Kyandoghere Kyamakya, Wolfgang A. Halang, Herwig Unger,
Jean Chamberlain Chedjou, Nikolai F. Rulkov, and Zhong Li (Eds.)

Recent Advances in Nonlinear Dynamics and Synchronization

Studies in Computational Intelligence, Volume 254

Editor-in-Chief
Prof. Janusz Kacprzyk
Systems Research Institute
Polish Academy of Sciences
ul. Newelska 6
01-447 Warsaw
Poland
E-mail: kacprzyk@ibspan.waw.pl

Further volumes of this series can be found on our homepage: springer.com

Vol. 234. Aruna Chakraborty and Amit Konar
Emotional Intelligence, 2009
ISBN 978-3-540-68606-4

Vol. 235. Reiner Onken and Axel Schulte
System-Ergonomic Design of Cognitive Automation, 2009
ISBN 978-3-642-03134-2

Vol. 236. Natalio Krasnogor, Belén Melián-Batista, José A. Moreno-Pérez, J. Marcos Moreno-Vega, and David Pelta (Eds.)
Nature Inspired Cooperative Strategies for Optimization (NICSO 2008), 2009
ISBN 978-3-642-03210-3

Vol. 237. George A. Papadopoulos and Costin Badica (Eds.)
Intelligent Distributed Computing III, 2009
ISBN 978-3-642-03213-4

Vol. 238. Li Niu, Jie Lu, and Guangquan Zhang
Cognition-Driven Decision Support for Business Intelligence, 2009
ISBN 978-3-642-03207-3

Vol. 239. Zong Woo Geem (Ed.)
Harmony Search Algorithms for Structural Design Optimization, 2009
ISBN 978-3-642-03449-7

Vol. 240. Dimitri Plemenos and Georgios Miaoulis (Eds.)
Intelligent Computer Graphics 2009, 2009
ISBN 978-3-642-03451-0

Vol. 241. János Fodor and Janusz Kacprzyk (Eds.)
Aspects of Soft Computing, Intelligent Robotics and Control, 2009
ISBN 978-3-642-03632-3

Vol. 242. Carlos Artemio Coello Coello, Satchidananda Dehuri, and Susmita Ghosh (Eds.)
Swarm Intelligence for Multi-objective Problems in Data Mining, 2009
ISBN 978-3-642-03624-8

Vol. 243. Imre J. Rudas, János Fodor, and Janusz Kacprzyk (Eds.)
Towards Intelligent Engineering and Information Technology, 2009
ISBN 978-3-642-03736-8

Vol. 244. Ngoc Thanh Nguyen, Rados law Piotr Katarzyniak, and Adam Janiak (Eds.)
New Challenges in Computational Collective Intelligence, 2009
ISBN 978-3-642-03957-7

Vol. 245. Oleg Okun and Giorgio Valentini (Eds.)
Applications of Supervised and Unsupervised Ensemble Methods, 2009
ISBN 978-3-642-03998-0

Vol. 246. Thanasis Daradoumis, Santi Caballé, Joan Manuel Marquès, and Fatos Xhafa (Eds.)
Intelligent Collaborative e-Learning Systems and Applications, 2009
ISBN 978-3-642-04000-9

Vol. 247. Monica Bianchini, Marco Maggini, Franco Scarselli, and Lakhmi C. Jain (Eds.)
Innovations in Neural Information Paradigms and Applications, 2009
ISBN 978-3-642-04002-3

Vol. 248. Chee Peng Lim, Lakhmi C. Jain, and Satchidananda Dehuri (Eds.)
Innovations in Swarm Intelligence, 2009
ISBN 978-3-642-04224-9

Vol. 249. Wesam Ashour Barbakh, Ying Wu, and Colin Fyfe
Non-Standard Parameter Adaptation for Exploratory Data Analysis, 2009
ISBN 978-3-642-04004-7

Vol. 250. Raymond Chiong and Sandeep Dhakal (Eds.)
Natural Intelligence for Scheduling, Planning and Packing Problems, 2009
ISBN 978-3-642-04038-2

Vol. 251. Zbigniew W. Ras and William Ribarsky (Eds.)
Advances in Information and Intelligent Systems, 2009
ISBN 978-3-642-04140-2

Vol. 252. Ngoc Thanh Nguyen and Edward Szczerbicki (Eds.)
Intelligent Systems for Knowledge Management, 2009
ISBN 978-3-642-04169-3

Vol. 253. Akitoshi Hanazawa, Tsutom Miki, and Keiichi Horio (Eds.)
Brain-Inspired Information Technology, 2009
ISBN 978-3-642-04024-5

Vol. 254. Kyandoghere Kyamakya, Wolfgang A. Halang, Herwig Unger, Jean Chamberlain Chedjou, Nikolai F. Rulkov, and Zhong Li (Eds.)
Recent Advances in Nonlinear Dynamics and Synchronization, 2009
ISBN 978-3-642-04226-3

Kyandoghere Kyamakya, Wolfgang A. Halang,
Herwig Unger, Jean Chamberlain Chedjou,
Nikolai F. Rulkov, and Zhong Li (Eds.)

Recent Advances in Nonlinear Dynamics and Synchronization

Theory and Applications

Prof. Kyandoghere Kyamakya
Institute for Smart-Systems Technologies
Alpen-Adria Universität Klagenfurt
Universitätsstraße 65-67
9020 Klagenfurt
Austria
E-mail: kyandoghere.kyamakya@uni-klu.ac.at

Dr. Jean Chamberlain Chedjou
Institute for Smart-Systems Technologies
Alpen-Adria Universität Klagenfurt
Universitätsstraße 65-67
9020 Klagenfurt
Austria
E-mail: jean.chedjou@uni-klu.ac.at

Prof. Wolfgang A. Halang
FernUniversität in Hagen
Lehrgebiet Informationstechnik
Universitätsstraße 27 - PRG
58097 Hagen
Germany
E-mail: wolfgang.halang@fernuni-hagen.de

Prof. Nikolai F. Rulkov
Institute for Nonlinear Science
University of California, San Diego
9500 Gilman Drive
La Jolla, CA 92093-0402
USA
E-mail: nrulkov@ucsd.edu

Prof. Herwig Unger
FernUniversität in Hagen
Lehrgebiet Kommunikationsnetze
Universitätsstraße 27 - PRG
58097 Hagen
Germany
E-mail:herwig.unger@fernuni-hagen.de

Prof. Zhong Li
FernUniversität in Hagen
Lehrgebiet Informationstechnik
Universitätsstraße 27 - PRG
58097 Hagen
Germany
E-mail: zhong.li@fernuni-hagen.de

ISBN 978-3-642-26057-5 e-ISBN 978-3-642-04227-0

DOI 10.1007/978-3-642-04227-0

Studies in Computational Intelligence ISSN 1860-949X

Softcover reprint of the hardcover 1st edition 2010

Typeset & Cover Design: Scientific Publishing Services Pvt. Ltd., Chennai, India.

Printed in acid-free paper

9 8 7 6 5 4 3 2 1

springer.com

Preface

In essence, the dynamics of real world systems (i.e. engineered systems, natural systems, social systesms, etc.) is nonlinear. The analysis of this nonlinear character is generally performed through both observational and modeling processes aiming at deriving appropriate models (mathematical, logical, graphical, etc.) to simulate or mimic the spatiotemporal dynamics of the given systems. The complex intrinsic nature of these systems (i.e. nonlinearity and spatiotemporal dynamics) can lead to striking dynamical behaviors such as regular or irregular, stable or unstable, periodicity or multi-periodicity, torus or chaotic dynamics. The various potential applications of the knowledge about such dynamics in technical sciences (engineering) are being intensively demonstrated by diverse ongoing research activities worldwide.

However, both the modeling and the control of the nonlinear dynamics in a range of systems is still not yet well-understood (e.g. system models with time varying coefficients, immune systems, swarm intelligent systems, chaotic and fractal systems, stochastic systems, self-organized systems, etc.). This is due amongst others to the challenging task of establishing a precise and systematic fundamental or theoretical framework (e.g. methods and tools) to analyze, understand, explain and predict the nonlinear dynamical behavior of these systems, in some cases even in real-time.

The full insight in systems' nonlinear dynamic behavior is generally achieved through approaches involving analytical, numerical and/or experimental methods. These approaches though complementary (since each of them can solve/depict some problems/aspects unsolved by its counterparts) are generally prone to some limitations when dealing with the analysis of complex, stiff and time varying dynamical systems. If it is generally tough/challenging to derive exact analytical solutions of models describing the dynamical behavior of such systems, numerical methods to analyze these systems are very time consuming and are exposed to accumulation of round-off errors during computations. Experimental studies are also limited by either the precision of physical components or calculators used or limitation in resources. These are some examples justifying the tremendous interest nowadays devoted to the development of a universal/unified framework which is robust to the underlined problems/drawbacks.

Due to the complex structural constitutions of real world systems which are generally made-up of autonomous and/or non-autonomous, coupled and/or uncoupled entities or sub-systems, the synchronization issue aims at achieving a general or global coordination/cooperation of the complete system as a whole. Basically, the realization and control of synchronization is a challenging task which appears far from being clearly understood and predicted in complex systems. Indeed, the synchronization phenomenon of is very widespread in nature as well as in complex technical systems. This phenomenon should be understood as the capacity of systems' entities (or sub-systems), either of the same or of different nature, to acquire a common operation regime which manifests itself by a general coordination (or global cooperation). This demonstrates the achievement of a given order or harmony in the behavior/motions of the systems or events. Some interesting applications can be found in urban area traffic control for instance, where phase synchronization of different traffic lights contribute to realize the so-called green waves and to optimize the overall throughput of the road network. In the fields of mechanics or electro-mechanics, the complete synchronization is exploited to make different sub-systems or coupled systems perform identical tasks. In the field of electronics, chaotic synchronization has been intensively exploited, for example in secure digital communications and in chaos-based cryptography. In medicine, synchronization has been exploited to regulate the cardiac beat. In biology, synchronization has been shown to be an important part of the function or dysfunction of a complex biological system. For instance, epileptic seizures correspond to a particular state of the brain in which too many neurons are desynchronized. Synchronization between these neurons leads to a correct functioning of the brain. These few applications of synchronization reveal the multiple interesting potential applications of synchronization and therefore justify the tremendous interest devoted on the investigation of this phenomenon.

The selected contributions of this book shed light on a series of interesting aspects related to nonlinear dynamics and synchronization with the aim of demonstrating some of their interesting applications in a series of selected disciplines. This book contains thirteenth chapters which are organized around five main parts. The first part (containing five chapters) does focus on theoretical aspects and recent trends of nonlinear dynamics and synchronization. The second part (two chapters) presents some modeling and simulation issues through concrete application examples. The third part (two chapters) is focused on the application of nonlinear dynamics and synchronization in transportation. The fourth part (two chapters) presents some applications of synchronization in security-related system concepts. The fifth part (two chapters) considers further applications areas, i.e. pattern recognition and communication engineering. The following lines present a brief and concise summary of each of the chapters included in this book.

The first chapter, ***"Recent Advances in Complex Networks Synchronization"***, by Guanong Chen, Xiaofan Wang, Xiang Li, and Jinhu Lü, addresses the issues of synchronization in complex dynamical networks. A brief overview of the state-of-the-art on synchronization issues in these networks is presented and a particular emphasis is devoted to the field of technical sciences. Synchronization is investigated in time-varying networks, and some mathematical formulas are derived to predict its occurrence. The "pinning" control technique (i.e. a permanent control or active control in real-time) is applied to particular nodes of some complex networks which cannot achieve self-synchronization in order to bring them into synchronization regime and some mathematical criteria are derived to predict the achievement/occurrence of synchronization.

The second chapter, ***"Network of Limit Cycle Oscillators with Parametric Learning Capability"***, by Julio Rodriguez and Max-Olivier Hongler, proposes a novel concept based on the application of a parametric learning approach to achieve synchronization in networks of interacting limit cycle oscillators. The method is applied to two types of networks namely the network of stable limit cycle oscillators and the network of mixed canonical-dissipative systems. The learning process is characterized analytically in both cases. Interestingly, it is shown the possibility of controlling/adjusting the natural frequencies of the oscillators by the learning method in networks of coupled oscillators as well as in networks of uncoupled oscillators as well.

The third chapter, ***"Ragged Synchronizability and Clustering in a Network of Coupled Oscillators"***, by Przemyslaw Perlikowsky, Andrzej Stefanski, and Tomasz Kapitaniak, investigates the issues of ragged synchronization and clustering in a network of three coupled in-line van der Pol oscillators. The modeling of this network is carried out and sets of equations are proposed (mathematical model) to describe the dynamics of the coupled network. These equations are used to explain the occurrence of ragged synchronization and clustering. Numerical and experimental studies are carried out to show the occurrence of ragged synchronization in both regular and irregular (i.e. chaotic) states of the network of coupled van der Pol oscillators. The study is extended to the case of a network of eight coupled in-line van der Pol oscillators and the occurrence of ragged synchronization is observed again.

The fourth chapter, ***"Cumulant Analysis of Strange Attractors -- Theory and Applications"***, by V. Kontorovitch and Z. Lovtchikova, presents both a theory and applications of the cumulant analysis of strange attractors. A theoretical description of a novel method/approach for the statistical analysis of strange attractors or chaos is carried out. The proposed method is applied to three classical and well-known self-sustained (or autonomous) nonlinear oscillators namely the Lorenz, Chua, and Rössler oscillators and various histograms are obtained for each of the oscillators. Theoretical results are compared with numerical results and a very good agreement is obtained as proof of concept of the proposed approach. Some applications of strange attractors are shown in the area of electrical engineering, specifically in communications.

The fifth chapter, ***"On Synchronization of Coupled Delayed Neural Networks"***, by Jinling Liang, Zidong Wang, and Xiaohui Liu, uses both continuous and discrete-time coupled neural networks schemes to investigate synchronization issues. Both linear and nonlinear couplings are considered in the case of a high dimensional system. The analytical study of the synchronization issues is carried out, and conditions are derived for the occurrence of both global synchronization and global exponential synchronization in the array of neural networks. Further, a numerical study is considered and the numerical results are compared with the analytical ones to illustrate the effectiveness (i.e. proof of concepts) of the proposed synchronization scheme.

The sixth chapter, ***"Multiset Agents in a Network for Simulation of Complex Systems"***, by V.K.Murthy and E.V.Krishnamurthy, analyses some key properties of complex systems and proposes models (mathematical and/or graphical) to understand the underlined properties. A concept based on the multiagent paradigm is proposed to simulate these properties. The methods agent based- modeling, simulation, and animation are considered and some advantages of these respective methods are described. Some application examples of these methods are further considered to investigate the swarm dynamics of ant colony, bacterial colonies, animal trails, etc. and the interest of developing the technology of agent modeling, simulation and animation to analyze biological phenomena is demonstrated.

The seventh chapter, ***"Simulation of Nonlinear Dynamics and Synchronization for Structural Control at Seismic Excitation"***, by Svetla Radeva addresses the modeling and simulation of earthquake phenomenon. The work relates to structural control systems in order to provide appropriate measurements necessary for reducing the destructive effects of seismic earthquakes on structures. The analysis of the nonlinear structural dynamics (or the structural behavior) during the vibrations caused by earthquakes is carried out. Using the modeling and computer simulation processes, some control algorithms are developed which are oriented to nonlinear structural dynamics. An elasto-mechanical model is proposed to describe the interaction between seismic waves, soil layers and the controlled structure. Some earthquake excitations are simulated, and the proposed model is exploited to study both structural control and synchronization mechanisms.

The eighth chapter, ***"Emergence of Synchronization in Transportation Networks with Biologically Inspired Decentralized Control"***, by Reik Donner, proposes a general concept for serving conflicting material flows in networks. Specifically, an approach is proposed which is based on a self-organized optimization for conflicting flows. The proposed approach is inspired by the oscillatory phenomena of pedestrian and animal flows at nodes/junctions. This approach is applied to some regular grid topologies and different synchronization regimes are obtained. The results obtained by the proposed approach are shown to be in very good agreement with those linked to the synchronization of the oscillatory service dynamics at different nodes in the network. The proposed approach is advantageous as it can easily be mapped to real-world traffic scenarios.

The ninth chapter, ***"Synchronization of Movements for Large-Scale Crowds"***, by Boris Goldengorin, Dmitry Krushinsky, and Alexander Makarenko, considers an approach based on cellular automata to analyze the movement of a large-scale crowd. Three main interdependent aspects/factors are considered. The first is the evacuation time, the second is the flow of pedestrians, and the last is the synchronization of a crowd. A framework based on a scenarios tree is proposed for modeling the optimal decision-making of pedestrians. Synchronization issues are investigated and it is shown that a more anticipating crowd might exhibit a less-complex behavior and might be more synchronized. The experimental study is considered and does show that increasing the probability characterizing a personal anticipation leads to maximizing the flow and minimizing the evacuation time of a crowd.

The tenth chapter, ***"Adaptive Synchronization of Chaotic Systems and Its Uses in Cryptanalysis"***, by Ying Liu and Wallace K. S. Tang, proposes a new design of an adaptive observer for adaptive synchronization of a class of chaotic systems with multiple unknown parameters. This design exploits a linear feedback control scheme and the dynamical minimization algorithm. Synchronization issues are investigated in chaotic systems and their stability as well. The exploitation of the new scheme designed in cryptanalysis for chaos-based communication systems is demonstrated. As proof of concept, the new scheme designed is used to solve diverse complex synchronization problems found in Lorenz, Rössler, and Genesio chaotic systems.

The eleventh chapter, ***"Trust-Based Collaborative Control for Teams in Communication Networks"***, by P. Ballal, F.L. Lewis, and G. R. Hudas, considers distributed networks in which issues related to trust establishment and consensus are addressed. The graphical modeling process is considered to represent collaboration between nodes through information exchange. Both continuous- and discrete-time trust update scheme for trust consensus are presented.

The twelfth chapter, ***"Investigating the Usability of SIFTfeatures in Biometrics"***, by Dakshina Ranjan Kisku, Ajita Rattani, Massimo Tistarelli, Jamuna Kanta Sing, addresses both usability and efficiency of the Scale Invariant Feature Transform (SIFT) features. A brief overview of biometric systems including face biometrics, fingerprint biometrics and multimodal biometrics is presented. Graph matching based face recognition approaches using SIFT features are discussed. The approach based on feature level fusion is proposed, where the SIFT-based face biometric and the minutiae based fingerprint biometric are fused at the feature extraction level.

The thirteenth chapter, ***"Coupled Oscillator Systems for Microwave Applications. Optimized Design Based on the Study and Control of the Multiple Coexisting Solutions in Systems with Symmetry"***, by Ana Collado and Apostolos Georgiadis, considers two oscillators architectures. The first architecture is made-up of arrays of linear coupled oscillators for beam steering applications. The second architecture is a structure of N-push oscillators for Radio frequency of High frequency (RF/HF) generation. The modeling process of the oscillators is described. These two oscillators architectures are examined and multiple coexisting solutions are identified as well as their stability analysis is carried out. The harmonic balance method is combined with the continuation method in order to optimize the coupling network. To ensure the appearance of only the desired traveling wave mode, stable regions of the modes are identified by simulation.

To finish, we would like to gratefully acknowledge and sincerely thank all the reviewers for their insightful comments and criticism of the manuscripts. Our thanks go also to the authors for their contributions and precious collaboration.

The Guest Editors

K. Kyamakya
W.A. Halang
H. Unger
J.C. Chedjou
N.F. Rulkov

Contents

Part I: Theoretical Aspects and Recent Trends

Part II: Some Modelling and Simulation Examples

Part III: Applications in Transportation

Part IV: Applications in Security Related System Concepts

Part V: Further Application Areas – Pattern Recognition and Communications Engineering

Part I
Theoretical Aspects and Recent Trends

Some Recent Advances in Complex Networks Synchronization*

Guanrong Chen, Xiaofan Wang, Xiang Li, and Jinhu Lü

Abstract. The current study of complex dynamical networks is pervading almost all kinds of science, engineering and technology, ranging from mathematics to computers, physics to biology, even to sociology. Its impacts on the modern high-tech industries, financial markets and human life are prominent and will be far-reaching. Research on fundamental properties and dynamical features of such complex networks has indeed become overwhelming.

This Chapter presents a brief overview of some past and current studies on the subject of complex dynamical network synchronization, particularly from an engineering and technological perspective. Some commonly concerned issues in the current research of network synchronization, mainly on

Guanrong Chen
Department of Electronic Engineering, City University of Hong Kong, Hong Kong SAR, China
e-mail: gchen@ee.cityu.edu.hk

Xiaofan Wang
Department of Automation, Shanghai Jiao Tong University, Shanghai 200240, China
e-mail: xfwang@sjtu.edu.cn

Xiang Li
Department of Electronic Engineering, Fudan University, Shanghai 200433, China
e-mail: lix@fudan.edu.cn

Jinhu Lü
Institute of Systems Science, Academy of Mathematics and Systems Science, Chinese Academy of Sciences, Beijing 100190, China
e-mail: jhlu@iss.ac.cn

* This work was supported by the NSFC-HKRGC Joint Research Scheme under Grant N-CityU 107/07, the Hong Kong Research Grants Council under the GRF Grant CityU 1117/08E, the National Natural Science Foundation of China under Grants 60821091 and 60772158, the National Basic Research (973) Program of China under Grant 2007CB310805, the Important Direction Project of Knowledge Innovation Program of Chinese Academy of Sciences under Grant KJCX3-SYW-S01, and the Scientific Research Foundation for the Returned Overseas Chinese Scholars, State Education Ministry, China.

K. Kyamakya (Eds.): Recent Adv. in Nonlinear Dynamics and Synchr., SCI 254, pp. 3–16.
springerlink.com

pinning-controlled network synchronization and time-varying network synchronization, will be addressed. Emphasis will be on some basic theories, methodologies, conditions and criteria for the network synchronizability as well as some possible relationships between the network topology and the network synchronizability.

Keywords: Complex dynamical network, pinning control, synchronizability, synchronization, time-varying network topology.

1 Introduction

Complex networks are ubiquitous in the real world [4], [22], [26]. A complex network typically refers to an ensemble of dynamical units with nontrivial topological features that do not occur in simple systems such as completely regular lattices and completely random structures [4]. Typical examples of complex dynamical networks in point include the Internet, the World Wide Web, power grids, biological neural networks, trading market trains, scientific citation networks, social relationship networks, among many others. Over the past two decades, the theoretical study of complex dynamical networks has become a very active area of scientific research, inspired largely by the empirical studies on a huge number of real-world networks.

Collective behavior of complex dynamical networks, in particular, is a focal topic of considerable interest within science and technology communities, where one of the most important subjects is the network synchronization [1]-[39]. Synchronization is a timekeeping dynamical behavior requiring coordination of events to operate the entire network in unison. Synchronization is one of the basic motions in nature where many connected systems are evolving in synchrony. In fact, the uniform synchrony of coupled oscillators serves as a platform for the study, which can well explain many natural and technological phenomena [22].

Given some recent comprehensive surveys [1, 3, 4, 26, 31, 33], this Chapter aims at providing only a very brief account of some recent advances in the field of complex network synchronization, mainly from an engineering and technological perspective. Some commonly concerned issues in the study of network synchronization, especially pinning-controlled network synchronization and time-varying network synchronization, will be addressed. Emphasis will be on some basic theories, methodologies, conditions and criteria for the network synchronizability, as well as some possible relationships between the network topology and the network synchronizability, rather than empirical studies or application investigation, which have been reviewed in, e.g., [1, 3, 4].

The rest of the Chapter is organized as follows. Sections 2 introduces some basic concepts, theories and methodologies of complex network synchronization. Section 3 then addresses the pinning-controlled network synchronization problem. Section 4 further investigates the time-varying version of network synchronization. Some concluding remarks are finally given in Section 5.

2 Problems of Network Synchronization

Only linearly and diffusively coupled networks are considered in this Chapter for simplicity of presentation. Such a network consists of N identical nodes, each being an n-dimensional autonomous dynamical system, described by [15]

$$\dot{\mathbf{x}}_i = \mathbf{f}(\mathbf{x}_i) + \sum_{\substack{j=1 \\ j \neq i}}^{N} c_{ij}(t)\, \Gamma(t)\, (\mathbf{x}_j - \mathbf{x}_i), \qquad i = 1, 2, \cdots, N, \tag{1}$$

where $t \geq t_0 = 0$ is the time variable, $\mathbf{f} : \mathbf{R}^n \to \mathbf{R}^n$ is continuously differentiable, $\mathbf{x}_i := \mathbf{x}_i(t) \in \mathbf{R}^n$ is the state vector of node i, $\Gamma(t)$ is the inter-state coupling matrix between node j and node i ($j \neq i$; $1 \leq i, j \leq N$) at time t, $C(t) = [\, c_{ij}(t) \,]_{N \times N}$ is the configuration coupling matrix representing the topological structure of the network at time t, in which the coupling strengths $c_{ij}(t)$ are defined as follows: if there is a connection between node i and node j ($j \neq i$) at time t, then $c_{ij}(t) \neq 0$; otherwise, $c_{ij}(t) = 0$; and the diagonal elements of $C(t)$ are defined by the diffusive coupling conditions:

$$c_{ii}(t) = -\sum_{\substack{j=1 \\ j \neq i}}^{N} c_{ij}(t), \qquad i = 1, 2, \cdots, N. \tag{2}$$

The above time-varying network can also be written in a compact form as [15]

$$\dot{\mathbf{x}}_i = \mathbf{f}(\mathbf{x}_i) + \sum_{j=1}^{N} c_{ij}(t)\, \Gamma(t)\, \mathbf{x}_j, \qquad i = 1, 2, \cdots, N. \tag{3}$$

As a special case, $\Gamma(t)$ can be a constant diagonal matrix of the form $\Gamma = \mathrm{diag}\{\gamma_1, \gamma_2, \cdots, \gamma_n\}$, where $\gamma_i \geq 0$ are not all zero, $i = 1, 2, \cdots, N$, and $C(t) = c\,[\, a_{ij} \,]_{N \times N}$ for all time t, where c is a constant coupling strength and a_{ij} satisfy the following connectivity condition: if there is a connection between node i and node j ($i \neq j$), then $a_{ij} = a_{ji} = 1$; otherwise, $a_{ij} = a_{ji} = 0$; and the diagonal elements are defined by the diffusive coupling conditions:

$$a_{ii} = -\sum_{\substack{j=1 \\ j \neq i}}^{N} a_{ij}, \qquad i = 1, 2, \cdots, N. \tag{4}$$

In this case, the time-varying network (1) or (3) reduces to a simple time-invariant model [25, 26, 27]:

$$\dot{\mathbf{x}}_i = \mathbf{f}(\mathbf{x}_i) + c\sum_{j=1}^{N} a_{ij}\,\Gamma\,\mathbf{x}_j, \qquad i = 1, 2, \cdots, N. \tag{5}$$

Hereafter, assume that networks (3) and (5) are connected in the sense that there are no isolated clusters at all times. Thus, the coupling matrices $C(t)$ and $c\,[\,a_{ij}\,]$ are irreducible for all $t \geq 0$.

A mathematical definition of network synchronization is first given. Only complete synchronization is discussed here, again for simplicity of presentation.

Definition 1. [20] Network (3) is said to achieve synchronization if

$$\lim_{t\to\infty} \|\mathbf{x}_i(t) - \mathbf{x}_j(t)\| = 0 \qquad \text{for all} \quad i, j = 1, 2, \cdots, N, \tag{6}$$

where $\|\cdot\|$ is the Euclidean norm.

In other words, by defining the synchronization manifold in the phase space as

$$\{\,\mathbf{x}_1 = \mathbf{x}_2 = \cdots = \mathbf{x}_N\,\}, \tag{7}$$

the network synchronization is achieved if all state trajectories $\mathbf{x}_i(t)$ $(1 \leq i \leq N)$ converge into this manifold as $t \to \infty$.

Strictly based on this definition of network synchronization, some rigorous mathematical analysis on the network synchronizability and synchronization conditions can be carried out (see, for example, [7, 11, 12, 19, 35]), which will be further reviewed below.

At this point, it is noted that Definition 1 is pure mathematical, namely, it does not concern about the ultimate synchronous behavior of the synchronized state vectors, wherever they arrive at. In practice, however, the purpose of studying network synchronization is to achieve something useful, for example to know where the network evolves to, when the synchronizing time is long enough, or better off, to ensure the network self-synchronize to a preferable and attainable target state. Of course, if this target state has no relations with the network, then this becomes a typical target-tracking problem in control theory. As further indicated in Remark 1 below, however, a self-synchronizing network will naturally evolve to one of its own solutions throughout the synchronization process continuously. Thus, from an engineering and technological application point of view, the above mathematical definition is very often being slightly twisted, as follows.

Definition 2. [2] Network (3) is said to achieve synchronization if

$$\lim_{t\to\infty} \|\mathbf{x}_i(t) - \mathbf{s}(t)\| = 0, \qquad i = 1, 2, \cdots, N, \tag{8}$$

for some $\mathbf{s}(t) \in \mathbf{R}^n$.

Remark 1. It may be possible that the network self-synchronize to a solution of its node system: $\dot{\mathbf{s}}(t) = \mathbf{f}(\mathbf{s}(t))$. For example, as a special case, when $\mathbf{s}(t) = \mathbf{s}$

is a constant vector, under some mild conditions to take a time limit on both sides of network (5) with the assumption that the network synchronization is indeed achieved, namely, if the network is able to self-synchronize to a constant vector: $\mathbf{x}_i \to \mathbf{s}$ $(i = 1, 2, \cdots, N)$, then due to the diffusive condition (4), one has

$$0 = \dot{\mathbf{s}} = \mathbf{f}(\mathbf{s}). \tag{9}$$

This means that the constant state $\mathbf{s}$ is an equilibrium solution of the node equation of the network. Of course, if the network is unable to self-synchronize to a constant vector, then external control input is needed, a topic to be further discussed below.

In a more general situation, $\mathbf{s}(t)$ may even be a periodic or chaotic trajectory. But in the non-constant case the above time limiting process cannot be directly carried out; hence some rigorous working arguments are still needed to develop. As pointed out in [19], both variational analysis and linearization can be done only near the trajectory $\mathbf{s}(t)$ therefore this trajectory must contain an attracting set. Nevertheless, if external control input is allowed, then network controlled-synchronization to a pre-assigned target trajectory $\mathbf{s}(t) \in \mathbf{R}^n$ is often possible, as further discussed below.

Remark 2. In some previous publications, the above two definitions of network synchronization were not clearly distinguished, and often messed up. For instance, in [17, 18], the discussed network synchronization by nature belongs to the second type specified by Definition 2 above, but it was stated there as the first type described by Definition 1 instead.

3 Basic Approaches and Results

For simplicity of presentation, consider the complex dynamical network (5) and define $\mathbf{x} = [\mathbf{x}_1, \mathbf{x}_2, \cdots, \mathbf{x}_N]$, $\mathbf{F}(\mathbf{x}) = [\mathbf{f}(\mathbf{x}_1), \mathbf{f}(\mathbf{x}_2), \cdots, \mathbf{f}(\mathbf{x}_N)]$, $\mathbf{A} = [a_{ij}]_{N\times N}$, and $\overline{\Gamma} = \text{diag}\,\{\Gamma, \Gamma, \cdots, \Gamma\}$. Then, the network can be rewritten as in the following matrix equation form:

$$\dot{\mathbf{x}} = \mathbf{F}(\mathbf{x}) + c\,\mathbf{A} \otimes \overline{\Gamma}\,, \tag{10}$$

where $\otimes$ is the Kronecker product.

To achieve network synchronization under Definition 1, a common approach is to apply the variational principle. Following [20], let ξ_i be the variation of the state vector of the ith node, $\mathbf{x}_i$, and introduce matrix $\xi = [\xi_1, \xi_2, \cdots, \xi_N]$. Then, one can obtain the following variational matrix equation:

$$\dot{\xi} = \left[\, \mathbf{I}_N \otimes [D\mathbf{f}(\mathbf{x})] + c\,\mathbf{A} \otimes \overline{\Gamma}\,\right] \xi\,, \tag{11}$$

where $[D\mathbf{f}(\mathbf{x})]$ is the Jacobian matrix at $\mathbf{x}$ and $\mathbf{I}_N$ is the N-dimensional identity matrix, in which each block is

$$\dot{\xi}_k = [[Df(\mathbf{x})] + c\lambda_k \Gamma]\, \xi_k, \tag{12}$$

with λ_k being the kth eigenvalue of $\mathbf{A}$, $k = 1, 2, \cdots, N$. Obviously, the Jacobian matrix $[Df(\mathbf{x})]$ here depends on the state vector $\mathbf{x}(t)$ for all $t \geq 0$, which is unknown beforehand and continuously evolves throughout the whole process.

To achieve network synchronization under Definition 2, a common approach is again to apply the variational principle, or directly linearize the network and then evaluate it at a state trajectory of the network. Following [2], one can obtain

$$\dot{\xi} = \left[\, \mathbf{I}_N \otimes [Df(\mathbf{s})] + c\, \mathbf{A} \otimes \overline{\Gamma} \,\right] \xi, \tag{13}$$

where $[Df(\mathbf{s})]$ is the Jacobian matrix evaluated at the trajectory $\mathbf{s}(t)$ which, as mentioned above, may be an equilibrium, periodic, or even chaotic trajectory. Similarly, each block of (13) is

$$\dot{\xi}_k = [[Df(\mathbf{s})] + c\lambda_k \Gamma]\, \xi_k, \tag{14}$$

with λ_k being the kth eigenvalue of $\mathbf{A}$, $k = 1, 2, \cdots, N$.

Let $\mathbf{A}_L = -\mathbf{A}$ be the Laplacian matrix of the network. Then, since the network is connected and diffusively coupled, 0 is an eigenvalue of $\mathbf{A}_L$ with multiplicity one and all the other eigenvalues of $\mathbf{A}_L$ are strictly positive, which are denoted by

$$0 = \lambda_1 < \lambda_2 \leq \lambda_3 \leq \cdots \leq \lambda_N. \tag{15}$$

Since the eigenvalue sets of $\mathbf{A}_L$ and $\mathbf{A}$ differ only by a minus sign, these two matrices will not be precisely distinguished below.

Now, consider the largest Lyapunov exponent $L_{\max}$ of network (5). This $L_{\max}$ is generally a function of the components and parameters of the network equation, such as $f(\cdot)$, A, Γ and c. This $L_{\max}$ is usually referred to as the master stability function [20]. Furthermore, given $f(\cdot)$ and Γ, the region S of negative real λ_k where $L_{\max}$ is also negative is called the synchronized region [2]. The synchronized solution of dynamical network (5) is locally asymptotically stable if

$$-c\lambda_k \in S\,, \qquad k = 2, 3, \cdots, N. \tag{16}$$

For a given and fixed coupling strength $c > 0$, if the synchronized region S is unbounded, in the form of $(-\infty, \alpha]$ for some real constant α, then the eigenvalue λ_2 of $\mathbf{A}$ characterizes the network synchronizability with respect to its topology. On the other hand, if the synchronized region S is bounded, in the form of $[\alpha_1, \alpha_2]$ for some real constants $\alpha_1 \leq \alpha_2$, then the eigenratio $r(A) = \lambda_2/\lambda_N$ of $\mathbf{A}$ characterizes the synchronizability. According to condition (16), the larger the $|\lambda_2|$ or the $r(A)$ is, the better the synchronizability will be, depending on the type of the synchronized region.

The following discussions are based on Definition 1.

Consider network (5), where the coupling matrix $\mathbf{A}$ is symmetrical. Let

$$\bar{\mathbf{x}}(t) = \frac{1}{N}\sum_{i=1}^{N} \mathbf{x}_i(t). \tag{17}$$

Then, the following result can be obtained.

Theorem 1. [19] Consider network (5). If all the following variational equations:

$$\dot{\mathbf{w}} = [\,[D\mathbf{f}(\bar{\mathbf{x}}(t))] + \lambda_k \Gamma\,]\,\mathbf{w}\,, \qquad k = 2, 3, \cdots, N\,, \tag{18}$$

are exponentially stable, then the synchronization manifold (7) is locally exponentially stable, namely, the network synchronization is achieved locally and exponentially.

Actually, this result holds for networks in a slightly more general form, e.g., with a non-autonomous node function $\mathbf{f}(\cdot, t)$ [19]. Moreover, a corresponding global result can also be obtained [19].

Next, the following discussions are based on Definition 2.

As mentioned above, under Definition 2, in a general situation, $\mathbf{s}(t)$ may be (i) an equilibrium, (ii) a periodic trajectory, or (iii) a chaotic trajectory, as discussed respectively below.

(i) For an equilibrium $\mathbf{s}$

As mentioned above, under the assumption that the network synchronization is indeed achieved, namely, self-synchronizing to s constant vector, then this constant vector will be an equilibrium of the node system (see (9)). In this case, notice that in network model (18), the eigenvalues of the Jacobian matrix $[D\mathbf{f}(\bar{\mathbf{x}}(t))]$ are time functions, therefore the eigenvalue-region type of synchronization criterion (16) cannot be directly applied. For networks with an equilibrium $\mathbf{s}(t) = \mathbf{s} \in \mathbf{R}^n$, this synchronization criterion (16) turns out to be convenient to use, as shown by the following result.

Theorem 2. [10] In network (5), if the inter-state coupling matrix Γ is anti-stable, namely, all its eigenvalues have positive real parts, then the network has an unbounded synchronized region about an equilibrium, in the form of $S = (-\infty, -\alpha)$ for some $\alpha > 0$.

Note that the synchronized region S may be bounded in the form of $(\alpha_1, \alpha_2) \subset (-\infty, \infty)$ [2], or a union of some bounded and unbounded regions [6].

(ii) For a periodic trajectory $\mathbf{s}(t)$

For this case, the following concepts are useful.

Definition 3. [15] Let $\mathbf{s}(t)$ be a T-periodic solution ($T > 0$) of the node system $\dot{\mathbf{x}} = \mathbf{f}(\mathbf{x})$, with $\mathbf{x}(0) = \mathbf{x}_0 \in \mathbf{R}^n$, and let γ represent the closed orbit of $\mathbf{s}(t)$ in the state space:

$$\gamma = \{ \mathbf{x} \mid \mathbf{x} = \mathbf{s}(t), \ 0 \le t < T \} .$$

If, for any $\varepsilon > 0$, there exits a constant $\delta = \delta(\varepsilon) > 0$ such that for any $\mathbf{x}_0$ satisfying

$$d(\mathbf{x}_0, \gamma) = \inf_{\mathbf{x} \in \gamma} ||\mathbf{x}_0 - \mathbf{x}|| < \delta ,$$

a solution of the system, $\mathbf{s}(t)$, satisfies

$$d(\mathbf{s}(t), \gamma) < \varepsilon , \qquad \text{for all } t \ge 0 ,$$

then this T-periodic solution trajectory, $\mathbf{s}(t)$, is said to be *orbitally stable.* Moreover, if there exist positive constants α, β and a real constant h such that $\|\mathbf{x}(t-h) - \mathbf{s}(t)\| \le \alpha e^{-\beta t}$ for $t \ge 0$, then $\mathbf{s}(t)$ is said to be *orbitally asymptotically stable with an asymptotic phase h.*

Definition 4. [15] Suppose that $\mathbf{s}(t)$ is a periodic solution of the node system $\dot{\mathbf{x}} = \mathbf{f}(\mathbf{x})$. Let $\gamma_1 = 1, \gamma_2, \cdots, \gamma_n$ be the Floquet multipliers of the variational equation of $\mathbf{s}(t)$, $\dot{\mathbf{x}} = [D\mathbf{f}(\mathbf{s}(t))] \, \mathbf{x}$. Then, $\mathbf{s}(t)$ is said to be a *hyperbolic periodic solution* if $|\gamma_j| \ne 1$ for all $j = 2, \cdots, n$. Moreover, $\mathbf{s}(t)$ is said to be a *hyperbolic synchronous periodic solution* of network (5) if all the Floquet multipliers of the variational equation of $\mathbf{s}(t)$ have absolute values less than 1, except one multiplier which equals 1.

Now, consider network (5) again.

Theorem 3. [15] Suppose that $\mathbf{s}(t)$ is a hyperbolic periodic solution of the node system $\dot{\mathbf{x}} = \mathbf{f}(\mathbf{x})$, and is orbitally asymptotically stable with a certain asymptotic phase. Suppose also that the network configuration coupling matrix $A = [a_{ij}]$ is diagonalizable. Then, $\mathbf{x}_i = \mathbf{s}(t)$ $(i = 1, 2, \cdots, N)$ is a hyperbolic synchronous periodic solution of network (5) and is orbitally asymptotically stable with a certain asymptotic phase, if and only if all the following linear time-varying systems:

$$\dot{\mathbf{w}} = [[D\mathbf{f}(\mathbf{s}(t))] + \lambda_k \Gamma] \, \mathbf{w} , \quad k = 2, \cdots, N , \tag{19}$$

are asymptotic stable about their zero solutions, where $\{\lambda_k\}$ are eigenvalues of the network coupling matrix A.

(iii) For a chaotic trajectory $\mathbf{s}(t)$

This case is difficult to analyze theoretically, due to the complexity of chaos. Nevertheless, under some "brute-force" types of conditions, such as the conservative LMI-type of conditions, some results can still be established, such as in [17] where however the approach of fixing $\mathbf{x}_1(t) = \mathbf{s}(t)$ is only heuristic but not completely rigorous.

There does not seem to be any rigorous result in the current literature on network synchronization to a chaotic trajectory, a solution of the chaotic node system of the network. A promising approach will be based on the profound chaos theory such as the ergodicity, an important and yet challenging topic for future research.

4 Pinning-Controlled Network Synchronization

Once again, consider network (5) for simplicity of presentation. The objective here is to let the network synchronize to some desired state, for example an equilibrium state $\mathbf{s} \in \mathbf{R}^n$.

If, under certain conditions as reviewed above, the network can self-synchronize to $\mathbf{s} \in \mathbf{R}^n$, then no external control input is needed.

If, however, the network is unable to self-synchronize to $\mathbf{s} \in \mathbf{R}^n$, then one may apply some control input to the network to force it to achieve the objective. Since a complex network typically has a large number of nodes, so applying one controller to each node is clearly impractical or often impossible. A common practice is to apply pinning control onto only a small fraction δ ($0 < \delta \ll 1$) of the nodes in the network. Here, the name "pinning" means to let the controller stay there forever once it is applied at a node, in order to distinguish itself from others like self-organizing control and switching control.

Without loss of generality, suppose that the first l nodes $(1, 2, \cdots, l)$ are selected to pin, where $l = \lfloor \delta N \rfloor$ is the smaller but nearest integer to the real number δN. Then, referred to (5), the controlled network can be described by

$$\begin{aligned} \dot{\mathbf{x}}_k &= \mathbf{f}(\mathbf{x}_k) + c \sum_{j=1}^{N} a_{kj} \Gamma \mathbf{x}_j + \mathbf{u}_k \,, \qquad k = 1, 2, \cdots, l, \quad (1 \le l < N) \\ \dot{\mathbf{x}}_k &= \mathbf{f}(\mathbf{x}_k) + c \sum_{j=1}^{N} a_{kj} \Gamma \mathbf{x}_j \qquad\quad k = l+1, l+2, \cdots, N \,. \end{aligned} \tag{20}$$

For simplicity here, use linear state-feedback control:

$$\mathbf{u}_k = -c\, d_k \Gamma (\mathbf{x}_k - \mathbf{s}) \,, \qquad k = 1, 2, \cdots, l, \tag{21}$$

where $\{d_k\}$ are positive constant feedback gains to be determined. Let

$$D = \text{diag}\{ d_1, d_2, \cdots, d_l, 0, \cdots, 0 \}$$

Theorem 4. [24] Suppose that there exists a constant $\rho > 0$ such that $[\,[D\mathbf{f}(\mathbf{s})] + \rho \Gamma\,]$ is Hurwitz stable. Let $\lambda_{\max}$ be the largest eigenvalue of matrix $A - D$. If

$$c \lambda_{\max} \le \rho \,, \tag{22}$$

then the equilibrium state $\mathbf{s}$ of the controlled network (5) is locally exponentially stable.

Theorem 5. [9] Assume that $\mathbf{f}(\mathbf{x})$ is Lipschitz continuous in $\mathbf{x}$ with Lipschitz constant $L_c^f > 0$, and the node $\dot{\mathbf{x}}_i = \mathbf{f}(\mathbf{x}_i)$ is chaotic for all $i = 1, 2, \cdots, N$, with the maximum positive Lyapunov exponent $h_{max} > 0$ and λ_{min} being the minimum eigenvalue of the positive definite matrix $[-A + D]$:

(i) if

$$c > L_c^f / \lambda_{min}, \tag{23}$$

then the controlled network (5), with $\Gamma = I$, is globally stable about the equilibrium $\mathbf{s}$;

(ii) if

$$c > h_{max} / \lambda_{min}, \tag{24}$$

then the controlled network (5) with $\Gamma = I$ is locally asymptotically stable about the equilibrium $\mathbf{s}$.

Remark 3. There were several closely-related papers on this approach to pinning control of network, e.g., [13].

For a more general form of networks, there are a couple of results reported in [9] where, however, Theorem 1 and its proof need some corrections: some involved matrices need to be symmetrical and the matrix U therein needs to be a positive-definite diagonal matrix (as in [7]).

For some even more general cases, e.g., with time-varying topologies and (or) time-delayed couplings, see [28, 29], for which the network may even be pinned to a rather arbitrary trajectory (not limited to an equilibrium).

For networks with unknown topological parameters, adaptive pinning controllers may be employed, as further discussed below.

Consider a general complex dynamical network model of N identical nodes, with linear and diffusive couplings, in the following form [36, 37]:

$$\dot{\mathbf{x}}_i = \mathbf{f}(\mathbf{x}_i) + c \sum_{j=1}^{N} a_{ij} \Gamma \mathbf{x}_j + \mathbf{u}_i(\mathbf{x}_1, \cdots, \mathbf{x}_N), \tag{25}$$

where the control inputs satisfy $\mathbf{u}_i(\mathbf{x}, \cdots, \mathbf{x}) = 0$. Let λ_i^m be the maximum eigenvalues of matrix A_i, where A_i is the ith minor of the coupling matrix $A = [\, a_{ij} \,]$, obtained by removing the first $i-1$ rows and the first $i-1$ columns from A, $i = 1, 2, \cdots, N$.

Theorem 6. [36] Suppose that the Jacobian matrix of the network (25) is bounded, $||\,[D\mathbf{f}(\mathbf{s})]\,|| \le \alpha$ at a trajectory $\mathbf{s}(t) \in \mathbf{R}^n$ for all $t \ge 0$. If there is an integer l, $1 \le l < N$, such that $\lambda_{l+1}^m < -\alpha / ||\Gamma||$, then the controlled network (25) is locally asymptotically stable about the trajectory $\mathbf{s}(t)$ under the following adaptive pinning controllers:

$$\begin{cases} \mathbf{u}_i = -p_i\,(\mathbf{x}_i - \mathbf{s}), & \dot{p}_i = q_i \|\mathbf{x}_i - \mathbf{s}\|^2, 1 \le i \le l, \\ \mathbf{u}_i = 0, & l+1 \le i \le N, \end{cases}$$

where q_i are some positive constants, $1 \le i \le l$.

A more general setting of this problem and results can be found in [36].

The simplest pinning control scheme is to use only one single controller (with $l = 1$), namely, in (25) with $\mathbf{u}_1 = c\varepsilon\,(\mathbf{x}_1(t) - \mathbf{s}(t))$ and $\mathbf{u}_i(t) = 0$ for all $i = 2, 3, ..., N$. Under some conditions (e.g., strong coupling strength c), this is possible, as shown in [7].

5 Time-Varying Complex Network Models

In this section, a network synchronization criterion for the time-varying network (3) is reviewed.

In network (3), assume that $\mathbf{f} : \Omega \to \mathbf{R}^n$ is continuously differentiable on $\Omega = \{\mathbf{x} \in \mathbf{R}^n \mid \|\mathbf{x} - \mathbf{s}(t)\| < r\}$. Suppose that the Jacobian matrix $[D\mathbf{f}(\mathbf{x})]$ is bounded and Lipschitz on Ω uniformly in t. Suppose also that there exists a real matrix, $\Phi(t)$, nonsingular for all $t \geq 0$, such that

$$\Phi^{-1}(t)\,A(t)\,\Phi(t) = \mathrm{diag}\,\{\lambda_1(t), \lambda_2(t), \cdots, \lambda_N(t)\} \tag{26}$$

and $\dot{\Phi}^{-1}(t)\,\Phi(t) = \mathrm{diag}\{\beta_1(t), \beta_2(t), \cdots, \beta_N(t)\}$.

Theorem 7. [15] Let $\mathbf{s}(t)$ be an exponentially stable solution of the node system $\dot{\mathbf{x}} = \mathbf{f}(\mathbf{x})$ in network (3). Then, all state vectors of the network will synchronize to $\mathbf{s}(t)$ if and only if all the following linear time-varying systems:

$$\dot{\mathbf{w}} = [D\mathbf{f}(\mathbf{s}(t)) + \lambda_k(t)\,\Gamma(t) - \beta_k(t)\,\mathbf{I}_n]\,\mathbf{w}\,, \qquad k = 2, \cdots, N\,, \tag{27}$$

are exponentially stable about their zero solutions.

Remark 4. Theorem 7 indicates that synchronization of the time-varying dynamical network (3) is completely determined by its inter-state coupling matrix $\Gamma(t)$, and the eigenvalues (26) as well as the corresponding eigenvectors of the configuration coupling matrix $A(t)$, where $\beta_k(t)$ are functions of these eigenvectors. In comparison, the synchronization of the time-invariant dynamical network (5) is completely determined by its inter-state coupling matrix Γ and the eigenvalues of the network-configuration coupling matrix A [18, 26].

6 Conclusions

This Chapter has presented a very brief overview of some progressive results on complex dynamical network synchronization from an engineering and technological perspective. Some commonly concerned issues in the current research on network synchronization, particularly on pinning-controlled network synchronization and time-varying network synchronization, have been addressed, with emphasis on some basic theories, methodologies, conditions and criteria for the network synchronizability as well as some possible relationships between the network topology and the network synchronizability.

In this Chapter, however, due to space limitation many other types of even more general networks have not been discussed, such as weighted, evolutionary, switching, delayed, clustering, and nonlinearly coupled networks, and many other interesting and important results have not been reviewed. Moreover, only complete synchronization has been discussed. Many other types of synchronization such as partial, cluster, projective, generalized, and phase synchronization have been left out. The interested readers may search for them in the following list of closely-related good references, therefore will not be further discussed here for brevity.

Acknowledgement

The authors thank Professor Tianping Chen for his valuable discussions and comments on some related technical issues.

References

1. Arenas, A., Diaz-Guilera, A., Kurths, J., Moreno, Y., Zhou, C.: Synchronization in complex networks. Phys. Rep. 469, 93–153 (2008)
2. Barahona, M., Pecora, L.M.: Synchronization in small-world systems. Phys. Rev. Lett. 89(5) art. 054101 (2002)
3. Boccaletti, S., Kurths, J., Osipov, G., Valladares, D.L., Zhou, C.S.: The synchronization of chaotic systems. Phys. Rep. 366, 1–101 (2002)
4. Boccaletti, S., Latora, V., Moreno, Y., Chavez, M., Hwang, D.-U.: Complex networks: Structure and dynamics. Phys. Rep. 424, 175–308 (2006)
5. Cai, S., Zhou, J., Xiang, L., Liu, Z.: Robust impulsive synchronization of complex delayed dynamical networks. Phys. Lett. A 372, 4990–4995 (2008)
6. Chen, G., Duan, Z.: Network synchronizability analysis: a graph-theoretic approach. Chaos 18, 037102 (2008)
7. Chen, T.P., Liu, X., Lu, W.: Pinning complex network by a single controller. IEEE Trans. Circ. Syst. I 54, 1317–1326 (2007)
8. Li, X.: Sync in complex dynamical networks: stability, evolution, control, and application. Int. J. Comput. Cognition 3, 16–26 (2005)
9. Li, X., Wang, X., Chen, G.: Pinning a complex dynamical network to its equilibrium. IEEE Trans. Circ. Syst.-I 51, 2074–2087 (2004)
10. Liu, C., Duan, Z., Chen, G., Huang, L.: Analyzing and controlling the network synchronization regions. Physica A 386, 531–542 (2007)
11. Liu, X., Chen, T.P.: Exponential synchronization of the linearly coupled dynamical networks with delays. Chin. Ann. Math., Ser. B 28, 737–746 (2007)
12. Liu, X., Chen, T.P.: Synchronization analysis for nonlinearly-coupled complex networks with an asymmetrical coupling matrix. Physica A 387, 4429–4439 (2008)
13. Lü, J.: Mathematical models and synchronization criterions of complex dynamical networks. Syst. Eng. Theory Pract. 24, 17–22 (2004)

14. Lü, J.: Synchronization of complex networks: Theories, approaches, applications and outlook. Adv. Mech. 38, 713–722 (2008)
15. Lü, J., Chen, G.: A time-varying complex dynamical network model and its controlled synchronization criteria. IEEE Trans. Autom. Control. 50, 841–846 (2005)
16. Lü, J., Leung, H., Chen, G.: Complex dynamical networks: Modelling, synchronization and control. Dyn. Continuous Discret. Impul. Syst., Ser. B 11a, 70–77 (2004)
17. Lü, J., Yu, X., Chen, G.: Chaos synchronization of general complex dynamical networks. Physica A 334, 281–302 (2004)
18. Lü, J., Yu, X., Chen, G., Cheng, D.: Characterizing the synchronizability of small-world dynamical networks. IEEE Trans. Circ. Syst.-I 51, 787–796 (2004)
19. Lu, W., Chen, T.P.: New approach to synchronization analysis of linearly coupled ordinary diferential systems. Physica D 213, 214–230 (2007); A new approach to synchronization analysis of linearly coupled map lattices. Chin. Ann. Math. Ser. B 28, 149–160 (2007)
20. Pecora, L., Carroll, T.: Master stability functions for synchronized coupled systems. Phys. Rev. Lett. 80, 2109–2112 (1998)
21. Sorrentino, F., di Bernardo, M., Garofalo, F., Chen, G.: Controllability of complex networks via pinning. Phys. Rev. E 75, 046103 (2007)
22. Strogatz, S.H.: Sync: The Emerging Science of Spontaneous Order. Hyperion, New York (2003)
23. Strogatz, S.H., Stewart, I.: Coupled oscillators and biological synchronization. Sci. Amer. 12, 102–109 (1993)
24. Wang, X., Chen, G.: Pinning control of scale-free dynamical networks. Physica A 310, 521–531 (2002)
25. Wang, X., Chen, G.: Synchronization in scale-free dynamical networks: robustness and fragility. IEEE Trans. Circ. Syst.-I 49, 54–62 (2002)
26. Wang, X., Chen, G.: Complex networks: small-world, scale-free, and beyond. IEEE Circ. Syst. Mag. 3(1), 6–20 (2003)
27. Wang, X., Li, X., Chen, G.: Complex Networks Theory and Its Applications. Tsinghua Univ. Press, Beijing (2006)
28. Xiang, L., Chen, Z.Q., Liu, Z., Chen, F., Yuan, Z.: Pinning control of complex dynamical networks with heterogeneous delays. Computers Math. Appl. 56, 1423–1433 (2008)
29. Xiang, L., Liu, Z., Chen, Z.Q., Yuan, Z.: Pinning weighted complex networks with heterogeneous delays by a small number of feedback controllers. Sci. China, Ser. F 51, 511–523 (2008)
30. Yu, W., Cao, J., Lü, J.: Global synchronization of linearly hybrid coupled networks with time-varying delay. SIAM J. Appl. Dyn. Syst. 7, 108–133 (2008)
31. Wu, C.W.: Synchronization in Coupled Chaotic Circuits and Systems. World Scientific, Singapore (2002)
32. Wu, C.W.: Synchronization and convergence of linear dynamics in random directed networks. IEEE Trans. Autom. Control 51, 1207–1210 (2006)
33. Wu, C.W.: Synchronization in Complex Networks of Nonlinear Dynamical Systems. World Scientific, Singapore (2007)
34. Zhang, Q., Lu, J.A., Lü, J., Tse, C.K.: Adaptive feedback synchronization of a general complex dynamical network with delayed nodes. IEEE Trans. Circ. Syst.-II 55, 183–187 (2008)

35. Zhou, J., Chen, T.P.: Synchronization in general complex delayed dynamical networks. IEEE Trans. Circ. Syst.-I 53, 733–744 (2006)
36. Zhou, J., Lu, J.A., Lü, J.: Adaptive synchronization of an uncertain complex dynamical network. IEEE Trans. Autom. Control 51, 652–656 (2006)
37. Zhou, J., Lu, J.A., Lü, J.: Pinning adaptive synchronization of a general complex dynamical network. Automatica 44, 996–1003 (2008)
38. Zhou, J., Lu, J.A., Lü, J.: Erratum to: Pinning adaptive synchronization of a general complex dynamical network. Automatica 45, 598–599 (2009); Automatica 44, 996–1003 (2008)
39. Zhou, J., Xiang, L., Liu, Z.: Synchronization in complex delayed dynamical networks with impulsive effects. Physica A 384, 684–692 (2007)

Networks of Limit Cycle Oscillators with Parametric Learning Capability

Julio Rodriguez and Max-Olivier Hongler

Abstract. The fundamental idea that synchronized patterns emerge in networks of interacting oscillators is revisited by allowing a parametric learning mechanism to operate on the local dynamics. The local dynamics consist of stable limit cycle oscillators which, due to mutual interactions, are allowed, via an adaptive process, to permanently modify their frequencies. Adaptivity is made possible by conferring to each oscillator's frequency the status of an additional degree of freedom. The network of individual oscillators is ultimately driven to a stable synchronized oscillating state which, once reached, survive even if mutual interactions are removed. Such a permanent, plastic type deformation of an *initial* to a *final consensual* state is realized by a dissipative mechanism which vanishes once a consensus is established. By considering diffusive couplings between position- and velocity-dependent state variables, we are able to analytically explore the resulting dynamics and in particular to calculate the resulting consensual state. The ultimate consensual state is topology network-independent. However, the interplay between the graph connectivity and the local dynamics does strongly influence the learning rate.

Julio Rodriguez and Max-Olivier Hongler
Faculté des Sciences et Techniques de l'Ingénieur (STI)
Institut de Microtechnique (IMT)
Laboratoire de Production Microtechnique (LPM)
Ecole Polytechnique Fédérale de Lausanne (EPFL)
CH-1015 LAUSANNE
e-mail: julio.rodriguez@epfl.ch, max.hongler@epfl.ch

Julio Rodriguez
Fakultät für Physik
Universität Bielefeld
D-33501 BIELEFELD
e-mail: jrodrigu@physik.uni-bielefeld.de

K. Kyamakya (Eds.): Recent Adv. in Nonlinear Dynamics and Synchr., SCI 254, pp. 17–48.
springerlink.com

1 Introduction

Tal como un péndulo

Así, el ser que ha despertado, como un péndulo viviente, ha de sostenerse en movimiento incesante, sostenido por un punto remoto, transformando el desfallecimiento en pausa, y la pausa, en lugar de más honda y obediente oscilación, revelando así su secreto de ser un diapasón del imperceptible fluir musical del interior del tiempo vivo.

María Zambrano

The *synchronization capabilities* of coupled oscillators and their relation to to a time and frequency distribution over a network of clocks is an old and fundamental problem with direct relevance for chronology and positioning applications. Already discussed in the seventeenth century by the genus contributions of Christiaan Huyghens, it is truly remarkable and quite unusual that after more than three centuries, this general problematic continues to trigger a strong and interdisciplinary research activity. Indeed numerous contributions ranging from applied mathematics to most up-to-date technology devices continue to feed the present literature. Besides these applications, such a timelessness of synchronization phenomena relies on close connections with modern fundamental perspectives jointly involving nonlinear dynamical systems and recent advances in the algebraic classification of networks.

The starting point of our study is a general class of dynamical system based on what was presented in [15]. It consists of N local dynamical systems, mutually coupled via time- and state-variable-dependent connections. The dynamical system is

$$\begin{aligned} \dot{x}_k(t) &= \mathsf{F}_k(x_k(t); \Omega_k) + \mathsf{C}_k(x_1(t), \ldots, x_N(t), \Gamma(t)), \qquad k = 1, \ldots, N, \\ \dot{\Gamma}(t) &= \mathsf{D}_\Gamma(x_1(t), \ldots, x_N(t), \Gamma(t)), \end{aligned} \tag{1}$$

where:

$x_k(t) = (x_k^1(t), \ldots, x_k^{n_k}(t)) \in \mathbb{R}^{n_k}$ are the `state variables` of the kth local system,

$\mathsf{F}_k : \mathbb{R}^{n_k} \longrightarrow \mathbb{R}^{n_k}$ are vector fields determining the kth `local dynamics`,

$\Omega_k = (\omega_k^1, \ldots, \omega_k^{m_k}) \in \mathbb{R}^{m_k}$ are fixed parameters for the kth local system,

$\mathsf{C}_k : \mathbb{R}^{n_1+\cdots+n_N+p} \longrightarrow \mathbb{R}^{n_1+\cdots+n_N+p}$ are vector fields determining the mutual `coupling dynamics`,

$\Gamma(t) = (\gamma^1(t), \ldots, \gamma^p(t)) \in \mathbb{R}^p$ are **connection variables** for the interactions,

$\mathsf{D}^\Gamma : \mathbb{R}^{n_1+\cdots+n_N+p} \longrightarrow \mathbb{R}^{n_1+\cdots+n_N+p}$ is a vector field determining the **connection dynamics**.

Eqs.(1) encompass a vast class of models ranging from elementary mechanical devices of the original Huyghens' type to highly complex dynamical systems coupled via networks, as those presented in the seminal work of L. M. Pecora and T. L. Carroll [11]. In this contribution, we propose a natural extension of Eqs.(1) by enlarging the class of dynamical systems to:

$$\begin{aligned}
\dot{x}_k(t) &= \mathsf{F}_k(x_k(t), \Omega_k(t)) + \mathsf{C}_k(x_1(t), \ldots, x_N(t), \Gamma(t)), \qquad k = 1, \ldots, N, \\
\dot{\Gamma}(t) &= \mathsf{D}^\Gamma(x_1(t), \ldots, x_N(t), \Omega_1(t), \ldots, \Omega_N(t), \Gamma(t)), \\
\dot{\Omega}_k(t) &= \mathsf{D}^{\Omega_k}(x_1(t), \ldots, x_N(t), \Omega_1(t), \ldots, \Omega_N(t), \Gamma(t)), \qquad k = 1, \ldots, N,
\end{aligned} \tag{2}$$

where:

$\Omega_k(t) = (\omega_k^1(t), \ldots, \omega_k^{m_k}(t)) \in \mathbb{R}^{m_k}$ are now allowed to be time-dependent parameters influencing the kth local dynamics. These will, from now on, be referred to as **parametric variables**.

$\mathsf{D}^{\Omega_k} : \mathbb{R}^{n_1+\cdots+n_N+m_k} \longrightarrow \mathbb{R}^{n_1+\cdots+n_N+m_k}$ are vector fields governing the parametric variables $\Omega_k(t)$. This evolution law will be referred to as the **parametric dynamics**.

In words, Eqs.(2) effectively enlarge the dimensionality (i.e. the number of degrees of freedom) of the original system Eqs.(1). Let us now distinguish between several typical situations:

Trivial parametric dynamics. For $\Omega_k(t) := \Omega^0$ (i.e. $\mathsf{D}^{\Omega_k} \equiv 0 \ \forall \ k$), the resulting, already considered in [15], covers a wealth of actual mechanical and other synchronizing devices. The explicit time dependence given by $\Gamma(t)$, governed by the connection dynamics D^Γ, takes the time-dependent environment into account. A classical and paradigmatic illustration of this situation is the periodically *forced* Van der Pol oscillator:

$$\begin{aligned}
\dot{x}_1(t) &= \mathsf{F}_1(x_1(t); \Omega^0) && + \mathsf{C}(x_1(t), \Gamma(t)), & \dot{\Gamma}(t) &= \mathsf{D}^\Gamma(x_1(t), \Gamma(t)), \\
\dot{x}(t) &= \omega y(t), && & \dot{\gamma}(t) &= \kappa(t), \quad \dot{\epsilon}(t) = 0, \\
\dot{y}(t) &= -\omega x(t) + \alpha(1 - x(t)^2) y(t) && + \epsilon\gamma(t), & \dot{\kappa}(t) &= -\nu^2\gamma(t), \ \dot{\nu}(t) = 0.
\end{aligned} \tag{3}$$

Here, $\gamma(t) = \sin(\nu t)$ and for an appropriate choice of $\Omega^0 = \{\omega, \alpha\}$ and $\Gamma = (\gamma, \kappa, \epsilon, \nu)$, the resulting oscillation is "*caught*" by the forcing signal $\gamma(t)$.

While the generality of Eqs.(2) obviously precludes any exhaustive listing of actual applications, we mention here two important subclasses which exhibit a direct relevance for our present study.

i) *Network synchronization with time-dependent connections.*
An extension of [11] proposed by S. Bocaletti et al. [2] considers synchronization of a collection of identical systems (with fixed $\Omega_{k}(t) := \Omega^0 \ \forall\ k$) but with time-dependent networks, namely:

$$\begin{aligned} \dot{x}_k(t) &= \mathsf{F}(x_k(t), \Omega^0) - \sigma \sum_{j=1}^{N} \gamma_{k,j}(t) h(x_j(t)), \quad && k = 1, \ldots, N, \\ \dot{\gamma}_{k,j}(t) &= \mathsf{D}^{\Gamma}_{k,j}(\Omega^0, \Gamma(t)), && j,k = 1, \ldots, N, \\ \dot{\sigma}(t) &= 0, \end{aligned} \tag{4}$$

where $\mathsf{C}_k(x_1(t), \ldots, x_N(t), \Gamma(t)) = \sigma \sum_{j=1}^{N} \gamma_{k,j}(t) h(x_j(t))$, h is a common coupling function, $\Gamma(t)$ contains an $N \times N$ time-dependent connectivity matrix and σ, a constant coupling strength. The matrix entries $\gamma_{k,j}(t)$ of $\Gamma(t)$ have a specified time evolution governed by D^{Γ}.

ii) *Kuramoto dynamics*
Pioneered by Y. Kuramoto, this dynamics investigates the synchronization of phase oscillators [1] mutually coupled via a given network. A recent generalization including time-dependent network connections and environmental noise is given in [3], namely:

$$\begin{aligned} \dot{\theta}_k(t) &= \omega_k + \sum_{j=1}^{N} \sigma_{k,j} \gamma_{k,j}(t) \sin(\theta_j(t) - \theta_k(t)), \quad && k = 1, 2, \ldots, N, \\ \dot{\gamma}_{k,j}(t) &= \mathsf{D}^{\Gamma}_{k,j}(\Omega, \Gamma(t)), \quad \dot{\sigma}_{k,j}(t) = 0, && j,k = 1, 2, \ldots, N, \end{aligned} \tag{5}$$

where $\mathsf{F}_k(\theta_k, \Omega_k) = \omega_k$, ($\omega_k$ is a fixed scalar parameter here), $\mathsf{C}_k(\theta_1(t), \ldots, \theta_N(t), \Gamma(t)) = \sum_{j=1}^{N} \sigma_{k,j} \gamma_{k,j}(t) \sin(\theta_j(t) - \theta_k(t))$ for all k and $\sigma_{k,j}$ is a constant positive coupling strength.

Trivial connection dynamics. For $\Gamma(t) := \Gamma^0$ (i.e. $\mathsf{D}^{\Gamma} \equiv 0$), the mutual interactions are now allowed to affect the control parameters Ω_k of the local dynamics. The simplest and paradigmatic illustration is the *parametric oscillator*:

$$\begin{aligned} \dot{x}_1(t) &= \mathsf{F}_1(x_1(t), \Omega_1(t)), \qquad & \dot{\Omega}_1(t) &= \mathsf{D}^{\Omega_1}(\Omega_1(t)), \\ \dot{x}(t) &= (\omega_0 + \omega(t)) y(t), & \dot{\omega}(t) &= \kappa(t), \quad \dot{\omega}_0(t) = 0, \\ \dot{y}(t) &= -(\omega_0 + \omega(t)) x(t), & \dot{\kappa}(t) &= -\nu^2 \omega(t), \quad \dot{\nu}(t) = 0. \end{aligned} \tag{6}$$

Hence, here $\omega(t) = \sin(\nu t)$. This system is well known to exhibit a *parametric resonance* for ad-hoc tuning of the forcing and natural frequencies (i.e. $\nu = 2\omega_0$).

Non-trivial connection and parametric dynamics. When $\mathsf{D}^{\Omega_k} \not\equiv 0$ and $\mathsf{D}^{\Gamma} \not\equiv 0$ in Eqs.(2), we have a simultaneous interplay between parametric and connection dynamics. A simple illustration has recently been introduced in [12]:

$$\begin{aligned}
\dot{x}(t) &= \omega(t)y(t) + \eta(\mu - x(t)^2 - y(t)^2)x(t) + \epsilon\gamma(t), \\
\dot{y}(t) &= -\omega(t)\,x(t) + \eta(\mu - x(t)^2 - y(t)^2)y(t), \\
\ddot{\gamma}(t) &= -\nu^2\gamma(t), \\
\dot{\omega}(t) &= \epsilon\gamma(t)\frac{y(t)}{\sqrt{x^2(t)+y^2(t)}}.
\end{aligned} \tag{7}$$

For $\epsilon = 0$, the system Eq.(7) reduces to the Hopf oscillator with a stable limit cycle $\mathcal{L}(\mu) := \{(x, y) \in \mathbb{R}^2 \,|\, x^2 + y^2 = \mu\}$. For large $\eta > 0$ (i.e. *strong structural stability* of $\mathcal{L}(\mu)$) and small $\epsilon > 0$, the limit cycle orbit on $\mathcal{L}(\mu)$ is barely perturbed. However, the angular velocity on $\mathcal{L}(\mu)$ is affected and *catches* the driving frequency ν of $\gamma(t)$. This *adaptive matching* of the original system's frequency ω_0 to the driving signal $\gamma(t)$ is referred to in [12] as "*adaptive frequency learning*". Note that in Eqs.(7), the external signal $\gamma(t)$ *plastically* deforms the original dynamics; plasticity meaning here that once the frequency matching is realized, it subsists even if $\gamma(t)$ is removed. Directly inspired by Eqs.(7), we shall consider the class of dynamics given by:

$$\begin{aligned}
\dot{x}_k(t) &= \mathsf{F}_k(x_k(t), \Omega_k(t)) + \mathsf{C}_k(x_1(t), \ldots, x_N(t), \Gamma(t)), \quad k = 1, \ldots, N, \\
\dot{\Gamma}(t) &= \mathsf{D}^{\Gamma}(\Gamma(t)), \\
\dot{\Omega}_k(t) &= \mathsf{D}^{\Omega_k}(x_1(t), \ldots, x_N(t), \Gamma(t)), \qquad k = 1, \ldots, N,
\end{aligned} \tag{8}$$

A *learning mechanism* will be introduced via D^{Ω_k} to influence the time evolution parametric variables $\Omega_k(t)$ and, under suitable hypotheses, all $\Omega_k(t)$ will asymptotically converge toward a single common consensual $\Omega_c(t)$. Among the goals of this work, we shall give analytical answers to questions that can be raised, namely:

1) How to calculate the consensual Ω_c ?
2) How does the consensual Ω_c depend on the connectivity of the network ?
3) How does the network influence the convergence rate towards the consensual state ?

2 Network of Mixed Canonical-Dissipative Systems with Parametric Dynamics

To construct the class of systems on which we shall focus, we rewrite Eqs.(8) under the form:

$$\begin{aligned}
\dot{x}_k &= \mathsf{M}_k^x(x_k, y_k, \Omega_k) + \mathsf{C}_k^x(x_1, y_1, \ldots, x_N, y_N, \Gamma), \\
\dot{y}_k &= \mathsf{M}_k^y(x_k, y_k, \Omega_k) + \mathsf{C}_k^x(x_1, y_1, \ldots, x_N, y_N, \Gamma), \quad k = 1, \ldots, N, \\
\dot{\Gamma} &= \mathsf{D}^{\Gamma}(\Gamma), \\
\dot{\Omega}_k &= \mathsf{D}^{\Omega_k}(x_1, y_1, \ldots, x_N, y_N, \Gamma), \qquad k = 1, \ldots, N,
\end{aligned} \tag{9}$$

and in subsections 2.1 to 2.4, we will go into further detail of the four components occurring in Eqs.(9).

i) **Local dynamics**, specified by: $\mathsf{M}_k^x(x_k, y_k, \Omega_k)$ and $\mathsf{M}_k^y(x_k, y_k, \Omega_k)$, will belong to the class of *mixed canonical-dissipative* (MCD) sytems; this is discussed in subsection 2.1.
ii) **Connection dynamics**, specified by the D^Γ vector field, characterizes the time-dependent connections of the network; this is discussed in 2.2.
iii) **Coupling dynamics**, $\mathsf{C}_k^x(x_1, \dots, y_N, \Gamma)$ and $\mathsf{C}_k^y(x_1, \dots, y_N, \Gamma)$, characterize the interactions and is of diffusive type; this is discussed in 2.3.
iv) **Parametric dynamics**, specified by the D^{Ω_k} vector field, characterizes the learning processes; this is discussed in 2.4.

2.1 Local Dynamics: Mixed Canonical-Dissipative systems

We introduce a collection of `MCD oscillators` $\{\mathcal{O}_k\}_{k=1}^N$ with dynamics:

$$\mathcal{O}_k \begin{cases} \dot{x}_k = \underbrace{\omega_k \frac{\partial H_k}{\partial y_k}(x_k, y_k)}_{} + \underbrace{\eta_k\, g_k(H_k(x_k, y_k); \mu_k) \frac{\partial H_k}{\partial x_k}(x_k, y_k)}_{}, \\ \dot{y}_k = \underbrace{-\omega_k \frac{\partial H_k}{\partial x_k}(x_k, y_k)}_{\text{canonical evolution}} + \underbrace{\eta_k\, g_k(H_k(x_k, y_k); \mu_k) \frac{\partial H_k}{\partial y_k}(x_k, y_k)}_{\text{dissipative evolution}}, \end{cases} \tag{10}$$

where $H_k : \mathbb{R}^2 \longrightarrow \mathbb{R}^+$ and $g_k : \mathbb{R}^+ \longrightarrow \mathbb{R}$. The H_k's functions are $\mathfrak{C}^2$ and positive definite ($H_k(x, y) = 0 \Leftrightarrow (x, y) = (0, 0)$ for all k), independent of a set of parameters (i.e. dependence only on (x, y)) and play the role of Hamiltonians (i.e. energy). From now on, we assume that given $\mu_k > 0$, $H_k(x,y) = \mu_k$ uniquely defines a set of closed, concentric curves $\mathcal{L}_k(\mu_k)$ in $\mathbb{R}^2$ that surround the origin. The g_k function are $\mathfrak{C}^1$, parameterized by μ_k and $g_k(H_k(x,y); \mu_k)$ is a non-conservative term which, according to the value of H_k, feeds or dissipates energy from the purely Hamiltonian system. Indeed, if $g_k(H_k(x,y); \mu_k)$ vanishes for $H_k(x,y) = \mu_k$, the dynamics is purely Hamiltonian, (i.e.*canonical conservative* part of the evolution), and we therefore have:

$$H_k(x,y) = \mu_k \text{ defines the } \textit{limit cycle } \mathcal{L}_k(\mu_k)$$

with

$$\mathcal{L}_k(\mu_k) := \left\{(x, y) \in \mathbb{R}^2 \mid H_k(x,y) = \mu_k\right\}.$$

The stability of $\mathcal{L}_k(\mu_k)$'s is determined by the g_k function. If $g_k'(s; \mu_k) < 0$ for all s in $\mathbb{R}^+$, then $V_k : (x, y) \longmapsto \frac{1}{2}\left(g_k(H_k(x,y); \mu_k)\right)^2$ is a Ляпунов function guaranteeing that the compact set $\mathcal{L}_k(\mu_k)$ is asymptotically stable (see [14] for details). Therefore, for g_k vanishing at μ_k and when $\mathcal{L}_{k(\mu_k)}$ is stable, the energy-type control g_k drives all orbits towards the stable limit cycle $\mathcal{L}_{k(\mu_k)}$ which is hence an attractor. The rate at which the orbits converge to $\mathcal{L}_{k(\mu_k)}$ depends on $\eta_k > 0$: the larger η_k, the faster the convergence. The system defined by Eqs.(10) belongs to the general class of mixed canonical-dissipative systems (MCD) (c.f [9], [8] and [19]). In the sequel, we shall make use of the short-hand notation:

$$\mathsf{M}_k^x(x_k, y_k; \Omega_k) := \omega_k \frac{\partial H_k}{\partial y_k}(x_k,y_k) + g_k(H(x_k,y_k); \mu_k)\frac{\partial H_k}{\partial x_k}(x_k,y_k)$$

$$\mathsf{M}_k^y(x_k, y_k; \Omega_k) := -\omega_k \frac{\partial H_k}{\partial x_k}(x_k,y_k) + g_k(H(x_k,y_k); \mu_k)\frac{\partial H_k}{\partial y_k}(x_k,y_k)$$

Observe that in Eqs.(10), $\Omega_k = (\omega_k, \eta_k, \mu_k)$ are fixed parameters. Among the dynamical systems belonging to the MCD class, we will distinguish between **homogenous** and **heterogenous** oscillators:

Definition 1.
i) A collection $\{\mathcal{O}_k\}_{k=1}^N$ *of* **homogenous MCD oscillators** *satisfies:*

$$H_j(x,y) = H_k(x,y) \quad \forall j, k \text{ and } (x, y) \in \mathbb{R}^2.$$

ii) A collection $\{\mathcal{O}_k\}_{k=1}^N$ *of* **heterogenous MCD oscillators** *satisfies:*

$$\exists\ k, j \text{ and } (x^*, y^*) \in \mathbb{R}^2 \ \text{ such that } H_j(x^*,y^*) \neq H_k(x^*,y^*).$$

Examples MCD Dynamics

a) *Hopf oscillator.* For this classical example, the Hamiltonian is $H(x, y) = \frac{1}{2}(x^2 + y^2)$ and the function g ca be chosen as $g(s; \mu) = \mu - s$. In this case, Eqs.(10) are

$$\begin{aligned}\dot{x} &= \omega y + \eta(\mu - \tfrac{1}{2}(x^2 + y^2))x \\ \dot{y} &= -\omega x + \eta(\mu - \tfrac{1}{2}(x^2 + y^2))y\end{aligned}$$

where the stable limit cycle $\mathcal{L}_{k(\mu_k)}$ is the circle of radius $\sqrt{\mu}$ and the circulation on $\mathcal{L}_{k(\mu_k)}$ is a uniform rotation of angular velocity ω.
b) *Mathews-Lakshmanan oscillator.* Using the results derived in [18], we introduce the following Hamiltonian:

$$H(x, y) = \log(\cosh(y)) + \frac{1}{2}\log(1 + x^2)$$

Eqs.(10), with $g(s; \mu) = \mu - s$, become

$$\begin{aligned}\dot{x} &= \omega \tanh(y) + \eta(\mu - H(x, y))\tfrac{x}{1+x^2} \\ \dot{y} &= -\omega \tfrac{x}{1+x^2} + \eta(\mu - H(x, y)) \tanh(y)\end{aligned}$$

where $\mu = \frac{1}{2}\log(1+\mathsf{a}^2)$ ($\mathsf{a} > 0$ to be given) and the solution to the equation $g(H(x,y);\mu) = 0$ defines the stable limit cycle $\mathcal{L}(\mu)$. The circulation on $\mathcal{L}(\mu)$ (i.e. the solution of canonical dynamics) is explicitly given by

$$x(t) = \mathsf{a}\sin\left(\frac{\omega t}{\sqrt{1+\mathsf{a}^2}} + \phi_0\right)$$
$$y(t) = \tanh^{-1}\left(\frac{\mathsf{a}}{\sqrt{1+\mathsf{a}^2}}\cos\left(\frac{\omega t}{\sqrt{1+\mathsf{a}^2}} + \phi_0\right)\right)$$

Note that the harmonic solution $x(t)$ exhibits an amplitude-dependent frequency which is the signature of the underlying nonlinearity of the dynamics.

2.2 Connection Dynamics: Time-Dependent Network

The connections between the MCD's given by Eqs.(10) is realized via a *connected, undirected* and possibly *time-dependent* network $\mathcal{N}(t)$ with N vertices $\{v_1,\dots,v_N\}$ and where the kth vertex v_k is equipped with the kth MCD. The networks we shall consider are defined by:

Definition 2. *A connected, undirected and time-dependent network* $\mathcal{N}(t)$ *is :*

- *connected : at all time, there is a path from any vertex* v_k *to any other* v_j $(\forall\, j,k,t\ \exists$ *a none zero sequence* $\{a_{k,j_1}(t), a_{j_1,j_2}(t),\dots,a_{j_m,j}(t)\})$.
- *undirected :* $a_{k,j}(t) = a_{j,k}(t)\,\forall\, j,k,t,$ *(i.e.* $A(t) = A(t)^{\top}$*).*
- *time-dependent : a time-dependent adjacency matrix* $A(t)$ *with* $a_{k,j} : \mathbb{R}^+ \longrightarrow \mathbb{R}^+$ *are* $\mathfrak{C}^1$ *fonctions for all* k,j*, corresponding to the cumulative weights of all edges directly connecting vertex* v_k *to vertex* v_j *at time* t.

Note that in the context of Eqs.(9) the connecting variables (A,κ,ν) belong to Γ, and that the time-evolution of the adjacency matrix A, with entries $a_{k,j}$ and characterized by $\mathsf{D}^{\Gamma}(\Gamma)$, depends on the variables $\kappa(t) \in \mathbb{R}^s$ and on fixed parameters $\nu \in \mathbb{R}^d$.

2.3 Coupling Dynamics

The type of coupling to be considered reads as:

$$\begin{aligned} \mathsf{C}_k^x(x_1,\dots,y_N,\Gamma) &:= \sigma_k \sum_{j=1}^{N} l_{k,j}(t)\Pi_j^x(x_j,y_j),\\ \mathsf{C}_k^y(x_1,\dots,y_N,\Gamma) &:= \sigma_k \sum_{j=1}^{N} l_{k,j}(t)\Pi_j^y(x_j,y_j), \end{aligned} \tag{11}$$

with positive and constant *coupling strength* parameters $\sigma = (\sigma_1,\dots,\sigma_N) \in \Gamma$.

For a given matrix $A(t)$ as in Definition 2, we now define the *Laplacian matrix* $L(t) := A(t) - D(t)$ with entries $l_{k,j}(t)$ where $D(t)$ is the diagonal matrix with $d_{k,k}(t) := \sum_{j=1}^{N} a_{k,j}(t)$, ($l_{k,k}(t) := a_{k,k}(t) - d_{k,k}(t) = -\sum_{j\neq k}^{N} a_{k,j}(t)$). The functions $\Pi_k^x, \Pi_k^y : \mathbb{R}^2 \longrightarrow \mathbb{R}$ are $\mathfrak{C}^1$, only depending on state variables and act as *coupling fonctions*. We now distinguish between *homogenous coupling* and *heterogenous coupling*:

Definition 3.
i) A collection $\{\mathcal{O}_k\}_{k=1}^N$ *of MCD's will have* `homogenous coupling` *if the coupling functions* Π_k^x, Π_k^y *satisfy:*

$$\Pi_j^x(x,y) = \Pi_k^x(x,y) \quad \text{and} \quad \Pi_j^y(x,y) = \Pi_k^y(x,y) \quad \forall j,k \text{ and } (x,y) \in \mathbb{R}^2.$$

ii) A collection $\{\mathcal{O}_k\}_{k=1}^N$ *of MCD's will have* `heterogenous coupling` *if the coupling functions* Π_k^x, Π_k^y *satisfy:*

$$\exists\, k,j \text{ and } (x^*,y^*) \in \mathbb{R}^2 \text{ such that}$$

$$\Pi_j^x(x^*,y^*) \neq \Pi_k^x(x^*,y^*) \quad \text{or} \quad \Pi_j^y(x^*,y^*) \neq \Pi_k^y(x^*,y^*).$$

A few illustrations of homogenous coupling functions are:

	Π^x	Π^y
Identity	$\mathsf{Id}^x(x,y) = x$	$\mathsf{Id}^y(x,y) = y$
Gradient	$H^x(x,y) = \frac{\partial H}{\partial x}(x,y)$	$H^y(x,y) = \frac{\partial H}{\partial y}(x,y)$
Normalized Gradient	$\bar{H}^x(x,y) = \frac{H^x}{\sqrt{(H^x)^2+(H^y)^2}}$	$\bar{H}^y(x,y) = \frac{H^y}{\sqrt{(H^x)^2+(H^y)^2}}$

We shall use of the short-hand notation:

$$\mathsf{C}_k^{\Pi^x}(t) := \sum_{j=1}^N l_{k,j}(t)\Pi_j^x(x_j,y_j),$$

$$\mathsf{C}_k^{\Pi^y}(t) := \sum_{j=1}^N l_{k,j}(t)\Pi_j^y(x_j,y_j).$$

Since the Laplacian matrix is symmetric for all t, its eigenvalues are real. Furthermore, Гершгорин's theorem implies that all eigenvalues are negative for all t. For a connected network $\mathcal{N}(t)$, the dimension of $L(t)$'s kernel $Ker(L(t))$ is one. Since $L(t)$ is diagonalizable and $Ker(L(t))$ is an eigenspace for all t, then the geometric multiplicity of $Ker(L(t))$ equals the algebraic multiplicity of the zero eigenvalue. Thus, if $\mathcal{N}(t)$ is, for all t, connected and undirected, $L(t)$ has one and only one zero eigenvalue (see Appendix A for further details). When $Dim(Ker(L(t))) = 1$, the nature of the coupling dynamics is characterized by:

Lemma 1. *Let* $L(t)$ *be the Laplacian matrix for a given time-dependent, connected network* $\mathcal{N}(t)$. *We have:*

$$\left.\begin{array}{l} C_k^{\Pi^x}(t) = 0 \\ C_k^{\Pi^x}(t) = 0 \end{array} \quad \forall k,t \right\} \Longleftrightarrow \left\{ \begin{array}{l} \Pi_j^x(x_j(t),y_j(t)) = \Pi_k^x(x_k(t),y_k(t)) \\ \Pi_j^y(x_j(t),y_j(t)) = \Pi_k^y(x_k(t),y_k(t)) \end{array} \right. \quad \forall j,k,t.$$

Proof. [$\Rightarrow$] Since the network is connected, the dimension of the kernel of $L(t)$ is one. As $C_k^{\Pi^x}(t)$ (respectively $C_k^{\Pi^y}(t)$) is the product between the kth line of $L(t)$ with $(\Pi_1^x, \ldots, \Pi_N^x)^\top$ (respectively $(\Pi_1^y, \ldots, \Pi_N^y)^\top$), there exist $v(t)$ and $w(t)$ such that:

$$v(t) = \Pi_j^x(x_j(t), y_j(t)) \quad \text{and} \quad w(t) = \Pi_j^y(x_j(t), y_j(t)), \quad \forall j, t$$

and therefore

$$\Pi_j^x(x_j(t), y_j(t)) = \Pi_k^x(x_k(t), y_k(t)) \text{ and } \Pi_k^y(x_j(t), y_j(t)) = \Pi_k^y(x_k(t), y_k(t)), \quad \forall j, k, t\,.$$

[$\Leftarrow$] By definition of the matrix $L(t)$, we have:

$$\sum_{j=1}^{N} l_{k,j}(t)\, \Pi_j^x(x_j(t), y_j(t)) = \Pi_j^x(x_j(t), y_j(t)) \sum_{j=1}^{N} l_{k,j}(t) = 0.$$

Along the same lines for Π_j^y's, the proof follows. $\square$

2.4 Parametric Dynamics

The parametric dynamics in Eqs.(9) with $\Omega_k = \{\omega_k, \eta_k, \mu_k\}$ is given as :

$$\begin{aligned}
\dot{\omega}_k &= \mathsf{D}_1^{\Omega_k}(x_1, y_1, \ldots, x_N, y_N, \Gamma) = \mathsf{K}_k\big(\mathsf{C}_k^{\Psi^x}(t)\, \Psi_k^y - \mathsf{C}_k^{\Psi^y}(t)\, \Psi_k^x\big),\\
\dot{\eta}_k &= \mathsf{D}_2^{\Omega_k}(x_1, y_1, \ldots, x_N, y_N, \Gamma) = 0,\\
\dot{\mu}_k &= \mathsf{D}_3^{\Omega_k}(x_1, y_1, \ldots, x_N, y_N, \Gamma) = 0,
\end{aligned}$$

with positive *susceptibility constant* parameters $\mathsf{K} = (\mathsf{K}_1, \ldots, \mathsf{K}_N) \in \Gamma$. For $\mathsf{K}_k \gg 1$, the kth oscillator is strongly influenced by its neighbor, whereas a $\mathsf{K}_k \ll 1$ reflects its "stubbornness". From now on, η_k, μ_k are fixed positive parameters. The $\mathsf{C}_k^{\Psi^x}(t)$ and $\mathsf{C}_k^{\Psi^y}(t)$ are defined as in Eqs.(11) in which Ψ_j^x and Ψ_j^y are coupling functions to be chosen.

3 Learning Mechanism and Dynamics of the Network

We now look at the behavior of Eqs.(9) with the choices made in section 2.

3.1 Learning Mechanism

The main idea behind subsection 2.4 is to allow ω_k's to evolve in time by the mutual influences of the state variables. This will asymptotically drive the collection of oscillators to adopt a single common ω_c, from now on to be called the *consensual frequency*. Hence, all oscillators have a common set of parameters $\Omega_k = \{\omega_c, \eta_k, \mu_k\}$ for all k. As we shall show, our dynamics

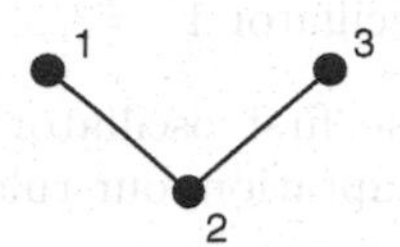

Fig. 1 Network of three oscillators

effectively introduces a *learning mechanism* in the sense that if the connections are removed at time t^* (i.e. $a_{k,j}(t) = 0$ for all $j, k, t > t^*$), then the individual oscillator frequency is ω_c, and no longer ω_k. To unveil the leaning process, let us first consider a simple illustration involving three stable limit cycle oscillators connected shown in Figure 1, which are first assumed to evolve independently (i.e. interactions are removed). We further assume that each oscillator has reached its limit cycle, $\mathcal{L}(\mu)$, (here assumed to be a circle), with frequency ω_k. Without loss of generality, assume that $\omega_1 < \omega_2 < \omega_3$. A snapshot of this situation is sketched in Figure 2, where the gradients H_k and vector fields F_k, $k = 1, 2, 3$ are represented, namely

$$\mathsf{H}_k = \begin{pmatrix} H_k^x(x_k, y_k) \\ H_k^y(x_k, y_k) \end{pmatrix}, \quad \mathsf{F}_k = \begin{pmatrix} H_k^y(x_k, y_k) \\ -H_k^x(x_k, y_k) \end{pmatrix}.$$

To construct a ω_k-learning rule, each oscillator has to adapt its ω_k to those of its connected neighbors. The rule depends on the coupling dynamics $C_k^{\Pi_x}$'s and $C_k^{\Pi_y}$'s. Let us first consider a discrete time reasoning.

Discrete-Time Learning Rule

Let $\{t_n\}_{n=1}^{\infty}$ be a discretization of the time, where $t_1 = 0$ and $t_{n+h} := t_n + h$ for a given small positive h. Let the snapshot in Figure 2 be at time $t = t_n$, when the interactions are turned on. We examine each oscillator behaviour individually:

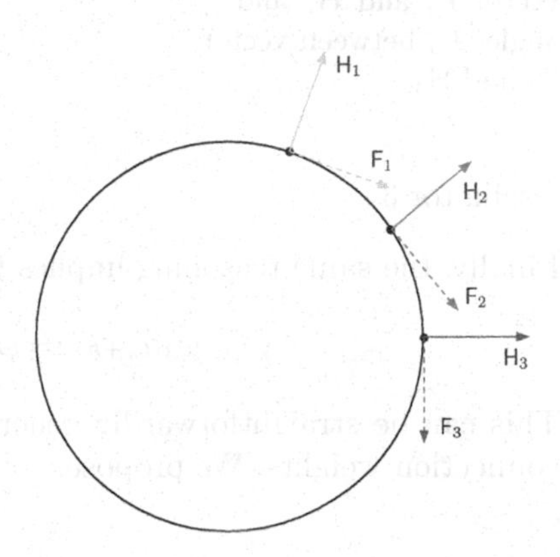

Fig. 2 Three oscillators on $\mathcal{L}(\mu)$

Oscillator 1

The first oscillator has a lower frequency than the second one, hence for adaptation, our rule implies:

at $t = t_n + h$ oscillator 1 must go "faster" to adjust with oscillator 2

Since $\langle \mathsf{H}_2|\mathsf{F}_1\rangle = \|\mathsf{H}_2\|\|\mathsf{F}_1\| \cos(\beta_{12}) > 0$, we propose

$$\omega_{1}(t_n+h) := \omega_{1}(t_n) + h\langle \mathsf{H}_2|\mathsf{F}_1\rangle.$$

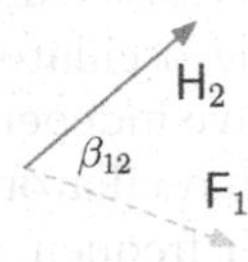

Fig. 3 Representation of the angle β_{12} between vector F_1 and H_2

Oscillator 2

For the second oscillator, adaptation implies:

at $t = t_n + h$ oscillator 2 must go "slower" to adjust with oscillator 1
oscillator 2 must go "faster" to adjust with oscillator 3

Since $\langle \mathsf{H}_1|\mathsf{F}_2\rangle = \|\mathsf{H}_1\|\|\mathsf{F}_2\| \cos(\beta_{21}) < 0$ and $\langle \mathsf{H}_3|\mathsf{F}_2\rangle = \|\mathsf{H}_3\|\|\mathsf{F}_2\| \cos(\beta_{23}) > 0$, we propose:

$$\omega_{2}(t_n+h) := \omega_{2}(t_n) + h\langle \mathsf{H}_1|\mathsf{F}_2\rangle + h\langle \mathsf{H}_3|\mathsf{F}_2\rangle.$$

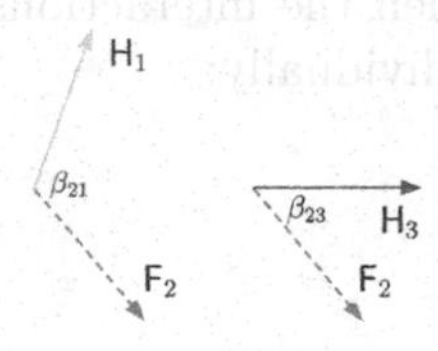

Fig. 4 Representation of the angle β_{21} between vector F_2 and H_1 and angle β_{23} between vector F_2 and H_3

Oscillator 3

Finally, the same reasoning implies for the third oscillator:

$$\omega_{3}(t_n+h) := \omega_{3}(t_n) + h\langle \mathsf{H}_2|\mathsf{F}_3\rangle.$$

This can be straightforwardly generalized to N oscillators by including the connection weights. We propose:

$$\omega_{k(t_n+h)} = \omega_{k(t_n)} + h\sum_{j\neq k}^{N} a_{k,j}\langle \mathsf{H}_j|\mathsf{F}_k\rangle + l_{k,k}\langle \mathsf{H}_k|\mathsf{F}_k\rangle = \omega_{k(t_n)} + h\sum_{j=1}^{N} l_{k,j}\langle \mathsf{H}_j|\mathsf{F}_k\rangle,$$

The second equality is true, simply because $\langle \mathsf{H}_j|\mathsf{F}_j\rangle = 0$ for all j. The continuous time version of this procedure is as follows:

$$\frac{\omega_{k(t_n+h)} - \omega_{k(t_n)}}{h} = \sum_{j=1}^{N} l_{k,j}\langle \mathsf{H}_j|\mathsf{F}_k\rangle$$

and therefore if we let h tend to zero ($h \to 0$), we have

$$\dot{\omega}_{k(t)} = \sum_{j=1}^{N} l_{k,j}\langle \mathsf{H}_j|\mathsf{F}_k\rangle = C_k^{H^x}\, H^y_{k(x_k,y_k)} - C_k^{H^y}\, H^x_{k(x_k,y_k)}.$$

Observe that this learning mechanism depends only on the $C_k^{H^x}$ and $C_k^{H^y}$ which belong to the homogeneous coupling class. However, we may generalize this learning mechanism for arbitraries Ψ_j^x and Ψ_j^y, and for time-dependent networks $\mathcal{N}(t)$.

3.2 Dynamics of the Network

Let us recall the specific dynamical system of interest:

$$\begin{aligned}
\dot{x}_k &= \mathsf{M}_k^x(x_k, y_k, \omega_k; \{\eta_k, \mu_k\}) + \sigma_k \mathsf{C}_k^{\Pi^x}(t),\\
\dot{y}_k &= \mathsf{M}_k^y(x_k, y_k, \omega_k; \{\eta_k, \mu_k\}) + \sigma_k \mathsf{C}_k^{\Pi^y}(t),\\
\dot{\omega}_k &= \mathsf{K}_k\big(\mathsf{C}_k^{\Psi^x}(t)\,\Psi_k^y - \mathsf{C}_k^{\Psi^y}(t)\,\Psi_k^x\big),\\
\dot{\Gamma} &= \mathsf{D}^\Gamma(\Gamma).
\end{aligned} \qquad k = 1, \dots, N, \qquad (12)$$

From now on, the solution for the connection dynamics (i.e. $\dot{\Gamma} = \mathsf{D}^\Gamma(\Gamma)$) is supposed to be given (i.e. time-dependent entries of the adjacency matrix are specified). We will study two cases separately:

1) constant Laplacian matrix, $L(t) := L^0$ which corresponds to $\mathsf{D}^\Gamma \equiv 0$,

2) time-dependent adjacency matrix.

While in case 1) we consider any Laplacian matrix L that fulfills Definition 2 with constant entries, in case 2) we focus on network topologies undergoing *commutative evolution*, i.e. $L(t)\,L(s) = L(s)\,L(t)$ for all times s and t. This subclass of networks offers the possibility of simultaneously diagonalizing all $L(t)$ with an orthogonal matrix V with real entries (i.e. $V \in \mathrm{Mat}_{N,N}(\mathbb{R})$) (c.f. [17]).

A class of illustrations of such matrices are *circulant matrices* (c.f. [16]) which have the form:

$$C := \mathsf{circ}(c_1, \dots, c_N),$$
$$= \begin{pmatrix} c_1 & c_2 & \cdots & c_N \\ c_N & c_1 & \cdots & c_{N-1} \\ \vdots & \vdots & \ddots & \vdots \\ c_2 & c_3 & \cdots & c_1 \end{pmatrix},$$

and where $c_1, \dots, c_N$ will here be considered here real numbers. Since we are working with Laplacian matrices related to undirected networks, we restrict symmetric to circulant matrices with $c_j \geqslant 0$ for $j = 2, \dots, N$, and $c_1 = -\sum_{j=2}^{N} c_j$. The symmetry of the matrix implies:

$$\begin{array}{ll} N = 0 \mod 2 & N = 1 \mod 1 \\ c_{2+j} = c_{N-j} \quad j = 1, \dots, \frac{N}{2} - 2 & c_{2+j} = c_{N-j} \quad j = 1, \dots, \frac{N+1}{2} - 2. \\ c_{\frac{N}{2}+1} \text{ no restrictions} & \end{array}$$

A famous property of circulant matrices is precisely that they are diagonalizable by the Fourrier matrix, so we can analytically express their eigenvalues (c.f. [16] for details). Eigenvalues of a Laplacian matrix and in particular the non-vanishing closest-to-zero eigenvalue known as the *algebraic connectivity* or *Fiedler number* (c.f. [10], [4] and [6]), characterize the connectivity of the network. For symmetric circulant matrices, we have:

$$\begin{aligned} & N = 0 \mod 2 \\ & \lambda_j = c_1 + c_{\frac{N}{2}+1} \cos(\pi(j-1)) + \sum_{k=2}^{\frac{N}{2}} 2c_k \cos(\frac{2\pi}{N}(j-1)(k-1)), \\ & N = 1 \mod 2 \\ & \lambda_j = c_1 + \sum_{k=2}^{\frac{N+1}{2}} 2c_k \cos(\frac{2\pi}{N}(j-1)(k-1)). \end{aligned} \tag{13}$$

For simple, no loop, connected networks with $a_{k,j} \in \{0, 1\}$, an important subclass with a symmetric circulant matrix as their Lapalcian matrix are the (N, k) *regular lattices* defined in [10]:

Definition 4. *An* (N, k) `regular lattice` *denoted as* $\mathcal{C}_{N,k}$ *is*

- *k-regular network (i.e. all vertices have degree k)*
- *the N vertices are evenly spaced on a ring in which each vertex is connected to its k nearest neighbors.*

Every $\mathcal{C}_{N,k}$ can be straightforwardly generalized by introducing time-dependent weights $a_{k,j}(t)$ while preserving the *commutative evolution* of the Laplacian matrix.

3.3 Synchronization of the Network

In Eqs.(12), the specific type of synchronization mechanism that we will consider fulfills:

Definition 5. *For a given initial state* $\mathcal{S}(0)$ *with state variables lying on* $\mathcal{L}_k(\mu_k)$*, the pair* $(\mathcal{S}(t), \mathcal{S}(0))$ *is the* **synchronized limit cycle solution** *(SLC) of Eqs.(12) if*

- $\mathsf{C}_k^{\Psi^x}(t) = \mathsf{C}_k^{\Psi^y}(t) = 0$ *for all* k *and time* t.
- $\mathcal{S}(t) = (x_1^{(s)}(t), y_1^{(s)}(t), \omega_c, \ldots, x_N^{(s)}(t), y_N^{(s)}(t), \omega_c) \in \mathbb{R}^{3N}$ *with,* $\forall\, k$, $(x_k^{(s)}(t), y_k^{(s)}(t))$ *T-periodic functions solving the* k^{th} *canonical MCD evolution.*

In particular, Definition 4 implies that for a SLC solution:

i) $\omega_k(t)$ is a common constant $\omega_c > 0$ imposed by the initial conditions $\mathcal{S}(0)$, ($\omega_c > 0$ will be chosen, by convention, to be positive),
ii) there is no coupling dynamics $\mathsf{C}_k^{\Pi^x}(t) = \mathsf{C}_k^{\Pi^y}(t) = 0$ for all k and all time t.

Now we can establish the existence of a constant of motion:

Proposition 1. *For* $\omega_k(t)$ *solutions of the dynamical system given in Eqs.(12), then*

$$\mathsf{J}^0 := \sum_{k=1}^{N} \frac{\omega_k(t)}{\mathsf{K}_k} \tag{14}$$

is a constant of motion.

Proof. By direct calculation, we have:

$$\begin{aligned}
&\sum_{k=1}^{N} \frac{1}{\mathsf{K}_k}\Big(\mathsf{K}_k\big(C_k^{\Psi^x}(t)\Psi_k^y - C_k^{\Psi^y}(t)\Psi_k^x\big)\Big) \\
&= \sum_{k=1}^{N}\Big[\Big(\sum_{j=1}^{N} l_{k,j}(t)\Psi_j^x(x_j,y_j)\Big)\Psi_k^y(x_k,y_k) - \Big(\sum_{j=1}^{N} l_{k,j}(t)\Psi_j^y(x_j,y_j)\Big)\Psi_k^x(x_k,y_k)\Big] \\
&\underbrace{=}_{l_{k,j}=l_{j,k}} \sum_{j=1}^{N}\Big[\sum_{k=1}^{N} l_{j,k}(t)\Psi_k^y(x_k,y_k)\Psi_j^x(x_j,y_j)\Big] - \sum_{k=1}^{N}\Big[\sum_{j=1}^{N} l_{k,j}(t)\Psi_j^y(x_j,y_j)\Psi_k^x(x_k,y_k)\Big] \\
&\underbrace{=}_{s=j,l=k} \sum_{s=1}^{N}\Big[\sum_{l=1}^{N} l_{s,l}(t)\Psi_l^y(x_l,y_l)\Psi_s^x(x_s,y_s)\Big] - \sum_{k=1}^{N}\Big[\sum_{j=1}^{N} l_{k,j}(t)\Psi_j^y(x_j,y_j)\Psi_k^x(x_k,y_k)\Big] \\
&= 0\,.
\end{aligned}$$

□

Illustration. Existence of a synchronized solution for heterogeneous oscillators

Let us explicitly illustrate the possibility of a *SLC* involving heterogeneous oscillators.

$$H(x,y) = \frac{1}{2}(x^2+y^2) \qquad H_{ML}(x,y) = \log(\cosh(y)) + \frac{1}{2}\log(1+x^2)$$

$$\begin{aligned}
\dot{x}_k &= \omega_k y_k && + \eta_k(\tfrac{\mathsf{a}^2}{2} - H(x_k,y_k))x_k && + \sigma_k \mathsf{C}_k^{\mathsf{Id}^x}(t)\\
\dot{y}_k &= -\omega_k x_k && + \eta_k(\tfrac{\mathsf{a}^2}{2} - H(x_k,y_k))y_k && + \sigma_k \mathsf{C}_k^{ML^y}(t)\\
\dot{\omega}_k &= \mathsf{K}_k\big(\mathsf{C}_k^{\mathsf{Id}^x}(t)\, y_k - \mathsf{C}_k^{ML^y}(t)\, x_k\big)\\
\dot{x}_s &= \omega_s\sqrt{1+\mathsf{a}^2}\tanh(y_s) + \eta_s(\mu - H_{ML}(x_s,y_s))\tfrac{x_s}{1+x_s^2} && && + \sigma_s \mathsf{C}_s^{\mathsf{Id}^x}(t)\\
\dot{y}_s &= -\omega_s\sqrt{1+\mathsf{a}^2}\tfrac{x_s}{1+x_s^2} && + \eta_s(\mu - H_{ML}(x_s,y_s))\tanh(y_s) && + \sigma_s \mathsf{C}_s^{ML^y}(t)\\
\dot{\omega}_s &= \mathsf{K}_s\big(\mathsf{C}_s^{\mathsf{Id}^x}(t)\sqrt{1+\mathsf{a}^2}\tanh(y_s) - \mathsf{C}_s^{ML^y}(t)\, x_s\big)
\end{aligned} \tag{15}$$

where $\mu = \frac{1}{2}\log(1+\mathsf{a}^2)$ ($\mathsf{a} > 0$ to be given)

$$\mathsf{C}_k^{ML^y}(t) := \sum_{j=1}^{n} l_{k,j}(t) y_j + \sum_{j=n+1}^{N} l_{k,j}(t)\sqrt{1+\mathsf{a}^2}\tanh(y_j)$$

The *SLC* solution:

initial condition

$$(x_k(0), y_k(0), \omega_k(0)) = (\mathsf{a}, 0, \omega_c) \quad \forall\, k = 1, \ldots, N,$$

orbit

$$\begin{aligned}
x_k(t) &= \mathsf{a}\sin(\omega_c t + \tfrac{\pi}{2}), & x_s(t) &= \mathsf{a}\sin(\omega_c t + \tfrac{\pi}{2}),\\
y_k(t) &= \mathsf{a}\cos(\omega_c t + \tfrac{\pi}{2}), & y_s(t) &= \tanh^{-1}(\tfrac{\mathsf{a}}{\sqrt{1+\mathsf{a}^2}}\cos(\omega_c t + \tfrac{\pi}{2})).
\end{aligned}$$

In general, heterogeneous dynamics of the type exposed in this examples offers little possibility for deriving further general conclusions. For collections of homogeneous oscillators, however, additional properties can be derived as shown next.

3.4 Homogenous Collections MCD

Now we focus on a collection of homogenous MCD with, in Eqs.(12), common normalized gradient couplings functions $\Pi^x \equiv \bar{H}^x$ (c.f. 2.3) and $\Pi^y \equiv \bar{H}^y$ and $\Psi^x \equiv \bar{H}^x$ and $\Psi^y \equiv \bar{H}^y$ (c.f. 2.4). For this configuration, we have:

Lemma 2. *Let $\{\mathcal{O}_k\}_{k=1}^N$ be a collection of homogenous MCD's and $\mathcal{L}_{k(\mu_k)}$ their respective stable limit cycle. For $k = 1, \ldots, N$, let*

$$\begin{aligned} \psi_k : \mathbb{R}^+ &\longrightarrow \mathcal{L}_{k(\mu_k)} \\ t &\longmapsto (x_{k(t)}, y_{k(t)}), \end{aligned}$$

be the solution of

$$\begin{cases} \dot{x} = & H^{y_k}(x,y) \\ \dot{y} = & -H^{x_k}(x,y) \end{cases} \quad (x_{k(0)}, y_{k(0)}) \in \mathcal{L}_{k(\mu_k)}$$

where the initial conditions $\psi_{j(0)}$ $j = 1, \ldots, N$ are two by two collinear (i.e. $\forall\, k\, \exists\, \{b_{k,j}\}_{j=1}^N$ such that $\psi_{k(0)} = b_{k,j}\psi_{j(0)}\, \forall\, j$). Then, for any k, we have

$$\psi_{k(s)} = b_{k,j}\psi_{j(s)} \Longleftrightarrow \begin{pmatrix} H^{x_k}(\psi_{k(s)}) \\ H^{y_k}(\psi_{k(s)}) \end{pmatrix} = b_{k,j} \begin{pmatrix} H^{x_j}(\psi_{j(s)}) \\ H^{y_j}(\psi_{j(s)}) \end{pmatrix}$$

for all j.

Proof. If we have $\dot{\psi}_{k(s)} = b_{k,j}\dot{\psi}_{j(s)}$ then $\psi_{k(s)} = T_{k,j} + b_{k,j}\psi_{j(s)}$, $T_{k,j}$ a constant and since $\psi_{k(0)} = b_{k,j}\psi_{j(0)}$ then

$$\psi_{k(s)} = b_{k,j}\psi_{j(s)} \Leftrightarrow \dot{\psi}_{k(s)} = b_{k,j}\dot{\psi}_{j(s)} \Leftrightarrow$$

$$\begin{pmatrix} H^{y_k}(\psi_{k(s)}) \\ -H^{x_k}(\psi_{k(s)}) \end{pmatrix} = b_{k,j} \begin{pmatrix} H^{y_j}(\psi_{j(s)}) \\ -H^{x_j}(\psi_{j(s)}) \end{pmatrix} \Leftrightarrow \begin{pmatrix} H^{x_k}(\psi_{k(s)}) \\ H^{y_k}(\psi_{k(s)}) \end{pmatrix} = b_{k,j} \begin{pmatrix} H^{x_j}(\psi_{j(s)}) \\ H^{y_j}(\psi_{j(s)}) \end{pmatrix}$$

□

Lemma 2 exhibit the necessary and sufficient conditions for $\mathsf{C}_k^{\bar{H}^x}(t) = 0$ and $\mathsf{C}_k^{\bar{H}^y}(t) = 0$ to hold for all t, and this even if not all individual limit cycles are identical. This is now exploited.

Proposition 2 (Existence of a synchronized solution). *For a given constant $\omega_c > 0$ and with $(\psi_{k(t)}, \psi_{k(0)})$ satisfying Lemma 2 for all k, the pair $(\mathcal{S}(t), \mathcal{S}(0))$ with $\mathcal{S}(0)$ as initial condition and where*

$$\mathcal{S}(t) := (\omega_c\psi_{1(t)}, \omega_c, \ldots, \omega_c\psi_{N(t)}, \omega_c) \in \mathbb{R}^{3N},$$

is the SLC solving Eqs.(12).

Proof. Since $(\psi_{k(t)}, \psi_{k(0)})$ satisfies Lemma 2, then, for any k, $\bar{H}^{x_k}(x_k,y_k) = \bar{H}^{x_j}(x_j,y_j)$ and $\bar{H}^{y_k}(x_k,y_k) = \bar{H}^{y_j}(x_k,y_k)$ for all j. Due to the diffusive nature of the coupling, the terms $\mathsf{C}_k^{\bar{H}^x}(t) = 0$ and $\mathsf{C}_k^{\bar{H}^y}(t) = 0$ for all k and times t. Hence, $\omega_{k(t)} = \omega_c$. The $(\psi_{k(t)}, \psi_{k(0)})$ solve the canonical part of the kth MCD, they have, due to Lemma 2, the same period.

□

3.5 Stability of the SLC Solutions

Let us consider a `perturbed` SLC `solution` PLC defined as:

$$\check{\mathcal{S}}(t) = \mathcal{S}(t) + \mathcal{E}(t)$$

with $\mathcal{S}(t)$ the SLC and $\mathcal{E}(t)$ small perturbations introduced to perform a linear stability analysis. We raise the question: do the perturbations $\mathcal{E}(t)$ decay in time?

To make head on, assume decay occurs implying that $\lim_{t\to\infty} \omega_k(t) = \omega_c$ for all k. In this case the resulting frequency ω_c can be analytically expressed. Given initial ferquencies $\omega_k(0)$, (k^{th} frequency when the coupling is turned on) and due to Eq.(14), $\sum_{k=1}^{N} \omega_k(t) = J^0$ for all t and in particular for the initial time $t = 0$. This implies that:

$$\text{if} \quad \lim_{t\to\infty} \omega_k(t) = \omega_c \qquad \text{then} \qquad \sum_{k=1}^{N} \frac{\omega_c}{\mathsf{K}_k} = \sum_{k=1}^{N} \frac{\omega_k(0)}{\mathsf{K}_k}.$$

Therefore, we conclude that a consensual frequency is reached and

$$\text{Consensual frequency:} \qquad \omega_c = \frac{\sum_{j=1}^{N} \frac{\omega_j(0)}{\mathsf{K}_j}}{\sum_{j=1}^{N} \frac{1}{\mathsf{K}_j}}. \tag{16}$$

The convergence of PLC towards SLC can be derived explicitly for specific cases which are now discussed.

4 Network of Hopf Oscillators (NHO)

In what follows, we focus on MCD with $H(x,y) = \frac{1}{2}(x^2 + y^2)$ with $\eta_k > 0$, and $g(s;\mu_k) = \mu_k - s$.

4.1 NHO with Time-Dependent Network Connections

Here the coupling functions are chosen to be of Normalized Gradient type (for both, the coupling and parametric dynamics) (see 2.3 and 2.4). Furthermore, we chose $\mathsf{K}_k := \mathsf{K}$ and $\sigma_k := \sqrt{2\mu_k}\,\mathsf{K}$ and the Laplacian matrices are such that $L(t)L(s) = L(s)L(t)$ (i.e. commutative evolution). Written in polar coordinates, we obtain:

$$\begin{aligned}
\dot{r}_k &= \eta_k(\mu_k - \frac{1}{2}r_k^2)r_k + \mathsf{K}\sum_{j=1}^{N} l_{kj}(t)\cos(\phi_k - \phi_j), \\
\dot{\phi}_k &= -\omega_k - \frac{\sqrt{2\mu_k}\mathsf{K}}{r_k}\sum_{j\neq k}^{N} l_{kj}(t)\sin(\phi_k - \phi_j), \\
\dot{\omega}_k &= \mathsf{K}\sum_{j\neq k}^{N} l_{kj}(t)\sin(\phi_k - \phi_j).
\end{aligned} \tag{17}$$

At initial time $t = 0$, we assume given:

- $\{\omega_k(0)\}_{k=1}^N$ the frequencies of the kth oscillator,
- $\{\theta_k(0)\}_{k=1}^N$ the phases of the kth oscillator,
- $\{r_k(0)\}_{k=1}^N$ the radii of the kth oscillator.

Define:

$$\omega_c := \frac{1}{N}\sum_{j=1}^{N}\omega_j(0) \qquad \text{and} \qquad \theta^0 := \frac{1}{N}\sum_{j=1}^{N}\theta_j(0).$$

The SLC solution is:

$$\mathcal{S}(t) = (\sqrt{2\mu_1}, \theta_s(t), \omega_c, \ldots, \sqrt{2\mu_N}, \theta_s(t), \omega_c) \in \mathbb{R}^{3N} \tag{18}$$

with $\theta_s(t) := -\omega_c t + \theta^0$. Perturbations read:

$$\begin{aligned}
\rho_k(t) &:= r_k(t) - \sqrt{2\mu_k} &\Leftrightarrow r_k(t) &= \sqrt{2\mu_k} + \rho_k(t) \\
\tau_k(t) &:= \theta_k(t) - \theta_s(t) &\Leftrightarrow \theta_k(t) &= \theta_s(t) + \tau_k(t) \\
\epsilon_k(t) &:= \omega_k(t) - \omega_c &\Leftrightarrow \omega_k(t) &= \omega_c + \epsilon_k(t).
\end{aligned}$$

The associated PLC solutions is:

$$\check{\mathcal{S}}(t) = (r_1(t), \theta_1(t), \omega(t), \ldots, r_N(t), \theta_N(t), \omega_N(t)) \in \mathbb{R}^{3N}. \tag{19}$$

Theorem 1. *With the previous hypothesis, the PLC $\check{\mathcal{S}}(t)$ solution given in (19) converges towards the $\mathcal{S}(t)$ solution in (18), i.e.*

$$\rho_k(t),\ \tau_k(t) \text{ and } \epsilon_k(t) \xrightarrow[t\to\infty]{} 0, \quad \forall k.$$

The convergence rate is controlled by the algebraic connectivity of the network.

Proof. We linearize the vector field given by Eqs.(17) around the SLC solution in (18). Without loss of generalization we arrange the variables as:

$$\begin{pmatrix}\dot{\rho}\\ \dot{\tau}\\ \dot{\epsilon}\end{pmatrix} = \begin{pmatrix} -2\,[\eta]\,[\mu] & \mathbb{O} & \mathbb{O} \\ \mathbb{O} & \mathsf{K}L(t) & -\mathrm{Id} \\ \mathbb{O} & -\mathsf{K}L(t) & \mathbb{O}\end{pmatrix}\begin{pmatrix}\rho\\ \tau\\ \epsilon\end{pmatrix}, \tag{20}$$

with $\rho := (\rho_1, \ldots, \rho_N)$, $\tau := (\tau_1, \ldots, \tau_N)$ and $\epsilon := (\epsilon_1, \ldots, \epsilon_N)$ and where $[\eta]$ and $[\mu]$ are diagonal matrices with entries η_k and μ_k and Id is the identity.

The upper left $(N \times N)$-bloc in Eqs.(20) has N real negative eigenvalues, thus the radial perturbations decay exponentially.

Since $L(t)$ is symmetric and commutes with all other Laplacian $L(s)$, there exists an orthogonal matrix V with real entries, time-independent such that $V^\top L(t) V$ is a diagonal matrix $\Lambda(t)$ with its spectrum $\{\lambda_k(t)\}_{k=1}^N$. As V is time-independent, for a change of variable $(\tilde{\tau}, \tilde{\epsilon}) := (V^\top \tau, V^\top \epsilon)$ we have $(\dot{\tilde{\tau}}, \dot{\tilde{\epsilon}}) = (V^\top \dot{\tau}, V^\top \dot{\epsilon})$. Therefore, changing the basis for the remaining $(2N \times 2N)$-bloc system by a $(2N \times 2N)$-bloc matrix with $V^\top$ on its diagonal, we have:

$$\begin{pmatrix} \dot{\tilde{\tau}} \\ \dot{\tilde{\epsilon}} \end{pmatrix} = \begin{pmatrix} \mathsf{K}\Lambda(t) & -\mathrm{Id} \\ -\mathsf{K}\Lambda(t) & \mathbb{O} \end{pmatrix} \begin{pmatrix} \tilde{\tau} \\ \tilde{\epsilon} \end{pmatrix}$$

which is reducible to N 2-dimensional systems of the form:

$$\begin{pmatrix} \dot{\tilde{\tau}}_k \\ \dot{\tilde{\epsilon}}_k \end{pmatrix} = \begin{pmatrix} \mathsf{K}\lambda_k(t) & -1 \\ -\mathsf{K}\lambda_k(t) & 0 \end{pmatrix} \begin{pmatrix} \tilde{\tau}_k \\ \tilde{\epsilon}_k \end{pmatrix}.$$

Without loss of generality, we assume $\lambda_1(t) = 0$, (basic property for connected networks). For $k = 1$, we have:

$$\dot{\tilde{\tau}}_1 = -\tilde{\epsilon}_1, \qquad \dot{\tilde{\epsilon}}_1 = 0.$$

and so $\tilde{\epsilon}_1(t) = \tilde{\epsilon}_1(0)$ which itself is zero because the first orthonormal base vector is $\frac{1}{\sqrt{N}}(1, \ldots, 1)^\top$ and the first coordinate of the product $V^\top \epsilon$ is

$$\tilde{\epsilon}_1(0) = \frac{1}{\sqrt{N}} \sum_{j=1}^{N} \epsilon_j(0) = \frac{1}{\sqrt{N}} \Big(\sum_{j=1}^{N} \omega_j(0) - N\omega_c \Big) = 0.$$

Along the same lines we have $\tilde{\tau}_1(t) = \tilde{\tau}_1(0) = 0$.

For $k \neq 1$, (below we omit the index),

$$\ddot{\tilde{\tau}}(t) - \mathsf{K}\lambda(t)\dot{\tilde{\tau}}(t) \underbrace{-\mathsf{K}\big(\dot{\lambda}(t) + \lambda(t)\big)}_{+a_0(t)} \tilde{\tau}(t) = 0. \tag{21}$$

Despite the linearity of Eq.(21) the stability of the $\tilde{\tau}(t) = 0$ solution requires care, as Eq. (21) exhibits time-dependent coefficients. For our specific case we have $-M < \lambda(t) < 0$ with M finite. This enables us to invoke Theorem 1.3 from [5], implying that $a_0(t) > 0$ is a necessary condition for $\tilde{\tau} = 0$ to be asymptotically stable. □

Remark. Rapid time variations of the underlying network, occurring when $\left|\frac{d \ln}{dt}(-\lambda(t))\right| > 1$, alter the positivity of $a_0(t)$, and stability of the $\tilde{\tau} = 0$ can no longer be guaranteed. This opens the possibility for *parametric pumping* induced by the network configuration - an issue we shall address separately.

4.2 NHO with Constant Network Connections

Remember that for NHO we have $H_{(x,y)} = \frac{1}{2}(x^2 + y^2)$, $\eta_k > 0$, and $g_{(s;\mu_k)} = \mu_k - s$. The coupling type is the Normalized Gradient for both the coupling and parametric dynamics (see 2.3 and 2.4). Furthermore, $\mathsf{K}_k > 0$ and $\sigma_k := \sqrt{2\mu_k}\,\mathsf{K}_k$. The network is connected and undirected with Lapalcian matrix L. We define, as in subsection 4.1, the *SLC* and *PLC* solutions with here the consensual frequency and initial phase shift as:

$$\omega_c := \frac{\sum_{j=1}^{N} \frac{\omega_j(0)}{\mathsf{K}_j}}{\sum_{j=1}^{N} \frac{1}{\mathsf{K}_j}}, \qquad \theta^0 := \frac{\sum_{j=1}^{N} \frac{\theta_j(0)}{\mathsf{K}_j}}{\sum_{j=1}^{N} \frac{1}{\mathsf{K}_j}}.$$

Theorem 2. *With the previous hypotheses we have:*

$$\rho_k(t), \tau_k(t) \text{ and } \epsilon_k(t) \xrightarrow[t\to\infty]{} 0 \quad \forall k$$

implying that $\check{\mathcal{S}}(t) \xrightarrow[t\to\infty]{} \mathcal{S}(t)$. *The convergence rate depends on* $\lambda_{2(K)}$ *where* $\lambda_{N(K)} \leqslant \cdots \leqslant \lambda_{2(K)} < 0$ *and* $\lambda_{1(K)} = 0$ *are the eigenvalues of*

$$\mathsf{K}^{1/2} L \mathsf{K}^{1/2} \quad \text{with } \mathsf{K}_{kk}^{1/2} = \sqrt{\mathsf{K}_k}\ .$$

Proof. As before, linear analysis implies:

$$\begin{pmatrix} \dot{\rho} \\ \dot{\tau} \\ \dot{\epsilon} \end{pmatrix} = \begin{pmatrix} -2\,[\eta]\,[\mu] & \mathbb{O} & \mathbb{O} \\ \mathbb{O} & [\mathsf{K}]\,L & -\mathrm{Id} \\ \mathbb{O} & -[\mathsf{K}]\,L & \mathbb{O} \end{pmatrix} \begin{pmatrix} \rho \\ \tau \\ \epsilon \end{pmatrix} \tag{22}$$

where $[\mathsf{K}]$ is the diagonal matrix with entries $\mathsf{K}_1, \mathsf{K}_2, \ldots, \mathsf{K}_N$. Again, as in Theorem 1, radial perturbations decay exponentially. Let $L_{(\mathsf{K})} := \mathsf{K}^{\frac{1}{2}}\, L\, \mathsf{K}^{\frac{1}{2}}$ and consider S an orthogonal matrix such that $S^{\top} L_{(\mathsf{K})} S = \Lambda_{(\mathsf{K})}$, a diagonal matrix. The matrix $L_{(\mathsf{K})}$ is a left and right multiplication of L by an identical positive definite diagonal matrix $\mathsf{K}^{\frac{1}{2}}$, hence the sign of its spectrum coincides with L, namely $\lambda_{k(\mathsf{K})} < 0$ for $k = 2, \ldots, N$ and $\lambda_{1(\mathsf{K})} = 0$. Changing the basis of the system by means of a $(2N \times 2N)$-bloc matrix with $S^{\top}\mathsf{K}^{-\frac{1}{2}}$ on its diagonal gives us:

$$\begin{pmatrix} \dot{\tilde{\tau}} \\ \dot{\tilde{\epsilon}} \end{pmatrix} = \begin{pmatrix} \Lambda_{(\mathsf{K})} & -\mathrm{Id} \\ -\Lambda_{(\mathsf{K})} & \mathbb{O} \end{pmatrix} \begin{pmatrix} \tilde{\tau} \\ \tilde{\epsilon} \end{pmatrix}$$

which is reducible to N 2-dimensional systems:

$$\begin{pmatrix} \dot{\tilde{\tau}}_k \\ \dot{\tilde{\epsilon}}_k \end{pmatrix} = \begin{pmatrix} \lambda_{k(\mathsf{K})} & -1 \\ -\lambda_{k(\mathsf{K})} & 0 \end{pmatrix} \begin{pmatrix} \tilde{\tau}_k \\ \tilde{\epsilon}_k \end{pmatrix}. \tag{23}$$

For $k = 1$, we have:

$$\dot{\tilde{\tau}}_1 = -2\tilde{\epsilon}_1 \quad \dot{\tilde{\epsilon}}_1 = 0$$

and so $\tilde{\epsilon}_1(t) = \tilde{\epsilon}_{1,0}$ which is zero. This is because the first orthonormal base vector is $N_1(\mathsf{K}_1^{\frac{1}{2}}, \ldots, \mathsf{K}_N^{\frac{1}{2}})^\top$ ($N_1 := \sum_{j=1}^{N} \frac{1}{\mathsf{K}_j}$) and the first coordinate of the product $S^\top \mathsf{K}^{\frac{1}{2}} \epsilon$ is

$$\tilde{\epsilon}_1(t) = N_1 \sum_{j=1}^{N} \frac{\epsilon_k(t)}{\mathsf{K}_j} = N_1 (\sum_{j=1}^{N} \frac{\omega_k(t)}{\mathsf{K}_j} - \omega_c \sum_{j=1}^{N} \frac{1}{\mathsf{K}_j}) = 0.$$

We then have $\tilde{\tau}_1(t) = \tilde{\tau}_{1,0}$. The same reasoning enables us to conclude that $\tilde{\tau}_{1,0} = 0$. For $k \neq 1$, the eigenvalues of the Eq.(23) are:

$$\alpha_{k,\pm} = \frac{1}{2}\lambda_k(\mathsf{K}) \pm \frac{1}{2}\sqrt{\lambda_k(\mathsf{K})^2 + 8\mu\lambda_k(\mathsf{K})}.$$

For $\lambda_k(\mathsf{K}) \in]-\infty, -8\mu]$ then $\alpha_{k,\pm} < 0$. For $\lambda_k(\mathsf{K}) \in]-8\mu, 0[$, then $\Re(\alpha_{k,\pm}) < 0$. This ensures an exponential asymptotic convergence towards $\mathcal{S}(t)$. □

4.3 NHO with Constant Network and Identical Limit Cycle

In this case we have $H_{(x,y)} = \frac{1}{2}(x^2 + y^2)$ and $\eta_k > 0$ with here $g_{(s;\mu)} = \mu - s$ (identical stable limit cycle for all oscillators). The coupling is to be of Gradient type (for both the coupling and parametric dynamics) and the coupling strength σ_k are given by $\mathsf{L}_k > 0$, not necessarily equal to the susceptibility constants $\mathsf{K}_k > 0$. The connections are defined via a constant Laplacian matrix L. This dynamics allows to explicitly construct a Ляпунов function, to establish that the compact set $\mathcal{A} := \bigcup_{k=1}^{N} \{(\mathcal{L}_{k(\mu)}, \{\omega_c\}_k)\}_k$ is asymptotically stable (c.f. [14]). The Ляпунов function reads:

$$Л_{(x,y,\omega)} := \sum_{j=1}^{N} \frac{\eta_j}{2\mathsf{L}_j}(g(H_j))^2 + \sum_{j=1}^{N} \frac{1}{2\mathsf{K}_j}(\omega_j - \omega_c)^2 - \frac{1}{2} H_x \, L \, H_x^\top - \frac{1}{2} H_y \, L \, H_y^\top$$

with $x := (x_1, \ldots, x_N), y := (x_1, \ldots, y_N)$, $\omega := (\omega_1, \ldots, \omega_N)$ and with $H_x := (H_1^x, \ldots, H_N^x)$ and $H_y := (H_1^y, \ldots, H_N^y)$. Since $g'(s) = -1$, $\frac{\partial^2 H}{\partial x^2}_{(x_k, y_k)} = \frac{\partial^2 H}{\partial y^2}_{(x_k, y_k)} = 1$ and $\frac{\partial^2 H}{\partial x \partial y} = \frac{\partial^2 H}{\partial y \partial x} = 0$, the partial derivatives are:

$$\frac{\partial Л}{\partial x_k} = -\Big(\frac{\eta_k g(H_k) H_k^x}{\mathsf{L}_k} + C_k^{H^x}\Big), \qquad \frac{\partial Л}{\partial y_k} = -\Big(\frac{\eta_k g(H_k) H_k^y}{\mathsf{L}_k} + C_k^{H^y}\Big),$$

$$\frac{\partial Л}{\partial \omega_k} = \frac{\omega_k - \omega_c}{\mathsf{K}_k}.$$

Computing the time derivatives:

$$\begin{aligned}\frac{\partial Л}{\partial x_k}\dot{x}_k &= -\Big(\tfrac{\eta_k}{\mathsf{L}_k} g(H_k)H_k^x + C_k^{H^x}\Big)\Big(\omega_k H_k^y + \eta_k g(H_k)H_k^x + \mathsf{L}_k C_k^{H^x}\Big)\\ &= -\tfrac{\eta_k}{\mathsf{L}_k} g(H_k)H_k^x\omega_k H_k^y - C_k^{H^x}\omega_k H_k^y - \mathsf{L}_k\Big(\tfrac{\eta_k}{\mathsf{L}_k} g(H_k)H_k^x + C_k^{H^x}\Big)^2,\end{aligned}$$

$$\begin{aligned}\frac{\partial Л}{\partial y_k}\dot{y}_k &= -\Big(\tfrac{\eta_k}{\mathsf{L}_k} g(H_k)H_k^y + C_k^{H^y}\Big)\Big(-\omega_k H_k^x + \eta_k g(H_k)H_k^y + \mathsf{L}_k C_k^{H^y}\Big)\\ &= \tfrac{\eta_k}{\mathsf{L}_k} g(H_k)H_k^y\omega_k H_k^x + C_k^{H^y}\omega_k H_k^x - \mathsf{L}_k\Big(\tfrac{\eta_k}{\mathsf{L}_k} g(H_k)H_k^y + C_k^{H^y}\Big)^2,\end{aligned}$$

$$\frac{\partial Л}{\partial \omega_k}\dot{\omega}_k = \tfrac{1}{\mathsf{K}_k}(\omega_k - \omega_c)\mathsf{K}_k\Big(C_k^{H^x} H_k^y - C_k^{H^y} H_k^x\Big)$$

and therefore

$$\begin{aligned}&\sum_{j=1}^{N} \tfrac{\partial Л}{\partial x_j}\dot{x}_j + \tfrac{\partial Л}{\partial y_j}\dot{y}_j + \tfrac{\partial Л}{\partial \omega_j}\dot{\omega}_j\\ &= -\sum_{j=1}^{N} \mathsf{L}_j\Big[\big(\tfrac{\eta_j}{\mathsf{L}_j} g(H_j)H_j^x + C_j^{H^x}\big)^2 + \big(\tfrac{\eta_j}{\mathsf{L}_j} g(H_j)H_j^y + C_j^{H^y}\big)^2\Big]\\ &\quad + \sum_{j=1}^{N}\omega_j\big(-C_j^{H^x}H_j^y + C_j^{H^y}H_j^x\big) + \sum_{j=1}^{N}\omega_j\Big[C_j^{H^x}H_j^y - C_j^{H^y}H_j^x\Big]\\ &\quad \underbrace{-\omega_c\sum_{j=1}^{N}\Big[C_j^{H^x}H_j^y - C_j^{H^y}H_j^x\Big]}_{=0,\ \text{constant of motion}}\\ &= -\sum_{j=1}^{N} \mathsf{L}_j\Big[\big(\tfrac{\eta_j}{\mathsf{L}_j} g(H_j)H_j^x + C_j^{H^x}\big)^2 + \big(\tfrac{\eta_j}{\mathsf{L}_j} g(H_j)H_j^y + C_j^{H^y}\big)^2\Big] < 0.\end{aligned}$$

Note that having explicitly constructed a Ляпунов function allows us to conclude that the attractor $\mathcal{A}$ is globally stable.

5 Numerical Simulations

We report numerical simulations preformed with oscillators defined in 2.1. In all numerical simulations, parameter $\eta_k = 1$ for all k. Three different topologies are considered: "(4,2) regular lattice ", "All to All "and "Königsberg "(c.f. Figure 5).

5.1 Homogenous Hopf Oscillators

a) Time-dependent connections

We consider a time-dependent network switching from "(4,2) regular lattice "to "All to All "defined by

$$\begin{pmatrix} -2-\cos(\nu t)^2 & 1 & \cos(\nu t)^2 & 1 \\ 1 & -2-\cos(\nu t)^2 & 1 & \cos(\nu t)^2 \\ \cos(\nu t)^2 & 1 & -2-\cos(\nu t)^2 & 1 \\ 1 & \cos(\nu t)^2 & 1 & -2-\cos(\nu t)^2 \end{pmatrix}$$

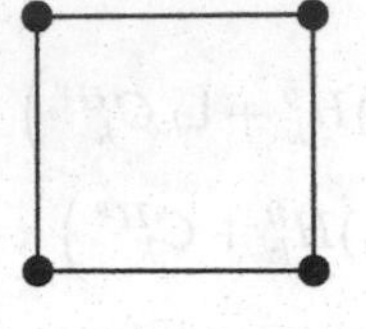

$(4,2)$ regular lattice $\mathcal{C}_{4,2}$

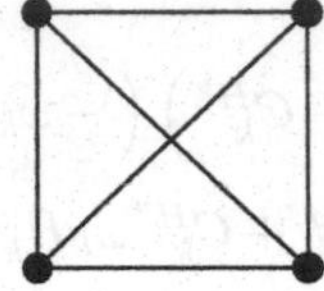

All to All

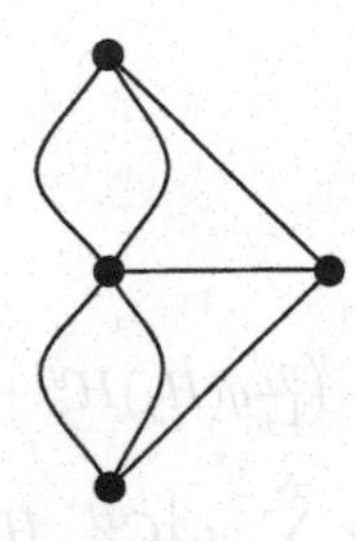

Königsberg

Fig. 5 Three types of network topologies: "(4,2) regular lattice ", "All to All "and "Königsberg "

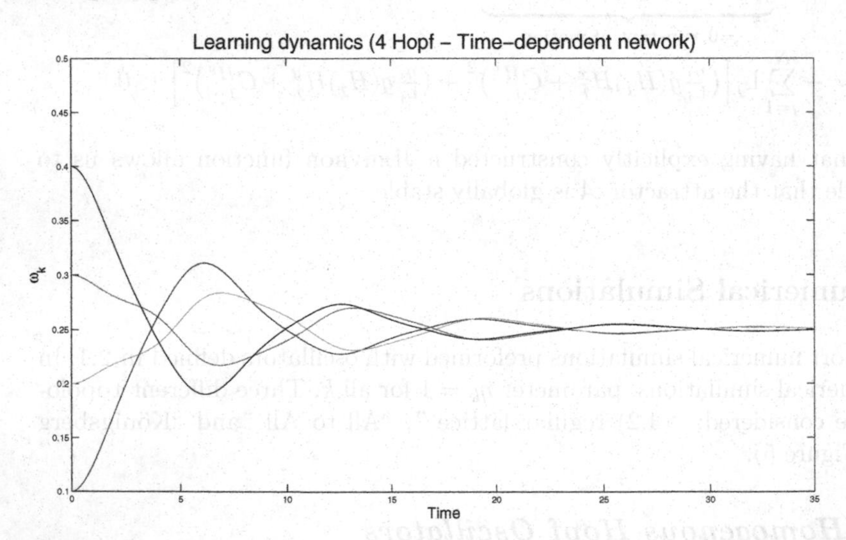

Fig. 6 Time evolution of the parametric variables $\omega_k(t)$ for four Hopf oscillators with $\mathsf{K}_k = 0.1$ for all k, $\omega_1(0) = 0.1$, $\omega_2(0) = 0.2$, $\omega_3(0) = 0.3$, $\omega_4(0) = 0.4$, $\mu_1 = \frac{1}{2}$, $\mu_2 = \frac{1.25^2}{2}$, $\mu_3 = \frac{1.5^2}{2}$, $\mu_4 = \frac{1.75^2}{2}$ and with time-dependent network topology ($\nu = 0.49$). The consensual frequency, given by Eq.(16), is here $\omega_c = 0.25$.

In this case, the relevant spectrum can be explicitly given and reads:

$$\lambda_1(t) = 0, \quad \lambda_2(t) = -4, \quad \lambda_2(t) = \lambda_3(t) = -2 - 2\cos(\nu t)^2.$$

The stability sufficient condition given in Theorem 1 is fulfilled provided $\dot{\lambda}(t) + \lambda(t) < 0$. For suitable small ν this is obviously true and an explicit illustration is given in Figures 6, 7.

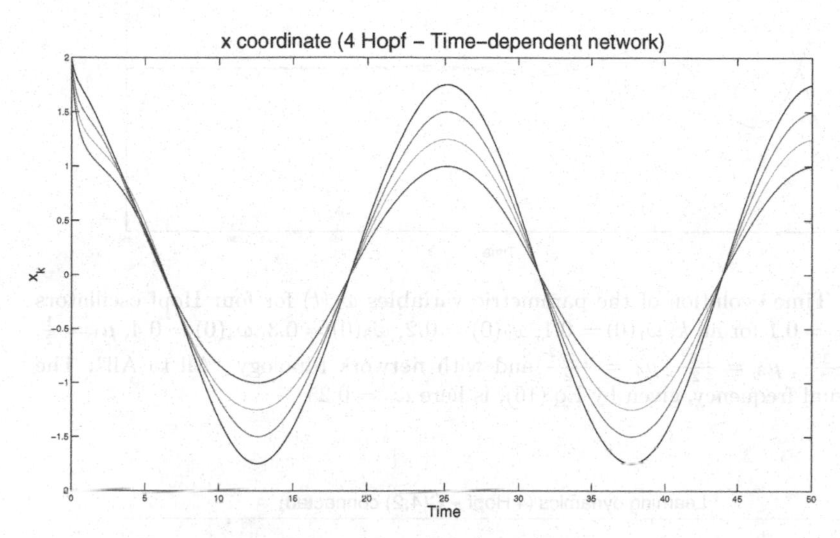

Fig. 7 Time evolution of the state variables $x_k(t)$ for four Hopf oscillators with $\mathrm{K}_k = 0.1$ for all k, $\omega_1(0) = 0.1$, $\omega_2(0) = 0.2$, $\omega_3(0) = 0.3$, $\omega_4(0) = 0.4$, $\mu_1 = \frac{1}{2}$, $\mu_2 = \frac{1.25^2}{2}$, $\mu_3 = \frac{1.5^2}{2}$, $\mu_4 = \frac{1.75^2}{2}$ and with time-dependent network topology $(\nu = 0.49)$

b) Constant connections in time

Illustrations are obtained for the "All to All"connecting topology in Figure 8 and with "$\mathcal{C}_{4,2}$"connecting topology in Figure 9. Note that the time axis is chosen to be identical in both Figures. This allows us to appreciate the modification of the convergence rate.

We further explore the case involving constant topologies connecting Hopf oscillators and the result obtained with the "Königsberg"topology is reported in Figures 10, 11. Connections where turned on and off with smooth functions (see Appendix B for further details).

5.2 Networks with Heterogeneous Oscillators

Experiments conducted with networks with heterogeneous oscillators are reported in Figures 12 and 13. Here, we couple two harmonic oscillators with two non-harmonic oscillators of the type given in Eq.(15).

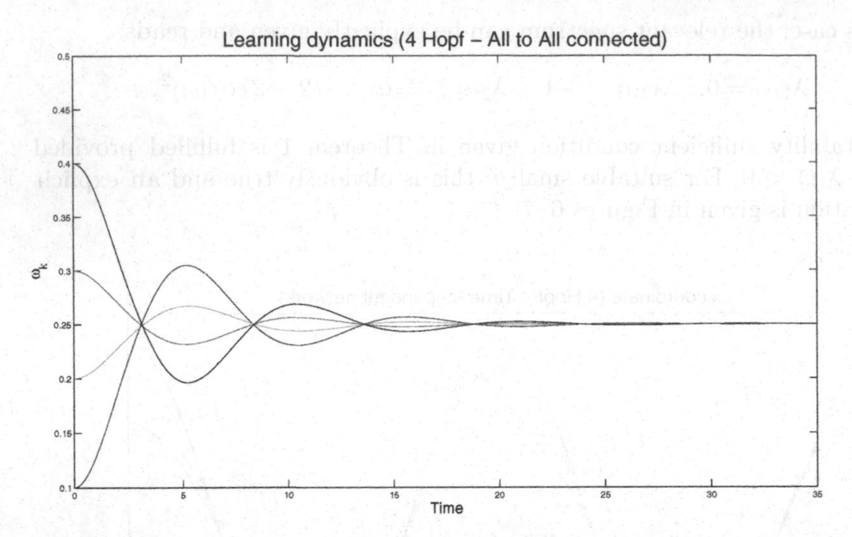

Fig. 8 Time evolution of the parametric variables $\omega_k(t)$ for four Hopf oscillators with $\mathsf{K}_k = 0.1$ for all k, $\omega_1(0) = 0.1$, $\omega_2(0) = 0.2$, $\omega_3(0) = 0.3$, $\omega_4(0) = 0.4$, $\mu_1 = \frac{1}{2}$, $\mu_2 = \frac{1.25^2}{2}$, $\mu_3 = \frac{1.5^2}{2}$, $\mu_4 = \frac{1.75^2}{2}$ and with network topology "All to All". The consensual frequency, given by Eq.(16), is here $\omega_c = 0.25$

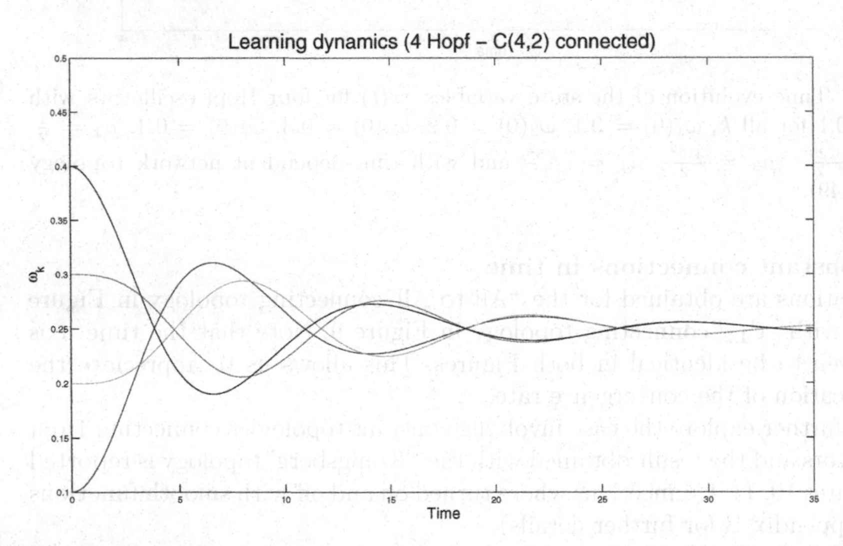

Fig. 9 Time evolution of the parametric variables $\omega_k(t)$ for four Hopf oscillators with $\mathsf{K}_k = 0.1$ for all k, $\omega_1(0) = 0.1$, $\omega_2(0) = 0.2$, $\omega_3(0) = 0.3$, $\omega_4(0) = 0.4$, $\mu_1 = \frac{1}{2}$, $\mu_2 = \frac{1.25^2}{2}$, $\mu_3 = \frac{1.5^2}{2}$, $\mu_4 = \frac{1.75^2}{2}$ and with network topology "$\mathcal{C}_{4,2}$". The consensual frequency, given by Eq.(16), is here $\omega_c = 0.25$.

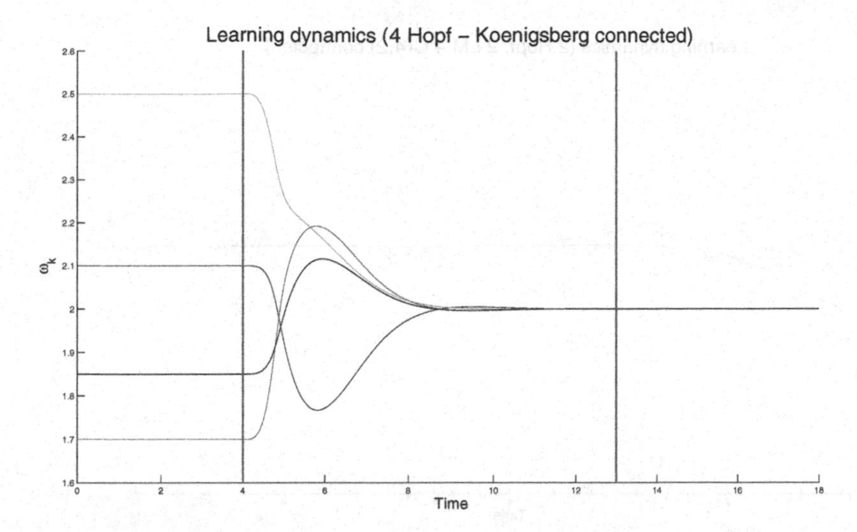

Fig. 10 Time evolution of the parametric variables $\omega_k(t)$ for four Hopf oscillators with $\mathsf{K}_1 = \frac{1}{2}$, $\mathsf{K}_2 = \frac{5}{3}$, $\mathsf{K}_3 = 2$, $\mathsf{K}_4 = \frac{3}{7}$, $\omega_1(0) = 2.1$, $\omega_2(0) = 2.5$, $\omega_3(0) = 1.7$, $\omega_4(0) = 1.85$ and with the network topology "Königsberg". Connections are turned on at $t = 4$, and turned off at $t = 13$, shown by thick red line. The consensual frequency, given by Eq.(16), is here $\omega_c = 2$.

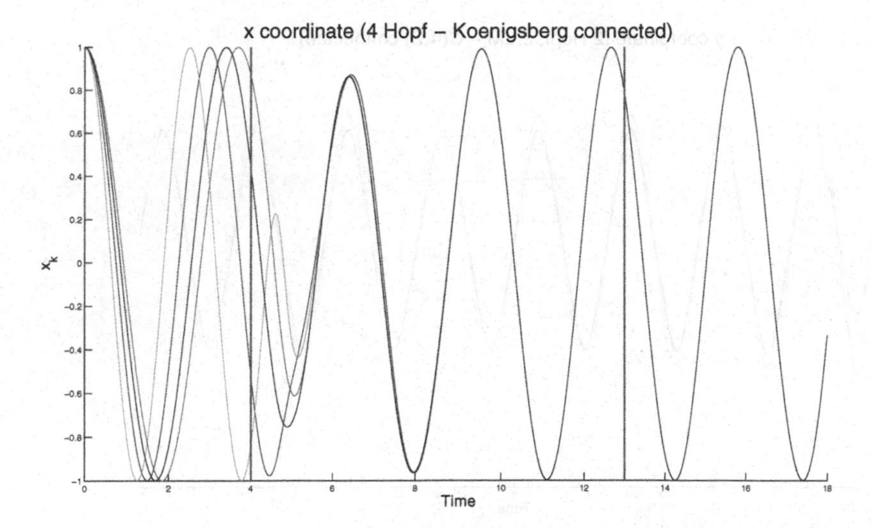

Fig. 11 Time evolution of the state variables $x_k(t)$ for four Hopf oscillators with $\mathsf{K}_1 = \frac{1}{2}$, $\mathsf{K}_2 = \frac{5}{3}$, $\mathsf{K}_3 = 2$, $\mathsf{K}_4 = \frac{3}{7}$, $\omega_1(0) = 2.1$, $\omega_2(0) = 2.5$, $\omega_3(0) = 1.7$, $\omega_4(0) = 1.85$ and with the network topology "Königsberg". Connections are turned on at $t = 4$, and turned off at $t = 13$, shown by thick red line.

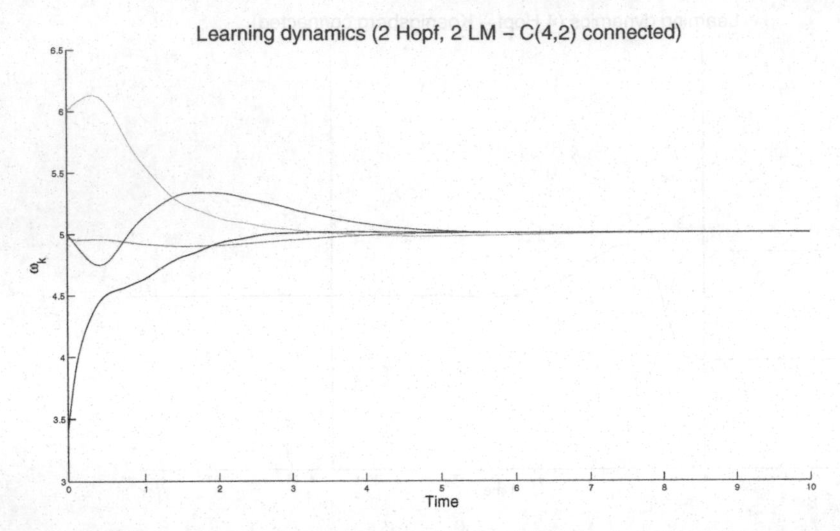

Fig. 12 Time evolution of the parametric variables $\omega_k(t)$ for two Hopf and two LM (with $a = 1$) oscillators with $K_1 = 1$, $K_2 = 2$, $K_3 = \frac{1}{4}$, $K_4 = 4$, $\omega_1(0) = 5$, $\omega_2(0) = 6$, $\omega_3(0) = 4.975$, $\omega_4(0) = 3.4$ and with the network topology "$\mathcal{C}_{4,2}$". The consensual frequency, given by Eq.(16), is here $\omega_c = 5$.

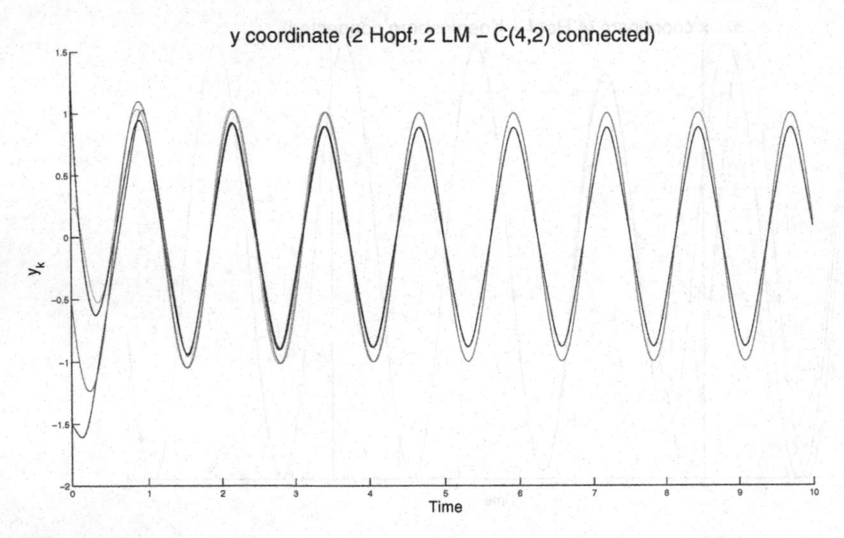

Fig. 13 Time evolution of the state variables $y_k(t)$ for two Hopf and two LM oscillators (with $a = 1$) with $K_1 = 1$, $K_2 = 2$, $K_3 = \frac{1}{4}$, $K_4 = 4$, $\omega_1(0) = 5$, $\omega_2(0) = 6$, $\omega_3(0) = 4.975$, $\omega_4(0) = 3.4$ and with the network topology "$\mathcal{C}_{4,2}$".

6 Conclusions and Perspectives

An abundance of recent literature is devoted to the synchronization of interacting oscillators located at vertices of complex networks. However, we note a complete absence of learning issues in this general context, an absence which motivates our present address. While in the "classical approaches" individual oscillators adjust their frequencies and produce a synchronized state, such adjustments vanish whenever mutual interactions are removed. Under a learning mechanism, mutual interactions permanently affect individual frequencies, even after removal of interactions. Such a definitive (i.e. *plastic*) deformation of the individual dynamics is nothing but the natural tuning process that any watch maker would perform before delivering a clock to his customer. In a first attempt devoted to learning issues in these complex networks, we study an array of individual clocks (i.e. *stable limit cycles oscillators*) in which parametric frequency learning on individual oscillators drives the system towards a consensual state. For an appropriate choice of individual clock dynamics (i.e.*mixed canonical-dissipative systems*) and for the general class of diffusive coupling between network vertices, we are able to analytically characterize the learning process and the resulting consensual state. While the constant of motion and the diffusive coupling used imply the consensual state to be topology network-independent, its algebraic connectivity strongly affects transient behavior. In our models, as long as learning operates, the global dynamics is non-conservative (i.e. the underlying Liouville volume element in phase space is not conserved), but once the consensual state is reached the global dynamics becomes purely Hamiltonian.

As a perspective for further research, the analytical tractability of our dynamics seems to be perfectly suitable to address the question "Can you hear the connectivity of the network?" this directly inspired by Mark Kac's famous original question "Can you hear the shape of the drum?". In the network context, this issue could indeed find an answer by inferring the transient response obtained after perturbation of a consensual state.

Acknowledgment

The authors thank Martin Anderegg for valuable advice and discussions that contributed to the elaboration of this work.

A Eigenvalues of Laplacian Matrix

We recall Гершгорин's circle theorem which implies that all eigenvalues of a Laplacian matrix $L(t)$ associated to a time-dependent network $\mathcal{N}(t)$ (not necessarily connected nor undirected but with $a_{k,j}(t) > 0$ for all t as entries of its adjacency matrix) lie on the left side of the $\mathbb{C}$ plan.

Theorem 3 (Гершгорин (1931)). *Let L be a $N \times N$ matrix (with elements in $\mathbb{R}$ or $\mathbb{C}$). If λ is an eigenvalue of L, then there exist k such that*

$$|\lambda - l_{k,k}| \leqslant \sum_{j \neq k}^{N} |l_{k,j}|,$$

that is, all eigenvalues of A are in the union of the discs

$$D_k := \Big\{ \lambda \mid |\lambda - l_{k,k}| \leqslant \sum_{j \neq k}^{N} |l_{k,j}| \Big\}.$$

For undirected $\mathcal{N}(t)$, $L(t)$ is symmetric for all t and by the spectral theorem in $\mathbb{R}$ all its eigenvalues are real. By Гершгорин's circle theorem they are all negative numbers. A matrix is diagonalizable over $\mathbb{R}$ if, and only if, its characteristic polynomial has real roots (which is the case for $L(t)$ when $\mathcal{N}(t)$ is undirected) and the algebraic multiplicity (i.e. the multiplicity of the corresponding root of the characteristic polynomial) of each eigenvalue is equal to the geometric multiplicity (i.e. the dimension of the associated eigenspace). By definition of $L(t)$, its kernel $Ker(L(t))$ is an eigenspace for all t. Therefore, by knowing its dimension, we know how many eigenvalues of $L(t)$ are strictly negative at time t. This is determined by the topology of the network $\mathcal{N}(t)$ as it is shown in the following lemma.

Lemma 3. *Let $L(t)$ be the Laplacian matrix of a time-dependent network $\mathcal{N}(t)$. We have:*

$$\text{The network } \mathcal{N}(t) \text{ is connected } \forall\, t \Longleftrightarrow Dim(Ker(L(t))) = 1\ \forall\, t$$

Proof. [$\Rightarrow$] Let $\mathbf{1} := (1,\dots,1)^\top$. The definition of $L(t)$ implies: $L(t)\mathbf{1} = 0$, therefore $Dim(Ker(L(t))) \geqslant 1$ for all t. Suppose that there exists t_0 such that $Dim(Ker(L(t_0))) \geqslant 2$. This implies the existence of a vector $x(t_0) := (x_1(t_0),\dots,x_N(t_0))^\top \in \mathbb{R}^N$ such that

I $Lx(t_0) = 0$
II $\exists\, j,k$ such that $x_j(t_0) \neq x_k(t_0)$
III $\mathbf{1}$ and $x(t_0)$ are linearly independent.

Let $x^*(t_0) := \min_{j=1,\dots,N}\{x_j(t_0)\}$ and define $z := x(t_0) - x^*(t_0)\mathbf{1}$. By definition, $z_j \geqslant 0$ for all j and there exist k such that $z_k = 0$ and $L(t_0)z = 0$. Let $I_0 := \{j_1,\dots,j_k\}$ such that $z_s = 0$ for $s \in I_0$ and $I_+ := \{j_{k+1},\dots,j_N\}$ such that $z_s > 0$ for $s \in I_+$. We have for all k

$$\sum_{j=1}^{N} l_{k,j}(t_0)z_j = \sum_{j\neq k}^{N} a_{k,j}(t_0)z_j + (a_{k,k}(t_0) - \sum_{j=1}^{N} a_{k,j}(t_0))z_k$$

$$= \sum_{j=1}^{N} a_{k,j}(t_0)z_j - z_k \sum_{j=1}^{N} a_{k,j}(t_0) = 0$$

For $k \in I_0$, $z_k = 0$ and so

$$\sum_{j=1}^{N} a_{k,j}(t_0)z_j = \sum_{j\in I_+}^{N} a_{k,j}(t_0)z_j = 0$$

Since $a_{k,j}(t_0) \geqslant 0$ for all j, k and $z_j > 0$ for $j \in I_+$, then $a_{k,j}(t_0) = 0$ for $k \in I_0$ and $j \in I_+$ and this implies that $\mathcal{N}$ is not connected at $t = t_0$, which is in contradiction with the hypothesis.

[$\Leftarrow$] Suppose that there exist t_0 such that $\mathcal{N}(t_0)$ is not connected, which means that there exist two sets of indices I and I_* such that $a_{k,j}(t_0) = 0$ for $k \in I$ and $j \in I_*$. Define a vector $x := (x_1, \ldots, x_N) \in \mathbb{R}^N$ such that $x_s = x$ for all $s \in I$ and $x_s = x_*$ for all $s \in I_*$ with $x \neq x_*$. The product $L(t_0)x$ gives, for all k, $\sum_{j=1}^{N} a_{k,j}(t_0)x_j - x_k \sum_{j=1}^{N} a_{k,j}(t_0)$. For $k \in I$

$$\sum_{j=1}^{N} a_{k,j}(t_0)x_j - x_k \sum_{j=1}^{N} a_{k,j}(t_0) = \sum_{j\in I}^{N} a_{k,j}(t_0)x - x\sum_{j\in I}^{N} a_{k,j}(t_0) = 0$$

since $a_{k,j}(t_0) = 0$ for all $j \in I_*$ with $k \in I$. With the same argument for $k \in I_*$, we conclude that $L(t_0)x = 0$, which is in contradiction with the fact that $Dim(Ker(L(t_0))) = 1$. This concludes the proof. □

The lemma is of course true for the particular case when the network is undirected (which is needed for $L(t)$ to be diagonalizable and thus allowing to conclude that the algebraic multiplicity of an eigenvalue is equal to the geometric multiplicity). Therefore, for a time-dependent, connected and undirected network the eigenvalues $\{\lambda_j(t)\}_{j=1}^{N}$ of $L(t)$ are all real negative numbers except for one (say $j = 1$) which is zero ($\lambda_1(t) = 0$) and this for all t.

B Turning Connections *on* and *off*

With the help of Fresnel Integrals defined as

$$\mathfrak{c}(s) := \int_0^s \cos(\tfrac{1}{2}\pi z^2)dz \quad \text{and} \quad \mathfrak{s}(s) := \int_0^s \sin(\tfrac{1}{2}\pi z^2)dz,$$

one can construct $\mathfrak{C}^1$ functions that take on the value 0 (respectively 1) until t^* and 1 (respectively 0) after t^*+1, namely

$$\mathcal{O}_{\uparrow(t;t^*)} = \begin{cases} \xi\, \mathfrak{s}(t-t^*) & t \in [t^*, t^* + \frac{1}{\sqrt{2}}] \\ \xi\left(\mathfrak{c}(t-t^*) - \left(\mathfrak{c}(\frac{1}{\sqrt{2}}) - \mathfrak{s}(\frac{1}{\sqrt{2}})\right)\right) & t \in [t^* + \frac{1}{\sqrt{2}}, t^*+1] \end{cases}$$

$$\mathcal{O}_{\downarrow(t;t^*)} = \begin{cases} -\xi\, \mathfrak{s}(t-t^*) + 1 & t \in [t^*, t^* + \frac{1}{\sqrt{2}}] \\ -\xi\left(\mathfrak{c}(t-t^*) - \left(\mathfrak{c}(\frac{1}{\sqrt{2}}) - \mathfrak{s}(\frac{1}{\sqrt{2}})\right)\right) + 1 & t \in [t^* + \frac{1}{\sqrt{2}}, t^*+1] \end{cases}$$

with $\xi := \left(\mathfrak{c}(1) - \left(\mathfrak{c}(\frac{1}{\sqrt{2}}) - \mathfrak{s}(\frac{1}{\sqrt{2}})\right)\right)^{-1}$.

References

1. Acebrón, J.A., Bonilla, L.L., Vicente, C.J.P., Ritort, F., Spigler, R.: Reviews of Modern Physics 77, 137 (2005)
2. Boccaletti, S., Hwang, D.-U., Chavez, M., Amann, A., Kurths, J., Pecora, L.M.: Physical Review E 74, 016102 (2006)
3. Chen, M., Shang, Y., Zou, Y., Kurths, J.: Physical Review E 77, 027101-1 (2008)
4. Cucker, F., Smale, S.: IEEE Transactions on Automatic Control 52, 852 (2007)
5. Duc, L.H., Ilchmann, A., Siegmund, S., Taraba, P.: Quarterly of Applied Mathematics 1, 137 (2006)
6. Fiedler, M.: Czechoslovak Mathematical Journal 23, 298 (1973)
7. Gersho, A., Karafin, B.J.: The Bell System Technical Journal 45, 1689 (1966)
8. Schweitzer, F., Ebeling, W., Tilch, B.: Physical Review E 64, 211101 (2001)
9. Hongler, M.-O., Ryter, D.M.: Zeitschrift für Physik B 31, 333 (1978)
10. Olfati-Saber, R.: Proceedings of the 2007 American Control Conference, p. 4619 (2007)
11. Pecora, L.M., Carroll, T.L.: Physical Review Letters 80, 2109 (1998)
12. Righetti, L., Buchli, J., Ijspeert, A.: Physica D 216, 269 (2006)
13. Baruh, H.: Analytical Dynamics. McGraw-Hill, New York (1999)
14. Bhatia, N.P., Szegö, G.P.: Dynamical Systems: Stability Theory and Applications. Springer, Heidelberg (1967)
15. Blekhman, I.I.: Synchronization in science and technology. ASME Press, New York (1988)
16. Davis, P.J.: Circulant Matrices. John Wiley & Sons, New York (1979)
17. Horn, R.A., Johnson, C.R.J.: Matrix analysis. Cambridge University Press, Cambridge (1985)
18. Lakshmanan, M., Rajasekar, S.: Nonlinear dynamics: integrability, chaos, and patterns. Springer, Heidelberg (2003)
19. Schweitzer, F.: Brownian Agents and Active Particles. Springer, Heidelberg (2003)

Ragged Synchronizability and Clustering in a Network of Coupled Oscillators

Przemyslaw Perlikowski, Andrzej Stefanski, and Tomasz Kapitaniak

Abstract. In this chapter we describe phenomenon of ragged synchronizability. First we give a theoretical introduction about detecting of the complete synchronization in the network of coupled systems. As an example we choose a network of non-diagonally coupled forced van der Pol oscillators. We present construction of the single electrical van der Pol oscillator and array of such systems. We show numerical and experimental results of the existence of the complete synchronization in the array and we discuss it in the very detailed way. We also consider the synchronization in clusters, experimentally in the small array and numerically in the larger network. Our numerical simulations show good agreement with the experimental observations.

1 Introduction

Over the last decade, chaotic synchronization in the networks of coupled dynamical systems has been intensively investigated, e.g.,[1, 2, 3, 4, 5, 6, 7, 8, 9, 10]. An issue, the most often appearing during the study of any

Przemyslaw Perlikowski
Division of Dynamics, Technical University of Lodz, Stefanowskiego 1/15, 90-924 Lodz, Poland
Institute of Mathematics, Faculty of Mathematics and Natural Sciences II, Humboldt University of Berlin, Rudower Chaussee 25, 12489, Berlin Post: Unter den Linden 6, 10099 Berlin, Germany
e-mail: przemyslaw.perlikowski@p.lodz.pl

Andrzej Stefanski
Division of Dynamics, Technical University of Lodz, Stefanowskiego 1/15, 90-924 Lodz, Poland
e-mail: steve@p.lodz.pl

Tomasz Kapitaniak
Division of Dynamics, Technical University of Lodz, Stefanowskiego 1/15, 90-924 Lodz, Poland
e-mail: tomaszka@p.lodz.pl

K. Kyamakya (Eds.): Recent Adv. in Nonlinear Dynamics and Synchr., SCI 254, pp. 49–75.
springerlink.com

synchronization problem, is to determine a synchronization threshold, i.e. the strength of coupling which is required for the appearance of synchronization. In the case of identical systems (the same set of ODEs and values of the system parameters) a complete synchronization [3] can be obtained. The first analytical condition for the complete synchronization of regular sets (all-to-all or nearest-neighbour types of coupling) of completely diagonally coupled identical dynamical systems has been formulated in [11, 12, 13, 14]. A complete diagonal (CD) coupling is realized by all diagonal components of output function for each pair of subsystems. Such a type of coupling induces a situation, when the condition of synchronization is determined only by the largest Lyapunov exponent of a node system and the coupling coefficient [11, 12, 13, 14, 15, 16]. This property of CD coupling causes, that a synchronous range of a coupling parameter for time-continuous subsystems is only bottom-limited by a value of coupling coefficient being a linear function of the largest Lyapunov exponent [14]. If the coupling is partly diagonal (PD, i.e. realized by not all diagonal components of output function or non-diagonal (ND - also or only non-diagonal components of output function are used in the coupling, then more advanced techniques like a concept called Master Stability Function (Sec. II) have to be applied [16]. This approach allows to solve the networks synchronization problem for any set of coupling weights, connections or number of coupled oscillators. Generally, in the literature dealing with PD or ND coupling problems dominate the works, where the synchronization ranges of a coupling parameter are only bottom-limited or they are double-limited, i.e. there exists one window of synchronization (interval) in desynchronous regime [11, 12, 13, 14, 15, 16, 17, 18, 19].

In real systems it is hard to find networks with regular structure, usually one can observe short–cuts, high clustering coefficient and others effects. Due to this properties a new view on topology of connection between single nodes in network was introduced by applying a graph theory [18, 20, 21, 22, 23, 24, 25]. There are two most frequently meet examples of description of the topology in real systems, first is small–world systems [26, 27], there are many paper reporting synchronization in such type of network [28, 29, 30, 31, 32]. Second scheme is called scale–free networks [33, 34] and as well as one can fine many publication about synchronization in it [35, 36, 37, 38, 39].

In the previous work [40], we presented an example of ND coupled oscillators array, in which more than one separated ranges of synchronization occur when the coupling strength increases. We observe the appearance or disappearance of desynchronous windows in coupling parameter space, when the number of oscillators in the array or topology of connections changes. This phenomenon has been called the ragged synchronizability (RSA). In [41] we showed first experimental proof of existence of the RSA in real coupled systems. We argue that RSA is generic and independent on small mismatch in values of system parameters. This chapter present more deeply view on RSA and let us demonstrate mechanism of synchronization and experimental results in very detail way.

The chapter is organized as follows. In Section 2 we presented some useful ideas which let us solve the problem of the synchronization in network of coupled oscillators. The main part of the chapter is Section 3 where we analyzed numerically and experimentally the complete synchronization in system of three coupled van der Pol oscillators. We also showed example of an appearance of clusters in this system. In Section 4 we showed clustering in larger network of van der Pol oscillators. Finally we concluded our work in Section 5.

2 Detecting of Complete Synchronization

2.1 Definition of Synchronization

Pecora and Carroll [42] defined the complete synchronization (CS) as a state when two state trajectories $\mathbf{x}(t)$ and $\mathbf{y}(t)$ converge to the same values and continue in such relation further in time. This phenomenon takes place between two identical dynamical systems and (when separated). If some kind of linking between them is introduced (direct diffusive or inertial coupling, common external signal, etc.), the CS, i.e., full coincidence of phases (frequencies) and amplitudes of their responses, becomes possible.

The CS can appear only in the case of identical coupled systems, i.e., defined by the same set of ODEs with the same values of system parameters, say $\dot{\mathbf{x}} = \mathbf{f}(\mathbf{x})$ and $\dot{\mathbf{y}} = \mathbf{f}(\mathbf{y})$. The CS takes place when all trajectories representing the coupled systems in the phase space, converge to the same value and remain in step with each other during further evolution. Hence, for two arbitrarily chosen trajectories $\mathbf{x}(t)$ and $\mathbf{y}(t)$, we have:

$$\lim_{t\to\infty} \parallel \mathbf{x}(t) - \mathbf{y}(t) \parallel = 0 \tag{1}$$

It is also described in the subject literature as identical or full synchronization [43, 3].

The CS state can be reached only when two identical dynamical systems are concerned, say, they are given with the same ODEs with identical system parameters. This condition of identity may not be fulfilled due to presence of an external noise or parameters mismatch what usually can happen in real systems. If scale of such disturbances is relatively small, then both systems may eventually reach a state called imperfect complete synchronization (ICS) [44], sometimes named as practical or disturbed synchronization.

In such a case ICS can be defined by following formula:

$$\lim_{t\to\infty} \parallel \mathbf{x}(t) - \mathbf{y}(t) \parallel \leq \epsilon \tag{2}$$

where ϵ is a vector of the small parameter.

The CS (or the ICS) of entire network or array of oscillators means a collective motion of them. However, if there is a set composed of $N > 2$

identical nodes then it can be divided into two or more subsets within which the motion of oscillators is collective while between subsets the dynamics is uncorrelated or at least a phase shift is observed, e.g. two group s of subsystems in anti-phase regime. Such subgroups of synchronized oscillators are called clusters [45, 46, 47, 48, 49].

2.2 Pecora and Caroll Approach

In this subsection we show the idea proposed by Pecora and Caroll [42, 3] nowadays called Pecora and Caroll approach.

The dynamical system is given by following equation:

$$\dot{\mathbf{x}} = \mathbf{F}(\mathbf{x}) \tag{3}$$

where $\mathbf{x} = (x_1, x_2, \ldots, x_m) \in R^m$ is a m-dimensional state vector and $\mathbf{F}(\mathbf{x}) = (f(x_1), , f(x_N))$ defining a vector field $\mathbf{F} : R^m \to R^m$. Assume that the system (3) can be divided in to subsystems:

$$\begin{array}{l} \left.\begin{array}{l} \dot{v} = f(v, u) \\ \dot{u} = g(v, u) \end{array}\right\} drive \\ \dot{w} = h(u, w) \,\}response \end{array} \tag{4}$$

where $\mathbf{v} = (v_1, v_2, \ldots, v_m) \in R^n$, $\mathbf{x} = (u_1, u_2, \ldots, u_k) \in R^k$, $\mathbf{w} = (w_1, w_2, \ldots, w_l) \in R^l$ and $m = n + k + l$. Such a property is called *drive decomposable.* The first two from Eqs. (3) define the drive system The last equation from Eqs. (3) correspond to the response system, whose evolution is guided by the driver trajectory given vector $\mathbf{u}$. Let us consider trajectory $\mathbf{w}$ and its replica $\mathbf{w}'$ forcing by the same chaotic signal $\mathbf{u}(t)$, whose starting in nearby point in the same basin of attraction. We can define complete synchronization by means of synchronization error $\mathbf{e}(t) = ||\mathbf{w} - \mathbf{w}'||$. When $\mathbf{e} \to 0$ for $t \to \infty$ the response systems synchronize with driven one.

The stability of Eqs. (4) can be determined by *Conditional Lyapunov Exponents* (CLE's) of response system [42, 3]. If all CLE's of subsystem $\mathbf{w}$ are negative, the trajectory is stable and the CS with driven system occurrence.

Now, let us show simple application of the Pecora and Caroll approach. Assume that the drive system is well known Rössler oscillator [50] given by equations:

$$\begin{array}{l} \dot{x} = -(y + x) \\ \dot{y} = x + ay \\ \dot{z} = b + z(x - c) \end{array} \tag{5}$$

where x, y, $z \in R^3$ are state vectors, $a = 0.15$, $b = 0.2$ and $c = 10.0$ are constants. For such parameters system (5) exhibits chaotic behavior determinate by following Lyapunov exponents: $\lambda_1 = 0.083$, $\lambda_2 = 0.000$, $\lambda_1 = -9.72$. The response system is driven by the variable y so its equation has following form:

$$\begin{aligned} \dot{x}' &= -(y + x') \\ \dot{z}' &= b + z'(x' - c) \end{aligned} \tag{6}$$

where x', $z' \in R^2$ are state vectors of the driven system.

In Figure 1 we presented numerical simulation of Eqs. (5) and (6). In Figure 1a we showed plot of trajectories $x(t)$ (black line) and $x'(t)$ (grey line) and in Figure 1b synchronization error $e = |x(t) - x'(t)|$ versus time. The CS between drive and systems occurs, after transition time, for for $\tau > 160$. We

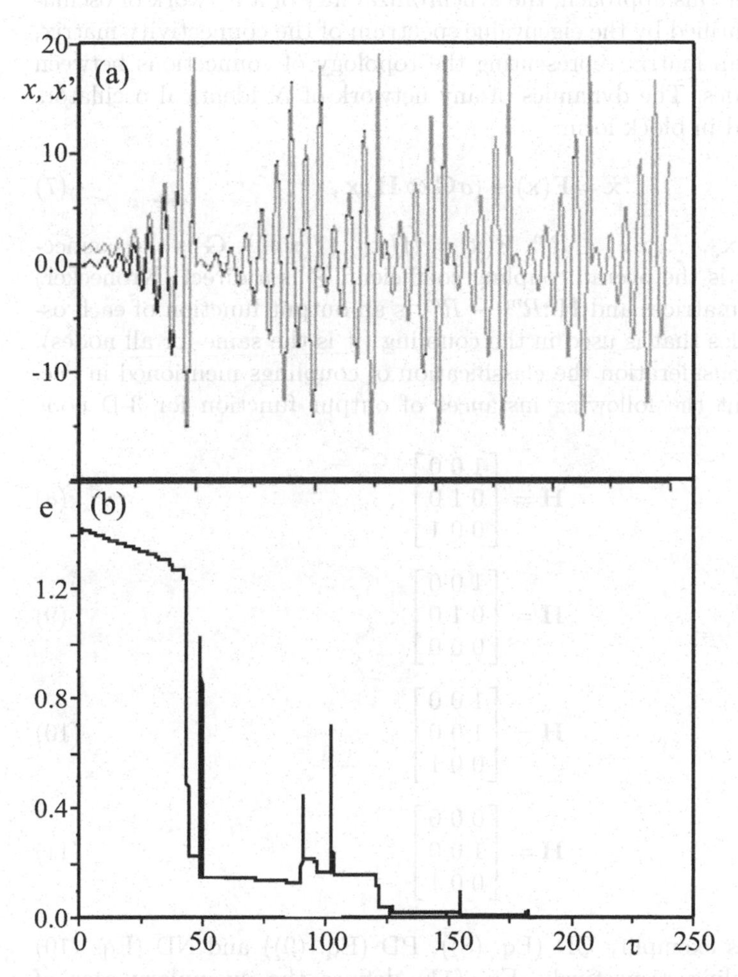

Fig. 1 Numerical simulation of Eqs. (5) and (6). Plot of (a) trajectories $x(t)$ (black line) and $x'(t)$ (grey line) and (b) synchronization error $e = |x(t)-x'(t)|$ versus time. For $\tau > 160$ CS between drive (5) and response (6) systems occurs. Parameters of system: $a = 0.2$, $b = 0.2$, $c = 0.2$, $x(0) = y(0) = z(0) = 0.1$, $x'(0) = 1.5$, $z'(0) = -0.1$.

calculated CLEs for response system: $\lambda_1^C = -0.026$ and $\lambda_2^C = -10.2$ which also proofed appearance of the CS.

The Pecora and Caroll approach was a very important idea which allowed to understand the CS between chaotic systems.

2.3 Master Stability Function

in 1998 Pecora and Caroll propose the idea called Master Stability Function MSF [16]. Under this approach, the synchronizability of a network of oscillators can be quantified by the eigenvalue spectrum of the connectivity matrix, i.e. the Laplacian matrix representing the topology of connections between the network nodes. The dynamics of any network of N identical oscillators can be described in block form:

$$\dot{\mathbf{x}} = \mathbf{F}(\mathbf{x}) + (\sigma \mathbf{G} \otimes \mathbf{H})\,\mathbf{x}\,, \tag{7}$$

where $\mathbf{x} = (\mathbf{x}_1, \mathbf{x}_2, \ldots, \mathbf{x}_N) \in R^m$, $\mathbf{F}(\mathbf{x}) = (\mathbf{f}(\mathbf{x}_1), , \mathbf{f}(\mathbf{x}_N))$, $\mathbf{G}$ is the connectivity matrix, σ is the overall coupling coefficient, $\otimes$ is a direct (Kronecker) product of two matrices and $\mathbf{H} : R^m \rightarrow R^m$ is an output function of each oscillator's variables that is used in the coupling (it is the same for all nodes). Taking under consideration the classification of couplings mentioned in Sec. I we can present the following instances of output function for 3-D node system:

$$\mathbf{H} = \begin{bmatrix} 1\,0\,0 \\ 0\,1\,0 \\ 0\,0\,1 \end{bmatrix}, \tag{8}$$

$$\mathbf{H} = \begin{bmatrix} 1\,0\,0 \\ 0\,1\,0 \\ 0\,0\,0 \end{bmatrix}, \tag{9}$$

$$\mathbf{H} = \begin{bmatrix} 1\,0\,0 \\ 1\,0\,0 \\ 0\,0\,1 \end{bmatrix}, \tag{10}$$

$$\mathbf{H} = \begin{bmatrix} 0\,0\,0 \\ 1\,0\,0 \\ 0\,0\,1 \end{bmatrix}. \tag{11}$$

The $\mathbf{H}$ matrices exemplify CD (Eq. (8)), PD (Eq. (9)) and ND (Eqs (10) and (11)) coupling respectively. Eq. (11) defines the exemplary case of pure ND coupling, because all the diagonal components are equal to zero. In accordance with the MSF concept, a tendency to synchronization of the network is a function of the eigenvalues γ_k of connectivity matrix $\mathbf{G}$, $k = 0, 1, 2, \ldots, N-1$. When oscillators are coupled bidirectionally in ring, which is a commonly meet scheme in mechanical and electrical systems, one can easy write an example of matrix $\mathbf{G}$ for four coupled systems:

$$G = \begin{pmatrix} -2 & 1 & 0 & 1 \\ 1 & -2 & 1 & 0 \\ 0 & 1 & -2 & 1 \\ 1 & 0 & 1 & -2 \end{pmatrix} \tag{12}$$

After block diagonalization of the variational equation of Eq. (7) there appear $N-1$ separated blocks $\dot{\gamma}_k = [\mathbf{Df} + \sigma\gamma_k D\mathbf{H}]$, (for $k = 0$, $\gamma_0 = 0$ is corresponding to the longitudinal mode), where γ_k represents different transverse modes of perturbation from synchronous state [16, 51]. Substituting $\sigma\gamma = \alpha + i\beta$, where $\alpha = Re(\gamma)$, $\beta = Im(\gamma)$ and γ represents an arbitrary value of γ_k, we obtain generic variational equation

$$\dot{\zeta} = [D\mathbf{f} + (\alpha + i\beta)D\mathbf{H}]\,\zeta \tag{13}$$

where ζ symbolizes an arbitrary transverse mode. The connectivity matrix $\mathbf{G} = \{G_{ij}\}$ satisfies $\sum_{j=1}^{N} G_{ij} = 0$ (zero row sum) so the synchronization manifold $\mathbf{x}_1 = \mathbf{x}_2 = \cdots = \mathbf{x}_N$ is invariant and all the real parts of eigenvalues γ_k associated with transversal modes are negative ($Re(\gamma_{k\neq 0} < 0)$. Hence, we obtain the following spectrum of the eigenvalues of $\mathbf{G}$: $\gamma_0 = 0 \geq \gamma_1 \geq \cdots \geq \gamma_{N-1}$. Now, we can define the MSF as a surface representing the largest transversal Lyapunov exponent (TLE) λ_T , calculated for generic variational equation, over the complex numbers plane (α, β). If all the eigenmodes corresponding to eigenvalues $\sigma\gamma_k = \alpha_k + i\beta_k$ can be found in the ranges of the negative TLE then the synchronous state is stable for the considered configuration of the couplings. If an interaction between each pair of nodes is mutual and symmetrical there exist only real eigenvalues of matrix $\mathbf{G}$ ($\beta_k = 0$). In such a case, which is called the real coupling [51, 17], the matrix $\mathbf{G}$ is symmetrical and the MSF is reduced to a form of a curve representing the largest TLE in function of a real number α fulfilling the equation

$$\alpha = \sigma\gamma\,. \tag{14}$$

In Figs 2a–c typical examples of the MSF for the CD coupling (Fig. 2a) and for PD or ND coupling (Figs 2b,c) are shown. Summarizing, one can calculate the MSF diagram basis on following equation:

$$\begin{aligned} \dot{\mathbf{x}} &= \mathbf{F}(\mathbf{x}) \\ \dot{\zeta} &= [D\mathbf{f} + (\alpha + i\beta)D\mathbf{H}]\,\zeta \end{aligned} \tag{15}$$

where first corresponding to dynamics of single node and second to its generic variational equation.

If the real coupling is applied to a set of oscillators with the MSF providing a single range of negative TLE as it is shown in Figs 2a, 2b and 2c, then the synchronous interval of a coupling parameter σ is simply

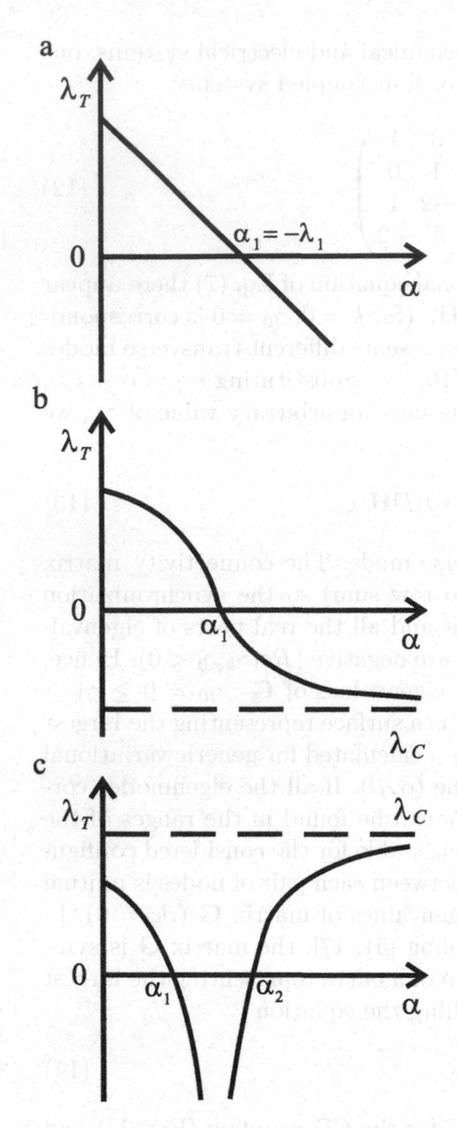

Fig. 2 Typical examples of MSF - $\lambda_T(\alpha)$ in the case of real coupling: (a, b) bottom-limited synchronous range (α_1, ∞), (c) double-limited synchronous interval (α_1, α_2)

reflected from the synchronous α-interval according to Eq. (14). For the case of MSF with double-limited α-interval of negative TLE (Fig. 2c) two transverse eigenmodes have an influence on the σ-limits of the synchronous regime: the longest spatial-frequency mode, corresponding to the largest eigenvalue γ_1, and the shortest spatial-frequency mode, corresponding to the smallest eigenvalue γ_{N-1}. These both eigenvalues determine the width of synchronous σ-range and two types of desynchronizing bifurcations can occur when the synchronous state loses its stability [17]. Decreasing σ leads to a long wavelength bifurcation (LWB), because the longest wavelength mode γ_1 becomes unstable. On the other hand, the increase of the coupling strength causes the shortest wavelength mode γ_{N-1} to become unstable, thus a short wavelength bifurcation (SWB) takes place [17, 16]. Another, characteristic feature of the coupled systems with double-limited synchronous interval is the array size limit, i.e. a maximum number of oscillators in an array which are able to synchronize. For the number of oscillators, which is larger then the size limit, the synchronous σ-interval does not exist. Such an interval exists if $\gamma_{N-1}/\gamma_1 < \alpha_2/\alpha_1$, where α_1 and α_2 are the boundaries of synchronous -α -interval (see Fig. 2c) [16]. If the synchronous range is only bottom-limited as it is depicted in Figs 2a and 2b, then the boundary (the smallest) value of σ, required for the appearance of synchronization, is determined only by the value of γ_1 and then desynchronizing LWB occurs with the decrease of σ. A type of single synchronous range appearing in the systems with PD coupling depends on CLEs of the remaining, uncoupled sub-block of node system. This property results from the asymptotic effect of the PD coupling [51]. An essence of this effect, depicted in Figs 2b and 2c, is that the largest TLE (MSF) tends asymptotically to the value of the largest CLE (λ_C) for strong coupling. Therefore, for negative λ_C the synchronous range is only bottom-limited (Fig. 2b) and for positive λ_C such a range is double-limited (Fig. 2c).

2.4 Probe of Three Oscillators

In last subsection we introduced MSF which base on calculation of the maximum TLE. In some cases it is hard to create accurate numerical model of systems coupled in network, but if one can construct experimental setup and control coupling weights of each node then it is possible to construct MSF experimentally. In real, working systems this idea is a good alternative method which take under consideration mismatches in parameters of subsystems, noise, ect. This idea is called *Universal probe of three oscillators* and was introduced by Fink et al. [17].

Let us introduced the original idea, consider the network of three coupled oscillators given by following equation:

$$\begin{aligned}\dot{\mathbf{x}}_i = \mathbf{F}(\mathbf{x}_i) + \tfrac{\alpha}{3}\left[\mathbf{H}(\mathbf{x}_{i+1}) + \mathbf{H}(\mathbf{x}_{i-1}) - 2\mathbf{H}(\mathbf{x}_i)\right] + \\ \tfrac{\beta}{\sqrt{3}}\left[\mathbf{H}(\mathbf{x}_{i+1}) - \mathbf{H}(\mathbf{x}_{i-1})\right],\end{aligned} \tag{16}$$

where $i = 1, 2, 3$ and i changes cyclically. Factors 3 and $\sqrt{3}$ coming from original article [51] and are added for simplicity. For Eq. (16) one can write a variational equation:

$$\frac{d\xi}{dt} = \begin{pmatrix} \mathbf{Df} & 0 & 0 \\ 0 & \mathbf{Df} & 0 \\ 0 & 0 & \mathbf{Df} \end{pmatrix} \xi + \begin{pmatrix} -2\frac{\alpha}{3} & \frac{\alpha}{3} + \frac{\beta}{\sqrt{3}} & \frac{\alpha}{3} - \frac{\beta}{\sqrt{3}} \\ \frac{\alpha}{3} - \frac{\beta}{\sqrt{3}} & -2\frac{\alpha}{3} & \frac{\alpha}{3} + \frac{\beta}{\sqrt{3}} \\ \frac{\alpha}{3} + \frac{\beta}{\sqrt{3}} & \frac{\alpha}{3} - \frac{\beta}{\sqrt{3}} & -2\frac{\alpha}{3} \end{pmatrix} \otimes D\mathbf{H}\xi. \tag{17}$$

The second matrix in Eq. 17 is the connectivity matrix $\mathbf{G}$, so according to MSF procedure showed in previous section, must be diagonalize:

$$\frac{d\zeta}{dt} = \begin{pmatrix} \mathbf{Df} & 0 & 0 \\ 0 & \mathbf{Df} & 0 \\ 0 & 0 & \mathbf{Df} \end{pmatrix} \zeta + \begin{pmatrix} 0 & 0 & 0 \\ 0 & (\alpha + i\beta)D\mathbf{H} & 0 \\ 0 & 0 & (\alpha + i\beta)D\mathbf{H} \end{pmatrix} \zeta, \tag{18}$$

where ζ is symbolizes an arbitrary transverse mode.

In the case of mutual coupled systems eigenvalues of connectivity matrix $\mathbf{G}$ have only real parts, hence $\beta = 0$ and one can simplify Eq. (18) to following form:

$$\frac{d\zeta}{dt} = \begin{pmatrix} \mathbf{Df} & 0 \\ 0 & \mathbf{Df} \end{pmatrix} \zeta + \begin{pmatrix} 0 & 0 \\ 0 & \alpha D\mathbf{H} \end{pmatrix} \zeta \tag{19}$$

where $\alpha = \sigma\gamma$ (see Eq. (14))

Summarizing, for systems with only bidirectional coupling (mutual) MSF can be create numerically and experimentally by investigation of the CS between two coupled systems.

3 Ragged Synchronizability and Clustering in Network of Van der Pol Oscillators – Experimental and Numerical Observation

3.1 Model of the System

To show experimental existence of RSA we choose the well know van der Pol (vdP) oscillator [52, 53]. In such a system one can observe self-excited oscillation as well as under forcing wild variety of complex dynamics [54]. In other hand to achieve chaotic behavior damping and excitation must have relatively large values and in our experimental circuit they are out of range. In Figure 3 one can see scheme of electric single vdP oscillator [41, 55, 56].

The circuit is composed of two capacitors $C1$ and $C2$, seven resistors $R(1-7)$ and two multiplicators AD-633JN (see Figure 4) [57] which introduce nonlinearity. Multiplicators have the following characteristic

$$W = (1/V_c)(X_1 - X_2)(Y_1 - Y_2) + Z, \tag{20}$$

where X_1, X_2, Y_1 and Y_2 are the input signals, W is an output signal, and $V_c = 10V$ is a characteristic voltage. The error caused by multiplicator is less

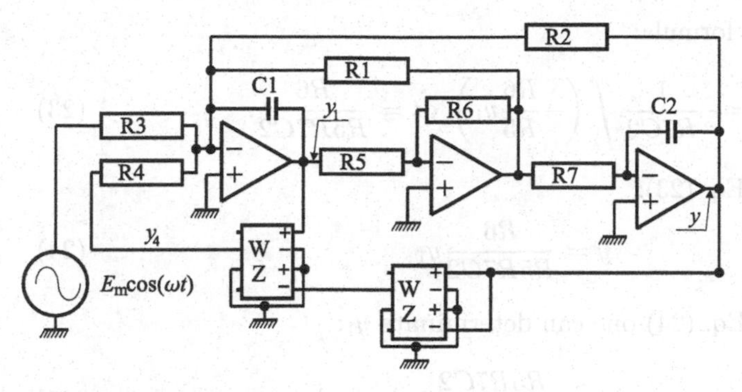

Fig. 3 Scheme of single electric vdP oscillator

Fig. 4 Scheme of multiplicator AD-633JN

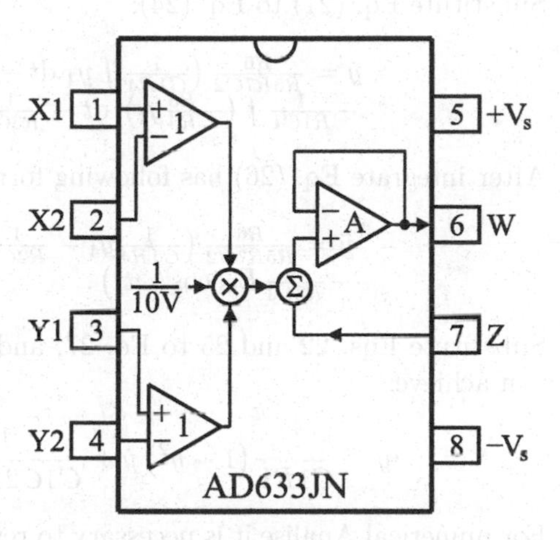

the 2% and the input resistance is (10 MΩ) and has negligible influence on input signal.

Oscillator is supplied by direct 18 V current from amplifier ZPA 81. The forcing signal $E_m \cos(\omega t)$ is generated in generator G 432 with maximum amplitude $E_m = 5$ V and frequency in range $\omega \in (0\text{Hz}, 1\text{MHz})$. The additional resistors $R8$ and R have been used to realize the coupling.

The node system y_1 is given by equation:

$$y_1 = \frac{1}{C1R4} \int y_4 \, \mathrm{dt} - \frac{1}{R2C1} \int y \, \mathrm{dt} - \frac{1}{R1C1} \int \left(-\frac{R6}{R5} y_1\right) \, \mathrm{dt} + \\ -\frac{1}{R3C1} \int E_m \cos(\omega t) \, \mathrm{dt}. \tag{21}$$

Basis on multiplicator property one can write:

$$y_4 = \frac{y_1 y^2}{100 \text{ V}^2}. \tag{22}$$

Signal y is given by formula:

$$y = -\frac{1}{R7C2}\int\left(-\frac{R6}{R5}y_1\right)\,\mathrm{dt} = \frac{R6}{R5R7C2}. \tag{23}$$

After derivation of Eq. (23):

$$\dot{y} = \frac{R6}{R5R7C2}y_1 \tag{24}$$

and rearranging of Eq. (24) one can determinate y_1:

$$y_1 = \frac{R5R7C2}{R6}\dot{y}. \tag{25}$$

Substitute Eq. (21) to Eq. (24):

$$\begin{aligned}\dot{y} = \tfrac{R6}{R5R7C2}\big(&\tfrac{1}{C1R4}\int y_4\,\mathrm{dt} - \tfrac{1}{R2C1}\int y\,\mathrm{dt} + \\ &-\tfrac{1}{R1C1}\int\left(-\tfrac{R6}{R5}y_1\right)\,\mathrm{dt} - \tfrac{1}{R3C1}\int E_m\cos\left(\omega t\right)\,\mathrm{dt}\big).\end{aligned} \tag{26}$$

After integrate Eq. (26) has following form:

$$\begin{aligned}\ddot{y} = \tfrac{R6}{R5R7C2}\big(&\tfrac{1}{C1R4}y_4 - \tfrac{1}{R2C1}y - \tfrac{1}{R1C1}\left(-\tfrac{R6}{R5}y_1\right) \\ &-\tfrac{1}{R3C1}E_m\cos\left(\omega t\right)\big).\end{aligned} \tag{27}$$

Substitute Eqs. 22 and 25 to Eq. 27, and use dependence $100R4 \approx R1$ one can achieve:

$$\ddot{y} - \frac{1}{C1R1}\left(1 - y^2\right)\dot{y} + \frac{1}{C1C2R2R7}y = \frac{E_m\cos\left(\omega t\right)}{C1C2R3R7}. \tag{28}$$

For numerical Analise it is necessary to rewrite Eq. (28) in two dimensionless first order equations:

$$\begin{aligned}\dot{z} &= x, \\ \dot{x} &= d\left(1 - x^2\right)z - x + \cos\left(\eta\tau\right),\end{aligned} \tag{29}$$

where $\omega_0^2 = \frac{1}{C1C2R2R7}$, $d = \frac{1}{C1R1\omega_0}$, $\eta = \frac{\omega}{\omega_0}$, $x = y\frac{R3}{E_mR2}$, $\dot{x} = \dot{y}\frac{R3}{E_mR2\omega_0}$, $\ddot{x} = \ddot{y}\frac{R3}{E_mR2\omega_0^2}$ and τ is a dimensionless time.

Consider an array consisting three coupled in-line vdP oscillators [41] given by equation:

$$\dot{x}_1 = z_1, \tag{30a}$$

$$\dot{z}_1 = d(1 - x_1^2)z_1 - x_1 + cos(\eta\tau) + \sigma(x_2 - x_1), \tag{30b}$$

$$\dot{x}_2 = z_2, \tag{30c}$$

$$\dot{z}_2 = d(1 - x_2^2)z_2 - x_2 + cos(\eta\tau) + \sigma(x_1 + x_3 - x_2), \tag{30d}$$

$$\dot{x}_3 = z_3, \tag{30e}$$

$$\dot{z}_3 = d(1 - x_3^2)z_3 - x_3 + cos(\eta\tau) + \sigma(x_2 - x_3), \tag{30f}$$

where $\sigma = \frac{R2}{R8}$ is a constant coupling coefficient.

In experiments we use an electronic implementation of this array shown in Figure 5.

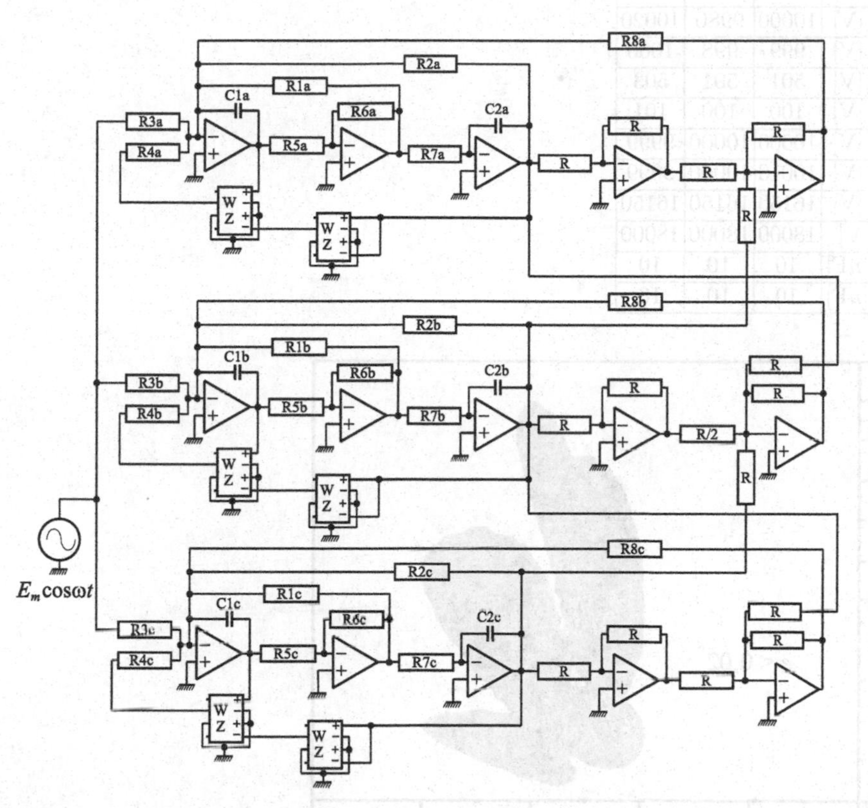

Fig. 5 Three coupled in-line electric vdP oscillators

In the Table 1 we present measurements of resistances and capacities of elements used in experimental circuits.

As it easy to see, the differences between each elements (bellow 1%) are small and can be negligible. Then for simplification, in numerical analysis, we assume mean value $d = 0.401$ and σ as a control parameter. Let us determinate the best value of η, for which in system (30) one can observe RSA phenomena.

In Figure 6 we presented the value of the synchronization error

$$e = \sum_{i=2}^{3} \sqrt{(x_1 - x_i)^2 + (z_1 - z_i)^2}, \tag{31}$$

versus the coupling coefficient σ and the frequency of external excitation η.

Table 1 Measurements of resistances and capacities of elements used in experimental circuits, where a, b, c are labels of vdP systems coupled in array

	a	b	c
$R1$ [V]	10000	9980	10020
$R2$ [V]	999	998	1000
$R3$ [V]	501	501	503
$R4$ [V]	100	100	101
$R5$ [V]	10000	10000	9999
$R6$ [V]	10010	10000	9999
$R7$ [V]	16150	16150	16150
R [V]	18000	18000	18000
$C1$ [nF]	10	10	10
$C2$ [nF]	10	10	10

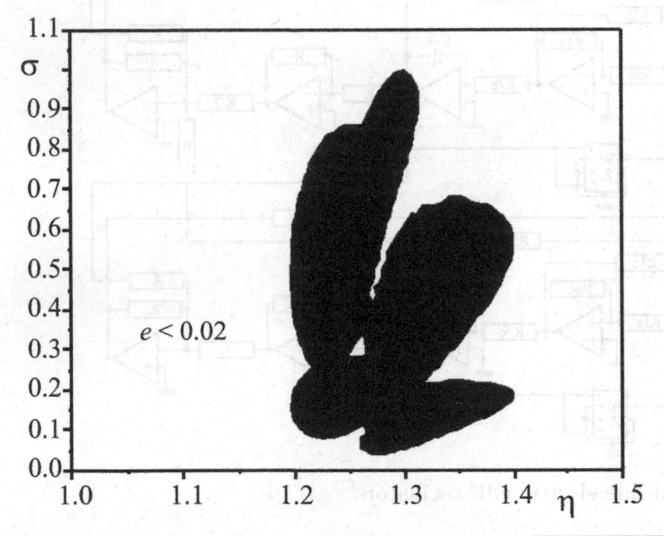

Fig. 6 The synchronization error $e = \sum_{i=2}^{3} \sqrt{(x_1 - x_i)^2 + (z_1 - z_i)^2}$ versus coupling coefficient σ and the frequency of external excitation η for Eqs. (9): $d = 0.401$. White region correspond to $e < 0.02$, while others to desynchronization rangers.

In the white region $e < 0.02$ and we assumed that systems are synchronized (calculations were consuming a lot of processor time, so we shorten them as much it was possible). It is easy to see that RSA can appear for $\eta \in (1.2, 1.22) \cup (1.32, 1.4)$.

As an example we consider $\eta = 1.207$ (its real equivalent $\omega = 4834$Hz in experiment). Then, in the absence of coupling each oscillator exhibits periodic behavior (the largest LE $\lambda_1 = -0.19$) with the period equal to the period of excitation. The phase plots of comparison of numerical calculated and experimental trajectories are shown in Figure 7.

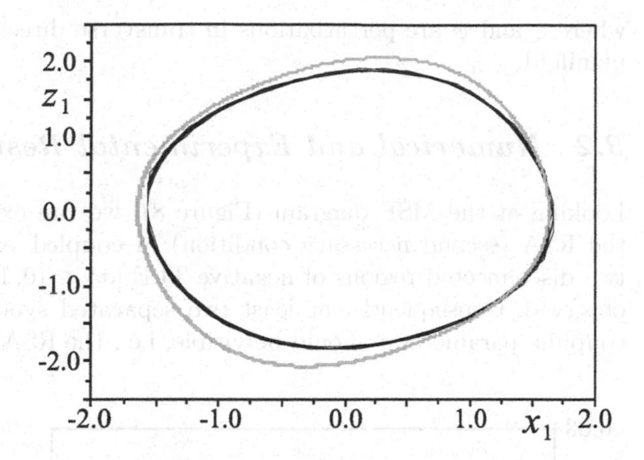

Fig. 7 Phase plots of single system Eqs. (29), gray color (experimental results), black color (numerical results). Parameters: $d = 0.401$, $\eta = 1.207$.

Black trajectory correspond to numerical calculation and gray one to experimental measurements, taking in to account all errors and unknowns existing in circuit and appearing during measurements, the similarity is satisfactory. We have done the same plots for different values of η and we also had the good comparison which proof correctness of mathematical model.

The coupling scheme presented in Figure 5 is a well known nearest-neighbor configuration. Such a connection of oscillators can be classified as the case of pure (diagonal components are equal to zero) ND coupling due to the form of output function

$$\mathbf{H} = \begin{bmatrix} 0 & 0 \\ 1 & 0 \end{bmatrix}. \tag{32}$$

The structure of the nearest-neighbor connections and also in this case mutual coupling of array nodes is described by the following connectivity matrix

$$\mathbf{G} = \begin{bmatrix} -1 & 1 & 0 \\ 1 & -2 & 1 \\ 0 & 1 & -1 \end{bmatrix}. \tag{33}$$

Matrix $\mathbf{G}$ has the following eigenvalues: $\gamma_0 = 0, \gamma_1 = -1, \gamma_2 = -3$. Since nonzero eigenvalues are not equal to each other one can expect the appearance of RSA (this is one of necessary condition). Applied idea of MSF to this case we obtain the generic variational equation for calculating the MSF, i.e., $\lambda_T(\alpha)$ in the form

$$\dot{\zeta} = \psi, \tag{34a}$$

$$\dot{\psi} = d(1 - x^2)\zeta - 2dx\psi\zeta - \psi + \alpha\psi. \tag{34b}$$

where ζ and ψ are perturbations in transverse direction to synchronization manifold.

3.2 Numerical and Experimental Results

Looking at the MSF diagram (Figure 8), we can expect an appearance of the RSA (second necessary condition) in coupled vdP oscillators, because two disconnected regions of negative TLE $[\sigma\gamma \in (0,1-)$ or $(1+,\infty)]$ can be observed. Consequently, at least two separated synchronous ranges of the coupling parameter σ should be visible, i.e., the RSA effect takes place.

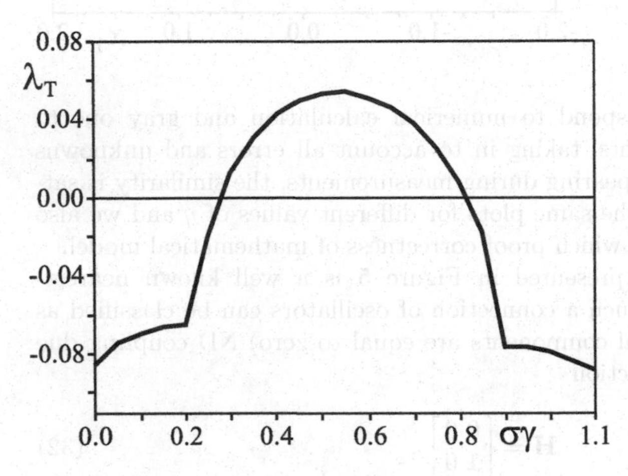

Fig. 8 Diagram of MSF for vdP oscillator; parameters of system: $d = 0.401$ and $\eta = 1.207$

However, synchronous intervals of the coefficient σ not always are an exact reflection of the MSF intervals, where the TLE is positive. The RSA mechanism is explained in Figure 9(a–e), where a projection from the MSF diagram (Figure 9(a)), via eigenvalues of connectivity matrix **G** (Figure 9(b)), to the bifurcation diagram of synchronization error

$$e = \sum_{i=2}^{3} \sqrt{(x_1 - x_i)^2 + (z_1 - z_i)^2}, \tag{35}$$

versus coupling strength σ (Figure 9(c)) is shown. Complete synchronization takes place in the σ-ranges where e approaches zero value (Figure 9(c)).

We can observe third synchronous $(\sigma_{1+}^2, \sigma_{1-}^1)$ and second desynchronous $(\sigma_{1-}^2, \sigma_{1+}^2)$ σ-intervals in comparison with only two synchronous and one desynchronous MSF-ranges, respectively. Additional desynchronous interval

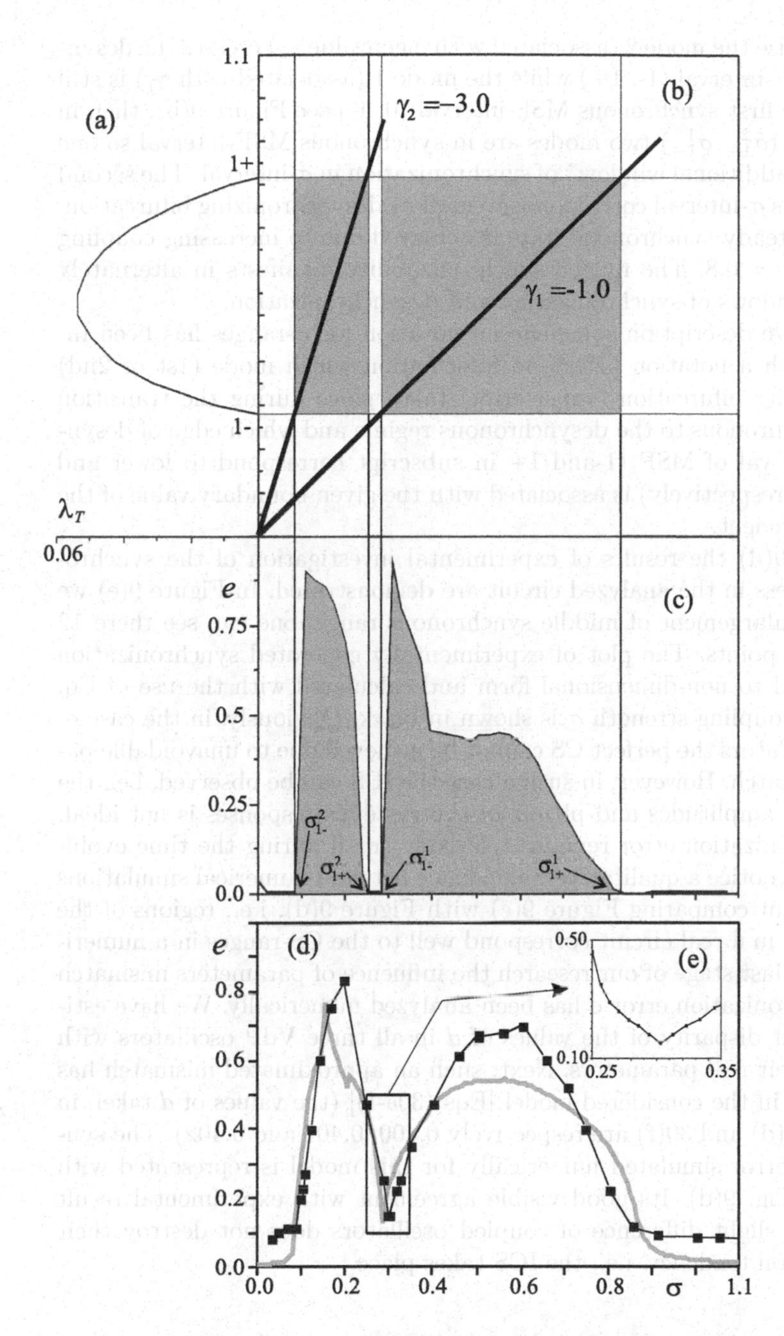

Fig. 9 Projection from the MSF $\lambda_T(\sigma\gamma)$ (a) via eigenvalues of the connectivity matrix (b) to the bifurcation diagram of the synchronization error (c) calculated according to Eq. (30). Desynchronization intervals connected with eigenvalues γ_1 and γ_2 are shown in grey; $d = 0.401$, $\eta = 1.207$. Comparative diagram of experimentally detected synchronization error (d) and its enlargement (e).

appears because the mode 2 (associated with eigenvalue γ_2) crosses the desynchronous MSF-interval (1-, 1+) while the mode 1 (associated with γ_1) is still located in the first synchronous MSF-interval (0, 1-) see Figure 9(b). then in narrow range $(\sigma^2_{1+}, \sigma^1_{1-})$ two modes are in synchronous MSF-interval so one can observe "additional window" of synchronization in σ-interval. The second desynchronous σ-interval corresponds to mode 1 desynchronizing bifurcation. Finally, the steady synchronous state is achieved due to increasing coupling strength at $\sigma = 0.8$. The ragged synchronizability manifests in alternately appearing windows of synchronization and desynchronization.

In the above description some special notation for σ-ranges has been introduced. Such a notation brings an information which mode (1st or 2nd) desynchronizing bifurcation (superscript) takes place during the transition from the synchronous to the desynchronous regime and which edge of desynchronous interval of MSF (1-and 1+ in subscript correspond to lower and higher edges, respectively) is associated with the given boundary value of the coupling coefficient.

In Figure 9(d) the results of experimental investigation of the synchronization process in the analyzed circuit are demonstrated. In Figure 9(e) we showed the enlargement of middle synchronous range, one can see there 12 measurement points. The plot of experimentally generated synchronization error (reduced to non-dimensional form and calculated with the use of Eq. (30)) versus coupling strength σ is shown in black. Obviously, in the case of real vdP oscillators the perfect CS cannot be achieved due to unavoidable parameter mismatch. However, in such a case the ICS can be observed, i.e., the correlation of amplitudes and phases of the system's responses is not ideal, but a synchronization error remains relatively small during the time evolution. One can notice a qualitative coincidence between numerical simulations and experiment comparing Figure 9(c) with Figure 9(d), i.e., regions of the ICS tendency in a real circuit correspond well to the CS-ranges in a numerical model. In last stage of our research the influence of parameters mismatch on the synchronization error e has been analyzed numerically. We have estimated a slight disparity of the values of d in all three VdP oscillators with measuring their real parameters. Next, such an approximated mismatch has been realized in the considered model [Eqs (30a-f)] (the values of d taken in Eqs 30(b), 30(d) and 30(f) are respectively 0.400, 0.401 and 0.402). The synchronization error simulated numerically for this model is represented with grey line in Fig. 9(d). Its good visible agreement with experimental result shows that a slight difference of coupled oscillators does not destroy their synchronization tendency, i.e., the ICS takes place.

3.3 Clustering

In considered network one can observe the phenomenon of clustering [45, 46, 58]. Such behavior corresponding to existence one or more groups of

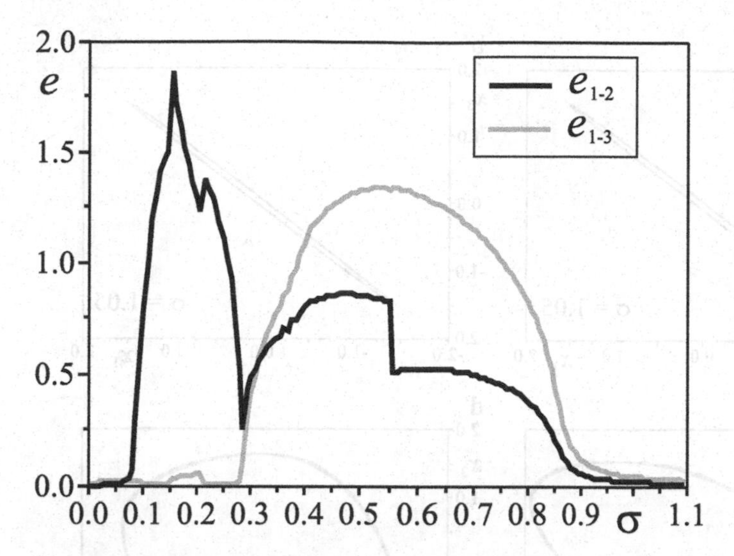

Fig. 10 Experimental synchronization error e_{1--2} (black line) e_{1--3} (gray line) and versus coupling coefficient σ, cluster between first and third oscillator for $\sigma \in (0.07, 0.27)$ is clearly visible ;paraments $d = 0.401$, $\eta = 1.207$

synchronized oscillators although the whole network is in the desynchronized state. In our case we can obviously observe only (2,1) cluster, i.e, two nodes have common behavior and one node is independent. We defined the synchronization errors between first and second ($e_{1-2} = \sqrt{(x_1 - x_2)^2 + (z_1 - z_2)^2}$) and first and third ($e_{1-3} = \sqrt{(x_1 - x_3)^2 + (z_1 - z_3)^2}$) oscillator. In Figure 3.3 we present results of numerical calculation of synchronization error e_{1-2} (black line) and e_{1-3} (grey line).

As it easy to see in range $\sigma = (0.07, 0.27)$ one can observed a cluster between fist and third oscillator, while second system is in the desynchronized state with them. This phenomenon is confirmed by calculation of eigenvectors [46] of connectivity matrix **G**. The synchronization in range $\sigma = (0.07, 0.27)$ is governed by eigenvalue $\gamma_2 = -3$ with corresponding eigenvector $v_2 = [1, -2, 1]$. This values of v_2 leads to existence of cluster shown in Figure 10.

Let us present experimental phase plots which proofed existence of the cluster in real system. In Figures 11 and 12 are shown phase plots of x_1 versus x_2 in left row and x_1 versus x_3 for chosen values of the coupling coefficient σ.

According to theoretical investigation there is no clustering for desynchronization governed by mode 2 (γ_2). One can see it in experiment (Figures 11 and 12) where for $\sigma = 0.502$ ((Figure 11c,d)) desynchronization is visible. Two other values of σ: $\sigma = 1.05$ ((Figure 11a,b)) and $sigma = 0.29$ ((Figure 11e,f)) correspond to synchronization rages. When mode 1 is responsible for

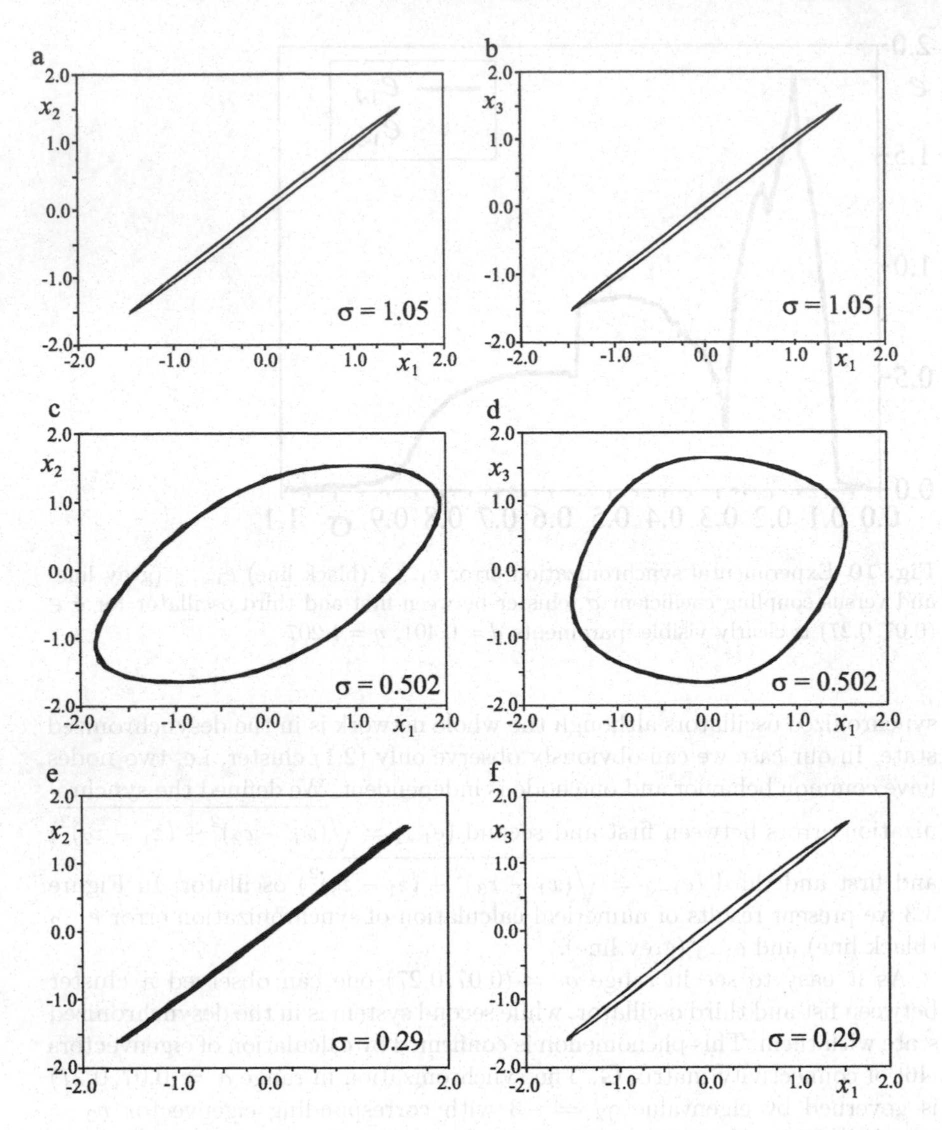

Fig. 11 Phase plot of synchronization manifolds (x_1, x_2) and (x_1, x_3) for different values of coupling coefficient σ. Parameters of system: $d = 0.401$, $\eta = 1.207$ and $\sigma = 1.05$ (a,b), $\sigma = 0.502$ (c,d), $\sigma = 0.29$ (d,e).

desynchronization one can expected cluster between first and third oscillators ($x_1 = x_3$) and desynchronization between first and second. This behavior is shown in Figure 12a–d. In last plots (Figure 12e,f) two mode are in range (0,1-) of MSF and synchronization occurs.

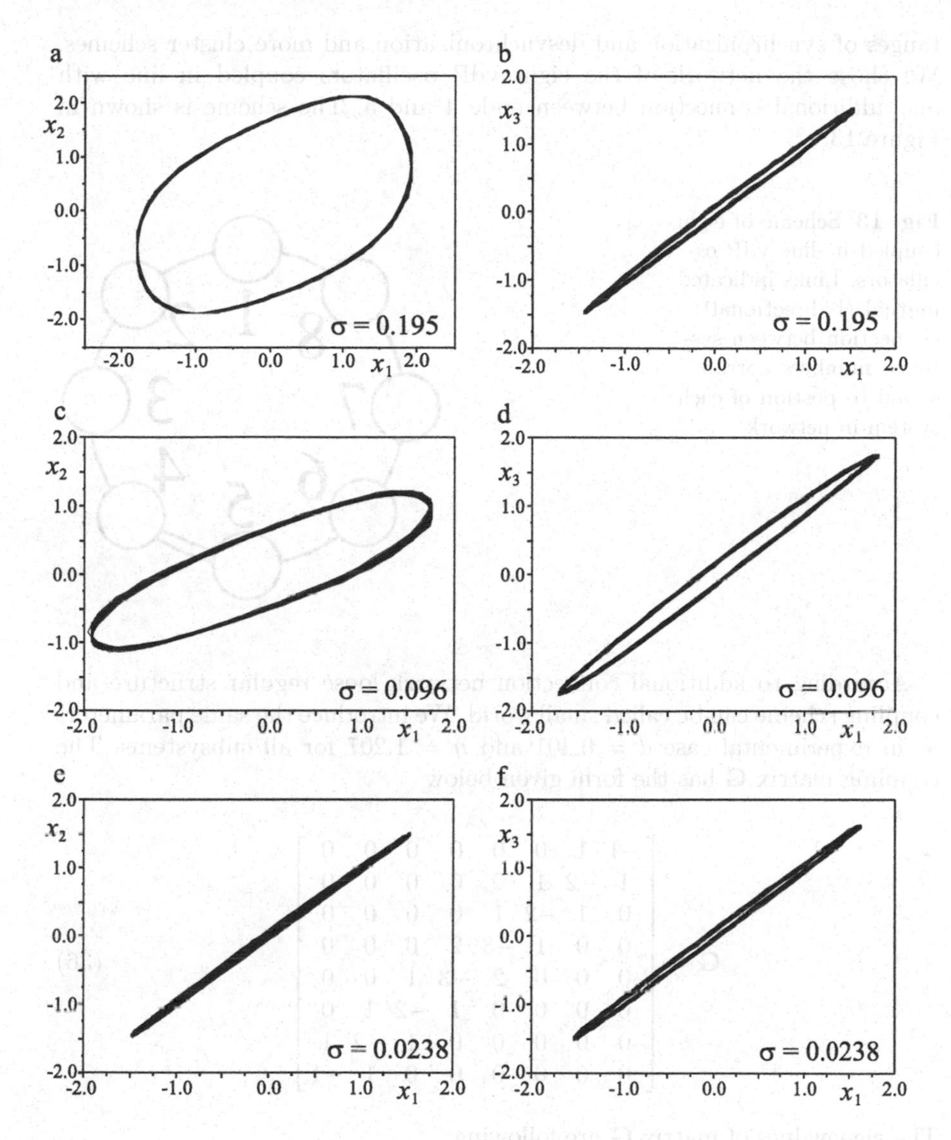

Fig. 12 Phase plot of synchronization manifolds (x_1, x_2) and (x_1, x_3) for different values of coupling coefficient σ. Parameters of system: $d = 0.401$, $\eta = 1.207$ and $\sigma = 0.195$ (a,b), $\sigma = 0.096$ (c,d), $\sigma = 0.0238$ (d,e).

4 Numerical Example of Clustering in Network with More Oscillators

In previous section we show simple example of clustering and RSA. Let us consider some more advanced case, where one can observe more altering ranges of synchronization and desynchronization and more cluster schemes. We chose the network of the eight vdP oscillators coupled in–line with one additional connection between node 4 and 5. The scheme is shown in Figure 13.

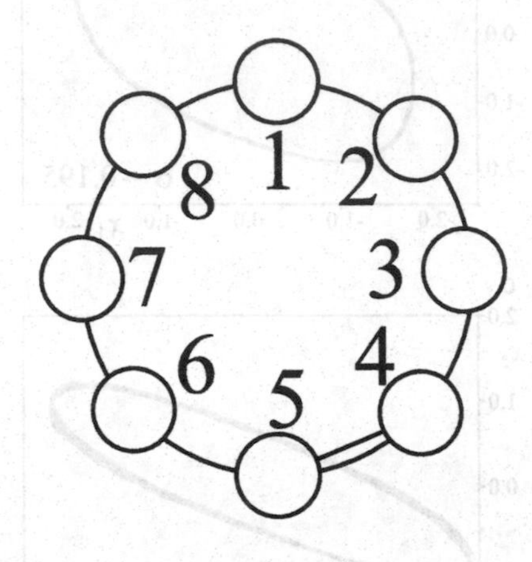

Fig. 13 Scheme of eight coupled in–line vdP oscillators. Links indicated mutual (bidirectional) connection between systems, numbers correspond to postion of each system in network

According to additional connection network loose regular structure and coupling scheme can be called small world. We introduce the same parameters as in experimental case $d = 0.401$ and $\eta = 1.207$ for all subsystems. The coupling matrix $\mathbf{G}$ has the form given below

$$\mathbf{G} = \begin{bmatrix} -1 & 1 & 0 & 0 & 0 & 0 & 0 & 0 \\ 1 & -2 & 1 & 0 & 0 & 0 & 0 & 0 \\ 0 & 1 & -2 & 1 & 0 & 0 & 0 & 0 \\ 0 & 0 & 1 & -3 & 2 & 0 & 0 & 0 \\ 0 & 0 & 0 & 2 & -3 & 1 & 0 & 0 \\ 0 & 0 & 0 & 0 & 1 & -2 & 1 & 0 \\ 0 & 0 & 0 & 0 & 0 & 1 & -2 & 1 \\ 0 & 0 & 0 & 0 & 0 & 0 & 1 & -1 \end{bmatrix}. \tag{36}$$

The eigenvalues of matrix $\mathbf{G}$ are following:

$$\lambda_{0-7} = 0.0, -0.1732, -0.5858, -1.3998, 2.0000, -3.0948, -3.4142, -5.3322, \tag{37}$$

and one can expect not only clustering but also more complicated synchronization mechanism which can cause appearance of RSA phenomena.

Before we analyze clustering let us explain mechanism of synchronization that is shown in Figure 14.

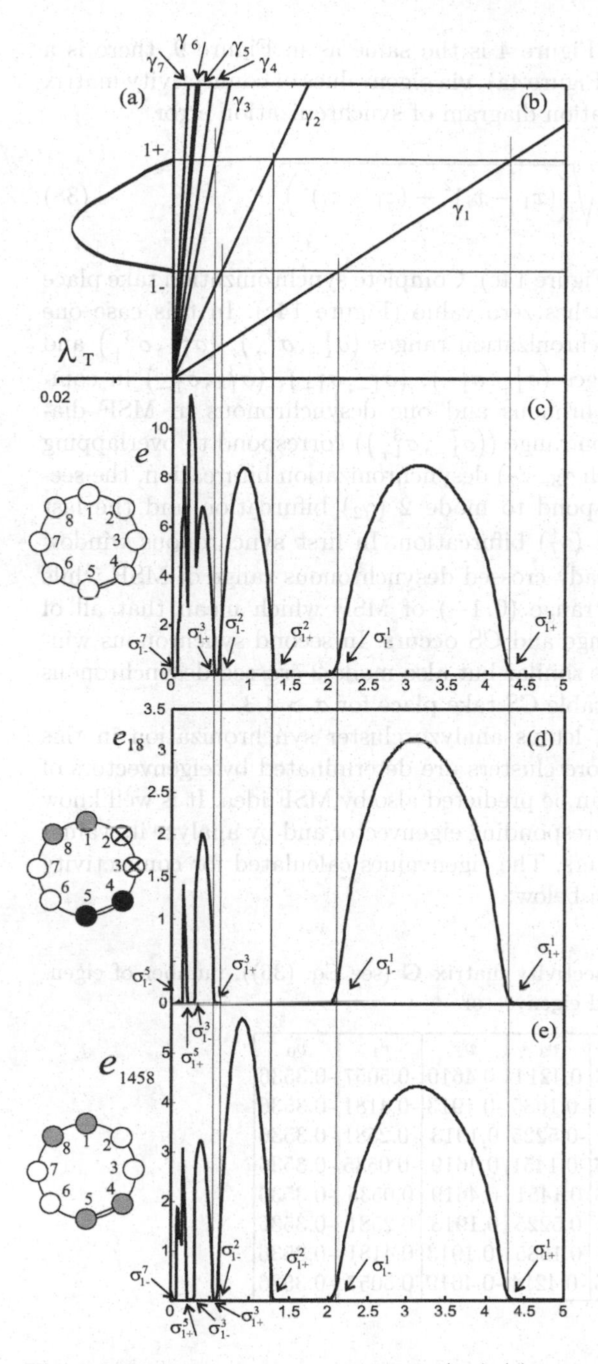

Fig. 14 Projection from the MSF $\lambda_T(\sigma\gamma)$ (a) via eigenvalues of the connectivity matrix (b) to the bifurcation diagram of the synchronization error (c). Clustering in pairs (d) and in four (e); $d = 0.401$, $\eta = 1.207$.

The idea of construction Figure 4 is the same as in Figure 9, there is a projection of MSF diagram (Figure 4a), via eigenvalues of connectivity matrix G (Figure 4b), to the bifurcation diagram of synchronization error

$$e = \sum_{i=2}^{8} \sqrt{\left((x_1 - x_i)^2 + (z_1 - z_i)^2\right)} \tag{38}$$

versus coupling strength σ (Figure 14c). Complete synchronization take place in σ ranges where e approaches zero value (Figure 14c). In this case one can observe two middle synchronization ranges $(\sigma^1_{1-}, \sigma^2_{1+})$, $(\sigma^2_{1-}, \sigma^3_{1+})$ and three desynchronization ranges $(\sigma^1_{1+}, \sigma^1_{1-})$, $(\sigma^2_{1-}, \sigma^2_{1+})$, $(\sigma^3_{1+}, \sigma^7_{1-})$ in comparison with only two synchronous and one desynchronous in MSF diagram. First desynchronization range $((\sigma^7_{1-}, \sigma^3_{1+}))$ correspond to overlapping of modes 3–7 (associated with γ_{3--7}) desynchronization bifurcation, the second one $((\sigma^2_{1-}, \sigma^2_{1+}))$ correspond to mode 2 (γ_2) bifurcation and the last one $((\sigma^1_{1-}, \sigma^1_{1+}))$ to mode 1 (γ_7) bifurcation. In first synchronous window $((\sigma^3_{1+}, \sigma^2_{1-}))$ modes 3–7 already crossed desynchronous range of MSF while modes 1 and 2 are still in range $(0, 1-)$ of MSF which mean that all of them are in synchronous range and CS occurs. In second synchronous window $((\sigma^2_{2+}, \sigma^1_{1-}))$ situation is similar but also mode 2 crossed desynchronous range of MSF. Finally the stable CS take place for $\sigma > 4.3$.

After description of RSA let us analyze cluster synchronization in this network. As we mention before clusters are determinated by eigenvectors of connectivity matrix G and can be predicted also by MSF idea. It is well know that each eigenvalue have corresponding eigenvector and by analyze its values one can find scheme of clusters. The eigenvalues calculated for connectivity matrix G (Eq. 36) are shown below:

Table 2 Eigenvectors of connectivity matrix **G** (see Eq. (36)), number of eigenvector correspond to associated eigenvector

v_7	v_6	v_5	v_4	v_3	v_2	v_1	v_0
0.0165	0.1913	0.2578	-0.3536	0.4214	-0.4619	-0.5057	-0.3536
-0.0715	-0.4619	-0.5401	0.3536	-0.1685	-0.1913	-0.4181	-0.3536
0.2217	0.4619	0.3335	0.3536	-0.5225	0.1913	-0.2581	-0.3536
-0.6674	-0.1913	0.1750	-0.3536	-0.1451	0.4619	-0.0535	-0.3536
0.6674	-0.1913	-0.1750	-0.3536	0.1451	0.4619	0.0535	-0.3536
-0.2217	0.4619	-0.3335	0.3536	0.5225	0.1913	0.2581	-0.3536
0.0715	-0.4619	0.5401	0.3536	0.1685	-0.1913	0.4181	-0.3536
-0.0165	0.1913	-0.2578	-0.3536	-0.4214	-0.4619	0.5057	-0.3536

In the first row we numbered eigenvectors and this number correspond to associated eigenvector. Is is clearly see the for γ_0 and v_0 so for direction along synchronization manifold all values in eigenvector v_0 are the same which mean the case of the CS.

When desynchronization in network is governed only by $\gamma_2 - v_2$ (all other modes ale in the synchronous range of MSF) one can observe four synchronized pairs of oscillators (first: oscillators 1 and 8, second: 2 and 7, third: 3 and 6, fourth: 4 and 5) show in Figure 14d. The same situation take place for $\gamma_6 - v_6$ also show in Figure 14d. If desynchronization is caused only by mode 6 ($\gamma_4 - v_4$) one can observe two clusters with four oscillators in each (first: 1, 8, 4, 5 oscillators and second: 2,3,6,7 oscillators) show in Figure 4f.

5 Conclusion

This chapter present in very detailed way idea of ragged synchronizability (RSA), as well as, its theoretical backgrounds. We have confirmed and explained the phenomenon of the RSA in the networks of van der Pol's oscillators with ND coupling between the nodes. Its occurrence is independent of the motion character (periodic or chaotic) of an isolated node system. We have shown the mechanism responsible for the appearance or disappearance of the windows of synchronizability is the same as the previously studied network of Duffing oscillators. It seems that the phomenonon of RSA is common for the systems with non-diagonal coupling and not sensitive for the small parameter mismatch, i.e., can be observed in real experimental systems. We also showed the appearance of cluster in experimental systems and we presented more complicated case by numerical simulation. There are still open topics for future investigation for example: confirmation of RSA existence in chaotic systems and in large networks (more then eight or ten systems).

Acknowledgements. This study has been supported by the Polish Department for Scientific Research (DBN) under Projects N N501 0710 33, N501 033 31/2490 and 293/N-DAAD/2008/0.

References

1. Afraimovich, V.S., Verichev, N.N., Rabinovich, M.: Radiophys. and Quantum Electron 29, 795 (1986)
2. Blekman, I.I.: Synchronization in Science and Technology. ASME Press, New York (1988)
3. Pecora, L., Carroll, T.: Physical Review A 44, 2374 (1991)
4. Boccaletti, S., Kurths, J., Osipov, G., Valladares, D.L., Zhou, C.S.: Physics Reports 366(1-2), 1 (2002)
5. Yanchuk, S.: Physical Review E, Statistical, Nonlinear, and Soft Matter Physics 72(3), 036205 (2005), `http://link.aps.org/abstract/PRE/v72/e036205`
6. Arenas, A., Dáz-Guilera, A., Kurths, J., Moreno, Y., Zhou, C.: Physics Reports 469(3), 93 (2008), `http://www.science-direct.com/science/article/B6TVP-4THJGHT-1/2/b218ffd0fb6a28204e1277ca59f086cb`

7. Kapitaniak, T.: Physical Review E 50, 1642 (1994)
8. West, B.J., Geneston, E.L., Grigolini, P.: Physics Reports 468(1-3), 1 (2008)
9. Pyragas, K., Pyragienė, T.: Physical Review E (Statistical, Nonlinear, and Soft Matter Physics) 78(4), 046217 (2008), `http://link.aps.org/abstract/PRE/v78/e046217`
10. Cao, L.Y., Lai, Y.C.: Physical Review E 58, 382 (1998)
11. Yamada, T., Fujisaka, H.: Prog. Theor. Phys. 70 (1983)
12. Fujisaka, H., Yamada, T.: Prog. Theor. Phys. 69 (1983)
13. Pikovsky, A.: Zeitschrift Phys. B 55, 149 (1984)
14. Stefański, A., Wojewoda, J., Kapitaniak, T., Yanchuk, S.: Physical Review E 70 (2004)
15. Dmitriev, A., Shirokov, M., Starkov, S.: IEEE Trans. Circuits Syst. I. Fund. Th. Appl. 44(9), 918 (1997)
16. Pecora, L., Carroll, T.: Physical Review Letters 80 (1998)
17. Pecora, L., Carroll, T., Johnson, G., Mar, D., Fink, K.S.: Int. J. Bifurcation Chaos 10(2) (2000)
18. Barahona, M., Pecora, L.M.: Physical Review Letters 89 (2002)
19. Nishikawa, T., Motter, A.E., Lai, Y.C., Hoppensteadt, F.C.: Phys. Rev. Lett. 91(1), 014101 (2003)
20. Boccaletti, S., Latora, V., Moreno, Y., Chavez, M., Hwang, D.-U.: Physics Reports 424(4-5), 175 (2006), `http://www.sciencedirect.com/science/article/B6TVP-4J0WTM2-1/2/739f254d99ff14ca96565d3d34a6d77a`
21. Belykh, I., de Lange, E., Hasler, M.: Physical Review Letters 94(18), 188101 (2005), `http://link.aps.org/abstract/PRL/v94/e188101`
22. Belykh, V.N., Belykh, I.V., Hasler, M.: Physica D: Nonlinear Phenomena 195(1-2), 159 (2004), `http://www.sciencedirect.com/science/article/B6TVK-4CNJD3M-3/2/13620fb652443d396fbfbb0a5fd16823`
23. Boccaletti, S., Hwang, D.-U., Chavez, M., Amann, A., Kurths, J., Pecora, L.M.: Physical Review E (Statistical, Nonlinear, and Soft Matter Physics) 74(1), 016102 (2006), `http://link.aps.org/abstract/PRE/v74/e016102`
24. Atay, F.M., Biyikoğlu, T.: Physical Review E (Statistical, Nonlinear, and Soft Matter Physics) 72(1), 016217 (2005), `http://link.aps.org/abstract/PRE/v72/e016217`
25. Chen, G., Duan, Z.: Chaos: An Interdisciplinary Journal of Nonlinear Science 18(3), 037102 (2008), `http://link.aip.org/link/?CHA/18/037102/1`
26. Watts, D.J., Strogatz, S.H.: Nature 393, 440 (1998)
27. Watts, D.J.: Small Worlds. Princeton University Press, Princeton (1999)
28. Barahona, M., Pecora, L.M.: Phys. Rev. Lett. 89(5), 054101 (2002)
29. Hong, H., Choi, M.Y., Kim, B.J.: Phys. Rev. E 65(2), 026139 (2002)
30. Belykh, I.V., Belykh, V.N., Hasler, M.: Physica D: Nonlinear Phenomena 195(1-2), 188 (2004)
31. Guclu, H., Korniss, G., Novotny, M.A., Toroczkai, Z., Rácz, Z.: Physical Review E (Statistical, Nonlinear, and Soft Matter Physics) 73(6), 066115 (2006)
32. Gade, P.M., Hu, C.-K.: Phys. Rev. E 62(5), 6409 (2000)
33. Barabási, A.L., Albert, R.: Science 286, 509 (1999)
34. Albert, R., Barabási, A.-L.: Rev. Mod. Phys. 74(1), 47 (2002)
35. Lee, D.-S.: Physical Review E (Statistical, Nonlinear, and Soft Matter Physics) 72(2), 026208 (2005), `http://link.aps.org/abstract/PRE/v72/e026208`

36. Fan, J., Wang, X.F.: Physica A: Statistical Mechanics and its Applications 349(3-4), 443 (2005)
37. Sorrentino, F., di Bernardo, M., Cuéllar, G.H., Boccaletti, S.: Physica D: Nonlinear Phenomena 224(1-2), 123 (2006); Dynamics on Complex Networks and Applications
38. Zhou, T., Zhao, M., Chen, G., Yan, G., Wang, B.-H.: Physics Letters A 368(6), 431 (2007)
39. Comellas, F., Rozenfeld, H.D., ben-Avraham, D.: Physical Review E (Statistical, Nonlinear, and Soft Matter Physics) 72(4), 046142 (2005)
40. Stefański, A., Perlikowski, P., Kapitaniak, T.: Physical Review E 75 (2007)
41. Perlikowski, P., Jagiello, B., Stefanski, A., Kapitaniak, T.: Physical Review E (Statistical, Nonlinear, and Soft Matter Physics) 78(1), 017203 (2008), http://link.aps.org/abstract/PRE/v78/e017203
42. Pecora, L., Carroll, T.: Physical Review Letters 64 (1990)
43. Rosenblum, M., Pikovsky, A., Kurths, J.: Physical Review Letters 76, 1804 (1996)
44. Kapitaniak, T., Sekieta, M., Ogorzalek, M.: Int. Journal of Bifurcation and Chaos 6, 211 (1996)
45. Kaneko, K.: Physica D 41, 137 (1990)
46. Yanchuk, S., Maistrenko, Y., Mosekilde, E.: Math. Comp. Simul. 54, 491 (2000)
47. Kestler, J., Kopelowitz, E., Kanter, I., Kinzel, W.: Physical Review E (Statistical, Nonlinear, and Soft Matter Physics) 77(4), 046209 (2008), http://link.aps.org/abstract/PRE/v77/e046209
48. Sorrentino, F., Ott, E.: Physical Review E (Statistical, Nonlinear, and Soft Matter Physics) 76(5), 056114 (2007)
49. Allefeld, C., Bialonski, S.: Physical Review E (Statistical, Nonlinear, and Soft Matter Physics) 76(6), 066207 (2007), http://link.aps.org/abstract/PRE/v76/e066207
50. Rössler, O.E.: Physics Letters A 57, 397–398 (1976)
51. Fink, K.S., Johnson, G., Carroll, T., Mar, D., Pecora, L.: Physical Review E 61(5) (2000)
52. Van der Pol, B., Van der Mark, J.: Nature 120, 363 (1927)
53. Van der Pol, B.: The London, Edinburgh, and Dublin Philosophical Magazine and Journal of Science Ser.7 3, 65 (1927)
54. Kennedy, M., Chua, L.: IEEE Transactions 33, 974 (1986)
55. Nana, B., Woafo, P.: Phys. Rev. E 74, 046213 (2006)
56. Perlikowski, P., Stefański, A., Kapitaniak, T.: Physical Review E 77 (2008)
57. Analog Devices, Inc. Low Cost Analog Multiplier (2002)
58. Belykh, V.N., Belykh, I.V., Mosekilde, E.: Phys. Rev. E 63(3), 036216 (2001)

Cumulant Analysis of Strange Attractors: Theory and Applications

V. Kontorovich and Z. Lovtchikova

Abstract. The material of this chapter presents a detailed description of the theoretical background of the new so-called "Degenerated Cumulant Equations Method for Statistical Analysis of Strange Attractors". The method is illustrated by applying the results to the analysis of concrete strange attractors of traditional interest: Lorenz, Chua, Rössler. Theoretical results are confirmed by computer simulations; it is shown how by using a limited set of the first cumulants, the PDF of the attractor components can be predicted applying orthogonal series representations or model distribution approach. Practical applications are presented through the modeling of Radio-Frequency Interferences (RFI) from digital interconnects of Laptops and Desktops computers as output signals of the strange attractors mentioned above[1].

1 Introduction – Basis of the Theory of Strange Attractors

Nowadays, the number of chaos applications in the electrical engineering field has grown considerably [17], [29]-[36], etc; moreover, recently some interesting reviews, related to the statistical analysis of strange attractors (continuous nonlinear dynamic systems generating chaos) (see [18, 19], and references therein) have been published. Despite of the comprehensive statistical study of the chaos, presented in the last two papers they were written "by physicist for physicist" and do not contain the material necessary for engineering statistical analysis of the chaotic attractors. For application purposes it is necessary to have a simple and rather accurate methodology for calculus of the statistical characteristics of attractors due to their circuit parameters, adjusting those parameters to the required features of the chaotic models, etc. Though, it is worth to mention, that there is still a lack of engineering methods

V. Kontorovich
CINVESTAV, Av. IPN #2508, Col San Pedro Zacatenco, C.P. 07360, Mexico D.F.
e-mail: valeri@cinvestav.mx

Z. Lovtchikova
Engineering and Advanced Technology Interdisciplinary Professional Unit UPIITA-IPN, Av. IPN #2580, Col Ticoman C.P. 07340, Mexico D.F.
e-mail: alovtchikova@ipn.mx

[1] This work was supported by the Intel Corporation through the grant "Intel VK"

K. Kyamakya (Eds.): Recent Adv. in Nonlinear Dynamics and Synchr., SCI 254, pp. 77–115.
springerlink.com

for the statistical analysis of attractors. Recently it was proposed by the authors the so-called "degenerate cumulant equations method" [20]-[22] for applied statistical analysis of the strange attractors, based on the parameters of the corresponding dynamic systems. It was shown, that by means of the proposed approach not only the expressions for the cumulants can be found, but also the so-called "model distributions" for each component of the attractors under analysis, etc. The attractive feature of the cumulants (instead of moments) for engineering purposes is explained in detail at [3, 4, 21, 22]; also a comprehensive and adequate method for the cumulant calculus is presented there: the cumulant brackets.

It is well known, that the "weight" of cumulants for applications diminishes as its order grows, so for engineering analysis it is often sufficient to consider only the first four cumulants: $\kappa_1 \div \kappa_4$, and the corresponding shape coefficients for PDF: γ_1 is the skewness coefficient and γ_2 is the kurtosis coefficient [4].Then, it was stated from the very beginning of the applications of the cumulants, that they can serve as a set of "independent statistical coordinates" of the random process and thus provide a "general view" of the statistical behavior of the system due to the system parameters [3, 4], etc. The cumulant analysis of the attractors is not only qualitative, but quantitative as well, and the "balance" between these two features highly depends on the type of attractors and its parameters and will be illustrated by examples presented at the Section 4. It will be shown in the following, that chaotic models are adequate to model several natural phenomena related to the communication field; for example, output signals of some components of the well known attractors: Lorenz, Chua, Rössler can successfully describe the probability density functions (PDFs) of the interferences from high speed digital interconnects, and the cumulant analysis provides a very good coincidence with the measurement data [20, 22]. Let us now introduce some basics of the chaotic (strange) attractor's theory. It is well known [1], that each dissipative continuous time dynamic system (strange attractor)[2]

$$\dot{\mathbf{x}} = \mathbf{f}(\mathbf{x}(t)), \quad \mathbf{x} \in \Re^n, \quad \mathbf{x}(t_0) = \mathbf{x_0}, \tag{1}$$

where $\mathbf{f}\left[f_1(\mathbf{x}),\ldots,f_n(\mathbf{x})\right]^T$ is a differentiable vector function. Chaotic attractors, described by (1) can be classified as hyperbolic, quasi-hyperbolic and non-hyperbolic [5]. All attractors, which were investigated beforehand (see, for example [1, 2, 25] are of the quasi-hyperbolic or non-hyperbolic type: (Lorenz, Chua, Rössler, etc.). The quasi-hyperbolic attractors (Lorenz attractor with the typically considered parameters [8]) do not actually differ from the hyperbolic (robust, ideal) attractors and existence of the invariant measure (see below) for them is practically guaranteed (see detail at [17, 18]). In the following, we shall explicitly follow the principles of ergodic theory for equations such as (1)[1], assuming the existence of an invariant measure for solutions of (1). As an invariant measure we will take a physical measure (see sections II-F and IV-H in [1]), applying the following idea of Kolmogorov: in (1) one has to consider a weak external noise $\xi(t)$, i.e.

[2] Attractors with both fractional and chaotic behavior will be considered in the Appendix. Please note that attractors are deterministic systems, but their statistical characterization is well explained at [1, 2, 7].

$$\dot{\mathbf{x}} = \mathbf{f}(\mathbf{x}(t) + \varepsilon\xi(t)), \tag{2}$$

where $\xi(t)$ is a vector of a weak external white noise with the related positive defined matrix of "intensities" $\varepsilon = [\varepsilon_{ij}]^{n\times n}$, [3].

Here it is time to give some comments regarding the concepts of the invariant and physical measures. An invariant measure is a measure for the "space" averaging of the chaos realizations: it is independent to initial distributions and completely specifies the statistical characteristics of the attractor, but it is not unique [1, 2]. A physical measure, which appears from the time averaging for each ergodic chaos realization, is obviously more reasonable from the engineering point of view and as mentioned before actually it follows from the Kolmogorov's idea (2). It is clear, that in the framework of the ergodic theory invariant and physical probabilistic measures might be synonyms, but physical measure is more relevant for the applications.

Therefore, it can be seen that (2) is a stochastic differential equation (SDE), generating the continuous Markov process with a physical measure (PDF) $W\varepsilon(\mathbf{x},t)$[3].

Note that relations between chaotic continuous processes and Markov processes were completely discussed in [2], etc; we will only stress here, that the linear Kolmogorov-Fokker-Plank (KFP) operator for $W\varepsilon(\mathbf{x},t)$ for SDE (2) has the same properties as the FPO operator [28] for discrete chaos and provides a stationary distribution: $\lim_{t\to\infty} W\varepsilon(\mathbf{x},t) = W^{\varepsilon}_{st}(\mathbf{x})$ when the vector function $\mathbf{f}(\cdot)$ does not depend on time t (sufficient conditions, see [3]).

Thus, the physical measure $W_{st}(\mathbf{x})$ has to be taken as

$$W_{st}(\mathbf{x}) = \lim_{\forall\varepsilon_{ij}\to 0} W^{\varepsilon}_{st}(\mathbf{x}). \tag{3}$$

Assumption (3) is important for all subsequent analysis, as we expect that when all elements of the matrix ε equally tend to zero, the solution of (1) and (2) has the same $W_{st}(\mathbf{x})$. In other words $W_{st}(\mathbf{x})$ is defined in asymptotic conditions of weak noise in (2).

Now, let us make some particular comments to this assumption for (2) if it describes the chaotic attractor. The Markov process, when noise intensities in SDE (2) tend to zero, is a so-called "quasi-deterministic" or "degenerated" Markov process and it is pretty hard to obtain the closed form expression for a physical measure in (3), if n in (1) is more than one. This issue was somehow briefly touched in [5]. Does it mean then, that the $W_{st}(\mathbf{x})$ does not exist? Of course not, it does exist, but the Markov description of the chaos simply might not provide us with the adequate closed form representation of the physical measure $W_{st}(\mathbf{x})$. In contrary, as it was shown in 22 and will be presented in the following, the "degenerate cumulant equations" approach provides us with the closed form solutions for the "degenerate" case. The non-hyperbolic attractors, in general, are rather sensitive to the variations of the parameters of attractors: the character of chaos can change drastically (from spiral to funnel [18, 19]), for example, for Rössler attractor when the c parameter (see below) changes from 5.7 to 10. Nevertheless the existence of the stationary measure for Rössler attractor was confirmed at [19]. The same matter was checked experimentally by the authors for the Lorenz and Chua attractor's cases.

Please note, that to the best of our knowledge, the detailed study of the existence of the stationary physical measure for the non-hyperbolic attractors is rather far from to be solved. Next, it is worth to mention here that the problem of the relaxation of output chaotic signals of the attractors to its stationary conditions is also far from to be settled [18, 19]. We propose in the following as a first approximation for the time constant of relaxation to use the inverse of the eigenvalues of the statistically linearized matrix of the attractor. Indeed, this approximation is only valid for the experimental evaluation of the first cumulants.

The $W_{st}(\mathbf{x})$, as well as its characteristic function, is totally defined by the complete set of cumulants [4] and we will show in the sequel, that conditions as in (3) are valid for all stationary cumulants of the solutions of (1) and (2). Besides, that in the following it will be dedicated a special section for the cumulants and moments here we'll introduce some basic ideas, related to the applications of cumulants for the statistical analysis of strange attractors. The concepts of cumulants used hereafter are well known [38].

The $W_{st}(\mathbf{x}) = F^{-1}\{\theta(j\nu)\}$, where $\theta(j\nu)$ is a characteristic function and $F\{\cdot\}$ and $F^{-1}\{\cdot\}$ are direct and inverse Fourier transforms respectively.

$$\theta(j\nu) = exp\left[\sum_{s=1}^{\infty}\frac{j^s}{s!}\sum_{m_1,m_2,\ldots,m_n}^{s}\kappa_{m_1,m_2,\ldots,m_n}^{\xi_1,\xi_2,\ldots,\xi_n}\nu_1^{m_1}\ldots\nu_n^{m_n}\right] \tag{4}$$

where $\{\xi_i\}_1^n$ are random variables, $m_1+m_2+\ldots+m_n = s$, $\kappa_{m_1,m_2,\ldots,m_n}^{\xi_1,\xi_2,\ldots,\xi_n}$ is the joint cumulant of the s-th order; $\nu = [\nu_1,\ldots,\nu_n]^T$.

As it was stated above and follows from (4), (see also [3, 4]) the full set of cumulants $\kappa_{m_1,m_2,\ldots,m_n}^{\xi_1,\xi_2,\ldots,\xi_n}$ completely represent $W_{st}(\mathbf{x})$, as long as the series (4) converges for all $\{\nu_i\}_{i=1}^n$.

Therefore, the analysis of cumulants renders the same results as the analysis of $W_{st}(\mathbf{x})$ and the advantages of cumulant versus moments for this goal are well known [3, 4].

It is worth to mention here that the formalism for cumulant calculus for random variables and stochastic processes was proposed by A. N. Malakhov ([3], see also [4]). The idea of applying the cumulant analysis for chaos dynamical systems was presented in [5, 21] (see also the references therein).

In the following, for $W_{st}(\mathbf{x})$ calculations we will evoke some ideas of Malakhov[4] such as model distributions, cumulant analysis and cumulant equations.

For example, for the characteristic function of one dimensional model distribution (or model approximation) the following representation is valid:

$$\hat{W}_{st}(\mathbf{x}) = F^{-1}\{\theta_m(j\nu)\}, \tag{5}$$

$$\theta_m(j\nu) = exp\left\{\sum_{s=1}^{m}\frac{(j\nu)^s}{s!}\kappa_s\right\},$$

where k_s is the cumulant of the s-th order. If one assumes that cumulants for all $s > m$ are equal to zero, then for the finite set of cumulants $\{\kappa_s\}_i^m$, we can introduce the "model distribution" $\hat{W}_{st}(\mathbf{x})$ of the m-th order and its characteristic function defined by $\theta_m(jv)$.

Note that the distribution $\hat{W}_{st}(\mathbf{x})$ is only an approximation of the true PDF and the model distributions provide better accuracy than the orthogonal series expansions [4] for the case $\gamma_2 < 0$.

Another option for analytical approximation for $W_{st}(\mathbf{x})$ is the orthogonal series representations, for example, Gramm-Charlier, Laguerre series, etc. [3, 38].

For example, the Gramm-Charlier series are defined by:

$$W(x) = W_G(x)\left[1 + \frac{\gamma_1}{3!}H_3(x) + \frac{\gamma_2}{4!}H_4(x)\right], \tag{6}$$

where

$$H_n(x) = (-1)^n exp\left\{\frac{x^2}{2}\right\}\frac{d^n}{dx^n}exp\left\{\frac{-x^2}{2}\right\},$$

is a Hermitian polynomial of n-order,

$$W_G(x) = \frac{1}{\sqrt{2\pi}}exp\left\{\frac{x^2}{2}\right\},$$

is a Gaussian distribution,

$$\gamma_1 = \frac{\kappa_3}{\sigma^3},$$

is the skewness coefficient and

$$\gamma_2 = \frac{\kappa_4}{\sigma^4},$$

is the kurtosis coefficient.

It is important to mention that for a symmetrical PDF those coefficients satisfies: $\sigma_3 = 0, \mu_3 = \kappa_3, \mu_4 = \kappa_4 + 3(\sigma^2)^2$, with κ_3 and κ_4 being the third and fourth cumulants; $\gamma_2 \geq -2$.

2 Moments and Cumulants – Cumulant Brackets

Now, let us give a formal presentation for moments and cumulants. Considering in the following one dimensional stationary distribution, one can introduce well known definitions for the moments and central moments [38]:

$$m_n = \int_{-\infty}^{\infty} x^n W_{st}(\mathbf{x})dx,$$
$$\mu_n = \int_{-\infty}^{\infty} (x-m)^n W_{st}(\mathbf{x})dx, \tag{7}$$

where m_n is a n th order moment and μ_n is a n th order central moments; $m(m_1)$ is an average (or expectation) of $W_{st}(\mathbf{x})$. The relations between the moments and the central moments are well known [3, 38]:

$$m_n = \sum_{k=0}^{n} C_n^k \mu_{n-k} m^k. \tag{8}$$

The moments also can be introduced through the characteristic function $\theta(jv)$ in the following way:

$$m_k = (-j)^k \frac{d^k}{dv^k} \theta(jv)|_{v=0}, \tag{9}$$

and the cumulants can be defined with the help of $\theta(jv)$ as well:

$$\ln \theta(jv) = \sum_{k=1}^{\infty} j^k \frac{\kappa_k}{k!} v^k.$$

The cumulant κ_k is:

$$\kappa_k = j^{-k} \left[\frac{d^k}{dv^k} \ln \theta(jv) \right]_{v=0}. \tag{10}$$

If moments m_n are known, the cumulants can be found from the following equations [3, 38]:

$$\begin{aligned} \kappa_1 &= m_1 = m \\ \kappa_2 &= m_2 - m^2 \\ \kappa_3 &= m_3 - 3m_1 m_2 + 2m_1^2 \\ \kappa_4 &= m_4 - 3m_2^2 - 4m_1 m_3 + 12 m_1 m_2 - 6m_1^4. \end{aligned} \tag{11}$$

Obviously, from (11), moments can be represented through cumulants as well [3, 38].

The following notation is also quite known: $\langle g(x) \rangle$, $\langle \xi_1 \cdot \xi_2 \rangle$, which denote the operator of statistical average of the $g(x)$ and for the product of two random variables ξ_1 and ξ_2 respectively. Following [4] we call this operator $\langle \cdot \rangle$ as *moment* bracket. In [4] the concept of *cumulant brackets* was introduced as an abbreviated representation for any cumulant, i.e.

$$\kappa_{m_1, m_2, \ldots, m_n}^{\xi_1, \xi_2, \ldots, \xi_n} = \langle \xi_1, \ldots, \xi_1, \xi_2, \ldots, \xi_2, \ldots, \xi_n, \ldots, \xi_n \rangle \equiv \langle \xi_1^{[m_1]}, \xi_2^{[m_2]}, \ldots, \xi_n^{[m_n]} \rangle, \tag{12}$$

where ξ_1 appear inside the brackets m_1 times, ξ_2 appears m_2 times, and so on; for example, the third cumulant is $\kappa_{1,2}^{\xi_1, \xi_2} = \langle \xi_1, \xi_2, \xi_2 \rangle$. Some useful features for cumulant brackets can be found at [3, 4] and are presented below. Please note, that m_i here are not the moments!

As it can be seen from (12) the formal difference between moment and cumulant brackets is, that the first one contains a "dot" between random variables (usually it is skipped), and the second one contains a "comma" between variables. The relations between moments and cumulants as well as the relations between moment and cumulant brackets were discussed in an exhaustive way in [3, 4]; the fact that by means of cumulants brackets it is possible to formalize the operations between random variables and their transformations (linear and non-linear, inertial and non-inertial), in a easy way was presented in [3, 4]. In order to continue, some features for cumulant brackets need to be taken into account [3, 4]:

1. $\langle \xi, \eta, \ldots, \omega \rangle$ is a symmetric function of its arguments.
2. $\langle a\xi, b\eta, \ldots, g\omega \rangle = a \cdot b \cdot \ldots \cdot g \langle \xi, \eta, \ldots, \omega \rangle$, where $a, b, \ldots, g$ are constants.
3. $\langle \xi, \eta, \ldots, \theta_1 + \theta_2, \ldots, \omega \rangle = \langle \xi, \eta, \ldots, \theta_1, \ldots, \omega \rangle + \langle \xi, \eta, \ldots, \theta_2, \ldots, \omega \rangle$.
4. $\langle \xi, \eta, \ldots, \theta, \ldots, \omega \rangle = 0$, if θ is independent of $\{\xi, \eta, \ldots, \omega\}$.
5. $\langle \xi, \eta, \ldots, a, \ldots, \omega \rangle = 0$.
6. $\langle \xi + a, \eta + b, \ldots, \omega + g \rangle = \langle \xi, \eta, \ldots, \omega \rangle$.

In addition to cumulant brackets it is necessary to introduce here the concept of Stratonovich symmetrization brackets which follows: symmetrization brackets together with the integer number in front of the brackets represent the sum of all possible permutations of the arguments inside the brackets. For example, the operator $3\{\langle \xi_1 \rangle \cdot \langle \xi_2, \xi_3 \rangle\}_s$ means that:

$$3\{\langle \xi_1 \rangle \cdot \langle \xi_2, \xi_3 \rangle\}_s = \langle \xi_1 \rangle \langle \xi_2, \xi_3 \rangle + \langle \xi_2 \rangle \langle \xi_1, \xi_3 \rangle + \langle \xi_3 \rangle \langle \xi_1, \xi_2 \rangle, \tag{13}$$

where $\{\cdot\}_s$ is the notation for the Stratonovich symmetrization brackets.

Rules for manipulations with cumulant brackets can be found in A.2 [3], or in Appendix II of [4]. Concrete examples for applications of cumulant and Stratonovich brackets will be done at the Section 4 with all necessary explanations. Thus, considering the formalism of cumulant and symmetrization brackets let us continue with the equations for the cumulants of the dynamical system represented by SDE (2).

3 Cumulant Equations – Degenerated Cumulant Equations

As it renders from the material presented above, (1) can be rewritten in the form of the stochastic differential equation (SDE); in other words $\mathbf{x}(t)$ is a continuous n-dimensional Markov process with kinetic coefficients, given by $\mathbf{K}_{1i}(\mathbf{x}) = \mathbf{f}_i(\mathbf{x})$ and $\mathbf{K}_2 = [\varepsilon_{ij}]^{n \times n}$ [3, 4]. Next, we assume (as it was proposed above), that $W_{st}(\mathbf{x})$ exists, and that it is a reasonable physical measure for (1) and (2), although there exists all cumulants that adequately represent $W_{st}(\mathbf{x})$.

For the SDE representation of the attractor the approach, named as cumulant equations for the SDE with the given $\mathbf{K}_{1i}((x))$ and $\mathbf{K}_2$ [3, 4], can be successfully applied [3].

Let us present here the procedure to develop the cumulant equations from SDE (2)[3, 4]. It is well known, that $W(\mathbf{x},t)$ for the solution of (2) follows the so-called Fokker-Plank-Kolmogorov Equation (FPK, or direct Kolmogorov Equation) which operator is self adjoint to the operator of the so-called inverse Kolmogorov Equation. Those direct and inverse Kolmogorov operators can be represented in the following way [3]:

$$L^+(\mathbf{x}) = \mathbf{K}_1(\mathbf{x}) \frac{\partial}{\partial x_i} + \mathbf{K}_{2ij}(\mathbf{x}) \frac{\partial^2}{\partial x_i \partial x_j} \tag{14}$$

[3] Here we apply the definition for kinetic coefficients in the Stratonovich form [3].

$$L^{-}(\mathbf{x}) = \frac{\partial}{\partial x_i}\mathbf{K}_{1i}(\mathbf{x}) + \frac{\partial^2}{\partial x_i \partial x_j}\mathbf{K}_{2ij}(\mathbf{x}),$$

where $L^{+}(\mathbf{x})yL^{-}(\mathbf{x})$ are direct and inverse Kolmogorov operators, which are self adjoint by definition [3]. Now with $L^{+}(\mathbf{x})$ from (15) it is possible to define the differential equation for the average of any function $g(\mathbf{x})$ as:

$$\frac{d\langle g(\mathbf{x})\rangle}{dt} = \langle L^{+}(\mathbf{x})g(\mathbf{x})\rangle \tag{15}$$

where $\langle\rangle$ is an averaging operator. Assuming that $g(\mathbf{x})$ is any product of $x_{\alpha}^{n_\alpha} x_{\beta}^{n_\beta} \dots x_{\omega}^{n_\omega}$, from (15) and (15) it is possible to obtain the equations for any moments of $\mathbf{x}(\mathbf{t})$. Then, in stationary conditions ($t \to \infty$), those differential equations tend to algebraic ones and applying the representation of cumulants through moments (11), one can finally get the so-called cumulant equations, which depend on $\mathbf{K}_{1i}(\mathbf{x})$ and $\mathbf{K}_{2ij}(\mathbf{x})$. The concrete algebra is not as complex as cumbersome (the interested reader can find it at [3, 4]). Here we will present this technique (for simplicity) only for one-dimensional case. Let $g(x) = x, x^2, x^3, \dots$ and substituting it into (15), one gets for $t \to \infty$:

$$\langle K_1(x)g'(x)\rangle + \langle K_2(x)g''(x)\rangle. \tag{16}$$

From (16) it immediately comes:

$$\begin{aligned} \langle K_1(x)\rangle &= 0 \\ 2\langle xK_1(x)\rangle + \langle K_2(x)\rangle &= 0 \\ &\vdots \\ s\langle x^{s-1}\cdot K_1(x)\rangle + \frac{s(s-1)}{2}\langle x^{s-2}\cdot K_2(x)\rangle &= 0. \end{aligned} \tag{17}$$

Now, with the help of (11) and applying cumulant brackets instead of moments we can get:

$$\begin{aligned} \langle K_1(x)\rangle &= 0 \\ 2\langle x, K_1(x)\rangle + \langle K_2(x)\rangle &= 0 \\ \langle x, x, K_1(x)\rangle + \langle x, K_2(x)\rangle &= 0 \\ &\vdots \end{aligned} \tag{18}$$

Generalizations for multidimensional case are straight forward [4]:

$$\begin{aligned} \langle K_{ij}(\mathbf{x})\rangle &= 0; \\ 2\left\{\langle x_i, K_{1i}(\mathbf{x})\rangle\right\}_s + \langle K_{2ij}\rangle &= 0; \\ 3\left\{\langle x_i, x_j, K_{1\beta}(\mathbf{x})\rangle\right\}_s + 3\left\{\langle x_1, K_{2j\beta}\rangle\right\}_s &= 0 \end{aligned} \tag{19}$$

$$\vdots$$

$$\sum_{l=1}^{n} C_n^l \left[\{\langle x_1,x_2,\ldots,x_{n-l},K_{1n-l+1}(\mathbf{x})\rangle\}_s + \langle x_1,x_2,\ldots,x_{n-l},K_{2n-l+1,l}\rangle \right] = 0,$$

where $i \neq j$, $\beta = \overline{1,n}$.

We can evidently see from (19), that if $\forall \varepsilon_{ij} \to 0$, then the second summand tends to zero and the equations in (19) tend to the so-called "*degenerated cumulant equations*". Therefore $W_{st}^{\varepsilon}(\mathbf{x})$ tends to $W_{st}(\mathbf{x})$ (when $\forall \varepsilon_{ij} \to 0$).

Hence, the degenerated cumulant equations have the following form:

$$\begin{aligned} \langle K_{1i}(\mathbf{x})\rangle &= 0; \\ 2\left\{\langle x_i, K_{1j}(\mathbf{x})\rangle\right\}_s &= 0; \\ 3\left\{\langle x_i, x_j, K_{1\beta}(\mathbf{x})\rangle\right\}_s &= 0, \\ \vdots \quad \vdots \quad & \vdots \end{aligned} \tag{20}$$

where $i,j,\beta = \overline{1,n}$.

Equations (18) and (20) are nonlinear algebraic equations and this set, in general, is not closed. However, it is always possible to "cut" the set of cumulants by neglecting all cumulants with order $s > n$. The distribution $\hat{\mathbf{W}}_{st}(\mathbf{x})$ which is created by applying all cumulants with order $s \leq n$ was earlier termed as a "model distribution".

Although the equations in (20) have to be consequently solved: first for each component of $\mathbf{x} = [x_1, x_2, \ldots, x_n]^T$ (first line), then for couples of components $\{x_i, x_j\}_{i,j=1}^{n}$ (second line), then for triples $\{x_i, x_j, x_\beta\}_{i,j,\beta=1}^{n}$ (third line), etc. The way to do it is to "open" the cumulant brackets (see the already mentioned references A.2 in [3] or Appendix II of [4]). We shall illustrate this technique with examples for the Lorenz, Chua and Rössler attractors.

4 Examples: Lorenz, Chua and Rössler Attractors

The parameters for the equations for Lorenz, Chua and Rössler attractors are based on the data presented in [30, Table 2], [18, 19], etc.

4.1 *Lorenz Attractor*

Equation (1) for Lorenz attractor case has the following form:

$$\begin{aligned} \dot{x} &= \sigma(y-x), \\ \dot{y} &= Rx - y - z, \\ \dot{z} &= xy - Bz, \end{aligned} \tag{21}$$

where σ, B and R are the parameters of the attractor, $\mathbf{x} = [x,y,z]^T$.

Thus the first kinetic coefficients for (20) are:

$$\begin{aligned} K_{11}(\mathbf{x}) &= \sigma(y-x), \\ K_{12}(\mathbf{x}) &= Rx - y - zx, \\ K_{13}(\mathbf{x}) &= xy - Bz. \end{aligned} \tag{22}$$

The PDF histograms for $x(t), y(t), z(t)$ components in normalized scale are presented at Fig. 1-3.

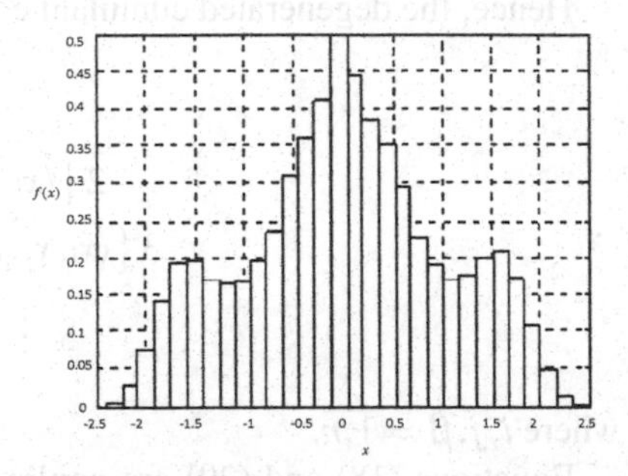

Fig. 1 PDF histogram of the "x" component of the Lorenz attractor

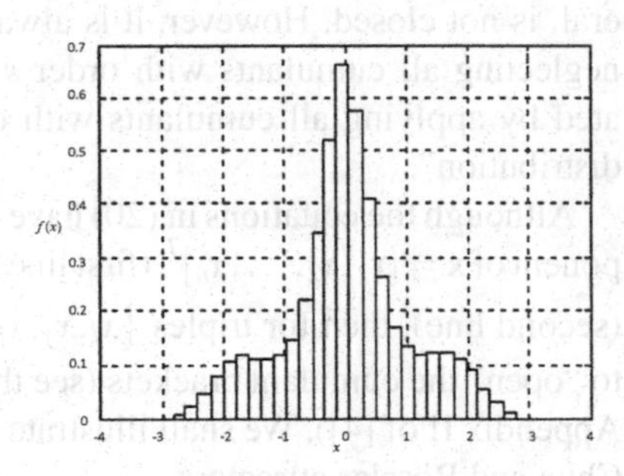

Fig. 2 PDF histogram of the "y" component of the Lorenz attractor

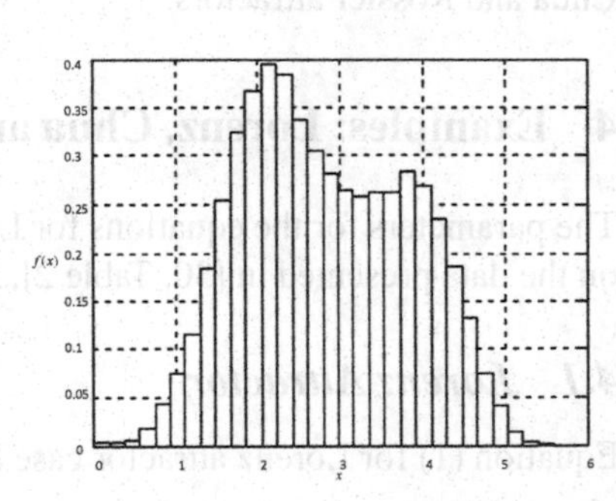

Fig. 3 PDF histogram of the "z" component of the Lorenz attractor

Note (see [7, 25]), that for the Lorenz attractor the mappings $x \rightarrow -x$ and $y \rightarrow -y$ are symmetrical so it is possible to take $\kappa_3^y = \kappa_3^x = 0$. From (18) and (20), using the first and the third property of cumulant brackets, we get:

-for component "x" [first line in (20)]:

$$\kappa_1^y = \langle y \rangle = \langle x \rangle = \kappa_1^x; \tag{23}$$
$$\kappa_{1,1}^{x,y} = \kappa_2^x; \qquad \kappa_{2,1}^{x,y} = \kappa_3^x = 0; \qquad \kappa_{3,1}^{x,y} = \kappa_{4j}^x; \ldots; \kappa_{5,1}^{x,y} = \kappa_6^x,$$

-for component "y" [first line in (20)]:

$$\kappa_{1,1}^{z,x} = 0;$$
$$R\kappa_{2,1}^{y,x} = \frac{\kappa_2^x}{B}\left(\kappa_{2,1}^{y,x} + \kappa_{2,1}^{y,z}\right); \tag{24}$$

-for component "z" [first line in (20)]:

$$\langle z \rangle = \kappa_1^z = \frac{\kappa_2^x}{B}; \qquad \kappa_{1,1,1}^{x,y,z} = B\kappa_2^z; \tag{25}$$

$$\kappa_3^z = 0; \qquad \kappa_{1,1}^{x,y} = \frac{\kappa_2^y + B\kappa_2^z + \frac{(\kappa_2^x)^2}{B}}{R}.$$

For two components [second line at (20)]:

$$-(x,y): \qquad \kappa_{1,2}^{x,y} = 0, \qquad \langle x \rangle = \langle y \rangle = 0;$$
$$-(y,z): \qquad \kappa_{1,1,1}^{x,y,z} = \frac{(2+B)\kappa_{2,1}^{y,z}}{\left(2B + \frac{\kappa_2^x}{B}\right)} \tag{26}$$
$$-(x,z): \qquad \frac{(R-1)}{B}\kappa_2^x = \kappa_{2,1}^{z,x}, \qquad \kappa_{2,1}^{z,y} = \frac{2B+\sigma}{\sigma}\kappa_{2,1}^{z,x}.$$

Note that in order to open cumulant brackets of the type $\langle x,x,zx \rangle$ in (25) and (26) we used the special case of the formula (A.26) from [3]:

$$\langle x, f(y), g(y) \rangle = \sum_{n=0}^{\infty} \frac{1}{n!}\left\{ \sum_{k=0}^{n} C_n^k \langle f^{(n-k)}(\cdot) g^{(k)} \rangle + \sum_{k=1}^{n-1} C_n^k \langle f^{(n-k)}(\cdot) \rangle \langle g^{(k)}(\cdot) \rangle \right\} \langle x, y^{[n]} \rangle.$$

In a similar way, we can perform the calculations for three components [third line at (15)], etc.

We can state some useful preliminary conclusions which follow from (23)-(26):

- The marginal distributions $W(x)$ and $W(y)$ are similar, but not identical, moreover both distributions are symmetrical with zero expectations.
- $\kappa_2^x \neq \kappa_2^y$.
- $\langle z \rangle = \frac{\kappa_2^x}{B} \neq 0$, but $W(z)$ is a symmetrical function as well.
- $\kappa_{1,1}^{x,z} = 0$, though components 'x' and 'z' are uncorrelated at the same time instant, higher order cumulants $\kappa_{2,1}^{y,z}$, $\kappa_{2,1}^{x,z}$ are non-negative.
- $\kappa_2^x \geq \frac{RB}{z}$, $\kappa_2^x + \kappa_2^z < \frac{BR^2}{4}$, etc.

Finally, one can summarize:

$$\kappa_1^x = \kappa_1^y = 0^*, \kappa_1^z = \langle z \rangle = \frac{\kappa_2^{x\,*}}{B}, \kappa_2^x = \kappa_{1,1}^{x,y*},$$

$$\kappa_{2,1}^{x,z} = \kappa_2^x \left(\frac{R-1}{B} \right), \kappa_3^y = \kappa_3^x = \kappa_{1,2}^{x,y} = 0^*, \kappa_{1,1,1}^{x,y,z} = B\kappa_2^z,$$

$$\kappa_4^x = \kappa_{3,1}^{x,y} = 0, \text{i.e.}, \gamma_4^x = 0^*, \kappa_2^y + \kappa_2^z < \frac{BR^2}{4}^*, \kappa_2^x \geq \frac{RB}{2}^*,$$

$$\kappa_3^z = \frac{2(B\kappa_2^z - \kappa_{1,2}^{z,y} - \kappa_1^z)}{B} \neq 0^*, \text{i.e.}, \gamma_3^z \neq 0, \kappa_{1,2}^{x,y} = \kappa_{1,1}^{z,x} = 0,$$

$$\kappa_{1,1}^{z,y} \approx 0, \text{and} \kappa_4^y = R\kappa_2^x - \kappa_2^y - 2B\kappa_2^z, \text{i.e.}, \gamma_4^z, \gamma_4^y \neq 0^*.$$

All the relations marked here by superscript "$*$" mean that such features were proved by numerical simulations (see also the material below).

4.2 Chua Attractor

$$\begin{aligned} \dot{x} &= \beta_1(y-x) - \alpha h(x), \\ \dot{y} &= \beta_2(x-y) + \beta_4 z, \\ \dot{z} &= -\beta_3 y \end{aligned} \tag{27}$$

where $\beta_1 \div \beta_4$ and α are the parameters of the attractor (see [18, 25, 28, 29] etc); $\mathbf{x} = [x, y, z]^T$.

An inertial nonlinearity $h(x)$ has the following form:

$$h(x) = \begin{cases} -L, & x < -L \\ x, & |x| < L \\ L, & x > L \end{cases}. \tag{28}$$

In the same way as for Lorentz attractor, histograms of the PDFs for $x(t)$, $y(t)$ and $z(t)$ components are presented for the Chua case at Fig 4-6.

Then, the first kinetic coefficients for (27) are:

$$K_{1,1}(\mathbf{x}) = \beta_1(y-x) - ah(x),$$

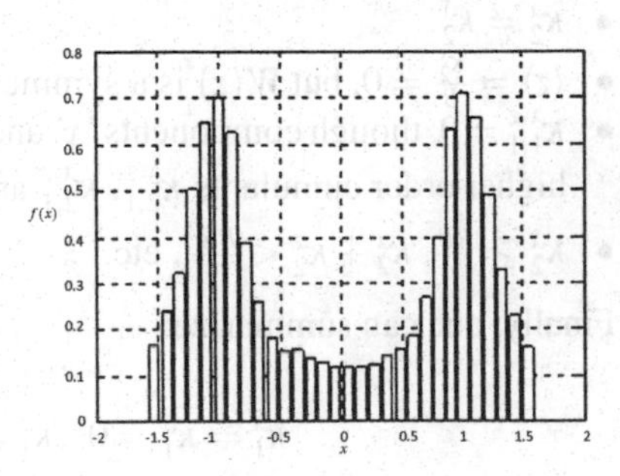

Fig. 4 PDF histogram of the "x" component of the Chua attractor

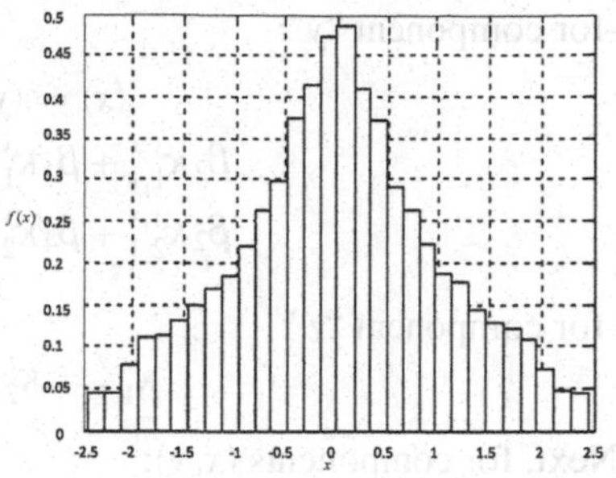

Fig. 5 PDF histogram of the "y" component of the Chua attractor

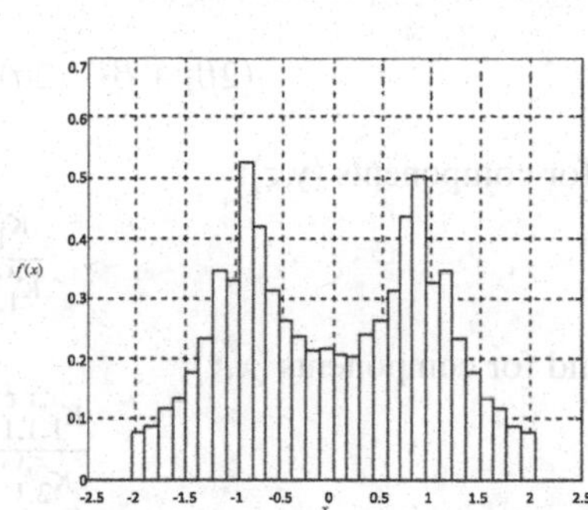

Fig. 6 PDF histogram of the "z" component of the Chua attractor

$$
\begin{aligned}
K_{1,2}(\mathbf{x}) &= \beta_2(x-y)+\beta_4 z \\
K_{1,3}(\mathbf{x}) &= -\beta_3 z.
\end{aligned}
\tag{29}
$$

Applying the same approach as in subsection 4.1 we get: - for component "x"

$$
\begin{aligned}
\frac{\beta_1}{a}(\langle y\rangle - \langle x\rangle) &= \langle h(x)\rangle, \\
\kappa_{1,1}^{x,y} &= \frac{a+\beta_1}{\beta_1}\kappa_2^x, \\
\kappa_{2,1}^{x,y} &= \frac{a+\beta_1}{\beta_1}\kappa_3^x
\end{aligned}
\tag{30}
$$

Note, that in order to open the cumulant brackets in (30) of the type $\langle x, h(x)\rangle$ or $\langle x, y, h(x)\rangle$ the following special cases of [3, Equation (A.26)] were applied:

$$
\langle x, f(x)\rangle = \sum_{s=1}^{\infty}\frac{1}{s!}\langle f^{(s)}(x)\rangle \kappa_{1+s}^x,
$$

$$
\langle x, y, f(x)\rangle = \sum_{s=1}^{\infty}\frac{1}{s!}\langle f^{(s)}(x)\rangle \left[\kappa_{1,1,s}^{x,y,z} + \sum_{L=1}^{s-1} C_s^L \kappa_{1,L}^{x,z}\kappa_{1,s-L}^{y,z}\right], \text{where } \langle h'(x)\rangle = 1,
$$

$$
\langle h^{(s)}(x)\rangle = 0 \text{ if } s \geq 2.
$$

-for component "y":

$$\begin{aligned} \langle x \rangle = \langle y \rangle = 0; \qquad \langle h(x) \rangle = 0, \\ \beta_2 \kappa_{1,1}^{y,x} + \beta_4 \kappa_{1,1}^{y,z} = \beta_2 \kappa_2^y \\ \beta_2 \kappa_{2,1}^{y,x} + \beta_4 \kappa_{2,1}^{y,z} = \beta_2 \kappa_3^y \end{aligned} \tag{31}$$

-for component "z":

$$\kappa_{1,1}^{z,y} = \kappa_{2,1}^{z,y} = 0; \qquad \langle z \rangle = 0. \tag{32}$$

Next, for components (x,y):

$$(2\beta_1 + \beta_2 + 2a)\kappa_{2,1}^{x,y} = 2\beta_1 \kappa_{1,2}^{x,y} + \beta_4 \kappa_{2,1}^{x,z}, \tag{33}$$

-for components (y,z):

$$\frac{\kappa_{1,2}^{y,z}}{\kappa_{1,1,1}^{x,y,z}} = -\frac{\beta_1}{\beta_4}, \tag{34}$$

and for components (x,z):

$$\frac{\kappa_{1,1,1}^{x,y,z}}{\kappa_{2,1}^{x,y}} = \frac{2(\beta_1 + 1)}{2\beta_1 - \beta_3}, \tag{35}$$

Similarly we can perform the calculations for the case of three components (x,y,z). If we assume, that $W(x)$ is symmetric, then formulas (33)-(35) provide us with the following features regarding the stochastic behavior of the Chua attractor:

$$\kappa_3^x \equiv 0, \qquad \kappa_{2,1}^{x,y} = 0, \qquad \kappa_{1,1}^{x,y} = C\kappa_2^x, \qquad \text{where } C > 1,$$

$$\kappa_{2,1}^{y,x} = -\frac{\beta_4}{\beta_2}\kappa_{2,1}^{y,z}, \qquad \kappa_{1,1}^{x,y} = \kappa_{1,1}^{y,x} = \kappa_2^y > \kappa_2^x,$$

$$\kappa_{1,2}^{x,y} = \frac{\beta_4}{2\beta_1}\kappa_{2,1}^{y,x}, \qquad \kappa_{2,1}^{y,x} = -\frac{\beta_4}{\beta_2}\kappa_{2,1}^{y,z}.$$

Finally, the set of relationships is:

$$\kappa_1^z = \kappa_1^x = \kappa_1^y = 0^*, \kappa_3^x = 0^*, \kappa_{1,1}^{x,y} = \frac{\alpha + \beta_1}{\beta_1}\kappa_2^x,$$

$$\kappa_4^x = \frac{\beta_1}{\beta_1 + \alpha}\kappa_{3,1}^{x,y}, \kappa_{3,1}^{x,y} = \left[\langle x^3 y \rangle - (\kappa_2^x)^2 \frac{\alpha + \beta_1}{\beta_1} \right] \approx -(\kappa_2^x)^2,$$

with the assumption that $\langle x^3 y \rangle \approx 0^*$, $\kappa_4^x \approx -(\kappa_2^x)^{2*}$, i.e., $\gamma_2 = \approx -1^*$.

This can be observed in Table 1.

4.3 Rössler Attractor

Let us now apply the approach for Rössler attractor. Equation (1) for the Rössler attractor has the following form:

Table 1 Experimental Data for Chua and Lorenz Attractors

	Chua			Lorenz		
	x	y	z	x	y	z
κ_1	0.0024	0	−0.0023	0.1275	0.1263	24.42
κ_2	2.0156	0.0278	2.4716	63.4806	75.4695	75.080
γ_1	−0.0037	0	0.0050	−0.0311	−0.0316	0.1386
γ_2	−1.6452	−0.3109	−1.1404	−0.6270	0.2190	0.8873

$$\begin{aligned} \dot{x} &= -y - z, \\ \dot{y} &= x + ay, \\ \dot{z} &= b + zx - zc, \end{aligned} \tag{36}$$

where a, b, c are the parameters of the attractor, $\mathbf{x} = [x, y, z]^T$.

In Figures 7-9 the histograms for PDFs of the "x", "y" and "z" components of the Rössler attractor are represented for normalized values.

One can see from figures 7-9, that the components "x" and "y" (contrary to Lorenz attractor) are "oppositely" asymmetric, and have unimodal PDFs with $\gamma_2 > 0$, i.e. the vertices of the distributions are "sharper", than the Gaussian ones. Next, the component "z" can be approximated by means of a delta-function.

Hence, the first kinetic coefficients for (36) are:

$$\begin{aligned} \mathbf{K}_{1,1}(\mathbf{x}) &= -y - z, \\ \mathbf{K}_{1,2}(\mathbf{x}) &= x + ay, \\ \mathbf{K}_{1,3}(\mathbf{x}) &= b + zx - zc. \end{aligned} \tag{37}$$

Introducing (36) and (37) into (20) for the first component "x", one gets:

$$\begin{aligned} \kappa_1^y = \langle y \rangle &= -\langle z \rangle = -\kappa_1^z, \\ \kappa_{1,1}^{x,z} &= -\kappa_{1,1}^{x,y}, \\ \kappa_{2,1}^{x,y} &= -\kappa_{2,1}^{x,z}, \\ \kappa_{3,1}^{x,y} &= -\kappa_{3,1}^{x,z}, \\ \kappa_{4,1}^{x,y} &= -\kappa_{4,1}^{x,z}. \end{aligned} \tag{38}$$

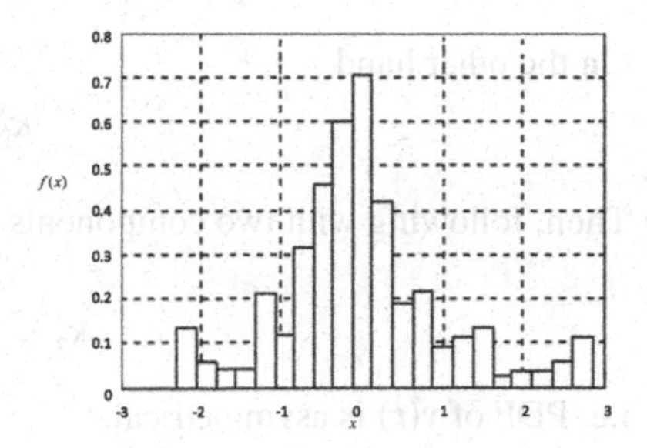

Fig. 7 PDF histogram of the "x" component of the Rössler attractor

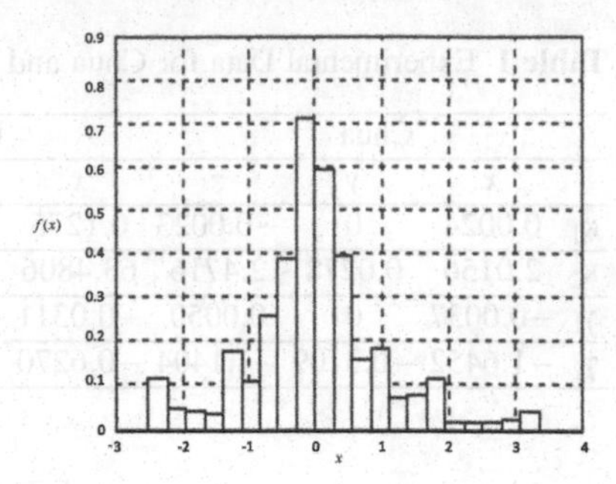

Fig. 8 PDF histogram of the "y" component of the Rössler attractor

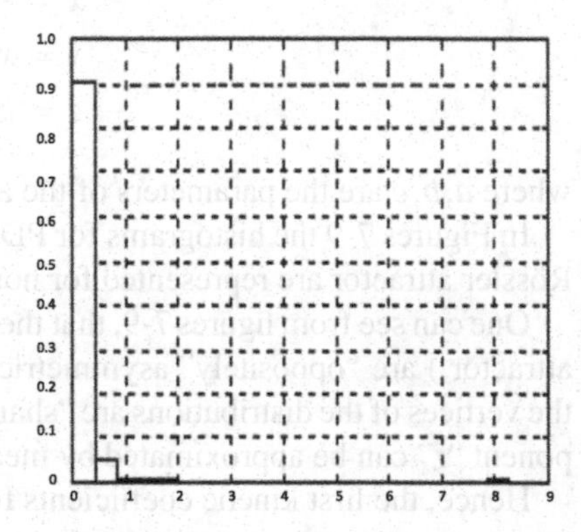

Fig. 9 PDF histogram of the "z" component of the Rössler attractor

For the second component "y":

$$\begin{aligned} \kappa_1^x &= \langle x \rangle = -a\langle y \rangle = -a\kappa_1^y, \\ \kappa_2^y &= -\frac{1}{a}\kappa_{1,1}^{x,y}, \\ \kappa_3^y &= -\frac{1}{a}\kappa_{2,1}^{y,x}, \\ \kappa_4^y &= -\frac{1}{a}\kappa_{3,1}^{y,x}. \end{aligned} \tag{39}$$

For the third component "z":

$$\begin{aligned} \kappa_{1,1}^{x,y} = \kappa_{1,1}^{z,x} &= c\langle z \rangle = -\langle z \rangle\langle x \rangle, \\ \kappa_2^y &= \frac{c\langle z \rangle - a\langle z \rangle^2}{a}. \end{aligned} \tag{40}$$

On the other hand

$$\kappa_2^y = \frac{\kappa_2^x}{\langle z \rangle}. \tag{41}$$

Then, following with two components $\{x, y\}$, we can obtain:

$$\kappa_3^y = \frac{2\kappa_{1,2}^{y,x}}{1+a} \neq 0, \tag{42}$$

i.e. PDF of $y(t)$ is asymmetrical.

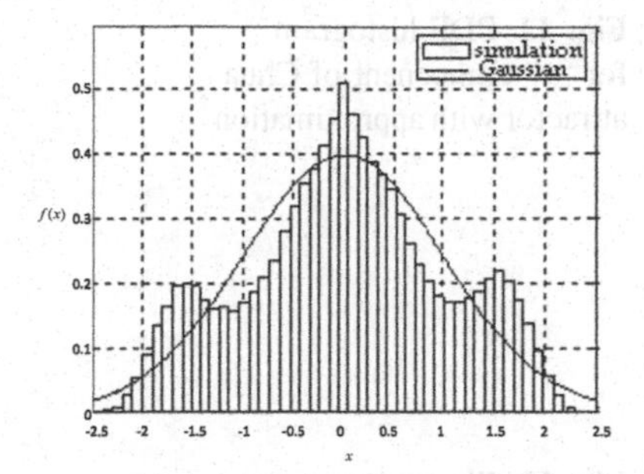

Fig. 10 PDF histogram of the Lorenz attractor and its approximations

With two components $\{x,z\}$ it follows: $\kappa_3^z \neq 0$, i.e. PDF of $z(t)$ is asymmetrical as well. The cumulants: $\kappa_{1,1}^{z,y}, \kappa_{1,1}^{z,x}$, are very important because they show the degree of "similarity" for the chaotic signals, $x(t)$, $y(t)$ and $z(t)$. Particularly they become

$$\kappa_{1,1}^{z,y} = a\langle y\rangle \kappa_2^y - \frac{\langle y\rangle}{a} \approx a\langle y\rangle \kappa_2^y, \qquad (43)$$
$$\kappa_{1,1}^{z,x} \cong \langle z\rangle c,$$

when $c >> a$.

Together with the previously defined $\kappa_{1,1}^{x,y}$, those two cumulants illustrate the measure of linear statistical dependence between the components of the attractor. At the same time, the mean value $\langle x\rangle$ is less than $\langle y\rangle$ and it is possible for the qualitative calculus to assume, that $\langle x\rangle \approx 0$. For the case of $\langle y\rangle$, the last assumption is not valid, as well as $\frac{\langle x\rangle}{a}$, etc.

Strictly speaking, as it follows from (38) and (39), all components of the Rössler attractor have an asymmetry in the PDFs, but the degree of its asymmetry is different: for "z" component: $\kappa_3^z \approx \frac{\kappa_2^z}{c}$, while $\kappa_3^x \approx 0$, i.e. $\gamma_3^z \neq \gamma_3^y \neq 0$. So, $x(t)$ has practically a symmetrical PDF, but the PDFs for $y(t)$ and $z(t)$ are significantly asymmetric.

4.4 *Analytical Approximations for the PDF of Strange Attractors*

Let us present here some analytical approximations for the PDFs of the components of the attractors, based on the results of the cumulant analysis. Some more data will be presented at Section 6 as well.

In Fig.10 we illustrate $W(x)$ in normalized scale for the Lorenz attractor: experimental (simulation) histogram; Gaussian distribution. The "model distribution" here is useless since the curtosis coefficient for $W(x)$ is zero (see 4.1).

For the Chua attractor let us present here an analytical approximation also for the x-component. An approximation of the PDF histogram that presents bimodal characteristics is chosen (Fig. 11) according to the following equation [3]:

$$W(x) = C(p,q)exp(px^2 - qx^4), \qquad (44)$$

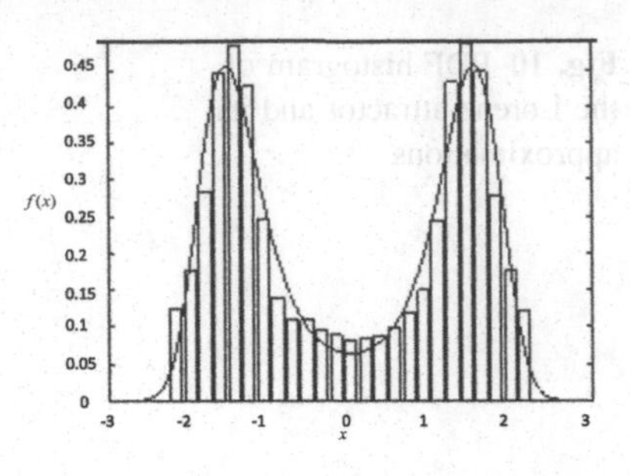

Fig. 11 PDF histogram for "x" component of Chua attractor with approximation

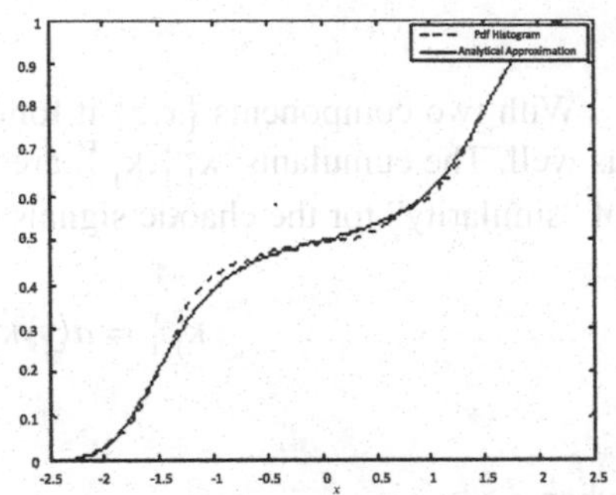

Fig. 12 Illustration of the KST application

where C represents a normalization constant, while p and q are approximation parameters.

The parameters p and q are obtained from the central moments equation for (44)[51]:

$$\overline{x^{2n}} = \frac{\Gamma\left(1+\frac{1}{2}\right) D_{-n-\frac{1}{2}}(-\delta)}{\sqrt{\pi} D_{-\frac{1}{2}}(-\delta)(2q)^{\frac{n}{2}}}, \tag{45}$$

where $n = 1,2,\ldots, \delta = \frac{p}{\sqrt{(2q)}}$ which can be found from κ_2^x and κ_4^x.

Once equations (45) are solved, parameters "p" and "q" are found to be $q = 1.5; p = 3.5$ while the normalization constant C is 0.063.

In order to confirm that the approximation is valid, the comparison between the histogram of the PDF obtained from the attractor and the analytical approximation was made using the Kolmogorov-Smirnoff goodness of fit test (KST) for the Cumulative Distribution Function (CDF) (Fig.12) with a level of significance: $\alpha = 0.05$. It can be concluded that the analytical approximation is valid.

For the Rössler attractor the following results are presented. For "x" and "y" components we describe the PDF histograms by means of the Laplace distribution defined by [38]:

$$f(x) = \frac{1}{2\lambda} exp\left\{-\frac{|x-\mu|}{\lambda}\right\}, \tag{46}$$

where μ is a location and λ is a scale. The use of Laplace distribution allows to make a right description for the "x" and "y" components of the Rössler attractor as it is observed in the Figures 13 and 14.

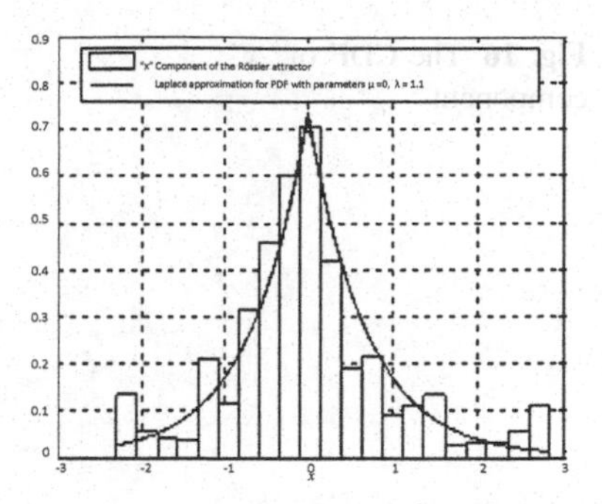

Fig. 13 Histogram of the "x" Component of the Rössler attractor and the Laplace PDF approximation

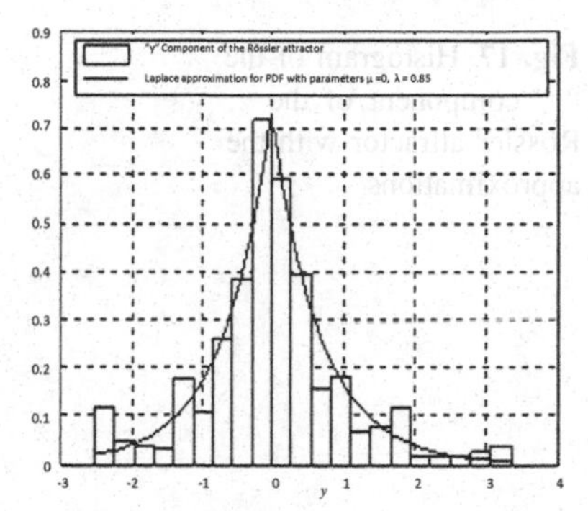

Fig. 14 Histogram of the "y" Component of the Rössler attractor and the Laplace PDF approximation

Then, as it follows from Figures 13 and 14, the PDF, for the "x" component of the Rössler attractor is approximated by a Laplace distribution with the local parameter $\mu = 0$ and scale parameter $\lambda = 1.1$ and for the "y" component with local parameter $\mu = 0$ and scale parameter $\lambda = 0.85$.

In Figures 15 and 16, we apply the Kolmogorov- Smirnoff goodness of fit test with a significance level $\alpha = 0.05$, in order to examine whether the accuracy of the PDF of the "x" and "y" components of the Rössler attractor and their approximations using the Laplace distribution are adequate or not. As it can be seen from the Figures 15 and 16, the approximation can be considered as acceptable. Utilization of the Chi-Square test gives the same results, which are not presented here.

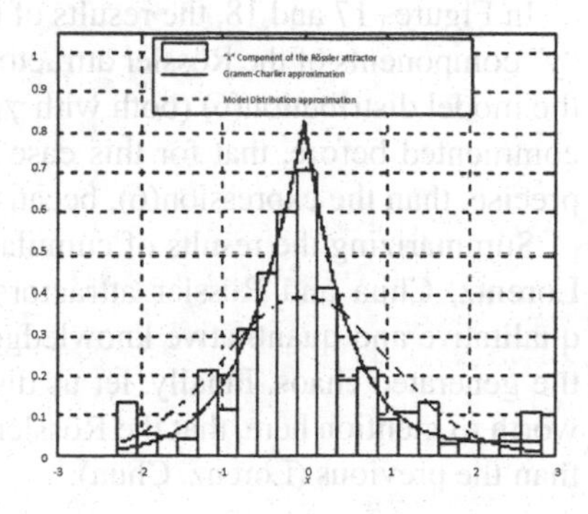

Fig. 15 Histogram of the "x" component of the Rossler attractor with approximations

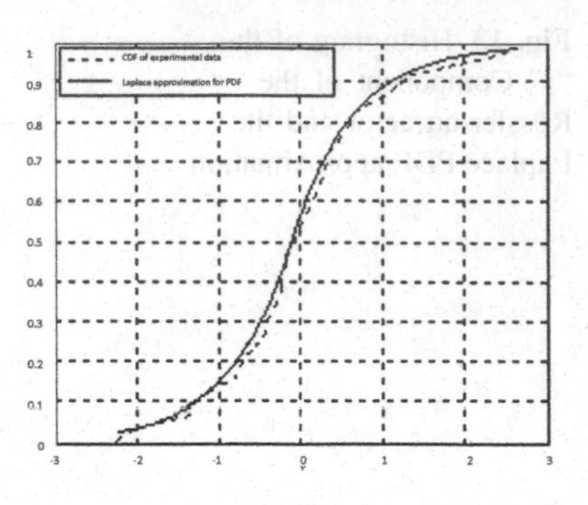

Fig. 16 The CDF of "x" component

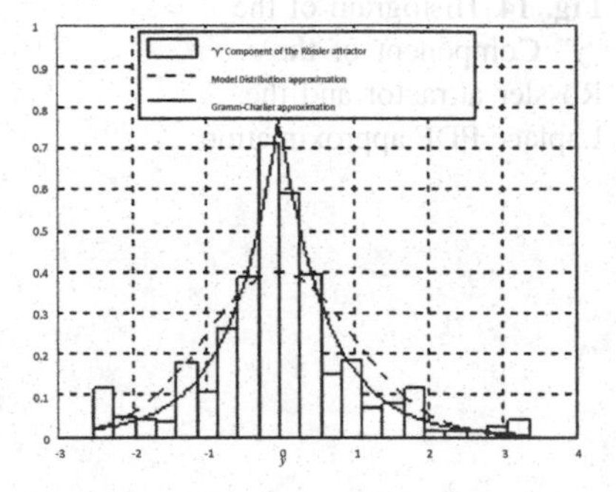

Fig. 17 Histogram of the "y" component of the Rössler attractor with the approximations

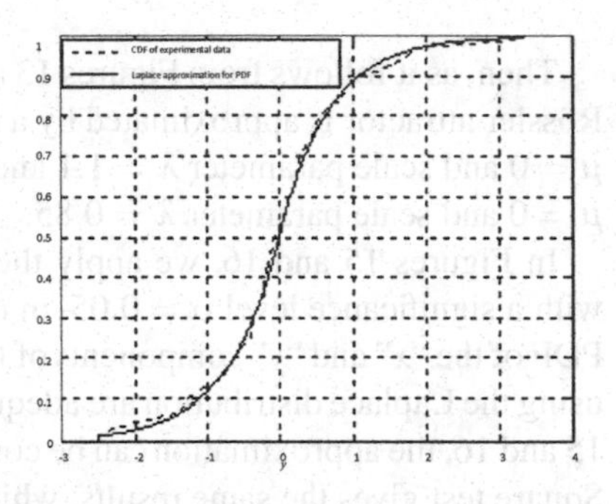

Fig. 18 The CDF of "y" component

In Figures 17 and 18, the results of the approximation of the PDF for the "x" and "y" components of the Rössler attractor by means of the Gramm-Charlier series and the model distribution(6) (both with $\gamma_4 > 0$) are represented. One can see, as it was commented before, that for this case the Gramm-Charlier approximation is more precise, than the expression(6), because of $\gamma_2 > 0$.

Summarizing the results of cumulant analysis even of the first cumulants of the Lorentz, Chua and Rössler attractors one can see that they give an interesting qualitative and quantitative knowledge regarding the statistical characterization of the generated chaos. Finally, let us discuss some features of Rössler attractor. It is worth to mention here, that the Rössler attractor is apparently a more complex case, than the previous (Lorenz, Chua).

Table 2 Data for Rössler attractor with variation of "c"

	$a=b=0.2; c=5.7$			$a=b=0.2; c=10$		
	x	y	z	x	y	z
κ_1	0.14	−0.78	0.76	0.131	−0.72	0.73
κ_2	20.91	23.17	7.83	35.22	36.48	21.2
γ_1	0.226	−0.198	5.22	0.218	−0.49	8.65
γ_2	0.59	−0.69	29	1	0.875	82.0

First, changing the "c" parameter from $c = 5.7$ to $c = 10$, one changes the type of chaos from spiral to funnel chaos. For us it is necessary to emphasize that by changing "c" we diminish the bandwidth of "smoothing" for "z" component (see equation (36)). The latter provokes a significant change for the variances $\{\kappa_2\}^{x,y}$ for "x" and "y" components of the attractor practically in a proportional way (see Table 2): variances are growing for the "continuous" components "x" and "y" proportionally to $10/5.7 = 1.75$. Moreover, the PDFs for "x" and "y" become in some sense more flat: the kurtosis γ_2 changes its sign and tends to one! Next, changing parameter "a" ("b" and "c" are fixed) one can observe the drastical change in the first two cumulants κ_1, κ_2 for all components, but not for γ_1 and γ_2.

The explanation of this effect is quite simple (see equation (36)): the parameter "a" is sort of "gain", which amplifies the absolute values, while it increases; as consequence κ_1 and κ_2 grow as well as their absolute values (see Table 3).

Table 3 Data for Rössler attractor with variation of "a"

	$a=0.35; b=0.2; c=5.7$			$a=b=0.2; c=5.7$		
	x	y	z	x	y	z
κ_1	1.02	−2.94	2.94	0.14	−0.78	0.76
κ_2	38.69	39.39	86.23	20.91	23.17	7.83
γ_1	0.105	0.016	4.6	0.226	−0.198	5.22
γ_2	−0.68	−0.9	23.7	−0.59	−0.69	29

Now, let us introduce some data (see Table 4) regarding the change of the number of applied samples; i.e. how fast the stationary regimen appears for different cumulants and shape coefficients of the components while the number of samples N grows.

One can see from Table 4, that $N \geq 5000$ is a sufficient number of samples for all components of the Rössler attractor for the first two cumulants and for the asymmetry coefficient γ_1 to be stationary, but not for kurtosis γ_2. The data, presented at Table 4 deserve to be commented in a more detailed way. First of all it demonstrates that while N grows, for Rössler attractor (in contrary to Lorenz and Chua cases), the PDFs for "x" and "y" components change drastically from significantly non-Gaussian to practically flat, nonetheless their asymmetry does not change significantly (see also [19, 20]). Next, for application purposes it is clear that by applying $N \leq 2000$, i.e. transient and stationary regimes all together, one can model

Table 4 Data for Rössler attractor with variation of number of samples

N	2000			5000			8000		
	x	y	z	x	y	z	x	y	z
κ_1	0.17	−0.4	0.4	0.15	−0.663	0.663	0.14	−0.76	0.76
κ_2	13.3	17.8	4.24	19.3	21.7	7	20.91	23.17	7.83
γ_1	0.3	−0.45	7.7	0.23	−0.24	5.7	0.226	−0.198	5.22
γ_2	0.89	0.83	63.43	−0.4	−0.4	34.46	0.59	0.69	29

non-Gaussian stochastic properties with rather "long tailed" PDFs. By increasing $N > 2000$, the PDF of "x" and "y" components tend to be practically flat and by adding "z" component to the "x" (or "y") output signals it is possible to model the mixture of pulse and "flat" noises. Moreover, one can use independently all three components of the Rössler attractor with different number of samples as a source of noises and interferences with different statistical features. For this reason in the following (see section 6) the results of analytical approximations of PDFs for Rössler attractor components for strongly non-Gaussian cases are presented. Note that from our experience Lorenz and Chua attractors do not provide us with those options; so the Rössler attractor can be considered as a promising object for more detailed future research. Finally, the results for analytical approximations of the PDFs for strange attractors show that the cumulant analysis can provide with rather accurate and simple tools for the "indirect" PDF calculus, which are quite difficult to achieve with the existing approaches [18, 19].

5 Approximate Approach for Variance Calculus

The results presented in the previous section show that the first cumulants usually depend on the evaluation of the variance κ_2, which cannot be done directly from the degenerate cumulant equations approach. This drawback is a clear shortcoming of the method and has to be solved. In the following, we present an approximate approach for the variance evaluation, based on the concept of the Kolmogorov-Sinai (H_{k-s}) entropy [25][4], with the conjecture that variance as a measure of the "fluctuations" of the process has to influence the information measure as well. The H_{k-s} entropy is defined by the sum of the Lyapunov exponents for the non-linear system of the attractor [25, page 122]; for a linear matrix **A** the Lyapunov exponents are defined by its eigen-values [25, pages 542-543]. In the proposed methodology we first compare the differential entropy obtained from the PDF approximation for an x-component for each attractor, with the H_{k-s} entropy computed through the parameters of the attractor (Lyapunov exponents). Note that in the framework of this analysis H_{k-s} coincides with the Kolmogorov differential entropy of the PDF [25, pages 542-543, 839]. The proposed methodology consists in the following steps:

[4] Considering that Kolmogorov-Sinai entropy can not be exactly calculated, it is impossible by using this method to obtain an exact result for variance. Nevertheless, assuming that for practical purposes an error about 10-20% is acceptable, we are able to apply this method.

1. The eigen-values must be found from the coefficients matrix formed by the linearized system (matrix **A**).
2. Once the eigen-values have been obtained, the H_{k-s} for the dynamic system can be estimated as follows:

$$\log|\lambda_{\max}| < H_{k-s} \leq \sum_{j=1}^{m} \log|\lambda_j|, \tag{47}$$

   where $j = \overline{1,m}$, λ_j is the j-th eigenvalue of the linear matrix and $\log|\lambda_j|$ is the j-th Lyapunov exponent for the linear matrix [25, page 542].
3. On the other hand the Kolmogorov differential entropy is:

$$h_{dif} = -\int_{-\infty}^{\infty} W(x)\log W(x)dx, \tag{48}$$

   where $W(x)$ is the PDF of the output signals whose parameters have to be represented through κ_2.
4. Then, one creates an algebraic equation depending on the variance according to steps 3 and 4.
5. Solving equation from steps 3 and 4, the solution for variance is obtained.

It is worth mentioning, that as the whole Kolmogorov-Sinai entropy is addressed to the given component of the attractor, the value of the variance obtained from the approach is actually its upper bound. The lower bound can be found applying the left hand side of the inequality (47).

Let us illustrate the proposed approach with the numerical examples for the already mentioned types of attractors, whose dates for convenience are summarized at the Table 5.

Table 5 Data of strange attractors

No.	Attractor	Equations	Parameters	Analytical PDF
1	Chua	$\dot{x} = \beta_1(y-x) - \alpha h(x)$	$\beta_{11} = 9.205$	
		$\dot{y} = \beta_2(x-y) + \beta_4 z$	$\beta_2 = 1$	$W(x) = Cexp(px^2 - qx^4)$
		$\dot{z} = -\beta_3 y$	$\beta_3 = 14.3,\ \beta_4 = 1$	
			$\alpha = 0.5$	
			$h(x) = \mathrm{SAT}\{x\}$	
2	Lorenz	$\dot{x} = \sigma(y-x)$	$\sigma = 10$	For x-component
		$\dot{y} = Rx - y - zx$	$B = \frac{8}{3}$	$W(x) = W_G(x)\left[1 + \frac{\gamma_4}{4!}H_4(x)\right]$
		$\dot{z} = xy - Bz$	$R = 28$	$H_4(x)$-fourth Hermite polynomial
3	Rössler	$\dot{x} = -y - z$	$c = 5.7$	For x-component
		$\dot{y} = x + ay$	$a = 0.2$	$W(x) = \frac{\lambda}{2}exp[-\lambda\|x-\mu\|]$
		$\dot{z} = b + zx - zc$	$b = 0.2$	
4	Duffing	$\dot{x} = y$	$\omega = 1$	For x-component
		$\dot{y} = x - x^3 - \delta y + \gamma\cos\omega t$	$\delta = 0.25$	$W(x) = Cexp(px^2 - qx^4)$
			$\gamma = 0.3$	

5.1 *Chua Attractor*

The equations for the Chua attractor and the corresponding PDF for the "x"-component can be found from Table 5. Applying the well known statistical linearization technique [25] with a symmetrical PDF one can get the linearized coefficients matrix **A** for the dynamic system (step 1) whose $\det \mathbf{A} \neq 0$:

$$\mathbf{A} = \begin{bmatrix} -\beta_1 + \alpha & \beta_1 & 0 \\ \beta_2 & -\beta_2 & \beta_4 \\ 0 & -\beta_4 & 0 \end{bmatrix}. \tag{49}$$

For (49), the characteristic equation, i.e. step 2 is:

$$\lambda^3 + a\lambda^2 + b\lambda + c = 0, \tag{50}$$

where $j^2 = -1$ and:

$$\begin{aligned} a &= \beta_2 + \beta_1 + \alpha, \\ b &= \beta_2 \alpha + \beta_3 \beta_4, \\ c &= \beta_3 \beta_4 (\beta_1 + \alpha). \end{aligned}$$

The eigenvalues from (50) form the H_{k-s} and so we need to represent its analytical solution depending on the parameters a, b, c. Using the well known procedure for solving cubic equations we obtain:

$$p_1 = \frac{3b - a^2}{3}, \qquad q_1 = \frac{2a^3}{27} - \frac{ab}{3} + c, \qquad D = \frac{p_1^3}{27} + \frac{q_1^3}{8}, \tag{51}$$

then:

$$n = \sqrt[3]{-q_1 + \sqrt{\frac{p_1^3}{27} + \frac{q_1^3}{8}}}, \qquad \nu = \sqrt[3]{-q_1 - \sqrt{\frac{p_1^3}{27} + \frac{q_1^3}{8}}}, \tag{52}$$

and

$$\begin{aligned} \lambda_1 &= n + \nu - \frac{a}{3}, \\ \lambda_2 &= -\frac{n+\nu}{2} + j\frac{n-\nu}{2}\sqrt{3} - \frac{a}{3}, \\ \lambda_3 &= -\frac{n+\nu}{2} - j\frac{n-\nu}{2}\sqrt{3} - \frac{a}{3}. \end{aligned} \tag{53}$$

Skipping here the detailed analysis of (53) for a, b and c, one can easily find, for a qualitative evaluation, that roughly all $\{\lambda_i\}_1^3$ depend on $a = \beta_2 + \beta_1$, (see Table 5), i.e. to the sum of the effective bandwidths of the "x" and "y" components. Note that for the linearized system **A** this kind of dependence for κ_2^x is rather predictable. Hence, the eigenvalues are:

$$\lambda_1 = -11,$$
$$\lambda_1 = -0.09 + j3.63, \quad (54)$$
$$\lambda_1 = -0.09 - j3.63. \quad (55)$$

Now, following with step 3, from (47) one gets $H_{k-s} \cong 2.16$ and from step 4 one gets the following algebraic equation:

$$q(\kappa_2^x)^2 - p\kappa_2^x - \log c = 2.16. \quad (56)$$

The solution of (56) is $\kappa_2^x = 2.24$. The exact solution (evaluated from simulation) is $\hat{\kappa}_2^x = 2.24$. As it is evident, the error is 11% and the solution is an upper bound of the real value of the variance.

5.2 *Lorenz Attractor*

From the second line of Table 5, we get the equations for the Lorenz attractor and the correspondent PDF for the x-component. Using the same methodology as in the previous case we get for $\mathbf{A}(\det \mathbf{A} \neq 0)$:

$$\mathbf{A} = \begin{bmatrix} -\sigma & \sigma & 0 \\ R & -1 & 0 \\ 0 & 0 & -B \end{bmatrix}. \quad (57)$$

Then, the characteristic equation is:

$$a\lambda^3 + b\lambda^2 + c\lambda + d = 0, \quad (58)$$

where $a = 1, b = B + \sigma + 1, c = B - R\sigma + \sigma + \sigma B, d = -R\sigma B + \sigma B$.

Repeating the same procedure, as it was done above for equation (50), one can see that for the Lorenz case κ_2^x depends mainly on "d" or on the product of $R\sigma B$ (see Table 5). The solution of (58) is $\lambda_1 = 11.82, \lambda_2 = -22.82, \lambda_3 = -2.66$. From (47) it is easy to find that $H_{k-s} \cong 1.35$ and the lower bound is:

$$\kappa_2^x = \frac{10^{2H_{k-s}}}{2\pi e} \quad (59)$$

Here, for simplicity, we assume that $W(x) \approx W_G(x)$ with $\hat{\kappa}_2^x = 36$. From (59) $\kappa_2^x = 30$ and so, the error for the worst case is 16.6%. It is easy to show that the upper bound does not have any sense here.

5.3 *Rössler Attractor*

For the Rössler attractor we take the third line from Table 5.

The $\mathbf{A}$ matrix in this case is:

$$\mathbf{A} = \begin{bmatrix} 0 & -1 & -1 \\ 1 & a & 0 \\ 0 & 0 & -c \end{bmatrix}. \tag{60}$$

Now, the characteristic equation becomes:

$$\lambda^3 - (a-c)\lambda^2 - (ac-1)\lambda + c = 0, \tag{61}$$

where $\lambda_1 = -5.7, \lambda_2 = 0.1 + j, \lambda_3 = 0.1 - j$.

Here it follows immediately that the main eigenvalue λ_1 is equal to the parameter c (see Table 5). For the entropy we have $H_{k-s} \approx \ln|\lambda_1| = 1.74$ from steps 3 and 4 it follows:

$$\kappa_2^x = \frac{exp(3.48)}{2e^2} = 2.1. \tag{62}$$

From the Table 5 it can be seen that $\hat{\kappa}_2^x = 2$ and thus, the error is 5%.

The number of examples can be extended not only to other types of strange attractors but also to the "y" and "z" components of the above-treated systems. So, for the concrete strange attractors it follows that for the variance calculation, the proposed approach performs rather good. Notice here that if instead of the strange attractors one analyzes chaotic attractors (like Hamiltonian systems, etc.), the straightforward use of the proposed methodology has to be done cautiously because erroneous or even absurd results may appear.

5.4 *Duffing Attractor (Counterexample)*

Equations for the Duffing attractor together with the analytical approximation for the PDF of the x-component come from the fourth line of Table 5. At first glance the x-component for the Duffing and Chua attractors has the same type of PDF.

Let us consider the matrix **A** for the linearized Duffing equation:

$$\mathbf{A} = \begin{bmatrix} 0 & 1 \\ 0 & -\delta \end{bmatrix}. \tag{63}$$

It is easy to observe that $\det \mathbf{A} = 0$, i.e. **A** is ill-posed. So, there is no way to take the eigenvalues of **A** as an approximation for the Lyapunov exponents [25, page 542] and the proposed method cannot be applied. If one pushes all this further and calculates the Lyapunov exponents following the exact procedure [25], the result becomes $\lambda_1 = 0.126, \lambda_2 = -0.376$, $H_{k-s} \approx 0.126$ and following steps 3 and 4, from the quadratic algebraic equation, it results that κ_2^x has complex values which is a clearly absurd outcome. Why it happened? The answer comes from [25, p.p. 122, 542-543, 839, etc]; for Hamiltonian systems the differential Kolmogorov entropy does not coincide with H_{k-s} as it is expressed beforehand. The variance κ_2^x for the Duffing attractor could be found if one substitutes the ordinary Duffing equation by its statistical analog (statistically equivalent SDE) and from there follow the approach detailed in [3, Ch. 7] to get finally κ_2^x. However, we do not address it here.

5.5 *Modified Approach*

In the previous material it was assumed that the parameters of the strange attractors are predefined as well as the PDFs for the output signals (see Table 5). Here we will consider a more general case. Let us suppose that the set of parameters of the strange attractors as well as their PDFs are not predefined, nevertheless the chaotic regime of the attractor is established. Thus, one can apply for the PDF choice the so-called "Maximum Entropy Method (MEM)" [38]. In the following, we consider only attractors with PDFs defined on $(-\infty,\infty)$ and so, a Gaussian PDF with fixed κ_2^x can be applied as a MEM distribution. It follows from the MEM principle that evaluated κ_2^x has to be always a lower boundary regarding its true value. It follows from (59) that the variance κ_2^x obeys the following inequality:

$$\kappa_2^x \geq \frac{10^{2H_{k-s}}}{2\pi e}. \tag{64}$$

As it was mentioned before, if one of the positive Lyapunov exponents predominates, it is possible to apply, as a lower boundary, the inequality $H_{k-s} > \log|\lambda_{\max}|$. If the values for all positive Lyapunov exponents are comparable then for the lower boundary evaluations, it might be applied the inequality:

$$\frac{1}{3}H_{k-s} \leq \sum_i \log(\lambda_i), \tag{65}$$

supposing that the strange attractor consist of three equations and assuming that all components are statistically independent. For simplicity let us take advantage of the numerical results early obtained at this section and use them to test the MEM approach application. For the Lorenz case, $\log|\lambda_{\max}| = 1.35$ and $\kappa_2^x \geq 29.34$ (see 5.2, the true value is 36). For the Chua case, supposing that $H_{k-s} \cong \frac{H_{k-s}}{3}$ (see λ_1, λ_2 and λ_3 in 5.1) one can get $\kappa_2^x \geq 1.8$ (the true value is 2.02). Finally, for the Rössler attractor, $\kappa_2^x \geq 1.8$ (the true value is 2).

As it can be seen, the lower boundaries of κ_2^x for all attractors under consideration are very close to their true values with an error of about 10%, so they are rather good. It is well known [38] that if the normalized chaotic signal is defined at $[0,\infty)$, the MEM distribution will be truncated Gaussian PDF and the variance evaluation can be done in the same way as above, but (64) has to be replaced by:

$$\kappa_2^x \geq \frac{10^{2H_{k-s}}}{\frac{\pi e}{2}}. \tag{66}$$

Finally, it is worth mentioning that the MEM approach is not the unique way to avoid the a priori knowledge of the PDF for the variance evaluation. Another way to find an analytical solution to the same problem can be found assuming orthogonal series representation for unknown PDF (see for example [3]) where unknown cumulants can be calculated applying the "degenerated cumulant equations" method.

5.6 Some Discussion

This material shows how, through the ordinary equations of strange attractors and their statistical linearization, it is possible to obtain analytical estimations of the Lyapunov exponents of the non-linear dissipative system describing the attractor. After the Lyapunov exponents are found it is possible to get a well known approximation for the Kolmogorov-Sinai entropy H_{k-s}. On the other side, applying the attractors PDFs for the components of interest (from computer simulations or from MEM approach) together with its analytical approximation, it is possible to find the Kolmogorov differential entropy (h_{dif}) which practically coincides with H_{k-s} for the strange attractors under consideration [7]. Finally, by solving a simple quadratic algebraic equation one can find $\kappa_2^x \geq 0$. The approach shows acceptable accuracy for engineering evaluation of κ_2^x (less than 20%) and it is completely analytical. Once more: why it happened? The reader must have already figured out that implicitly it has to depend on the details of the stochastic chaotic dynamics of the attractor such as the dimension of the stochastic set, dimension of the phase space, etc. (see [7, 25], etc. for details). For example, the dimension of the stochastic set D for the Lorenz strange attractor considered in the subsection 5.2 of this section is more than two, but has an almost zero volume. This means that almost all phase trajectories, that constitute the strange attractor are localized in a very thin layer, i.e. it can be approximately represented by a one dimensional Poincare mapping (details can be found in [17, 25]). It is actually true not only for Lorenz but also for Chua, Rössler, etc. strange attractors as well (see [25, 52]). So, any non-linear dissipative system with a dimension D equal or more than two represents a one-dimensional Poincare mapping (one dimensional dynamics) and one of the consequences of this is the similarity between H_{k-s} and the Kolmogorov differential entropy. Therefore, the acceptable accuracy of the proposed approach might be, in some sense, predicted.

Next, one should notice that the dependence of κ_2^x on the parameters of the strange attractors (even in the framework of the predefined chaotic regime) is not trivial and it depends on the attractor's type. As it was shown above, the κ_2^x for the Chua attractor depends mainly on the sum of the effective bandwidths β_1 and β_2; at the same time for the Lorenz case the dependence is on the product of $R\sigma B$. Finally, the Rössler attractor clearly shows the dependence of κ_2^x on the unique parameter c. Obviously, a thorough study of the relationship between the statistical behavior of the strange attractors and their non-linear dynamics might clarify the matter.

6 Modeling of the Radio Frequency Interference of Laptops and Desktops Computers

As it was mentioned before in section 1, here the results of the application of chaos signals to the modeling of the radio frequency interference (RFI) of laptops and desktops will be presented. Let us consider a PCIe bus, which is widely used in computers to communicate at high speed data rates. Specifically, the broadband emissions generated by high speed digital interconnects input-output (I/O) bus transactions associated with graphics benchmark (SW, 3Dmark) are the major contributors of the non-Gaussian radio interferences to the PCI'e platforms. Those interferences were experimentally investigated by the Intel measurement set up system. The measurement system consists of a high speed data capture system: RF gain and analog to digital converter (ADC). In addition, a low loss RF cable is used to connect a RF horn antenna. The methodology for the measurement can be summarized as follows:

1. The equipment used for measurements is put inside a Faraday cage, then the laptop is connected to another computer via wireless communication, which is located inside a compartment that allows isolation of the radio signals and the communication is intermittent if the door compartment opens or closes.
2. The RFI measure originated from a laptop, device under test (DUT), with all its systems assets (LCD/Graphics hardware/PCIe buses and wireless systems under normal traffic conditions).
3. The measuring of a radiated emission level is of -100dBm over a 20 MHz bandwidth, with an (ADC) 40 Msps. This is the minimum required level of the WiFi radio subsystem (i.e. 802.11a and 802.11g IEEE Standards). Therefore, the modeled statistics are representative of the conditions to be experienced by the platform's wireless subsystems.
4. The measurement conditions are: ON/OFF: The screen graphics of the laptop where the measurement was carried out is activated and deactivated. CLOSED/OPEN: The compartment door (where the computer with wireless card is) opens and closes.

At the PCI express (PCIe) bus there are various sources of deterministic or random jitter such as signal reflection, cross talk on transmission lines, and non-linear properties of electronic components which give rise to non-ideal PCI express signals, i.e. RFI under investigation. The experimental data (histograms) of the probability density function (PDF) of PCIe data at 5 Gbit/s and the PDF of "x" component of the Chua's attractor that models that interference are displayed in Figure 19. PCIe data and analytical results are presented along the normalized x-axis. One can see from those graphs that analytical and experimental results are practically identical, all necessary calculus of the parameters p, q and normalization constant "C" the reader can find at the Subsection 4.4. As it was previously mentioned, a Kolmogorov- Smirnoff (KST) test was applied as a criteria of goodness of fit.

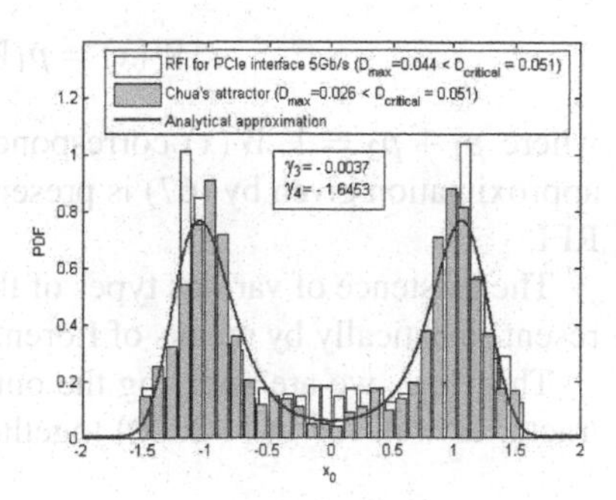

Fig. 19 "Histogram of the x" component of Chua´s attractor, experimental results and approximations

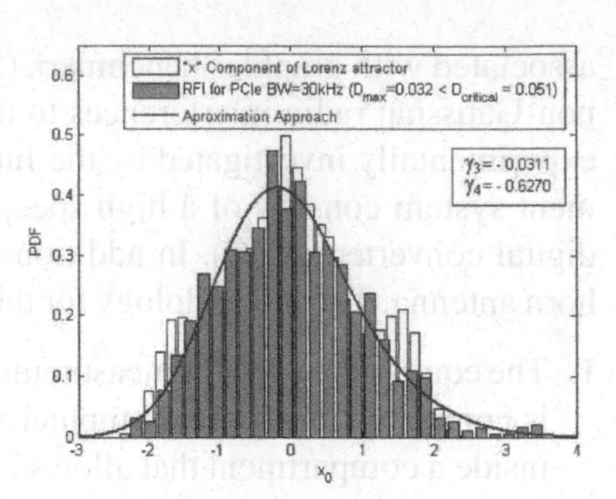

Fig. 20 Histogram of the "x" component of Lorenz attractor, experimental results and approximations

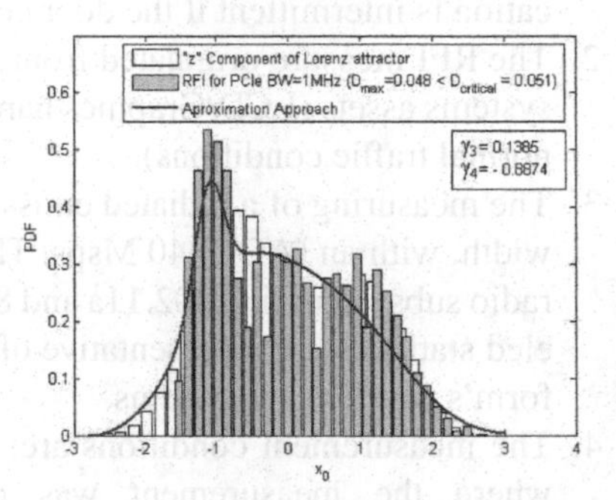

Fig. 21 Histogram of the "z" component of Lorenz attractor, experimental results and approximations

Let us take other examples of RFI from PCIe. For these examples, measurements were taken from PCIe buses (again) with data transmissions at 5 Gbit/s, but performing measurements of radiated interference at lower bandwidths; 30 kHz and 1 MHz. The corresponding distributions for the normalized random variable are presented in Figures 19 and 20, respectively. One can see that, for the case of measurements at 30 kHz, the PDF slightly differs from a Gaussian distribution, which is a consequence of the relatively narrow bandwidth. For the case of measurements at 1 MHz, the RFI is sufficiently non-Gaussian but with the peculiarity that the probability of the normalized component, equal to "-1", can be considered as the Gaussian PDF as well. For PCIe's PDF (BW 30 kHz) the approximation was done in the following way (see (6)): results of the approximation are shown in Fig. 19, and were tested by the Kolmogorov-Smirnoff test as well. For the RFI taken at 1 MHz bandwidth, the following representation for the distribution was assumed:

$$W(x) = p_1\tilde{W}(x) + p_2\delta(x+1), \tag{67}$$

where $p_1 + p_2 = 1$, $\tilde{W}(x)$ corresponds to (6) and $\delta(\cdot)$ is the delta- function. The approximation given by (67) is presented in Fig. 20 together with histogram of the RFI.

The existence of various types of the RFI shows, that not all of them can be represent statistically by means of Lorenz and Chua attractors signals.

Therefore, we are applying the output signals of the components of Rössler attractor as well for ($N < 2000$) together with their analytical approximations based on the cumulant analysis. So, the experimental data obtained from measurements of the interference of laptops and desktops computers at 20 MHz were analyzed using histograms, PDF and output signal of "x", "y" and "z" components of the Rössler attractor.

Considering the application of the "x" component of the Rössler attractor as a RFI model with the PDF in the Fig. 22, one can see that analytical and experimental results are practically identical, when the analytical approximation corresponds to the Laplace distribution function and their cumulants coincide as well.

Let us take another example of RFI from PCIe. For this example, measurements were taken from PCIe buses at 20 MHz (again). In this case the "y" component of the Rössler attractor was used as a RFI model with the PDF as shown in Fig. 23. One may see that analytical and experimental results are very similar, but it is necessary to examine whether the accuracy of RFI modeling by the PDF of the "x" and "y" component of the Rössler attractor and the analytical approximation of each of them by the Laplace distribution is adequate or not.

In the same way as in Section 4 this was achieved by applying the Kolmogorov-Smirnoff and Chi-Square goodness of fit test, both tests validate the given approximation, even though the Chi-Square test indicates a better accuracy. It is illustrated at Figures 22-24.

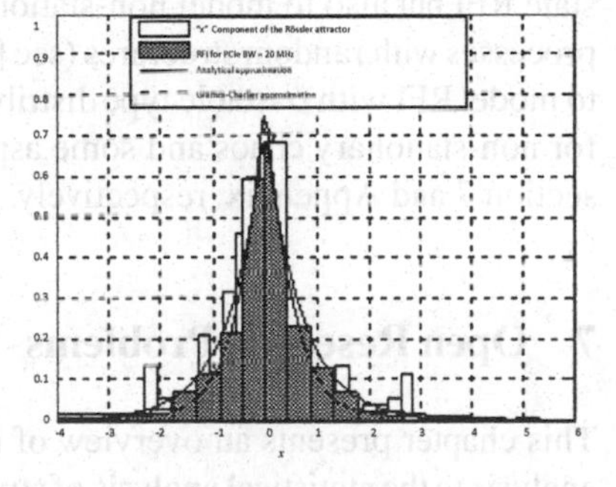

Fig. 22 Histogram of the "x" component of the Rössler attractor and experimental results

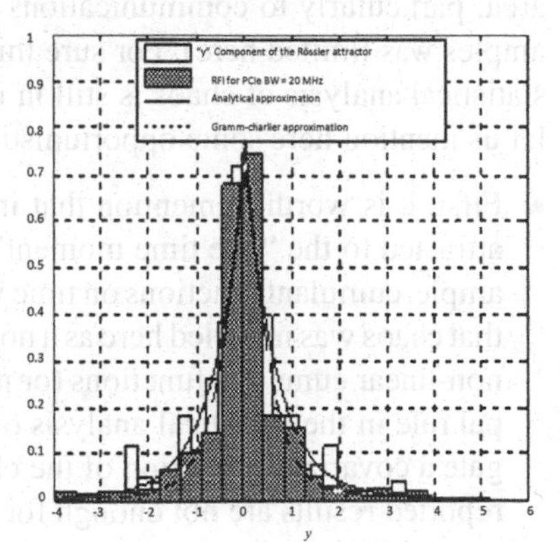

Fig. 23 Histogram of the "y" component of the Rössler attractor and experimental results

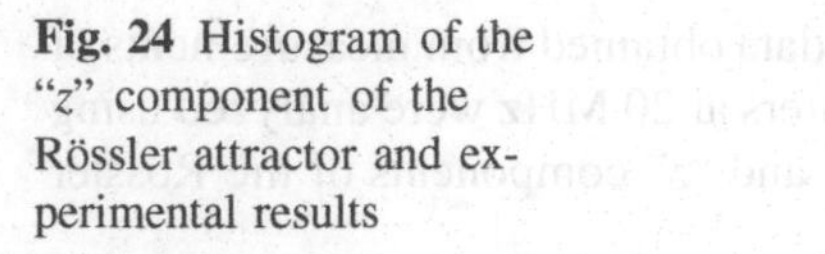

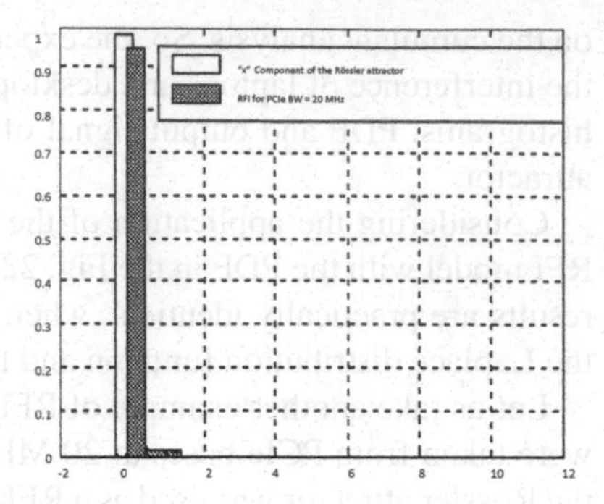

Fig. 24 Histogram of the "z" component of the Rössler attractor and experimental results

From the material of this section it follows that the output signals of the strange attractors can be successfully applied for statistical modeling of the RFI in high speed interconnects. This approach opens the opportunity to carry out practical implementations of interference mitigation techniques, since the attractors mentioned above can be represented in the form of Ordinary Differential Equations (ODE) or in state variable representation [3], which naturally implies an extended Kalman Filtering approach to their further filtering and mitigation (see [50], and references therein). The ODE or state variable representation of chaos allows to model not only stable-state RFI but also to model non-stationary interferences by applying methods of the processes with random structures (see [3]). Furthermore, fractal chaos can be applied to model RFI with α-stable type distribution as well [15, 16]. The cumulant approach for non-stationary chaos and some aspects of fractal chaos will be discussed at the section 7 and Appendix, respectively.

7 Open Research Problems

This chapter presents an overview of the results of the application of the cumulant analysis to the statistical analysis of strange attractors together with an example of the possible implementation in a practical problem, related to the electrical engineering area, particularly to communications (due to lack of the space, the number of examples was limited here). For sure this is only the top of the "iceberg", because he statistical analysis of chaos is still in its early stages of development. Nevertheless, let us mention here some opportunistic routes for the future research projects.

- First, it is worth to mention that in the present material the main attention was attracted to the "one time moment" cumulants and their generalizations, for example, cumulant functions on time were not even mentioned. Taking into account, that chaos was modeled here as a non-Gaussian stochastic processes the linear and non-linear cumulant functions (or non-linear covariance functions) play a principal role in the statistical analysis of strange attractors. Some attempts to investigate a covariance function of the chaos was presented at [18, 19], but, certainly, reported results are not enough for any complete study.

- Second, as it was shortly stated above, there is a clear lack of studies about the mutual influence of the non-linear dynamics and stochastic properties of the chaotic output signals of strange attractors. Despite of the fundamental results presented at [1, 2, 7, 25],etc. statistical properties of chaos are mainly experimentally investigated by computer simulations by changing the attractor parameters, i.e. in the same way as it was done in [18, 19] and subsection 4.3. From the authors' point of view, this way is rather "contemplative" and does not demonstrate the profound "ties" which indeed exist between the previously stated two parts of the strange attractor's theory. Strong research efforts in this field are really needed.
- Third, the theory of the fractal chaos, to the best of our knowledge, is still waiting for its specific approach for cumulants analysis. It was shown at the Appendix that the straightforward solution of the "fractional" FPKE is not a simple issue, so the alternative search of its solutions by cumulant methods can be rather useful.

The last, but not least problem is in applications of chaos in electrical engineering. Beforehand at the section 7, it was shown one of the possible applications: RFI modeling. Other applications can be found at [31, 44],[47]-[49], etc. but for sure, chaotic applications in electrical engineering could be as extensive ,as they are in physics, biology, chemistry, etc. Finally, we would like to stress, that the reader can find along the text of this chapter some useful comments regarding future research topics.

Acknowledgements. Authors are pleased to pass thanks to Dr. Fernando Ramos, M. Sc. Mario Mijangos, M. Sc. Luis Beltran and M. Sc. Oscar Filio for their valuable help and assistance in the preparation of this material.

A Appendix: Fractal Kinetics of Chaotic Dynamic Systems

As it was already stated above, we distinguish here between strange attractors for chaos generation and dynamic systems, both with fractional and chaotic behavior (see [1] for details). Currently, fractal models are widely used, for example in physics to represent different phenomena with "long-tail" PDFs and correlations [9]-[15].

Hereafter, some aspects of the statistical analysis for dynamic systems will be considered, particularly of Hamiltonian type, which exhibit both fractal and chaotic properties. To tackle this problem the methods of fractional calculus will be widely applied (see [8]).

One of the most effective tools of the aforementioned approach is an analytical solution of the fractional Fokker-Planck Equation (FFPE) to obtain not only the correlation function of the chaos, but its probability density function (characteristic function) as well [8]-[14].

Let us consider the generalized FFPE (see [8, pag.88,pag.193]).

$$\frac{\partial^{\alpha} P(x,t)}{\partial t^{\alpha}} = D^{\alpha}\left[A_{\alpha,\gamma}(x)P(x,t)\right] + D^{2\alpha}\left[B_{2\alpha,\gamma}(x)P(x,t)\right], \tag{68}$$

where

$$P(x,t) \simeq W\left(x,t|_{x_0}\right); x(t_0) = x_0; P(x,t_0) = \partial(x - x_0),$$

$$D^{2\alpha} = \left(D_+^\alpha + D_-^\alpha\right)^2, D^\alpha = \left(D_+^\alpha + D_-^\alpha\right),$$

D_+^α - left side fractional derivative, D_-^α - right side fractional derivative (see [9] for details).

α -fractal space dimension characteristic, γ -is the Holder exponent [8]-[10].

$A_{\alpha,\gamma}$ -drift, $B_{2\alpha,\gamma}(x)$ -diffusion coefficient. Actually, they are defined by (81), (82) (in p.192 of [9]) or (7.5) (in p.221 of [11]), in a similar way as local characteristics of the continuous Markov process [3].

When $t \to \infty, P(x,t) = W_{st}(x)$ - stationary distribution of x.

It is assumed in the sequel that $\gamma = 1$ and $A_{\alpha,1} = Ax, B_{2\alpha,1} = B$.

Then, introducing the characteristic function for the solution of (68) $\theta(\nu,t) = F\{P(x,t)\}$ and using [9, eq. (96)], one can get:

$$F\{D^\alpha(xP(x,t))\} = j|\nu|^\alpha sgn(\nu) \cdot A\nu \frac{\partial}{\partial \nu}\theta(\nu,t), \quad (69)$$

$$F\left\{D^{2\alpha}(P(x,t))\right\} = -|\nu|^{2\alpha}\theta(\nu,t);$$

Finally, with the help of (68), (69) it yields to the following equation for $\theta(\nu,t)$:

$$\frac{\partial}{\partial t}\theta(\nu,t) = -jA|\nu|^\alpha \nu \frac{\partial}{\partial \nu}\theta(\nu,t) - |\nu|^{2\alpha}B\theta(\nu,t). \quad (70)$$

Equation (70) can be called the FFPE for the characteristic function $\theta(\nu,t)$. For stationary conditions $t \to \infty$ (see for example [4]) the solution of (70) is:

$$\theta_{st}(\nu) = e^{-\frac{B}{A}\frac{|\nu|^\alpha}{\alpha}} \quad (71)$$

From [21, 22] it follows, that (71) is a special case of the characteristic function for α- stable distributions; its probability density function is symmetric and can be represented in the way [15, 16]:[5]

$$W_{st}(x,\alpha) = \begin{cases} \frac{1}{\pi x}\sum_{i=1}^{\infty}\frac{(-1)^{i-1}}{i!}\Gamma(\alpha i+1)x^{\alpha\kappa}\sin\left(\frac{i\pi\alpha}{2}\right), \ 0<\alpha<1 \\ \frac{1}{\pi x}\sum_{i=1}^{\infty}\frac{(-1)^{i-1}}{i!}\Gamma\left(\frac{i}{\alpha}+1\right)x^{i}\sin\left(\frac{i\pi}{2}\right); \ 1<\alpha<2 \end{cases} \quad (72)$$

for $x > 0$.

Note that in [10, 11] two cases were analyzed: $A = 0, B \neq 0$ and $A \neq 0, B = 0$ respectively, so (71) and (72) are the general results for FFPE (71). See also [15, 16] for applications of (71), (72) as models for impulsive noises, traffic, etc. The FFPE (70) can be also used to find a non-stationary $\theta(\nu,y)$. For that matter, let us represent $\theta(\kappa,t)$ in the following way [3]:

$$\theta(\nu,t) = e^{\phi(\nu,t)}$$

[5] The α -stable distributions with exception of $\alpha = 2$ do not have moments [15, 16].

With this representation, (70) can be easily transformed into:

$$\frac{\partial \phi(v,t)}{\partial t} + Av^{\alpha+1}\frac{\partial \phi}{\partial v} + v^{2\alpha}B = 0, \tag{73}$$

with boundary conditions: $t \to \infty$, $\phi(v,t) = \phi_{st}(v)$ and $t \to 0$ $\phi(v,0) = \phi_0(v)$.

The method for solution of (73) is well known [3]: it is necessary to look for a solution as a sum of the general solution for the homogeneous differential equation and the partial solution for non homogeneous equation (73). If one takes as a partial solution $\phi_{st}(v)$, assuming that for $t \to \infty$ $\phi_{st}(v)$ exists, and presents the general solution for homogeneous equation in (73) in the way:

$$\phi_{hom}(v,t) = \Phi\left(ve^{-A|v|^{\alpha t}}\right), \tag{74}$$

where $\Phi(\cdot)$ is an arbitrary function, which can be found from initial conditions for (68), then one can get an expression for $\theta(v,t)$ in the following form:

$$\theta(v,t) = \frac{\theta_{st}(v)}{\theta_{st}\left(ve^{-Av^{\alpha t}}\right)}\theta_0\left(ve^{-A|v|^{\alpha t}}\right), \tag{75}$$

where $\theta_{st}(v)$ stands for (71) and:

$$\theta_0\left(ve^{-Av^{\alpha t}}\right) = \int_{-\infty}^{\infty} \delta(x-x_0)\, e^{-A|v|^{\alpha t}} e^{jvx} dx = exp\left(jvx_0 e^{-A|v|^{\alpha t}}\right). \tag{76}$$

Finally (75) reduces to:

$$(v,t) = exp\left\{-\frac{B|v|^{\alpha}}{A\alpha}\left(1 - e^{\alpha A|v|^{\alpha t}}\right) + jvx_0 e^{-A|v|^{\alpha t}}\right\}. \tag{77}$$

It is easy to see that when $t \to \infty$ in (77) we obtain (71) and (72).

Therefore, from (71) and (77) we got a complete solution for the FFPE (70) both for the stationary and non-stationary cases, which generalizes the analysis presented in [9, 10].

Now, let us see how all these analytical results presented in this section and in sections 2 and 3 can be used for the statistical analysis of fractional chaotic systems. For this matter, we have to invoke some ideas of G. Zaslavsky [10], regarding to the fractional chaos in Hamiltonian systems. He demonstrated (see details in [10]), that it is conventional to write down a corresponding distribution function $P(x,t)$ in the form [6].

$$P(x,t) = P_n(x,t) + P_s(x,t), \tag{78}$$

where $P_n(x,t)$ - is the normal part of $P(x,t)$ and $P_s(x,t)$ -is the singular one. From [10] it follows, that only for the second term in (78) fractional kinetic equations are valid. The same representation can be done for the stationary distributions and characteristic functions, i.e:

[6] In the following, for simplicity, the weighting coefficients, which provide normalization of $P(x,t)$ are omitted in (78) and (79).

$$W_{st}(x) = W_n(x) + W_s(x), \tag{79}$$
$$\theta_{st}(x) = \theta_n(x) + \theta_s(x).$$

In what follows we will analyze only the (79) case. As it was shown in the section 1, the normal part can be represented by model distributions (6) created with the cumulants from the dynamic system equations. Actually the same idea, but without applying this terminology, was mentioned at [10].

Then it is necessary to tackle the problem of the singular part, supposing that $W_s(x)$ is -α stable (not indebtly symmetrical), one can use the more general representation, established in [8, 9], for its characteristic function [15, 16]:

$$\theta_s(\nu) = exp(j\alpha\nu - \gamma|\nu|^{\alpha}(1 + j\eta sgn(\nu)\xi(\nu,\alpha))) \tag{80}$$

where γ, α and η are parameters,

$$\xi(\nu,\alpha) = \begin{cases} \tan\frac{\alpha\pi}{2}, & \alpha \neq 1 \\ \frac{2}{\pi}\ln|\nu|, & \alpha = 1 \end{cases} \tag{81}$$

Considering the asymptotic: $|x| \to \infty, \nu \to 0$, from (80), we can get:

$$\theta_s(\nu) \sim 1 - |\nu|^{\alpha}\left(\frac{\gamma}{2} - j\frac{2}{\pi}\eta\gamma\right)(\alpha = 1), \tag{82}$$

and for $\alpha \neq 1$

$$\theta_s(\nu) \sim 1 - |\nu|^{\alpha}\left(\frac{\gamma}{2} - j\frac{2}{\pi}\nu\gamma\right)(\alpha \neq 1);$$

or from ((10.4),(10.6) [11]) it follows that:

$$\theta_s(\nu) \sim A_m|\nu|^{\alpha} + reminder,$$

where the coefficient A_m is complex.

The last equation together with the formula [4], $j^n < x^n = \frac{d^n}{d\nu^n}\theta_s(\nu)|_{\nu=0}$, where $< x^n >$ is an n-th moment, provides us with not only infinite, but also complex moments for the singular part. For α-stable symmetrical distributions $(\alpha = \eta = 0)$, $A_m = \frac{-\gamma}{2}$ the moments are "only" infinite (see [15, 16]). Then, for the singular part, one has all moments as infinite ones and possibly complex, which makes impossible to represent $W_s(x)$ as a model distribution.

To make the analysis more constructive, i.e. represent $W_s(x)$ or $\theta_s(\nu)$ in a closed form it is feasible to make a linearization of the dynamic system and use formulas (70), etc., for $W_s(x)$ $(\theta_s(\nu))$ representation. In this section we have analyzed the case $\gamma = 1$; though, we have not considered the long-tale correlations for the fractional chaos. Strictly speaking the case $\gamma \neq 1$, α-arbitrary require a solution of the FFPE (68) in the general way, which is hardly possible to be done analytically. But for $\alpha = 1$, $\gamma \neq 1$ the analytical solution can be found in [11]-[14].

References

1. Eckmann, J., Ruelle, D.: Ergodic theory and strange attractors. Review of Modern Physics 57, 617–656 (1985)
2. Horton, W., Ichikawa, Y.: Chaos and structures in non-linear plasmas. World Scientific, Singapore (1996)
3. Primak, S., Kontorovich, V., Lyandres, V.: Stochastic Methods and their Applications to Communications. Stochastic Differential Equations Approach. John Wiley & Sons, Chichester (2004)
4. Malakhov, A.: Cumulant analysis of random non-gaussian process and their transformation. Sovetskoe Radio (1976) (in Russian)
5. Kontorovich, V., Primak, S., Almustafa, K.: On limits of applicability of cumulant method to analysis of dynamics systems with random structure and chaotic behavior. In: Proc. of the III Int. DSDIS Conf. on Engineering Applications and Computer Algorithms, Guelph, Ontario, Canada (May 2003)
6. Cramer, H.: Mathematical Methods of Statistics. Princeton University Press, Princeton (1946)
7. Rabinovich, M., Trubetskov, D.: Introduction to Oscillation Theory. Nauka, Moscow (1984) (in Russian)
8. Hilfer, R.: Applications of fractional calculus in physics. World Scientific, Singapore (1998)
9. West, B.J., Grigolini, P.: From Applications of Fractional Calculus in Physics. Fractional differences, derivates and fractal time series, pp. 171–201. World Scientific, Singapore (1998)
10. Zaslavsky, G.M.: Fractional kinetics of hamiltonian chaotic systems. From Applications of Fractional Calculus in Physics, pp. 203–239. World Scientific, Singapore (1998)
11. Kontorovich, V., Lyandres, V., Primak, S.: Non-linear methods in information processing: modelling, numerical simulation and application. Dynamics of Continuous, Discrete and Impulsive Systems, Serie B: Aplications & Algorithms 10, 417–428 (2003)
12. Metzler, R., Barkai, E., Klafter, J.: Anomalous diffusion and relaxation close to thermal equilibrium: A fractional fokker-plank equation approach. Physical Review Letters 82, 3563–3567 (1999)
13. Aydmer, E.: Anomalous rotational relaxation: A fractional fokker-plank equation approach. Physical Review E71, 046103 (2005)
14. Khan, S., Reynolds, A.M., Morrison, I.E.G., Chouzy, R.J.: Stochastic modeling of protein motions within cell membrane. Physical Review E71, 041915 (2005)
15. Nicias, C.L., Shao, M.: Signal Processing with Alpha-Stable Distributions and Applications. N.Y. Wiley, Chichester (1995)
16. Middleton, D.: Non-gaussian noise models in signal processing for telecommunications: New methods and results for class a and class b noise models. IEEE Trans. on IT 15(4), 1129–1148 (1999)
17. Hasler, M., Mazzini, G., Ogorzalek, M.: Applications of nonlinear dynamics to electronics and information engineering. Special Issue, Proceedings of the IEEE 90 (May 2002)
18. Anischenko, V., et al.: Mixing and spectral-correlation properties of chaotic and stochastic systems: numerical and physical experiments. New Journal of Physics 7(76), 1–29 (2005)
19. Anischenko, V.S., et al.: Statistical properties of dynamical chaos. Physics-Uspehi 48(2), 151–166 (2005)

20. Mijangos, M., Kontorovich, V.: Non-gaussian perturbation modeling through chua's attractor. In: XVI International Conference on Electromagnetic Disturbances (EMD 2006), September 2006, pp. 27–29 (2006)
21. Kontorovich, V.: Applied statistical analysis for strange attractors and related problems. Mathematical Methods in Applied Science 30, 1705–1717 (2007)
22. Mijangos, M., Kontorovich, V., Aguilar-Torrentera, J.: Some statistical properties of strange attractors: Engineering view. Journal of Physics:Conference Series: 012147 96, 6 (2008)
23. Radev, R.: Full state observer for rössler's system. In: Int. IEEE Symp. Intelligent Systems (2002)
24. Kukharenco, B.G.: Detecting orbits of high periods from numerical solutions of the rössler system. In: Int. IEEE Sym. Phys. Com., pp. 866–871 (2005)
25. Scott, A. (ed.): Encyclopedia of nonlinear science. Routledge, New York (2005)
26. Nayfeh, A., Balanchandra, B.: Applied non-linear dynamics. John Wiley & Sons, Chichester (1995)
27. Badit, R., Polite, A.: Statistical description of chaotic attractors: The dimension function. Journal of Statistical Physics 40(5/6) (1985)
28. Lasota, A., Mackey, M.C.: Chaos, Fractals and Noise. Springer, N.Y (1994)
29. Delgado-Restituto, M., Rodríguez-Vásquez, A.: Integrated chaos generators. Proceedings of the IEEE 90, 747–767 (2002)
30. Abel, X., Schwartz, W.: Chaos communications principles, schemes, and system analysis. Special Issue, Proceedings of the IEEE 90, 691–710 (2002)
31. Costamagna, E., Favalli, L., Rizzardi, M.: Simulation of error process on mobile radio channels based on chaos equations. In: Proc. 52nd IEEE Vehicular Technology Conference, Boston, MA, September 2000, pp. 1866–1873 (2000)
32. Schimming, T., Goetz, M., Schwarz, W.: Signal modeling using piecewise linear chaotic generators. In: Proc. Eur. Signal Processing Conf., V. III, Island of Rhodes, Greece, September 1998, pp. 1377–1380 (1998)
33. Tannous, C., Davies, R., Angus, A.: Strange attractors in multipath propagation. IEEE Trans. on Communications 39, 629–631 (1991)
34. Costamagna, E., Schirru, A.: Channel error models derived from chaos equations. In: Proc. IEEE Int. Conf. Systems, Man, and Cybernetics, vol. 1, pp. 577–581 (October 1994)
35. Benelli, G., Costamagna, E., Favalli, L., Gamba: Satellite channels modeled by chaotic bit error generators. In: Vatalaro, F. (ed.) Mobile and Personal Communications 2., pp. 139–149. Springer, Berlin (1996)
36. Costamagna, E., Favalli, L., Gamba, P., Savazzi, P.: Block-error probabilities for mobile radio channels derived from chaos equations. IEEE Communications Letters 3, 66–68 (1999)
37. Chua, L.O.: Computer-aided analysis of nonlinear networks
38. Papoulis, A., Pillai, S.U.: Probability, Random Variables and Stochastic Processes. McGraw-Hill, New York (2002)
39. Wang, H., Dirik, C., Rodríguez, S., Gole, A., Jacob, B.: Radio frequency effects on the clock networks of digital circuits. In: Int. Symp. on Electromagnetic Compatibility, August 2004, vol. 1, pp. 93–96 (2004)
40. Stecher, M.: Possible effects of spread-spectrum-clock interference on wideband radiocommunication services. In: Int. Symp. on Electromagnetic Compatibility, vol. 1, pp. 60–63 (August 2005)
41. Matsumoto, Y., Fujii, K., Sugiera, A.: Estimating the amplitude reduction of clock harmonics due to frequency modulation. IEEE Trans. on Electromagnetic Compatibility 48(4), 734–741 (2006)

42. Hasler, M., Mazzini, G., Ogorzalek, M., Rovatti, R., Setti, G.: Applications of nonlinear dynamics to electronic and information engineering. Special Issue, Proceedings of the IEEE 90 (May 2002)
43. Lau, F.C.M., Tse, C.K.: Chaos-Based Digital Communications Systems. Springer, Heidelberg (2003)
44. Delgado-Restituto, M., Rodriguez-Vazquez, A.: Integrated chaos generators. Special Issue, Proceedings of the IEEE 90, 747–767 (2002)
45. Costamagna, E., Favelli, L., Gamba, P.: Multipath channel modeling with chaotic attractors. Special Issue, Proceeding of the IEEE 90, 842–859 (2002)
46. Patterson, E.K., Gowdy, J.N.: Modeling speech-interfering noise with applied chaos theory. In: IEEE Southeast Conf. 2001. Proceedings, April 2001, pp. 229–233 (2001)
47. Balestra, M., Bellini, A., Callegari, S., Rovatti, R., Setti, G.: Chaos-based generation pf pwm-like signals for low-emi induction motor drives: analysis and experimental results. In: IEICE Trans. on Electronics, January 2004, vol. E87-C, pp. 66–75 (2004)
48. Rovatti, R., Mazzini, G., Setti, G., Giovanardi, A.: Statistical modeling and design of discrete-time chaotic processes: advanced finite-dimensional tools and applications. Proceedings of the IEEE 90, 662–690 (2002)
49. Kontorovich, V.: A statistical analysis applied to strange attractors. In: Proc. of the Intern. Conf. on Numerical Analysis and Applied Math., ICNAAM 2005. Wiley-VCH, Weinheim (2005)
50. Chui, C., Chen, G.: Kalman filtering. Springer, Heidelberg (1987)
51. Gradshteyn, I.S., Ryzhik, I.M.: Table of Integrals, Series and Products. Academic Press, San Diego (2000)
52. Badii, R., Politi, A.: Statistical description of chaotic attractors: The dimension function. Journal of Statistical Physics 40(5/6), 725–749 (1985)
53. Fraser, A.: Information and entropy in strange attractors. IEEE Trans. on Information Theory 35, 245–261 (1989)

On Synchronization of Coupled Delayed Neural Networks

Jinling Liang, Zidong Wang, and Xiaohui Liu

Abstract. In this chapter, the global synchronization problem is investigated for both continuous- and discrete-time coupled neural networks. The neural networks appear in the form of coupled arrays, where both the linear and nonlinear couplings are taken into account. The activation functions include both the Lipchitz-type and the sector-type ones. Due to the high dimension of the system under consideration, the Kronecker product is utilized to facilitate the derivation and simplify the presentation. By resorting to the matrix functional method, we aim to establish sufficient conditions under which the considered array of neural networks is globally synchronized. It is shown that the globally exponential synchronization can be achieved by suitably designing the coupling matrix, the inner linking matrix and some free matrices representing the relationships between the system matrices. The sufficient conditions obtained that guarantee the synchronization are directly related to several matrix quantities describing the coupling topology. Furthermore, these conditions are expressed in terms of several linear matrix inequalities (LMIs) which can be easily verified by utilizing the numerically efficient Matlab LMI toolbox. Several illustrative examples are given to show the feasibility and applicability of the proposed synchronization scheme.

1 Introduction

Neural networks, especially Hopfield neural networks and cellular neural networks, have found successful applications in many areas, such as image processing, pattern

---

Jinling Liang
Department of Mathematics, Southeast University, Nanjing 210096, China
e-mail: jinlliang@seu.edu.cn

Zidong Wang and Xiaohui Liu
Department of Information Systems and Computing, Brunel University, Uxbridge, Middlesex, UB8 3PH, U.K
e-mail: Zidong.Wang@brunel.ac.uk

K. Kyamakya (Eds.): Recent Adv. in Nonlinear Dynamics and Synchr., SCI 254, pp. 117–149.
springerlink.com

recognition, associative memory and optimization problems. In both the biological and artificial neural networks, the interactions between neurons are generally asynchronous, which give rise to the inevitable signal transmission delays. Also, in electronic implementation of analog neural networks, time-delay is usually unavoidable due primarily to the finite switching speed of amplifiers. It is known that time-delays may cause undesirable dynamic network behaviors such as oscillation and instability. Consequently, the stability analysis problems for delayed neural networks have received considerable research attention [11, 12].

In the past decade, a large amount of results have appeared in the literature, where the delay type can be constant, time-varying, or distributed, and the stability criteria can be delay-dependent or delay-independent. Note that continuously distributed delays have recently gained particular attention, since a neural network usually has a spatial nature due to the presence of an amount of parallel pathways of a variety of axon sizes and lengths. Complex dynamics such as bifurcation and chaos phenomena have also been shown to exist in delayed neural networks [13, 26]. For recent exposures on neural networks that incorporate either discrete-time or distributed delays, see [6, 14, 18, 24, 35, 37, 45, 44] and the references cited therein. It is worth mentioning that most of the previous studies have predominantly concentrated on the stability analysis, periodic oscillations and dissipativity of various kinds of neural networks.

On the other hand, the synchronization problem has long been an attractive research topic. Both the mathematical analysis and practical applications of synchronized complex systems have gained considerable research attention. The synchronization analysis and synchronization control issues have been thoroughly investigated and successfully applied in many different areas such as secure communication, chaos generators design and harmonic oscillation generation [19, 30, 34]. Among others, the synchronization problems for large-scale and complex networks of chaotic oscillators have received particular research interests, see [10, 21, 33]. An interesting feature of complex networks is the coupling that leads to many interesting phenomena such as spatio-temporal chaos and autowaves, which are important in modelling populations of interacting biological systems [29, 31, 46]. It has also observed that coupling allows cells to synchronize to each other. In [42], the synchronization problem in an array of linearly coupled dynamical systems has been thoroughly analyzed, where the linear coupling configuration has been characterized by a diffusive coupling matrix, indicating that each dynamical system will behave in the same manner when the synchronization is achieved. Subsequently, the synchronization problem has attracted many researchers see [3, 9, 20, 22, 40, 47] for some representative results.

In the context of synchronization, among many coupled systems have been studied, neural networks have attracted a great deal of research focus. It has been revealed that the synchronization of coupled neural networks could have an important impact on both the basic science (e.g., the self-organization behavior in the brain) and the technological practice. Early theoretical research has mainly focused on the local synchronization problem, where the synchronization can be achieved if all the conditional Lyapunov exponents of the coupled system are negative. For the local nature of this kind of synchronization, the rate of convergence is very difficult to be determined and, furthermore, the computation of the conditional Lyapunov exponents is not an easy task as well. To address such an issue, much effort has been put forward. For example, in [8], by defining a distance between any point and the synchronization manifold, global synchronization is studied for the coupled connected neural networks in terms of the Lyapunov functional methods and the Hermitian matrix theory. In [27], without assuming the coupling matrix to be symmetric or irreducible, the authors studied local and global synchronization problem by referring to the variational method. Recently, by employing the Kronecker product technique, a general array model of coupled delayed neural networks with hybrid coupling, which is composed of constant coupling, discrete-delay coupling and distributed-delay coupling, was investigated in [2, 5].

In biological neural networks, cells interact with each other more likely through output-coupling, and the *nonlinear* nature of asynchronous coupling enables the rich, context-sensitive interactions that characterize real brain dynamics, suggesting that it plays a role in functional integration that may be as important as synchronous interactions. Unfortunately, in almost all literature mentioned previously, the coupling terms of the models are always assumed to be *linear*. To the best of our knowledge, up to now, there are very few results on the synchronization problem of an array of delayed neural networks with *nonlinear* coupling. It is, therefore, the main purpose of the present research to fill such a gap by extending the main results of [5] to the nonlinear coupling case.

In this chapter, we study the global synchronization problem for coupled neural networks with time delays, where both the linear and nonlinear couplings are simultaneously taken into account. By making use of the Kronecker product properties, we employ the matrix functional method and develop matrix inequality techniques to establish sufficient conditions under which the considered neural networks are globally synchronized. The conditions obtained are related to several matrix quantities describing the coupling topology, and they are expressed in terms of several linear matrix inequalities (LMIs) which can therefore be easily verified by utilizing the numerically efficient Matlab LMI toolbox. Several numerical examples are also given in this chapter to illustrate the effectiveness of the proposed synchronization scheme.

The rest of this chapter is organized as follows. In Section 2, global synchronization is investigated for an array of continuous-time neural networks. In Section 3, the discrete-time case is studied. Finally, in Section 4, we give the conclusions and some future studying directions.

Notations: Throughout this chapter, for real matrices X and Y, the notation $X \geq Y$ (respectively, $X > Y$) means that matrices X, Y are real, symmetric and the matrix $X - Y$ is positive semidefinite (respectively, positive definite). E is an identity matrix with appropriate dimension. For any real matrix X, $X^s = \frac{1}{2}(X + X^T)$ represents the symmetric part of X. The superscript "T" stands for matrix transposition. In symmetric block matrices, we use an asterisk "$*$" to represent a term that is induced by symmetry. R^n is the n-dimensional Euclidean space and the notation $\|\cdot\|$ refers to the Euclidean vector norm. We use $\lambda_{\min}(\cdot)$ and $\lambda_{\max}(\cdot)$ to denote the minimum and maximum eigenvalue of a symmetric matrix. The Kronecker product of an $n \times m$ matrix X and an $p \times q$ matrix Y is the $np \times mq$ matrix $X \otimes Y$ defined as

$$X \otimes Y = \begin{pmatrix} x_{11}Y & \cdots & x_{1m}Y \\ \vdots & & \vdots \\ x_{n1}Y & \cdots & x_{nm}Y \end{pmatrix}.$$

For integers α and β with $\alpha < \beta$, we use $N[\alpha,\beta]$ to denote the discrete interval given by $N[\alpha,\beta] = \{\alpha, \alpha+1, \ldots, \beta-1, \beta\}$. Let $C(N[\alpha,\beta], R^n)$ be the set of all functions $\phi : N[\alpha,\beta] \to R^n$. Matrices, if not explicitly stated, are assumed to have compatible dimensions.

2 Synchronization of Continuous-Time Coupled Neural Networks

In this section, we consider the following array of delayed neural networks with both linear and nonlinear couplings:

$$\begin{aligned} \frac{dx_i(t)}{dt} &= -Cx_i(t) + Af(x_i(t)) + Bf(x_i(t-\tau)) + I(t) \\ &\quad + \sum_{j=1}^{N} G_{ij}^{(1)} D_1 x_j(t) + \sum_{j=1}^{N} G_{ij}^{(2)} D_2 f(x_j(t)), \quad i = 1, 2, \ldots, N \end{aligned} \tag{1}$$

where $x_i(t) = (x_{i1}(t), x_{i2}(t), \ldots, x_{in}(t))^T$ is the state vector of the ith network at time t; $C = \mathrm{diag}\{c_1, \ldots, c_n\} > 0$ denotes the rate with which the cell i resets its potential to the resting state when isolated from other cells and inputs; A and B are the weight matrix and the delayed weight matrix, respectively; the activation function is given by $f(x_i(t)) = (f_1(x_{i1}(t)), f_2(x_{i2}(t)), \ldots, f_n(x_{in}(t)))^T$; $I(t) = (I_1(t), I_2(t), \ldots, I_n(t))^T$ is the external input and $\tau > 0$ represents the transmission delay; $G^{(1)} = (G_{ij}^{(1)})_{N\times N}$ and $G^{(2)} = (G_{ij}^{(2)})_{N\times N}$ denote, respectively, the linearly coupling configuration and the nonlinearly coupling configuration of the array and satisfy the following diffusive coupling connections

$$G_{ij}^{(k)} = G_{ji}^{(k)} \geq 0 \ (i \neq j), \quad G_{ii}^{(k)} = -\sum_{j=1, j\neq i}^{N} G_{ij}^{(k)} \quad (k = 1, 2;\ i, j = 1, 2, \ldots, N). \tag{2}$$

Furthermore, in (1), D_1 and $D_2 \in R^{n\times n}$ represent the linking matrix and the nonlinear linking matrix of the neural networks, respectively.

Remark 1. In model (1), the term $\sum_{j=1}^{N} G_{ij}^{(2)} D_2 f(x_j(t))$ describes the nonlinear coupling of the array of neural networks. For simplicity, the nonlinear function $f(x_j(t))$ is taken as the activation function. However, as can be seen later, our main results can be easily extended to the case where such a nonlinear coupling can be set as any norm-bounded function. Note that in almost all literature regarding synchronization of neural networks, the nonlinear coupling phenomenon has not been considered.

For presentation convenience, we let $x(t) = (x_1^T(t), x_2^T(t), \ldots, x_N^T(t))^T$, $F(x(t)) = (f^T(x_1(t)), f^T(x_2(t)), \ldots, f^T(x_N(t)))^T$, $J(t) = (I^T(t), I^T(t), \ldots, I^T(t))^T$. By using the symbol $\otimes$ of the Kronecker product, model (1) can be rewritten as

$$\begin{aligned}\frac{dx(t)}{dt} &= -(E \otimes C)x(t) + (E \otimes A)F(x(t)) + (E \otimes B)F(x(t-\tau)) \\ &\quad + J(t) + (G^{(1)} \otimes D_1)x(t) + (G^{(2)} \otimes D_2)F(x(t)). \qquad (3)\end{aligned}$$

The initial conditions associated with (3) are given by

$$x_i(s) = \phi_i(s) \in C([-\tau, 0], R^n), \quad i = 1, 2, \ldots, N. \qquad (4)$$

2.1 Some Preliminaries

For the activation functions $f_i(\cdot)$, we consider the following two assumptions:

(H) There exist constants $l_i > 0$ $(i = 1, 2, \ldots, n)$ such that

$$0 \le \frac{f_i(u) - f_i(v)}{u - v} \le l_i$$

for any different $u,\ v \in R$.

(H^*)There exist constants $l_i > 0$ $(i = 1, 2, \ldots, n)$ such that

$$|f_i(u) - f_i(v)| \le l_i |u - v|$$

for any $u,\ v \in R$.

Definition 1. *Model (3) is said to be globally exponentially synchronized, if there exist two constants $\varepsilon > 0$ and $M > 0$, such that for all $\phi_i(\cdot) \in C([-\tau, 0], R^n)$ and for sufficiently large $T_0 > 0$,*

$$\|x_i(t) - x_j(t)\| \le Me^{-\varepsilon t}$$

hold for all $t > T_0$, $i, j = 1, 2, \ldots, N$.

Lemma 1. *[15] For any constant matrix $X \in R^{n\times n}$, $X^T = X > 0$; a scalar $r > 0$ and a vector function $g:\ [0, r] \to R^n$ such that the integrations concerned are well defined, the following holds*

$$r\int_0^r g^T(t)Xg(t)dt \ge \left(\int_0^r g(t)dt\right)^T X \left(\int_0^r g(t)dt\right).$$

Lemma 2. *[1] The following linear matrix inequality*

$$\begin{bmatrix} Q(x) & S(x) \\ S^T(x) & R(x) \end{bmatrix} > 0,$$

where $Q(x) = Q^T(x)$, $R(x) = R^T(x)$ and $S(x)$ depend affinely on x, is equivalent to one of the following conditions:

(1) $R(x) > 0,\ Q(x) - S(x)R^{-1}(x)S^T(x) > 0$;
(2) $Q(x) > 0,\ R(x) - S^T(x)Q^{-1}(x)S(x) > 0$.

Lemma 3. *[7] Let $\alpha \in R$; X, Y, P and Q are matrices with appropriate dimensions, then*

(1) $(\alpha X) \otimes Y = X \otimes (\alpha Y)$;
(2) $(X+Y) \otimes P = X \otimes P + Y \otimes P$;
(3) $(X \otimes Y)(P \otimes Q) = (XP) \otimes (YQ)$;
(4) $(X \otimes Y)^T = X^T \otimes Y^T$.

Lemma 4. *Let e be the N-dimensional vector of all 1s, i.e., $e = (1,1,\ldots,1)^T$ and $U = NE - E_N$, where $E_N = ee^T$ is the $N \times N$ matrix of all 1s. Suppose P is an $n \times n$ matrix and $x = (x_1^T, x_2^T, \ldots, x_N^T)^T$ where $x_i = (x_{i1}, x_{i2}, \ldots, x_{in})^T \in R^n$. Then, the following relationships hold:*

(1) $UG^{(j)} = G^{(j)}U = NG^{(j)},\ j = 1,2$;
(2) $(U \otimes P)J = 0,\ J^T(G^{(j)} \otimes P) = 0,\ j = 1,2$;
(3) $x^T(U \otimes P)y = -\sum_{i=1}^{N-1} \sum_{j=i+1}^{N} u_{ij}(x_i - x_j)^T P(y_i - y_j)$;

where J, $G^{(1)}$ and $G^{(2)}$ are defined in model (3).

Proof. From the definition of U, we have

$$u_{ij} = u_{ji}, \quad u_{ii} = -\sum_{j=1, j\neq i}^{N} u_{ij}; \quad (i,j = 1,2,\ldots,N)$$

and therefore

$$\begin{aligned} x^T(U \otimes P)x &= \sum_{i=1}^{N}\sum_{j=1}^{N} u_{ij}x_i^T P x_j = \sum_{i=1}^{N} x_i^T P[u_{ii}x_i + \sum_{j=1, j\neq i}^{N} u_{ij}x_j] \\ &= \sum_{i=1}^{N} \sum_{j=1, j\neq i}^{N} u_{ij}x_i^T P(x_j - x_i) \\ &= \sum_{i=1}^{N} x_i^T P \sum_{j<i} u_{ij}(x_j - x_i) + \sum_{i=1}^{N} x_i^T P \sum_{j>i} u_{ij}(x_j - x_i); \end{aligned} \tag{5}$$

while

$$\sum_{i=1}^{N} x_i^T P \sum_{j<i} u_{ij}(x_j - x_i) = \sum_{i=2}^{N}\sum_{j=1}^{i-1} u_{ij} x_i^T P(x_j - x_i)$$
$$= \sum_{i=2}^{N}\sum_{j=1}^{i-1} u_{ji} x_i^T P(x_j - x_i) = \sum_{i=1}^{N-1}\sum_{j=i+1}^{N} u_{ij} x_j^T P(x_i - x_j); \tag{6}$$

$$\sum_{i=1}^{N} x_i^T P \sum_{j>i} u_{ij}(x_j - x_i)$$
$$= \sum_{i=1}^{N-1}\sum_{j=i+1}^{N} u_{ij} x_i^T P(x_j - x_i) = -\sum_{i=1}^{N-1}\sum_{j=i+1}^{N} u_{ij} x_i^T P(x_i - x_j). \tag{7}$$

It follows from (5)-(7) that

$$x^T(U \otimes P)x = -\sum_{i=1}^{N-1}\sum_{j=i+1}^{N} u_{ij}(x_i - x_j)^T P(x_i - x_j),$$

which completes the proof.

2.2 Main Results

In this subsection, a matrix functional method will be employed to investigate the global exponential synchronization of system (3).

Theorem 1. *Under Assumption (H), the dynamical system (3) with initial conditions (4) is globally exponentially synchronized if there exist matrices $P > 0$, $Q > 0$ and a diagonal matrix $S = \mathrm{diag}\{s_1, s_2, \ldots, s_n\} > 0$ such that the following LMIs hold for all $1 \le i < j \le N$:*

$$\Omega_{ij}^{(1)} = \begin{bmatrix} \Delta_{11} & PA - NG_{ij}^{(2)}PD_2 + LS & PB \\ SL + A^T P - NG_{ij}^{(2)} D_2^T P & Q - 2S & 0 \\ B^T P & 0 & -Q \end{bmatrix} < 0, \tag{8}$$

where $\Delta_{11} = -PC - CP - NG_{ij}^{(1)}(PD_1 + D_1^T P)$ and $L = \mathrm{diag}\{l_1, l_2, \ldots, l_n\}$.

Proof. Since condition (8) holds and $\tau > 0$, by Lemma 2, one can choose a positive definite matrix $R > 0$ and a scalar $\alpha \in (0,1)$ such that

$$\widetilde{\Omega}_{ij}^{(1)} = \begin{bmatrix} \Delta_{11} & PA - NG_{ij}^{(2)}PD_2 + LS & PB & 0 \\ SL + A^T P - NG_{ij}^{(2)} D_2^T P & Q - 2S + \tau R & 0 & 0 \\ B^T P & 0 & -Q & 0 \\ 0 & 0 & 0 & -\frac{1-\alpha}{\tau} R \end{bmatrix} < 0. \tag{9}$$

Denote $x_t(\theta) = x(t+\theta)$, $(\theta \in [-\tau, 0])$ and consider the following matrix functional candidate for system (3):

$$V(t,x_t) = e^{\varepsilon t}\left[V_1(t,x_t) + V_2(t,x_t) + V_3(t,x_t)\right], \tag{10}$$

where ε is a positive number to be determined later; U is the matrix defined in Lemma 4 and

$$\begin{aligned}
V_1(t,x_t) &= x^T(t)(U \otimes P)x(t), \\
V_2(t,x_t) &= \int_{t-\tau}^{t} F^T(x(s))(U \otimes Q)F(x(s))ds, \\
V_3(t,x_t) &= \int_{t-\tau}^{t}\int_{\theta}^{t} F^T(x(s))(U \otimes R)F(x(s))dsd\theta.
\end{aligned}$$

Bearing Lemma 4 in mind, we calculate the time derivative of $V_1(t)$ along the solutions of (3) and obtain

$$\begin{aligned}
\frac{dV_1(t)}{dt} &= 2x^T(t)(U \otimes P)\big[-(E \otimes C)x(t) + (E \otimes A)F(x(t)) + (E \otimes B)F(x(t-\tau)) \\
&\quad + J(t) + (G^{(1)} \otimes D_1)x(t) + (G^{(2)} \otimes D_2)F(x(t))\big] \\
&= -2x^T(t)(U \otimes (PC))x(t) + 2x^T(t)(U \otimes (PA))F(x(t)) \\
&\quad + 2x^T(t)(U \otimes (PB))F(x(t-\tau)) + 2x^T(t)(NG^{(1)} \otimes (PD_1))x(t) \\
&\quad + 2x^T(t)(NG^{(2)} \otimes (PD_2))F(x(t)) \\
&= 2\sum_{i=1}^{N-1}\sum_{j=i+1}^{N}(x_i(t)-x_j(t))^T\big[(-PC - NG_{ij}^{(1)}PD_1)(x_i(t)-x_j(t)) \\
&\quad + (PA - NG_{ij}^{(2)}PD_2)(f(x_i(t)) - f(x_j(t))) \\
&\quad + PB(f(x_i(t-\tau)) - f(x_j(t-\tau)))\big].
\end{aligned} \tag{11}$$

Again, calculating the time derivative of $V_2(t)$ along the trajectories of (3) gives

$$\begin{aligned}
\frac{dV_2(t)}{dt} &= F^T(x(t))(U \otimes Q)F(x(t)) - F^T(x(t-\tau))(U \otimes Q)F(x(t-\tau)) \\
&= \sum_{i=1}^{N-1}\sum_{j=i+1}^{N}\big[(f(x_i(t)) - f(x_j(t)))^T Q(f(x_i(t)) - f(x_j(t))) \\
&\quad - (f(x_i(t-\tau)) - f(x_j(t-\tau)))^T Q(f(x_i(t-\tau)) - f(x_j(t-\tau)))\big].
\end{aligned} \tag{12}$$

Similarly, by using Lemma 1, the time derivative of $V_3(t)$ along the solutions of (3) can be obtained as follows:

$$\begin{aligned}
\frac{dV_3(t)}{dt} &= -\int_{t-\tau}^{t} F^T(x(s))(U \otimes R)F(x(s))ds + \int_{t-\tau}^{t} F^T(x(t))(U \otimes R)F(x(t))d\theta \\
&= \tau F^T(x(t))(U \otimes R)F(x(t)) - \alpha\int_{t-\tau}^{t} F^T(x(s))(U \otimes R)F(x(s))ds \\
&\quad - (1-\alpha)\int_{t-\tau}^{t} F^T(x(s))(U \otimes R)F(x(s))ds
\end{aligned}$$

$$
\begin{aligned}
&\leq \tau F^T(x(t))(U\otimes R)F(x(t)) - \alpha\int_{t-\tau}^{t} F^T(x(s))(U\otimes R)F(x(s))ds \\
&\quad -\frac{1-\alpha}{\tau}\left(\int_{t-\tau}^{t} F(x(s))ds\right)^T (U\otimes R)\left(\int_{t-\tau}^{t} F(x(s))ds\right) \\
&= \sum_{i=1}^{N-1}\sum_{j=i+1}^{N}\Big[\tau(f(x_i(t))-f(x_j(t)))^T R(f(x_i(t))-f(x_j(t))) \\
&\quad -\frac{1-\alpha}{\tau}\left(\int_{t-\tau}^{t}(f(x_i(s))-f(x_j(s)))ds\right)^T R\left(\int_{t-\tau}^{t}(f(x_i(s))-f(x_j(s)))\right)\Big] \\
&\quad -\alpha\int_{t-\tau}^{t} F^T(x(s))(U\otimes R)F(x(s))ds. \qquad (13)
\end{aligned}
$$

Considering the Assumption (H), we can deduce

$$
\begin{aligned}
&[f(x_i(t))-f(x_j(t))]^T S[f(x_i(t))-f(x_j(t))] \\
&\leq [x_i(t)-x_j(t)]^T LS[f(x_i(t))-f(x_j(t))], \ \forall 1\leq i<j\leq N. \qquad (14)
\end{aligned}
$$

The relations (11)-(14) lead to

$$
\frac{d(\sum_{i=1}^{3}V_i(t))}{dt} \leq \sum_{i=1}^{N-1}\sum_{j=i+1}^{N}\xi_{ij}^T\widetilde{\Omega}_{ij}^{(1)}\xi_{ij} - \alpha\int_{t-\tau}^{t}F^T(x(s))(U\otimes R)F(x(s))ds, \quad (15)
$$

where $\xi_{ij}=[(x_i(t)-x_j(t))^T,(f(x_i(t))-f(x_j(t)))^T,(f(x_i(t-\tau))-f(x_j(t-\tau)))^T,(\int_{t-\tau}^{t}(f(x_i(s))-f(x_j(s)))ds)^T]^T$ and $\widetilde{\Omega}_{ij}^{(1)}$ is defined in (9).

On the other hand, from the definition of $V_i(t)$ $(i=1,2,3)$, it is easy to obtain the following inequalities:

$$
V_1(t,x_t) \leq \lambda_{\max}(P)\sum_{i=1}^{N-1}\sum_{j=i+1}^{N}(x_i(t)-x_j(t))^T(x_i(t)-x_j(t)); \qquad (16)
$$

$$
V_3(t,x_t) \leq \tau\int_{t-\tau}^{t}F^T(x(s))(U\otimes R)F(x(s))ds. \qquad (17)
$$

From (10) and (15)-(17), one obtains

$$
\begin{aligned}
\frac{dV(t,x_t)}{dt} &= e^{\varepsilon t}\left[\varepsilon(V_1(t)+V_2(t)+V_3(t)) + \frac{d(\sum_{i=1}^{3}V_i(t))}{dt}\right] \\
&\leq e^{\varepsilon t}\Bigg[\sum_{i=1}^{N-1}\sum_{j=i+1}^{N}(-\lambda_{\min}(-\widetilde{\Omega}_{ij}^{(1)})+\varepsilon\lambda_{\max}(P))\|x_i(t)-x_j(t)\|^2 \\
&\quad -\int_{t-\tau}^{t}F^T(x(s))(U\otimes(-\varepsilon Q-\varepsilon\tau R+\alpha R))F(x(s))ds\Bigg]. \qquad (18)
\end{aligned}
$$

Let

$$\varepsilon = \min\left\{\frac{\lambda_{\min}(-\widetilde{\Omega}_{ij}^{(1)})}{\lambda_{\max}(P)}, \frac{\alpha\lambda_{\min}(R)}{\lambda_{\max}(Q+\tau R)}\right\}. \tag{19}$$

Then, it follows from (18)-(19) that $dV(t)/dt \leq 0$, and therefore $V(t) \leq V(0)$, i.e., $V(t)$ is a bounded function. Subsequently, $e^{\varepsilon t}x^T(t)(U \otimes P)x(t)$ is also bounded, which yields

$$\lambda_{\min}(P)\|x_i(t)-x_j(t)\|^2 \leq \sum_{i=1}^{N-1}\sum_{j=i+1}^{N}(x_i(t)-x_j(t))^T P(x_i(t)-x_j(t)) = O(e^{-\varepsilon t}) \tag{20}$$

for all $1 \leq i < j \leq N$. From Definition 1, one can conclude that model (3) is globally exponentially synchronized, and the proof is completed.

Proposition 1. *Under Assumption (H*), the dynamical system (3) with initial conditions (4) is globally exponentially synchronized if there exist matrices $P>0$, $Q>0$ and a diagonal matrix S such that the following LMIs hold for all $1 \leq i < j \leq N$:*

$$\begin{bmatrix} \Delta_{11}^* & PA - NG_{ij}^{(2)}PD_2 & PB \\ A^TP - NG_{ij}^{(2)}D_2^TP & Q-S & 0 \\ B^TP & 0 & -Q \end{bmatrix} < 0 \tag{21}$$

where $\Delta_{11}^* = -PC - CP - NG_{ij}^{(1)}(PD_1 + D_1^TP) + LSL$ *and* $L = \mathrm{diag}\{l_1, l_2, \ldots, l_n\}$.

Proof. Proposition 1 can be easily proved by substituting (14) with

$$\begin{aligned} &[f(x_i(t)) - f(x_j(t))]^T S[f(x_i(t)) - f(x_j(t))] \\ &\leq [x_i(t)-x_j(t)]^T LSL[x_i(t)-x_j(t)] \quad (1 \leq i < j \leq N) \end{aligned} \tag{22}$$

when calculating the time derivative of $V(t)$ along the trajectories. The rest of the proof is similar to that of Theorem 1 and is therefore omitted.

In the following, a new criterion is obtained for the global exponential synchronization of system (3) by using a different matrix functional.

Theorem 2. *Under Assumption (H), the dynamical system (3) with initial conditions (4) is globally exponentially synchronized if there exist two positive definite matrices $P>0$, $Q>0$, two matrices T_1, T_2 and one positive diagonal matrix S such that the following LMIs hold for all $1 \leq i < j \leq N$:*

$$\Omega_{ij}^{(2)} = \begin{bmatrix} \Lambda_{11} & \Lambda_{12} & \Lambda_{13} & T_1B \\ \Lambda_{12}^T & -T_2 - T_2^T & \Lambda_{23} & T_2B \\ \Lambda_{13}^T & \Lambda_{23}^T & Q-2S & 0 \\ B^TT_1^T & B^TT_2^T & 0 & -Q \end{bmatrix} < 0, \tag{23}$$

where $\Lambda_{11} = -T_1C - CT_1^T - NG_{ij}^{(1)}(T_1D_1 + D_1^T T_1^T)$, $\Lambda_{12} = P - T_1 - CT_2^T - NG_{ij}^{(1)}D_1^T T_2^T$, $\Lambda_{13} = T_1A - NG_{ij}^{(2)}T_1D_2 + LS$, $\Lambda_{23} = T_2A - NG_{ij}^{(2)}T_2D_2$ *and* $L = \mathrm{diag}\{l_1, l_2, \ldots, l_n\}$.

Proof. Since condition (23) holds and $\tau > 0$, by Lemma 2, one can choose positive definite matrices $R > 0, M > 0$ and a constant $\alpha \in (0,1)$ such that

$$\widetilde{\Omega}_{ij}^{(2)} = \begin{bmatrix} \Lambda_{11} & \Lambda_{12} & \Lambda_{13} & T_1B & 0 & 0 \\ \Lambda_{12}^T & \tau M - T_2 - T_2^T & \Lambda_{23} & T_2B & 0 & 0 \\ \Lambda_{13}^T & \Lambda_{23}^T & \tau R + Q - 2S & 0 & 0 & 0 \\ B^T T_1^T & B^T T_2^T & 0 & -Q & 0 & 0 \\ 0 & 0 & 0 & 0 & -\frac{1-\alpha}{\tau}R & 0 \\ 0 & 0 & 0 & 0 & 0 & -\frac{1-\alpha}{\tau}M \end{bmatrix} < 0. \quad (24)$$

Consider the following matrix functional candidate for system (3):

$$V(t,x_t) = e^{\varepsilon t}[V_1(t,x_t) + V_2(t,x_t) + V_3(t,x_t) + V_4(t,x_t)], \quad (25)$$

where V_1, V_2, V_3 are defined as in Theorem 1, and

$$V_4(t,x_t) = \int_{-\tau}^{0}\int_{t+\theta}^{t} \dot{x}^T(s)(U\otimes M)\dot{x}(s)dsd\theta.$$

Then, the time derivative of $V_1(t,x_t)$ along the solutions of (3) can be obtained as

$$\frac{dV_1(t)}{dt} = 2x^T(t)(U\otimes P)\dot{x}(t) = 2\sum_{i=1}^{N-1}\sum_{j=i+1}^{N}(x_i(t)-x_j(t))^T P(\dot{x}_i(t)-\dot{x}_j(t)). \quad (26)$$

Also, calculating the time derivative of $V_4(t,x_t)$ along the solution of (3) and by Lemma 1, one has

$$\begin{aligned} \frac{dV_4(t,x_t)}{dt} &= \int_{-\tau}^{0} \dot{x}^T(t)(U\otimes M)\dot{x}(t)d\theta - \int_{-\tau}^{0} \dot{x}^T(t+\theta)(U\otimes M)\dot{x}(t+\theta)d\theta \\ &= \tau\dot{x}^T(t)(U\otimes M)\dot{x}(t) - \int_{t-\tau}^{t} \dot{x}^T(s)(U\otimes M)\dot{x}(s)ds \\ &\le \sum_{i=1}^{N-1}\sum_{j=i+1}^{N}\Big[\tau(\dot{x}_i(t)-\dot{x}_j(t))^T M(\dot{x}_i(t)-\dot{x}_j(t)) \\ &\quad -\frac{1-\alpha}{\tau}\left(\int_{t-\tau}^{t}(\dot{x}_i(s)-\dot{x}_j(s))ds\right)^T M\left(\int_{t-\tau}^{t}(\dot{x}_i(s)-\dot{x}_j(s))ds\right)\Big] \\ &\quad -\alpha\int_{t-\tau}^{t}\dot{x}^T(s)(U\otimes M)\dot{x}(s)ds. \end{aligned} \quad (27)$$

Moreover, from (3), it is not difficult to see that the following equation holds for any matrices T_1, T_2:

$$2[x^T(t)(U\otimes T_1)+\dot{x}^T(t)(U\otimes T_2)]\times\Big[-\dot{x}(t)-(E\otimes C)x(t)+(E\otimes A)F(x(t))$$
$$+(E\otimes B)F(x(t-\tau))+J(t)+(G^{(1)}\otimes D_1)x(t)+(G^{(2)}\otimes D_2)F(x(t))\Big]=0.\quad(28)$$

And then the relationships (2)-(14) and (26)-(28) ensure that

$$\frac{d\left(\sum_{i=1}^{4}V_i(t,x_t)\right)}{dt}\le\sum_{i=1}^{N-1}\sum_{j=i+1}^{N}\zeta_{ij}^T\widetilde{\Omega}_{ij}^{(2)}\zeta_{ij}-\alpha\int_{t-\tau}^{t}\dot{x}^T(s)(U\otimes M)\dot{x}(s)ds$$
$$-\alpha\int_{t-\tau}^{t}F^T(x(s))(U\otimes R)F(x(s))ds,\quad(29)$$

where $\zeta_{ij}=[(x_i(t)-x_j(t))^T,(\dot{x}_i(t)-\dot{x}_j(t))^T,(f(x_i(t))-f(x_j(t)))^T,(f(x_i(t-\tau))-f(x_j(t-\tau)))^T,(\int_{t-\tau}^{t}(f(x_i(s))-f(x_j(s)))ds)^T,(\int_{t-\tau}^{t}(\dot{x}_i(s)-\dot{x}_j(s))ds)^T]^T$. From (16), (17), (24), (25), (29) and the following inequality

$$V_4(t,x_t)\le\tau\int_{t-\tau}^{t}\dot{x}^T(s)(U\otimes M)\dot{x}(s)ds,\quad(30)$$

we have

$$\frac{dV(t,x_t)}{dt}=e^{\varepsilon t}\left[\varepsilon\sum_{i=1}^{4}V_i(t,x_t)+\frac{d}{dt}\Big(\sum_{i=1}^{4}V_i(t,x_t)\Big)\right]$$
$$\le e^{\varepsilon t}\left[\sum_{i=1}^{N-1}\sum_{j=i+1}^{N}(-\lambda_{\min}(-\widetilde{\Omega}_{ij}^{(2)})+\varepsilon\lambda_{\max}(P))\|x_i(t)-x_j(t)\|^2\right.$$
$$-\int_{t-\tau}^{t}F^T(x(s))(U\otimes(\alpha R-\varepsilon Q-\varepsilon\tau R))F(x(s))ds$$
$$\left.-\int_{t-\tau}^{t}\dot{x}^T(s)(U\otimes(\alpha M-\varepsilon\tau M))\dot{x}(s)ds\right].\quad(31)$$

Taking

$$\varepsilon=\min\left\{\frac{\lambda_{\min}(-\widetilde{\Omega}_{ij}^{(2)})}{\lambda_{\max}(P)},\ \frac{\alpha}{\tau},\ \frac{\alpha\lambda_{\min}(R)}{\lambda_{\max}(Q+\tau R)}\right\},\quad(32)$$

we can conclude that $dV(t)/dt\le 0$. The remaining proof is similar to that of Theorem 1 and hence omitted.

The following proposition is easily accessible from Theorem 2.

Proposition 2. *Under Assumption (H^*), the dynamical system (3) with initial conditions (4) is globally exponentially synchronized if there exist two positive definite matrices $P>0$, $Q>0$; two matrices T_1, T_2 and one positive diagonal matrix S such that the following LMIs hold for all $1\le i<j\le N$:*

$$\begin{bmatrix} \Lambda_{11}^* & \Lambda_{12} & \Lambda_{13}^* & T_1B \\ \Lambda_{12}^T & -T_2-T_2^T & T_2A-NG_{ij}^{(2)}T_2D_2 & T_2B \\ \Lambda_{13}^{*T} & A^TT_2^T-NG_{ij}^{(2)}D_2^TT_2^T & Q-S & 0 \\ B^TT_1^T & B^TT_2^T & 0 & -Q \end{bmatrix} < 0, \quad (33)$$

where $\Lambda_{11}^* = -T_1C - CT_1^T - NG_{ij}^{(1)}(T_1D_1 + D_1^TT_1^T) + LSL$, $\Lambda_{13}^* = T_1A - NG_{ij}^{(2)}T_1D_2$ *and the other symbols are defined in Theorem 2.*

Remark 2. In the proof of Theorem 2 and Proposition 2, some free matrices T_1 and T_2 are introduced to reduce the conservatism of the results.

If the linking matrix $D_1 = 0$, the system (1) is reduced to the following model:

$$\frac{dx_i(t)}{dt} = -Cx_i(t) + Af(x_i(t)) + Bf(x_i(t-\tau)) + I(t) + \sum_{j=1}^{N} G_{ij}Df(x_j(t)), \quad (34)$$

where $i = 1, 2, \ldots, N$. In this case, we can get the following corollary:

Corollary 1. *Under Assumption (H), the dynamical system (34) is globally exponentially synchronized if there exist two positive definite matrices $P > 0$, $Q > 0$, two matrices T_1, T_2 and one positive diagonal matrix S such that the following LMIs hold for all $1 \le i < j \le N$:*

$$\begin{bmatrix} -T_1C - CT_1^T & P - T_1 - CT_2^T & \Xi_{13} & T_1B \\ P - T_1^T - T_2C & -T_2 - T_2^T & T_2A - NG_{ij}T_2D & T_2B \\ \Xi_{13}^T & A^TT_2^T - NG_{ij}D^TT_2^T & Q - 2S & 0 \\ B^TT_1^T & B^TT_2^T & 0 & -Q \end{bmatrix} < 0, \quad (35)$$

where $\Xi_{13} = T_1A - NG_{ij}T_1D + LS$.

If the linking matrix $D_2 = 0$, the system (1) specializes to the following model:

$$\frac{dx_i(t)}{dt} = -Cx_i(t) + Af(x_i(t)) + Bf(x_i(t-\tau)) + I(t) + \sum_{j=1}^{N} G_{ij}Dx_j(t), \quad (36)$$

where $i = 1, 2, \ldots, N$. In this case, the following corollary can be obtained readily.

Corollary 2. *Under Assumption (H), the dynamical system (36) is globally exponentially synchronized if there exist two positive definite matrices $P > 0$, $Q > 0$; two matrices T_1, T_2 and one positive diagonal matrix S such that the following LMIs hold for all $1 \le i < j \le N$:*

$$\begin{bmatrix} \Pi_{11} & \Pi_{12} & T_1A + LS & T_1B \\ \Pi_{12}^T & -T_2 - T_2^T & T_2A & T_2B \\ A^TT_1^T + SL & A^TT_2^T & Q - 2S & 0 \\ B^TT_1^T & B^TT_2^T & 0 & -Q \end{bmatrix} < 0, \quad (37)$$

where $\Pi_{11} = -T_1C - CT_1^T - NG_{ij}(T_1D + D^TT_1^T)$ *and* $\Pi_{12} = P - T_1 - CT_2^T - NG_{ij}D^TT_2^T$.

Remark 3. In [8, 27], it has been assumed that the inner coupling matrix is diagonal. However, in this paper, by referring to the Kronecker product technique, this restriction can be removed, and hence our results improve the earlier works. Moreover, the results here are given in terms of LMIs, which can be easily checked by resorting to the Matlab LMI control toolbox.

Remark 4. If we make the following assumption on the activation functions:

(H^{**}) There exist constants l_i^- and l_i^+ such that

$$l_i^- \leq \frac{f_i(u) - f_i(v)}{u - v} \leq l_i^+$$

for any different $u, v \in R$.

By a similar method used in [24, 35], more general results could be obtained.

2.3 A Numerical Example

As in [13, 5], we consider a two dimensional neural network with delays:

$$\frac{dy(t)}{dt} = -Cy(t) + Af(y(t)) + Bf(y(t-0.95)) + I(t), \tag{38}$$

where $y(t) = (y_1(t), y_2(t))^T \in R^2$ is the state vector of the network, the activation function $f(y(t)) = (f_1(y_1(t)), f_2(y_2(t)))^T$ with $f_i(y_i) = 0.5(|y_i + 1| - |y_i - 1|)$ $(i = 1, 2)$, obviously, Assumption (H) is satisfied with $L = \mathrm{diag}\{1, 1\}$; the external input vector $I(t) = (0, 0)^T$; and the other matrices are as follows:

$$C = \begin{bmatrix} 1 & 0 \\ 0 & 1 \end{bmatrix}, \quad A = \begin{bmatrix} 1 + \frac{\pi}{4} & 20 \\ 0.1 & 1 + \frac{\pi}{4} \end{bmatrix}, \quad B = \begin{bmatrix} -\frac{1.3\pi\sqrt{2}}{4} & 0.1 \\ 0.1 & -\frac{1.3\pi\sqrt{2}}{4} \end{bmatrix}.$$

The dynamical chaotic behavior with initial conditions

$$y_1(s) = 0.2, \quad y_2(s) = 0.3, \qquad \forall s \in [-0.95, 0] \tag{39}$$

is shown in Fig.1.

Let us now turn to an array of three nonlinearly coupled identical models (38). The state equations of the entire array are

$$\begin{aligned} \frac{dx_i(t)}{dt} &= -Cx_i(t) + Af(x_i(t)) + Bf(x_i(t-0.95)) + I(t) \\ &\quad + \sum_{j=1}^{3} G_{ij}^{(1)} D_1 x_j(t) + \sum_{j=1}^{3} G_{ij}^{(2)} D_2 f(x_j(t)), \end{aligned} \tag{40}$$

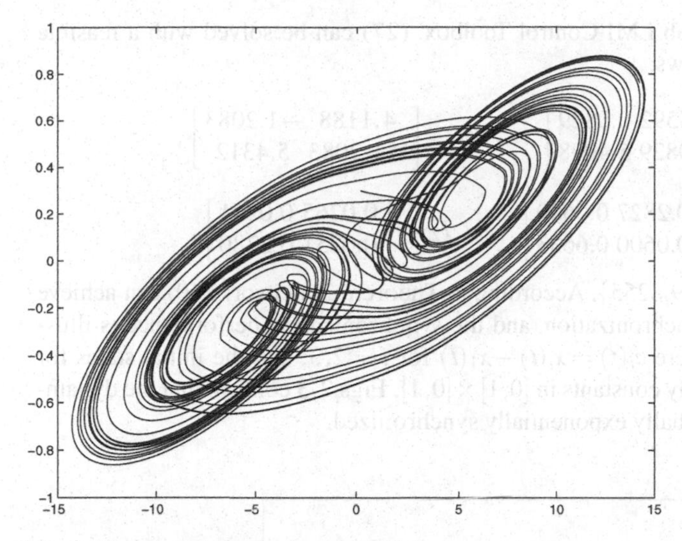

Fig. 1 Chaotic trajectory of model (38)

where $x_i(t) = (x_{i1}(t), x_{i2}(t))^T$ $(i = 1, 2, 3)$ is the state vector of the ith neural network. Choose the coupling matrices $G^{(1)}$, $G^{(2)}$ and the linking matrices D_1, D_2 as

$$G^{(1)} = G^{(2)} = \begin{bmatrix} -3 & 1 & 2 \\ 1 & -2 & 1 \\ 2 & 1 & -3 \end{bmatrix}, \quad D_1 = \begin{bmatrix} 4 & 0 \\ 0 & 4 \end{bmatrix}, \quad D_2 = \begin{bmatrix} 1 & 0 \\ 0 & 1 \end{bmatrix}.$$

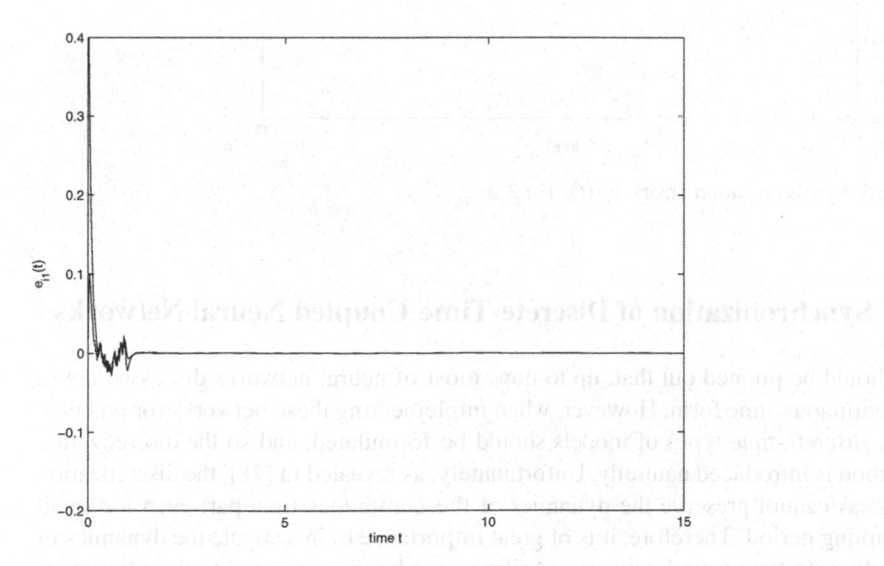

Fig. 2 Synchronization errors: $e_{i1}(t)$, $i = 2, 3$

By applying the Matlab LMI Control Toolbox, (27) can be solved with a feasible solution given as follows:

$$P = \begin{bmatrix} 0.5592 & 0.0829 \\ 0.0829 & 5.4098 \end{bmatrix}, \qquad Q = \begin{bmatrix} 4.1188 & -1.2083 \\ -1.2083 & 5.4312 \end{bmatrix},$$

$$T_1 = \begin{bmatrix} 0.2827 & 0.0600 \\ 0.0600 & 0.6633 \end{bmatrix}, \qquad T_2 = \begin{bmatrix} 0.0265 & 0.0251 \\ 0.0251 & 0.2780 \end{bmatrix},$$

and $S = \mathrm{diag}\{4.8447, 9.8255\}$. According to Theorem 2, network (40) can achieve global exponential synchronization, and the synchronization performance is illustrated in Figs.2-3, where $e_i(t) = x_i(t) - x_1(t)$ for $i = 2,3$; and the initial states for (37) are taken randomly constants in $[0,1] \times [0,1]$. Figs.2-3 confirm that the dynamical system (40) is globally exponentially synchronized.

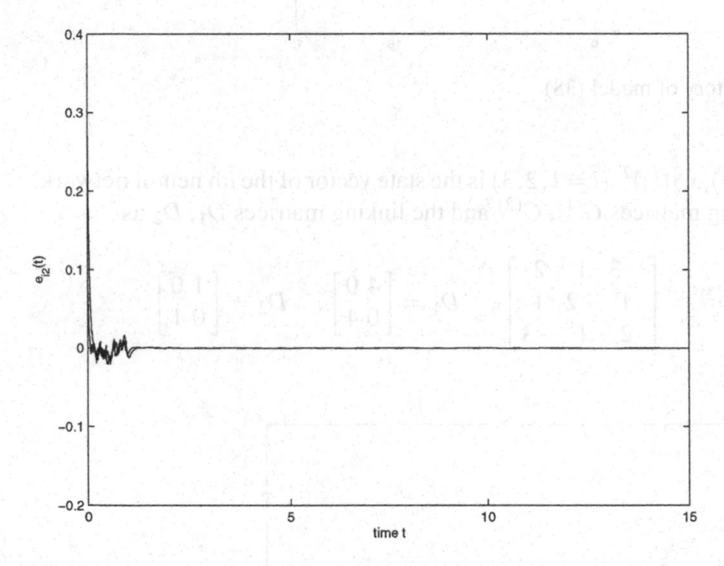

Fig. 3 Synchronization errors: $e_{i2}(t)$, $i = 2,3$

3 Synchronization of Discrete-Time Coupled Neural Networks

It should be pointed out that, up to now, most of neural networks discussed are in a continuous-time form. However, when implementing these networks for practical use, *discrete-time* types of models should be formulated, and so the discretization method is introduced naturally. Unfortunately, as revealed in [28], the discretization process cannot preserve the dynamics of the continuous-time part even for small sampling period. Therefore, it is of great importance to investigate the dynamics of the discrete-time neural networks. As far as we know, compared to the great number of results for continuous-time neural networks, relatively few results have been available for discrete-time models although they are important in practical applications such as time series analysis and system identification, especially where there exists time-varying delays [11, 23, 25, 32, 43]. In all the relevant synchronization papers we have discussed so far, the models are always assumed to be in a continuous-time form and the coupling terms of the system are mostly always assumed to be *constant* with or without time delays. However, in biological neural networks, cells interact with each other more likely through output-coupling, and the *time-varying* nature of asynchronous coupling enables the rich, context-sensitive interactions that characterize real brain dynamics, suggesting that it plays a role in functional integration that may be as important as synchronous interactions. To the best of the authors' knowledge, there has been very few results on the synchronization problem of an array of *discrete-time* neural networks, not to mention the mixed time-varying delays. Motivated by the above discussion, in this paper, we aim to deal with the global synchronization problem for an array of discrete-time neural networks with mixed couplings and time-varying delays. The activation functions used here are not required to be differentiable, bounded and monotonically increasing, and hence they are more appropriate to describe the neuron activation when describing and implementing an artificial neural networks.

3.1 Preliminaries

Consider the following array of discrete-time neural networks with mixed coupling and time-varying delays:

$$\begin{aligned} x_i(k+1) &= Ax_i(k) + Bf(x_i(k)) + Cf(x_i(k-d_k)) + I(k) \\ &\quad + \sum_{j=1}^{N} G_{ij}^{(1)} D_1 x_j(k) + \sum_{j=1}^{N} G_{ij}^{(2)} D_2 h(x_j(k-d_k)), \quad i = 1,2,\ldots,N \end{aligned} \tag{41}$$

where $k = 1,2,\ldots$; $x_i(k) = (x_{i1}(k), x_{i2}(k), \ldots, x_{in}(k))^T$ is the state vector of the ith network at time k; $A = \mathrm{diag}(a_1, a_2, \ldots, a_n)$ with $0 < a_i < 1$ $(i = 1,2,\ldots,n)$; d_k represents the time-varying transmission delay and satisfies:

$$d_m \le d_k \le d_M, \qquad k = 1,2,\ldots \tag{42}$$

where d_m and d_M are known positive integers; the constant coupling and the delayed coupling configuration $G^{(1)}$ and $G^{(2)}$ of the array still are assumed to satisfy the diffusive coupling connection conditions (2); the other symbols in (41) have the similar meaning as they are in (1).

Remark 5. The condition imposed on the time delay d_k is appropriate for reflecting the practical case in the discrete-time networks, where the transmission delays occur because of the finite propagation speed of signals along the axons in nervous systems or the finite switching speed of amplifier in neural circuits. Such delays are of a time-varying nature, and can be assumed to have both lower and upper bounds without loss of generality.

The initial conditions associated with model (41) are given by

$$x_i(s) = \varphi_i(s) \in C(N[-d_M,0],R^n), \qquad i = 1,2,\ldots,N. \tag{43}$$

For the activation functions, the following assumptions are made.
$(\widetilde{H})$ There exist constants k_i^-, k_i^+, l_i^-, l_i^+ such that

$$k_i^- \le \frac{f_i(u)-f_i(v)}{u-v} \le k_i^+, \quad l_i^- \le \frac{h_i(u)-h_i(v)}{u-v} \le l_i^+, \qquad i=1,2,\ldots,n$$

for any different u, $v \in R$.

Remark 6. In Assumption $(\widetilde{H})$, the constants k_i^-, k_i^+, l_i^- and l_i^+ are allowed to be negative, zero or positive. Hence, the resulting activation functions are more general than the usually used Lipschitz type or the mono-increasing ones. This description was firstly introduced in [24] and, since then, much attention has been drawn to this condition.

Denote $x(k) = (x_1^T(k), x_2^T(k), \ldots, x_N^T(k))^T$, $F(x(k)) = (f^T(x_1(k)), f^T(x_2(k)), \ldots, f^T(x_N(k)))^T$, $H(x(k-d_k)) = (h^T(x_1(k-d_k)), h^T(x_2(k-d_k)), \ldots, h^T(x_N(k-d_k)))^T$ and $J(k) = (I^T(k), I^T(k), \ldots, I^T(k))^T$. By making use of the Kronecker product, we can rewrite (41) as follows:

$$\begin{aligned}
x(k+1) &= (E_N \otimes A)x(k) + (E_N \otimes B)F(x(k)) + (E_N \otimes C)F(x(k-d_k)) \\
&\quad + J(k) + (G^{(1)} \otimes D_1)x(k) + (G^{(2)} \otimes D_2)H(x(k-d_k)) \\
&= (E_N \otimes A + G^{(1)} \otimes D_1)x(k) + (E_N \otimes B)F(x(k)) + (E_N \otimes C)F(x(k-d_k)) \\
&\quad + J(k) + (G^{(2)} \otimes D_2)H(x(k-d_k)).
\end{aligned} \tag{44}$$

Remark 7. The use of the Kronecker product would enable us to separate the effect caused by the connection topology (which is determined by the coupling matrices $G^{(1)}$ and $G^{(2)}$) from the effect due to the dynamics of each neuron and the couplings between them, see also [2, 39].

Definition 2. *Model (41) is said to be globally exponentially synchronized if there exist two constants $M > 0$ and $0 < \mu < 1$ such that, for all $\phi_i(\cdot) \in C(N[-d_M,0],R^n)$ and sufficiently large integer $T_0 > 0$, the inequality*

$$\|x_i(k) - x_j(k)\| \le M\mu^k \max_{s \in N[-d_M,0]} \|\phi_i(s) - \phi_j(s)\|$$

holds for all $k > T_0$ and $i, j = 1,2,\ldots,N$.

3.2 Global Synchronization Criteria

In this subsection, matrix functional method will be employed to investigate the global exponential synchronization of system (44). For simplicity, in the following, denote $G^{(1,1)} = G^{(1)}G^{(1)}$, $G^{(1,2)} = G^{(1)}G^{(2)}$, $G^{(2,2)} = G^{(2)}G^{(2)}$ and

$$K_1 = \mathrm{diag}(k_1^- k_1^+, k_2^- k_2^+, \ldots, k_n^- k_n^+), \quad L_1 = \mathrm{diag}(l_1^- l_1^+, l_2^- l_2^+, \ldots, l_n^- l_n^+),$$
$$K_2 = \mathrm{diag}(-\frac{k_1^- + k_1^+}{2}, -\frac{k_2^- + k_2^+}{2}, \ldots, -\frac{k_n^- + k_n^+}{2}),$$
$$L_2 = \mathrm{diag}(-\frac{l_1^- + l_1^+}{2}, -\frac{l_2^- + l_2^+}{2}, \ldots, -\frac{l_n^- + l_n^+}{2}).$$

Theorem 3. *Under Assumption ($\widetilde{H}$), the dynamical system (44) is globally exponentially synchronized if there exist matrices $P > 0$, $Q > 0$ and three diagonal matrices $R = \mathrm{diag}(r_1, r_2, \ldots, r_n) > 0$, $S = \mathrm{diag}(s_1, s_2, \ldots, s_n) > 0$, $W = \mathrm{diag}(w_1, w_2, \ldots, w_n) > 0$ such that the following LMIs hold for all $1 \le i < j \le N$:*

$$\Phi = \begin{bmatrix} \Pi_1^{(ij)} & 0 & \Pi_2^{(ij)} & APC - NG_{ij}^{(1)} D_1^T PC & \Pi_3^{(ij)} \\ * & \Pi_4 & 0 & -WK_2 & -SL_2 \\ * & * & B^T PB - R & B^T PC & -NG_{ij}^{(2)} B^T PD_2 \\ * & * & * & C^T PC - W & -NG_{ij}^{(2)} C^T PD_2 \\ * & * & * & * & -NG_{ij}^{(2,2)} D_2^T PD_2 - S \end{bmatrix} < 0, \quad (45)$$

where $\Pi_1^{(ij)} = APA - P - 2NG_{ij}^{(1)}(APD_1)^s - NG_{ij}^{(1,1)} D_1^T PD_1 - RK_1 + (d_M - d_m + 1)Q$, $\Pi_2^{(ij)} = APB - NG_{ij}^{(1)} D_1^T PB - RK_2$, $\Pi_3^{(ij)} = -NG_{ij}^{(2)} APD_2 - NG_{ij}^{(1,2)} D_1^T PD_2$, $\Pi_4 = -SL_1 - Q - WK_1$.

Proof. First of all, from Assumption ($\widetilde{H}$), we have

$$\begin{aligned} &[f_m(x_{im}(k)) - f_m(x_{jm}(k)) - k_m^+(x_{im}(k) - x_{jm}(k))] \\ \times &[f_m(x_{im}(k)) - f_m(x_{jm}(k)) - k_m^-(x_{im}(k) - x_{jm}(k))] \le 0, \quad (46) \\ &[f_m(x_{im}(k-d_k)) - f_m(x_{jm}(k-d_k)) - k_m^+(x_{im}(k-d_k) - x_{jm}(k -_k))] \\ \times &[f_m(x_{im}(k-d_k)) - f_m(x_{jm}(k-d_k)) - k_m^-(x_{im}(k-d_k) - x_{jm}(k-d_k))] \le 0 (47) \\ &[h_m(x_{im}(k-d_k)) - h_m(x_{jm}(k-d_k)) - l_m^+(x_{im}(k-d_k) - x_{jm}(k_k))] \\ \times &[h_m(x_{im}(k-d_k)) - h_m(x_{jm}(k-d_k)) - l_m^-(x_{im}(k-d_k) - x_{jm}(k-d_k))] \le 0 (48) \end{aligned}$$

where $m = 1, 2, \ldots, n$ and $1 \le i < j \le N$; which are equivalent to

$$\begin{bmatrix} x_i(k) - x_j(k) \\ f(x_i(k)) - f(x_j(k)) \end{bmatrix}^T \begin{bmatrix} k_m^- k_m^+ e_m e_m^T & -\frac{k_m^- + k_m^+}{2} e_m e_m^T \\ -\frac{k_m^- + k_m^+}{2} e_m e_m^T & e_m e_m^T \end{bmatrix} \times \begin{bmatrix} x_i(k) - x_j(k) \\ f(x_i(k)) - f(x_j(k)) \end{bmatrix} \le 0, \quad (49)$$

$$\begin{bmatrix} x_i(k-d_k) - x_j(k-d_k) \\ f(x_i(k-d_k)) - f(x_j(k-d_k)) \end{bmatrix}^T \begin{bmatrix} k_m^- k_m^+ e_m e_m^T & -\frac{k_m^- + k_m^+}{2} e_m e_m^T \\ -\frac{k_m^- + k_m^+}{2} e_m e_m^T & e_m e_m^T \end{bmatrix} \times \begin{bmatrix} x_i(k-d_k) - x_j(k-d_k) \\ f(x_i(k-d_k)) - f(x_j(k-d_k)) \end{bmatrix} \le 0, \quad (50)$$

$$\begin{bmatrix} x_i(k-d_k)-x_j(k-d_k) \\ h(x_i(k-d_k))-h(x_j(k-d_k)) \end{bmatrix}^T \begin{bmatrix} l_m^- l_m^+ e_m e_m^T & -\frac{l_m^-+l_m^+}{2} e_m e_m^T \\ -\frac{l_m^-+l_m^+}{2} e_m e_m^T & e_m e_m^T \end{bmatrix} \times \begin{bmatrix} x_i(k-d_k)-x_j(k-d_k) \\ h(x_i(k-d_k))-h(x_j(k-d_k)) \end{bmatrix} \leq 0, \quad (51)$$

where e_m denotes the n-dimensional unit column vector having "1" element on its mth row and zeros elsewhere.

Multiplying both sides of (49), (50) and (51) by r_m, w_m and s_m, respectively; and summing up from 1 to n with respect to m, we obtain

$$\begin{bmatrix} x_i(k)-x_j(k) \\ f(x_i(k))-f(x_j(k)) \end{bmatrix}^T \begin{bmatrix} RK_1 & RK_2 \\ RK_2 & R \end{bmatrix} \begin{bmatrix} x_i(k)-x_j(k) \\ f(x_i(k))-f(x_j(k)) \end{bmatrix} \leq 0, \quad (52)$$

$$\begin{bmatrix} x_i(k-d_k)-x_j(k-d_k) \\ f(x_i(k-d_k))-f(x_j(k-d_k)) \end{bmatrix}^T \begin{bmatrix} WK_1 & WK_2 \\ WK_2 & W \end{bmatrix} \times \begin{bmatrix} x_i(k-d_k)-x_j(k-d_k) \\ f(x_i(k-d_k))-f(x_j(k-d_k)) \end{bmatrix} \leq 0, \quad (53)$$

$$\begin{bmatrix} x_i(k-d_k)-x_j(k-d_k) \\ h(x_i(k-d_k))-h(x_j(k-d_k)) \end{bmatrix}^T \begin{bmatrix} SL_1 & SL_2 \\ SL_2 & S \end{bmatrix} \times \begin{bmatrix} x_i(k-d_k)-x_j(k-d_k) \\ h(x_i(k-d_k))-h(x_j(k-d_k)) \end{bmatrix} \leq 0. \quad (54)$$

Now, consider the following matrix functional candidate for system (44):

$$V(k) = V_1(k) + V_2(k) + V_3(k);$$

where

$$V_1(k) = x^T(k)(U \otimes P)x(k), \quad (55)$$

$$V_2(k) = \sum_{i=k-d_k}^{k-1} x^T(i)(U \otimes Q)x(i), \quad (56)$$

$$V_3(k) = \sum_{j=-d_M+1}^{-d_m} \sum_{i=k+j}^{k-1} x^T(i)(U \otimes Q)x(i), \quad (57)$$

with the matrix U defined in Lemma 4.

Calculating the difference of $V_1(k)$ along the trajectories of (44), we obtain

$$\begin{aligned} \Delta V_1(k) &= V_1(k+1) - V_1(k) \\ &= (x(k+1)+x(k))^T (U \otimes P)(x(k+1)-x(k)) \\ &= \Big[(E_N \otimes (A+E_n) + G^{(1)} \otimes D_1)x(k) + (E_N \otimes B)F(x(k)) \\ &\quad + (E_N \otimes C)F(x(k-d_k)) + J(k) + (G^{(2)} \otimes D_2)H(x(k-d_k)) \Big]^T (U \otimes P) \end{aligned}$$

$$\times \Big[(E_N \otimes (A-E_n)+G^{(1)} \otimes D_1)x(k)+(E_N \otimes B)F(x(k))$$
$$+(E_N \otimes C)F(x(k-d_k))+J(k)+(G^{(2)} \otimes D_2)H(x(k-d_k))\Big],$$

and it then follows from Lemma 3 and Lemma 4 that

$$\begin{aligned}
\Delta V_1(k) = &\Big[x^T(k)(E_N \otimes (A+E_n)+G^{(1)} \otimes D_1^T)+F^T(x(k))(E_N \otimes B^T) \\
&+F^T(x(k-d_k))(E_N \otimes C^T)+J^T(k)+H^T(x(k-d_k))(G^{(2)} \otimes D_2^T)\Big] \\
&\times \Big[(U \otimes (P(A-E_n))+NG^{(1)} \otimes (PD_1))x(k)+(U \otimes (PB))F(x(k)) \\
&+(U \otimes (PC))F(x(k-d_k))+(NG^{(2)} \otimes (PD_2))H(x(k-d_k))\Big] \\
= &\, x^T(k)\Big[U \otimes ((A+E_n)P(A-E_n))+NG^{(1)} \otimes (D_1^T P(A-E_n)) \\
&+NG^{(1)} \otimes ((A+E_n)PD_1)+NG^{(1,1)} \otimes (D_1^T PD_1)\Big]x(k) \\
&+2x^T(k)[U \otimes (APB)+NG^{(1)} \otimes (D_1^T PB)]F(x(k)) \\
&+2x^T(k)[U \otimes (APC)+NG^{(1)} \otimes (D_1^T PC)]F(x(k-d_k)) \\
&+2x^T(k)[NG^{(2)} \otimes (APD_2)+NG^{(1,2)} \otimes (D_1^T PD_2)]H(x(k-d_k)) \\
&+F^T(x(k))(U \otimes (B^T PB))F(x(k))+2F^T(x(k))(U \otimes (B^T PC))F(x(k-d_k)) \\
&+F^T(x(k-d_k))(U \otimes (C^T PC))F(x(k-d_k)) \\
&+2F^T(x(k))(NG^{(2)} \otimes (B^T PD_2))H(x(k-d_k)) \\
&+2F^T(x(k-d_k))(NG^{(2)} \otimes (C^T PD_2))H(x(k-d_k)) \\
&+H^T(x(k-d_k))(NG^{(2,2)} \otimes (D_2^T PD_2))H(x(k-d_k)).
\end{aligned} \tag{58}$$

Similarly, computing the difference of $V_2(k)$ and $V_3(k)$ along the solutions of (44), we have

$$\begin{aligned}
\Delta V_2(k) &= V_2(k+1)-V_2(k) \\
&= \sum_{i=k+1-d_{k+1}}^{k} x^T(i)(U \otimes Q)x(i) - \sum_{i=k-d_k}^{k-1} x^T(i)(U \otimes Q)x(i) \\
&= x^T(k)(U \otimes Q)x(k) - x^T(k-d_k)(U \otimes Q)x(k-d_k) \\
&\quad + \sum_{i=k+1-d_{k+1}}^{k-d_m} x^T(i)(U \otimes Q)x^T(i) \\
&\quad + \sum_{i=k-d_m+1}^{k-1} x^T(i)(U \otimes Q)x(i) - \sum_{i=k-d_k+1}^{k-1} x^T(i)(U \otimes Q)x(i) \\
&\le x^T(k)(U \otimes Q)x(k) - x^T(k-d_k)(U \otimes Q)x(k-d_k) \\
&\quad + \sum_{i=k+1-d_M}^{k-d_m} x^T(i)(U \otimes Q)x(i),
\end{aligned} \tag{59}$$

and

$$
\begin{aligned}
\Delta V_3(k) &= V_3(k+1) - V_3(k) \\
&= \sum_{j=-d_M+1}^{-d_m} \sum_{i=k+1+j}^{k} x^T(i)(U \otimes Q)x(i) - \sum_{j=-d_M+1}^{-d_m} \sum_{i=k+j}^{k-1} x^T(i)(U \otimes Q)x(i) \\
&= \sum_{j=-d_M+1}^{-d_m} [x^T(k)(U \otimes Q)x(k) - x^T(k+j)(U \otimes Q)x(k+j)] \\
&= (d_M - d_m)x^T(k)(U \otimes Q)x(k) - \sum_{i=k+1-d_M}^{k-d_m} x^T(i)(U \otimes Q)x(i). \qquad (60)
\end{aligned}
$$

Considering (58)-(60) and the fact that $u_{ij} = -1$ $(i \neq j)$, we obtain

$$
\begin{aligned}
\Delta V(k) \leq\; & x^T(k)\Big[U \otimes (APA - P) + NG^{(1)} \otimes (D_1^T PA + APD_1) + NG^{(1,1)} \otimes (D_1^T PD_1) \\
& +(d_M - d_m + 1)U \otimes Q\Big]x(k) + F^T(x(k))(U \otimes (B^T PB))F(x(k)) \\
& +2x^T(k)[U \otimes (APB) + NG^{(1)} \otimes (D_1^T PB)]F(x(k)) \\
& +2x^T(k)[U \otimes (APC) + NG^{(1)} \otimes (D_1^T PC)]F(x(k-d_k)) \\
& +2x^T(k)[NG^{(2)} \otimes (APD_2) + NG^{(1,2)} \otimes (D_1^T PD_2)]H(x(k-d_k)) \\
& +F^T(x(k-d_k))(U \otimes (C^T PC))F(x(k-d_k)) \\
& -x^T(k-d_k)(U \otimes Q)x(k-d_k) + 2F^T(x(k))(U \otimes (B^T PC))F(x(k-d_k)) \\
& +2F^T(x(k))(NG^{(2)} \otimes (B^T PD_2))H(x(k-d_k)) \\
& +2F^T(x(k-d_k))(NG^{(2)} \otimes (C^T PD_2))H(x(k-d_k)) \\
& +H^T(x(k-d_k))(NG^{(2,2)} \otimes (D_2^T PD_2))H(x(k-d_k)) \\
=\; & \sum_{i=1}^{N-1} \sum_{j=i+1}^{N} \Big\{(x_i(k) - x_j(k))^T [APA - P - NG_{ij}^{(1)}(D_1^T PA + APD_1) \\
& -NG_{ij}^{(1,1)} D_1^T PD_1 + (d_M - d_m + 1)Q](x_i(k) - x_j(k)) \\
& +(f(x_i(k)) - f(x_j(k)))^T (B^T PB)(f(x_i(k)) - f(x_j(k))) \\
& +2(x_i(k) - x_j(k))^T [APB - NG_{ij}^{(1)} D_1^T PB](f(x_i(k)) - f(x_j(k))) \\
& +2(x_i(k) - x_j(k))^T [APC - NG_{ij}^{(1)} D_1^T PC](f(x_i(k-d_k)) - f(x_j(k-d_k))) \\
& -(x_i(k-d_k) - x_j(k-d_k))^T Q(x_i(k-d_k) - x_j(k-d_k)) \\
& +2(f(x_i(k)) - f(x_j(k)))^T B^T PC(f(x_i(k-d_k)) - f(x_j(k-d_k))) \\
& -2(f(x_i(k)) - f(x_j(k)))^T NG_{ij}^{(2)} B^T PD_2(h(x_i(k-d_k)) - h(x_j(k-d_k))) \\
& +(f(x_i(k-d_k)) - f(x_j(k-d_k)))^T (C^T PC)(f(x_i(k-d_k)) - f(x_j(k-d_k))) \\
& +2(x_i(k) - x_j(k))^T [-NG_{ij}^{(2)} APD_2 - NG_{ij}^{(1,2)} D_1^T PD_2] \\
& \times (h(x_i(k-d_k)) - h(x_j(k-d_k))) - 2(f(x_i(k-d_k)) - f(x_j(k-d_k)))^T
\end{aligned}
$$

$$\times NG_{ij}^{(2)} C^T PD_2(h(x_i(k-d_k)) - h(x_j(k-d_k)))$$
$$-(h(x_i(k-d_k)) - h(x_j(k-d_k)))^T NG_{ij}^{(2,2)} D_2^T PD_2$$
$$\times (h(x_i(k-d_k)) - h(x_j(k-d_k)))\} . \tag{61}$$

Substituting (52), (53) and (54) into (61) gives

$$\Delta V(k) \le \sum_{i=1}^{N-1} \sum_{j=i+1}^{N} \xi_{ij}^T \Phi \xi_{ij} \le -\sum_{i=1}^{N-1} \sum_{j=i+1}^{N} \lambda_{\min}(-\Phi)\|x_i(k) - x_j(k)\|^2, \tag{62}$$

where $\xi_{ij} = ((x_i(k) - x_j(k))^T \ (x_i(k-d_k) - x_j(k-d_k))^T \ (f(x_i(k)) - f(x_j(k)))^T \ (f(x_i(k-d_k)) - f(x_j(k-d_k)))^T \ (h(x_i(k-d_k)) - h(x_j(k-d_k)))^T)^T$.

On the other hand, from the definition of $V(k)$, it is easy to see that

$$V(k) \le \sum_{i=1}^{N-1} \sum_{j=i+1}^{N} [\lambda_{\max}(P)\|x_i(k) - x_j(k)\|^2 + \rho \sum_{m=k-d_M}^{k-1} \|x_i(m) - x_j(m)\|^2], \tag{63}$$

$$V(k) \ge \sum_{i=1}^{N-1} \sum_{j=i+1}^{N} \lambda_{\min}(P)\|x_i(k) - x_j(k)\|^2, \tag{64}$$

where $\rho = \lambda_{\max}(Q)(d_M - d_m + 1)$.

For any $\varepsilon > 1$, inequalities (62) and (63) imply that

$$\begin{aligned}
&\varepsilon^{p+1} V(p+1) - \varepsilon^p V(p) \\
&= \varepsilon^{p+1} \Delta V(p) + \varepsilon^p(\varepsilon - 1)V(p) \\
&\le \sum_{i=1}^{N-1} \sum_{j=i+1}^{N} \{[-\varepsilon^{p+1}\lambda_{\min}(-\Phi) + \varepsilon^p(\varepsilon-1)\lambda_{\max}(P)]\|x_i(p) - x_j(p)\|^2 \\
&\quad + \rho\varepsilon^p(\varepsilon - 1) \sum_{m=p-d_M}^{p-1} \|x_i(m) - x_j(m)\|^2\}.
\end{aligned} \tag{65}$$

Summing up both sides of (65) from 0 to $T_0 - 1$ (T_0 is big enough to satisfy that $T_0 - 1 > d_M$) with respect to p, one has

$$\begin{aligned}
&\varepsilon^T V(T) - V(0) \\
&\le \sum_{i=1}^{N-1} \sum_{j=i+1}^{N} \{[-\varepsilon\lambda_{\min}(-\Phi) + (\varepsilon - 1)\lambda_{\max}(P)] \sum_{p=0}^{T_0-1} \varepsilon^p \|x_i(p) - x_j(p)\|^2 \\
&\quad + \rho(\varepsilon - 1) \sum_{p=0}^{T_0-1} \sum_{m=p-d_M}^{p-1} \varepsilon^p \|x_i(m) - x_j(m)\|^2\},
\end{aligned} \tag{66}$$

with

$$\sum_{p=0}^{T_0-1}\sum_{m=p-d_M}^{p-1}\varepsilon^p\|x_i(m)-x_j(m)\|^2$$
$$\leq (\sum_{m=-d_M}^{-1}\sum_{p=0}^{m+d_M}+\sum_{m=0}^{T_0-d_M-1}\sum_{p=m+1}^{m+d_M}+\sum_{m=T_0-d_M}^{T_0-1}\sum_{p=m+1}^{T_0})\varepsilon^p\|x_i(m)-x_j(m)\|^2$$
$$\leq d_M^2\varepsilon^{d_M}\max_{s\in N[-d_M,0]}\|x_i(s)-x_j(s)\|^2+d_M\varepsilon^{d_M}\sum_{m=0}^{T_0}\varepsilon^m\|x_i(m)-x_j(m)\|^2. \quad (67)$$

It follows from (63), (66) and (67) that

$$\varepsilon^{T_0}V(T_0)\leq\sum_{i=1}^{N-1}\sum_{j=i+1}^{N}\{L_1(\varepsilon)\max_{s\in N[-d_M,0]}\|x_i(s)-x_j(s)\|^2$$
$$+L_2(\varepsilon)\sum_{m=0}^{T_0}\varepsilon^m\|x_i(m)-x_j(m)\|^2\}, \quad (68)$$

where

$$L_1(\varepsilon)=\lambda_{\max}(P)+\rho d_M+\rho(\varepsilon-1)d_M^2\varepsilon^{d_M},$$
$$L_2(\varepsilon)=-\varepsilon\lambda_{\min}(-\Phi)+(\varepsilon-1)\lambda_{\max}(P)+\rho(\varepsilon-1)d_M\varepsilon^{d_M}.$$

Since $L_1(1)>0$ and $L_2(1)<0$, we can choose a $\mu>1$ such that $L_1(\mu)>0$ and $L_2(\mu)\leq 0$. From (64) and (68), we have

$$\sum_{i=1}^{N-1}\sum_{j=i+1}^{N}\mu^{T_0}\lambda_{\min}(P)\|x_i(T_0)-x_j(T_0)\|^2\leq\mu^{T_0}V(T_0)$$
$$\leq\sum_{i=1}^{N-1}\sum_{j=i+1}^{N}L_1(\mu)\max_{s\in N[-d_M,0]}\|x_i(s)-x_j(s)\|^2,$$

i.e.,

$$\|x_i(T_0)-x_j(T_0)\|\leq M(\frac{1}{\sqrt{\mu}})^{T_0}\max_{s\in N[-d_M,0]}\|x_i(s)-x_j(s)\|,$$

where M is a large positive number. According to Definition 2, we conclude that model (44) is globally exponentially synchronized, and the proof is then completed.

Remark 8. In Theorem 3, the synchronization problem is studied for an array of discrete-time neural networks with mixed coupling and time-varying delays. This is the first time that the synchronization of such kind of neural networks is analyzed. The obtained criteria are dependent on not only the upper bound but also the lower bound of the time-varying delays, hence are less conservative than the traditional delay-independent ones.

If $D_1 = 0$ in system (44), i.e., the system has only the time-delayed coupling terms, then the model reduces to the following:

$$\begin{aligned} x(k+1) = &(E_N \otimes A)x(k) + (E_N \otimes B)F(x(k)) + (E_N \otimes C)F(x(k-d_k)) \\ &+J(k) + (G \otimes D)H(x(k-d_k)), \end{aligned} \tag{69}$$

and the following corollary can be easily obtained from Theorem 3.

Corollary 3. *Under Assumption* $(\widetilde{H})$*, the dynamical system (69) is globally exponentially synchronized if there exist matrices* $P > 0$*,* $Q > 0$ *and three diagonal matrices* $R > 0$*,* $W > 0$ *and* $S > 0$ *such that the following LMIs hold for all* $1 \leq i < j \leq N$*:*

$$\Phi_1 = \begin{bmatrix} \Lambda & 0 & APB - RK_2 & APC & -NG_{ij}APD \\ & \Pi_4 & 0 & -WK_2 & -SL_2 \\ & * & B^T PB - R & B^T PC & -NG_{ij}B^T PD \\ & * & * & C^T PC - W & -NG_{ij}C^T PD \\ & * & * & * & -N\widetilde{G}_{ij}D^T PD - S \end{bmatrix} < 0, \tag{70}$$

where $\Lambda = APA - P + (d_M - d_m + 1)Q - RK_1$*,* $\widetilde{G} = GG = (\widetilde{G}_{ij})_{N\times N}$ *and* Π_4 *is defined in Theorem 3.*

Moreover, in system (69), if $H(x(k-d_k)) = x(k-d_k)$, i.e., the system has only linear delayed coupling terms, then it reduces to the following model:

$$\begin{aligned} x(k+1) = &(E_N \otimes A)x(k) + (E_N \otimes B)F(x(k)) + (E_N \otimes C)F(x(k-d_k)) \\ &+J(k) + (G \otimes D)x(k-d_k); \end{aligned} \tag{71}$$

and we have the following corollary easily.

Corollary 4. *Under Assumption* $(\widetilde{H})$*, the dynamical system (71) is globally exponentially synchronized if there exist matrices* $P > 0$*,* $Q > 0$ *and three diagonal matrices* $R > 0$*,* $W > 0$ *and* $S > 0$ *such that the following LMIs hold for all* $1 \leq i < j \leq N$*:*

$$\Phi_2 = \begin{bmatrix} \Lambda & 0 & APB - RK_2 & APC & -NG_{ij}APD \\ & \Pi_4' & 0 & -WK_2 & S \\ & * & B^T PB - R & B^T PC & -NG_{ij}B^T PD \\ & * & * & C^T PC - W & -NG_{ij}C^T PD \\ & * & * & * & -N\widetilde{G}_{ij}D^T PD - S \end{bmatrix} < 0, \tag{72}$$

where $\Pi_4' = -S - Q - WK_1$*,* Λ *and* $\widetilde{G}$ *are defined in Corollary 3.*

If $D_2 = 0$ in system (44), i.e., the system has only the linear coupling terms, then it specializes to the following model:

$$\begin{aligned} x(k+1) = &(E_N \otimes A)x(k) + (E_N \otimes B)F(x(k)) + (E_N \otimes C)F(x(k-d_k)) \\ &+J(k) + (G \otimes D)x(k), \end{aligned} \tag{73}$$

and from Theorem 3, the following corollary is readily established.

Corollary 5. *Under Assumption $(\widetilde{H})$, the dynamical system (73) is globally exponentially synchronized if there exist matrices $P > 0$, $Q > 0$ and two diagonal matrices $R > 0$, $W > 0$ such that the following LMIs hold for all $1 \leq i < j \leq N$:*

$$\Phi_3 = \begin{bmatrix} \Gamma^{(ij)} & 0 & APB - NG_{ij}D^T PB - RK_2 & APC - NG_{ij}D^T PC \\ & -Q - WK_1 & 0 & -WK_2 \\ & * & B^T PB - R & B^T PC \\ & * & * & C^T PC - W \end{bmatrix} < 0, \tag{74}$$

where $\Gamma^{(ij)} = APA - P - 2NG_{ij}(APD)^s - N\widetilde{G}_{ij}D^T PD + (d_M - d_m + 1)Q - RK_1$, *and* $\widetilde{G} = GG = (\widetilde{G}_{ij})_{N \times N}$.

Remark 9. We point out that it is possible to extend the main results developed in this paper to many other cases, for example, the synchronization problem of an array of neural network models with multiple delays, high-order terms and more complex couplings.

3.3 Numerical Examples

In this subsection, two illustrative examples are provided to show the effectiveness of our results.

Example 1. Consider a discrete-time neural network model with time-varying delays as follows:

$$y(k+1) = Ay(k) + Bf(y(k)) + Cf(y(k - d_k)) + I(k), \tag{75}$$

where $k = 1, 2, \ldots$; $y(k) = (y_1(k), y_2(k))^T$ is the state vector; the time-varying delay d_k is taken as $d_k = 4 + \cos\frac{k\pi}{2}$ and therefore $d_m = 2$, $d_M = 5$; the activation function is $f(y(k)) = (\tanh(-0.2y_1(k)), \tanh(0.4y_2(k)))^T$ and the system parameters are given by

$$A = \begin{bmatrix} 0.3 & 0 \\ 0 & 0.2 \end{bmatrix}, B = \begin{bmatrix} 0.2 & -0.1 \\ -0.5 & 0.3 \end{bmatrix}, C = \begin{bmatrix} -0.15 & -0.1 \\ -0.2 & -0.25 \end{bmatrix}, I(k) = \begin{bmatrix} 2\cos\frac{k\pi}{3} \\ -\sin\frac{k\pi}{3} \end{bmatrix}.$$

The state trajectory of neural model (75) is shown in Fig. 4 with initial conditions $(y_1(s), y_2(s))^T = (0.4, -0.5)^T$ for $s \in [-5, 0]$.

We now deal with the following dynamical system which is composed of three coupled identical models (75) with mixed coupling:

$$\begin{aligned} x_i(k+1) &= Ax_i(k) + Bf(x_i(k)) + Cf(x_i(k - d_k)) \\ &\quad + I(k) + \sum_{j=1}^{3} G_{ij}Dh(x_j(k - d_k)), \quad i = 1, 2, 3 \end{aligned} \tag{76}$$

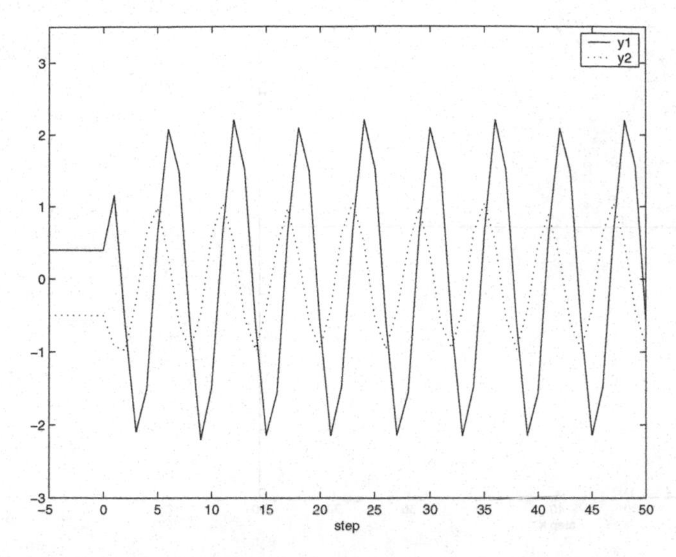

Fig. 4 State trajectory of model (75)

where $x_i(k) = (x_{i1}(k), x_{i2}(k))^T$ is the state variable of the ith neural network model and the activation function is $h(x_j(k-d_k)) = (\tanh(0.25x_{j1}(k-d_k)), \tanh(-0.23x_{j2}(k-d_k)))^T$. The coupling matrix and the linking matrix are given by

$$G = \begin{bmatrix} -3 & 1 & 2 \\ 1 & -2 & 1 \\ 2 & 1 & -3 \end{bmatrix}, \quad D = \begin{bmatrix} 0.2 & 0 \\ 0 & 0.1 \end{bmatrix}.$$

We can verify that Assumption $(\widetilde{H})$ holds for the activation functions with

$$K_1 = L_1 = \begin{bmatrix} 0 & 0 \\ 0 & 0 \end{bmatrix}, \quad K_2 = \begin{bmatrix} 0.1 & 0 \\ 0 & -0.2 \end{bmatrix}, \quad L_2 = \begin{bmatrix} -0.125 & 0 \\ 0 & 0.115 \end{bmatrix}.$$

By using the Matlab LMI Toolbox, We can solve the LMI (70) and have the feasible solutions as follows:

$$P = \begin{bmatrix} 1.1458 & 0.0807 \\ 0.0807 & 1.4719 \end{bmatrix}, \qquad Q = \begin{bmatrix} 0.1680 & 0.0077 \\ 0.0077 & 0.2065 \end{bmatrix},$$

$$R = \begin{bmatrix} 1.8936 & 0 \\ 0 & 1.4122 \end{bmatrix}, \quad W = \begin{bmatrix} 1.3404 & 0 \\ 0 & 1.2305 \end{bmatrix}, \quad S = \begin{bmatrix} 2.7883 & 0 \\ 0 & 1.8332 \end{bmatrix}.$$

According to Corollary 3, the array (76) can achieve synchronization, and the synchronization performance is illustrated by Fig. 5 and Fig. 6, where $e_i(t) = x_i(t) - x_1(t)$ for $i = 2, 3$, and the initial states for (76) are randomly taken as

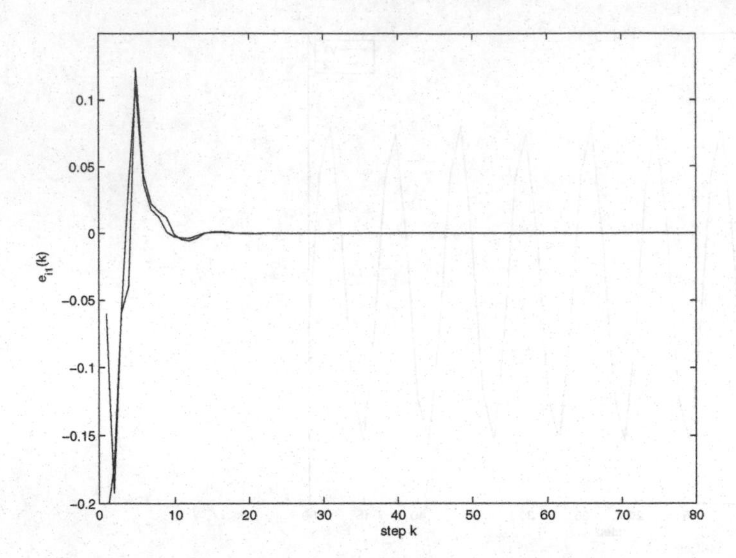

Fig. 5 Synchronization errors: $e_{i1}(t)$, $i = 2,3$

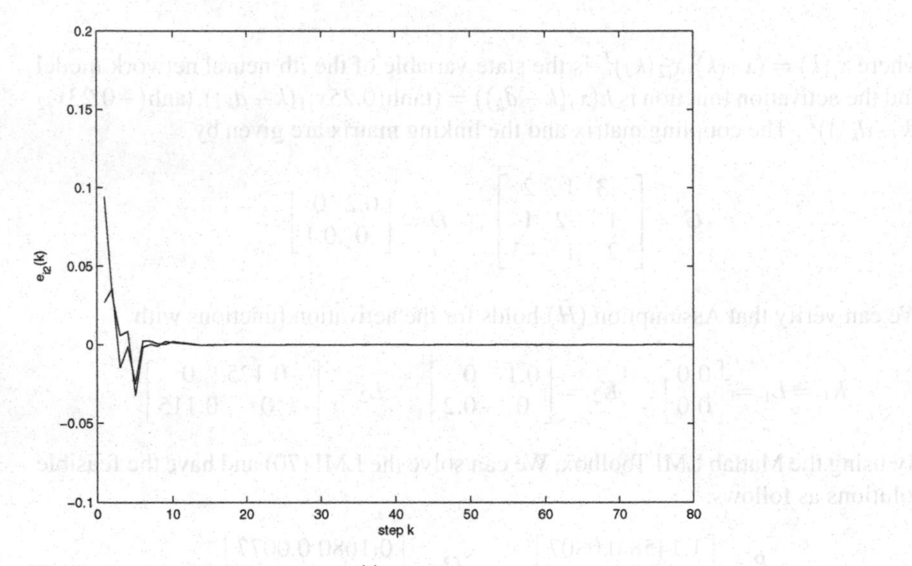

Fig. 6 Synchronization errors: $e_{i2}(t)$, $i = 2,3$

constants in $[0,1] \times [0,1]$. Figs. 5-6 confirm that the dynamical system (76) is globally exponentially synchronized.

Example 2. Consider the discrete-time neural network model with time-varying delays as follows:

$$y(k+1) = Ay(k) + Bf(y(k)) + Cf(y(k-d_k)) + I(k), \tag{77}$$

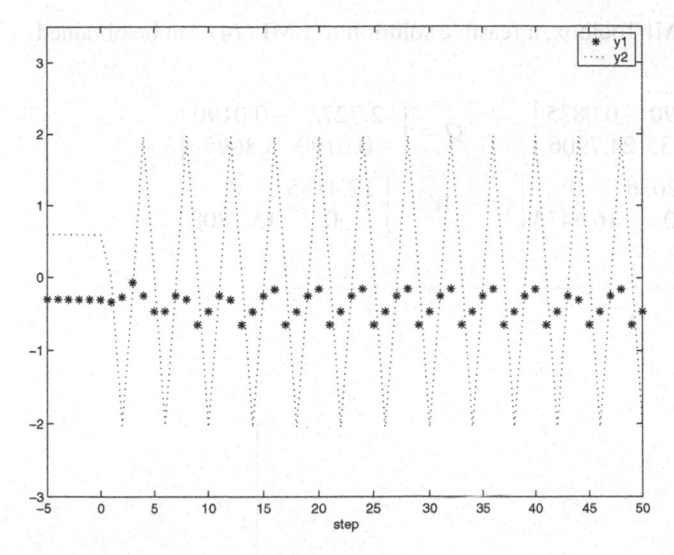

Fig. 7 State trajectory of model (77)

where $k = 1, 2, \ldots$; $y(k) = (y_1(k), y_2(k))^T$ is the state vector; the time-varying delay d_k is $d_k = 4 - \sin\frac{k\pi}{2}$ with $d_m = 2$ and $d_M = 5$; the activation function is $f(y(k)) = (\tanh(0.2y_1(k)), \tanh(-0.4y_2(k)))^T$ and the system parameters are given by

$$A = \begin{bmatrix} 0.06 & 0 \\ 0 & 0.1 \end{bmatrix}, \quad B = \begin{bmatrix} -0.25 & 0.3 \\ 0.2 & 0.1 \end{bmatrix}, \quad C = \begin{bmatrix} 0.1 & -0.2 \\ 0.3 & 0 \end{bmatrix}, \quad I(k) = \begin{bmatrix} -0.3 \\ 2\cos\frac{k\pi}{2} \end{bmatrix}.$$

It is clear that the activation functions in system (77) satisfy Assumption $(\widetilde{H})$ with $K_1 = 0$ and $K_2 = \mathrm{diag}\{-0.1, 0.2\}$. The state trajectory of the neural model (77) is shown in Fig. 7 with initial conditions as $(y_1(s), y_2(s))^T = (-0.3, 0.6)^T$ for $s \in [-5, 0]$.

Now, let us consider the following dynamical system consisting of three coupled identical models (77) with linear coupling:

$$\begin{aligned} x_i(k+1) &= Ax_i(k) + Bf(x_i(k)) + Cf(x_i(k - d_k)) \\ &\quad + I(k) + \sum_{j=1}^{3} G_{ij} D x_j(k), \quad i = 1, 2, 3 \end{aligned} \tag{78}$$

where $x_i(k) = (x_{i1}(k), x_{i2}(k))^T$ is the state variable of the ith neural network model. The coupling matrix and the linking matrix are taken as

$$G = \begin{bmatrix} -8 & 2 & 6 \\ 2 & -4 & 2 \\ 6 & 2 & -8 \end{bmatrix}, \quad D = \begin{bmatrix} 0.02 & 0 \\ 0 & 0.01 \end{bmatrix}.$$

By using the Matlab LMI Toolbox, a feasible solution to LMI (74) can be obtained as follows:

$$P = \begin{bmatrix} 22.7890 & -0.0835 \\ -0.0835 & 24.7906 \end{bmatrix}, \qquad Q = \begin{bmatrix} 2.7273 & -0.0190 \\ -0.0190 & 3.3695 \end{bmatrix},$$
$$R = \begin{bmatrix} 17.2636 & 0 \\ 0 & 16.6433 \end{bmatrix}, \qquad W = \begin{bmatrix} 17.4185 & 0 \\ 0 & 15.7808 \end{bmatrix}.$$

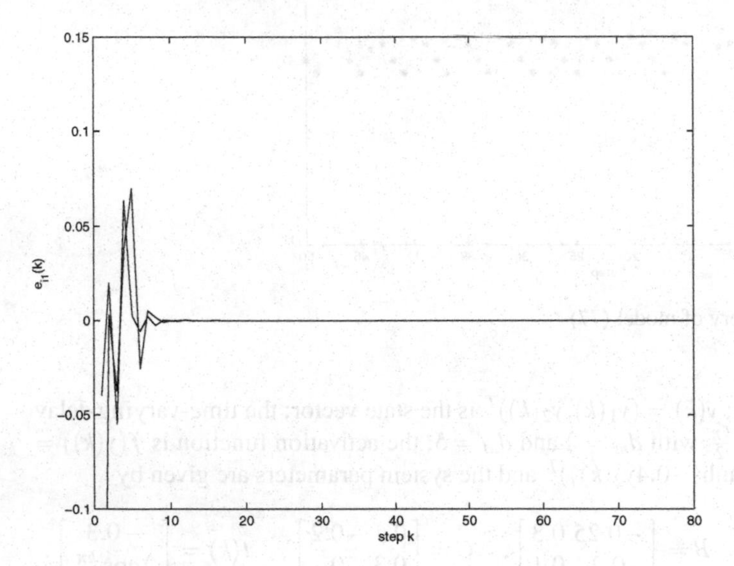

Fig. 8 Synchronization errors: $e_{i1}(t)$, $i = 2,3$

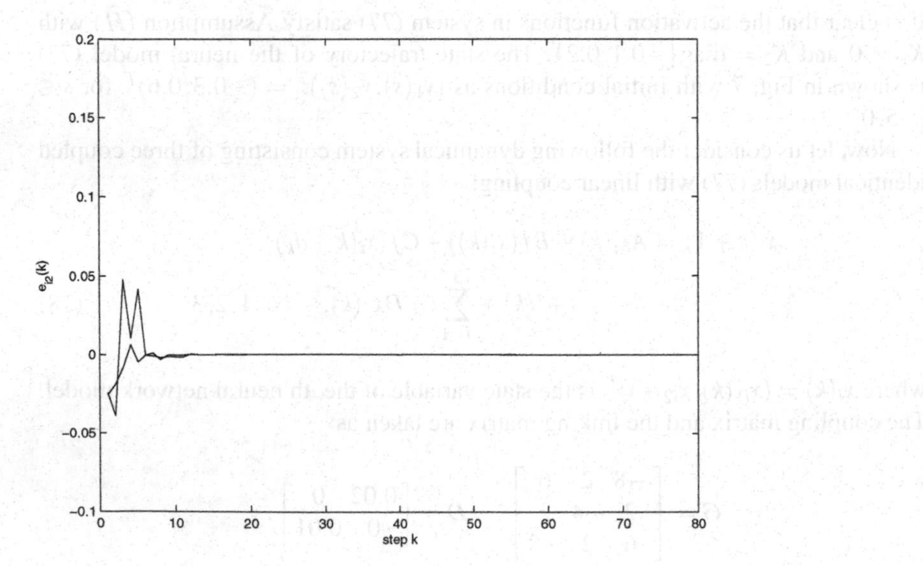

Fig. 9 Synchronization errors: $e_{i2}(t)$, $i = 2,3$

Let $e_i(t) = x_i(t) - x_1(t)$ for $i = 2,3$, and the initial states for (78) be randomly taken as constants in $[0,1] \times [0,1]$. Fig. 8 and Fig. 9 illustrate the synchronization performance, which confirm the conclusion from Corollary 5 that the dynamical system (78) is globally exponentially synchronized.

4 Conclusions

In this paper, the global exponential synchronization problem has been analyzed for an array of neural networks. Both the continuous-time type and the discrete-time analogue are investigated separately. In the complex system under study, both the cases of linear coupling and nonlinear coupling have been taken into account with time delays. By employing the inequality techniques, the properties of Kronecker product as well as the Lyapunov functional method, several sufficient criteria have been obtained to guarantee the coupled system to be globally exponentially synchronized. The main feature of results obtained is that we do not require the inner coupling matrices to be diagonal, and the mixed coupling term is introduced so as to account for the reality. The conditions obtained are related to several matrix quantities describing the coupling topology, moreover, they are described in terms of LMIs which can be easily verified by utilizing the numerically efficient Matlab LMI toolbox and therefore easy to be checked and applied in practice. Several commonly used examples have been given to illustrate the effectiveness of the proposed synchronization scheme.

We like to point out that our main results could be extended to more complex neural networks such as those with Markovian jumping parameters and/or stochastic disturbances. Also, similar problem can be addressed to and same methodology can be applied to complex networks, which are more general than the neural networks.

References

1. Boyd, S., Ghaoui, L.E., Feron, E., Balakrishnan, V.: Linear Matrix Inequalities in System and Control Theory. SIAM, Philadelphia (1994)
2. Cao, J., Li, P., Wang, W.W.: Global synchronization in arrays of delayed neural networks with constant and delayed coupling. Phys. Lett. A 353, 318–325 (2006)
3. Cao, J., Lu, J.: Adaptive synchronization of neural networks with or without time-varying delays. Chaos 16, 013133 (2006)
4. Cao, J., Yuan, K., Ho, D.W.C., Lam, J.: Global point dissipativity of neural networks with mixed time-varying delays. Chaos 16(1) 013105 (2006)
5. Cao, J., Chen, G., Li, P.: Global synchronization in an array of delayed neural networks with hybrid coupling. IEEE Trans. Syst. Man Cybern. B 38(2), 488–498 (2008)
6. Chen, Y.: Global stability of neural networks with distributed delays. Neural Networks 15, 867–871 (2002)
7. Chen, J.L., Chen, X.H.: Special Matrices. Tsinghua University Press, China (2001)
8. Chen, G., Zhou, J., Liu, Z.R.: Global synchronization of coupled delayed neural networks and applications to chaotic CNN models. Int. J. Bifurc. Chaos 14(7), 2229–2240 (2004)

9. Chen, M., Zhou, D.: Synchronization in uncertain complex networks. Chaos 16, 013101 (2006)
10. Chua, L.O.: CNN: a Paradigm for Complexity. World Scientific, Singapore (1998)
11. Gao, H., Chen, T.: New results on stability of discrete-time systems with time-varying state delay. IEEE Tran. Autom. Control 52(2), 328–334 (2007)
12. Gao, H., Lam, J., Chen, G.: New criteria for synchronization stability of general complex dynamical networks with coupling delays. Phys. Lett. A 360, 263–273 (2006)
13. Gilli, M.: Strange attractors in delayed cellular neural networks. IEEE Trans. Circuits Syst. -I 40(11), 849–853 (1993)
14. Gopalsamy, K., He, X.: Stability in asymmetric Hopfield nets with transmission delays. Physica D 76, 344–358 (1994)
15. Gu, K.Q., Kharitonov, V.L., Chen, J.: Stability of Time-Delay Systems. Birkhauser, Boston (2003)
16. He, Y., Liu, G., Rees, D.: New delay-dependent stability criteria for neural networks with time-varying delay. IEEE Trans. Neural Networks 18(1), 310–314 (2007)
17. He, Y., Wu, M., She, J.H.: Delay-dependent exponential stability of delayed neural networks with time-varying delay. IEEE Trans. Circuits Syst.-II 53(7), 553–557 (2006)
18. Hopfield, J.J.: Neurons with graded response have collective computational properties like those of two-stage neurons. Proc. Natl. Acad. Sci. USA 81, 3088–3092 (1984)
19. Hoppensteadt, F.C., Izhikevich, E.M.: Pattern recognition via synchronization in phase locked loop neural networks. IEEE Trans. Neural Networks 11(3), 734–738 (2000)
20. Huang, X., Cao, J.: Generalized synchronization for delayed chaotic neural networks: a novel coupling scheme. Nonlinearity 19(12), 2797–2811 (2006)
21. Jost, J., Joy, M.: Special properties and synchronization in coupled map lattices. Phys. Rev. E 65, 061201 (2002)
22. Li, Z., Chen, G.: Global synchronization and asymptotic stability of complex dynamical networks. IEEE Trans. Circuits Syst.-II 53(1), 28–33 (2006)
23. Liang, J., Cao, J., Lam, J.: Convergence of discrete-time recurrent neural networks with variable delay. Int. J. Birfurc. Chaos 15(2), 581–595 (2005)
24. Liu, Y., Wang, Z., Liu, X.H.: Global exponential stability of generalized recurrent neural networks with discrete and distributed delays. Neural Networks 19(5), 667–675 (2006)
25. Liu, Y., Wang, Z., Serrano, A., Liu, X.: Discrete-time recurrent neural networks with time-varying delays: exponential stability analysis. Phys. Lett. A 362, 480–488 (2007)
26. Lu, H.T.: Chaotic attractors in delayed neural networks. Phys. Lett. A 298, 109–116 (2002)
27. Lu, W.L., Chen, T.P.: Synchronization of coupled connected neural networks with delays. IEEE Trans. Circuits Syst.-I 51(12), 2491–2503 (2004)
28. Mohamad, S., Gopalsamy, K.: Exponential stability of continuous-time and discrete-time cellular neural networks with delays. Appl. Math. Comput. 135(1), 17–38 (2003)
29. Murray, J.D.: Mathematical Biology. Springer, Berlin (1989)
30. Pecora, L.M., Carroll, T.L.: Synchronization in chaotic systems. Phys. Rev. Lett. 64(8), 821–824 (1990)
31. Perez-Munuzuri, V., Perez-Villar, V., Chua, L.O.: Autowaves for image processing on a two-dimensional CNN array of excitable nonlinear circuits: flat and Wrinkled labyrinths. IEEE Trans. Circuits Syst. -I 40, 174–181 (1993)
32. Song, Q., Wang, Z.: A delay-dependent LMI approach to dynamics analysis of discrete-time recurrent neural networks with time-varying delays. Phys. Lett. A (2007)
33. Wang, X.F., Chen, G.: Synchronization in small-world dynamical networks. Int. J. Bifurc. Chaos 12(1), 187–192 (2002)

34. Wang, X.F., Chen, G.: Synchronization in scale-free dynamical networks: robustness and fragility. IEEE Trans. Circuits Syst.-I 49(1), 54–62 (2002)
35. Wang, Z., Liu, Y., Liu, X.: On global asymptotic stability of neural networks with discrete and distributed delays. Phys. Lett. A 345(4-6), 299–308 (2005)
36. Wang, Z., Liu, Y., Fraser, K., Liu, X.: Stochastic stability of uncertain Hopfield neural networks with discrete and distributed delays. Phys. Lett. A 354(4), 288–297 (2006)
37. Wang, Z., Liu, Y., Li, M., Liu, X.: Stability analysis for stochastic Cohen-Grossberg neural networks with mixed time delays. IEEE Trans. Neural Networks 17(3), 814–820 (2006)
38. Wang, Z., Liu, Y., Yu, L., Liu, X.: Exponential stability of delayed recurrent neural networks with Markovian jumping parameters. Phys. Lett. A 356(4-5), 346–352 (2006)
39. Wu, C.W.: Synchronization in coupled arrays of chaotic oscillators with nonreciprocal coupling. IEEE Trans. Circuits Syst.-I 50(2), 294–297 (2003)
40. Wu, C.W.: Synchronization in arrays of coupled nonlinear systems with delay and nonreciprocal time-varying coupling. IEEE Trans. Circuits Syst.-II 52(5), 282–286 (2005)
41. Wu, C.W.: Synchronization in networks of nonlinear dynamical systems coupled via a directed graph. Nonlinearity 18, 1057–1064 (2005)
42. Wu, C.W., Chua, L.O.: Synchronization in an array of linearly coupled dynamical systems. IEEE Trans. Circuits Syst.-I 42(8), 430–447 (1995)
43. Xiong, W., Cao, J.: Global exponential stability of discrete-time Cohen-Grossberg neural networks. Neurocomputing 64, 433–446 (2005)
44. Zhang, J., Suda, Y., Iwasa, T.: Absolutely exponential stability of a class of neural networks with unbounded delay. Neural Networks 17(3), 391–397 (2004)
45. Zhao, H.Y.: Global stability of neural networks with distributed delays. Phys. Rev. E 68, 051909 (2003)
46. Zheleznyak, A., Chua, L.O.: Coexistence of low- and high-dimensional spatio-temporal chaos in a chain of dissipatively coupled Chua's circuits. Int. J. Bifurc. Chaos 4(3), 639–674 (1994)
47. Zhou, J., Chen, T.P.: Synchronization in general complex delayed dynamical networks. IEEE Trans. Circuits Syst.-I 53(3), 733–744 (2006)

Part II
Some Modelling and Simulation Examples

Multiset of Agents in a Network for Simulation of Complex Systems

V.K. Murthy and E.V. Krishnamurthy

Abstract. A Complex System (CS) exhibits the four salient properties: (i) Collective, coordinated and efficient interaction among its components (ii) Self-organization and emergence (iii) Power law scaling under emergence (iv) Adaptation, fault tolerance and resilience against damage of its components. We describe briefly, three interrelated mathematical models that enable us to understand these properties: Fractal and percolation model, Stochastic / Chaotic (nonlinear) dynamical model and Topological (network) or graph model. These models have been very well studied in recent years and are closely related to the properties such as: self-similarity, scale-free, resilience, self-organization and emergence. We explain how these properties of CS can be simulated using the multi-set of agents-based paradigm (MAP) through random enabling, inhibiting, preferential attachment and growth of the multiagent network. We discuss these aspects from the point of view of geometric parameters-Lyapunov exponents, strange attractors, metric entropy, and topological indices-Cluster coefficient, Average degree distribution and the correlation length of the interacting network.

We also describe the advantages of agent-based modelling, simulation and animation. These are illustrated by a few examples in swarm dynamics- ant colony, bacterial colonies, human-animal trails, and graph-growth. We briefly consider the engineering of CS, the role of scales, and the limitations arising from quantum mechanics. A brief summary of currently available agent-tool kits is provided. Further developments of agent technology will be of great value to model, simulate and animate, many phenomena in Systems biology-cellular dynamics, cell motility, growth and development biology (Morphogenesis), and can provide for improved capability in complex systems modelling.

1 Introduction

The Science of Complex Systems (CS) is a highly interdisciplinary area and the definitions have not yet been well- formulated and vary between fields. From the

---

V.K. Murthy
University of Brunei Dausssalam, Brunei
e-mail: edayathuk@gmail.com

V.K. Murthy and E.V. Krishnamurthy
Computer Sciences Laboratory,
The Australian National University,
Canberra, ACT 0200, Australia
e-mail: Evk.Krishnamurthy@anu.edu.au

K. Kyamakya (Eds.): Recent Adv. in Nonlinear Dynamics and Synchr., SCI 254, pp. 153 – 200.
springerlink.com

point of view of Physics, a complex system consists of a large (possibly countably infinite) number of interacting inhomogeneous subsystems that are spatially and temporally structured on different scales and characterized by their own dynamics. A complex system often exhibits a collective ("Emergence") behaviour that is difficult to model deterministically based entirely on the properties of the individual subsystems. CS exchanges energy, entropy and matter with the environment. On an average CS exports entropy, it operates far from equilibrium, beyond a critical distance from the equilibrium state, positive and negative feedback are present and the system is nonlinear.

In Engineering, a CS is one that monitors itself and /or its environment in order to respond to changes in its conditions. This can result in (i) Adaptation (ii) Self-repair and (iii) Optimisation of resources.

Biosystems characterize CS by the properties:

1. The processing of information is distributed and is monitored
2. The communication system is reconfigurable.
3. The system is highly adaptive and acquires knowledge through interaction with its environment.
4. The energy consumption is optimized.
5. The system is self-organizing; it exhibits emergent properties which are not predictable, in advance from the properties of the individual local interactions among its components and the environment.

Thus it is reasonable to define CS with the following four interrelated properties that are reflected in each of the above viewpoints.

1. Interacting components,Interaction with Environment, Energy source

A CS consists of a large network of number of interacting components and an environment that is defined by a boundary to discriminate the inside (self) from the outside (nonself) of the CS. The components collectively and cooperatively perform actions, the actions being driven by an energy source. These actions are coordinated to obtain maximal efficiency to seek the optimum, under varying environmental conditions and constraints. Thus it is a nonlinear dynamic optimization problem in a nonstationary environment in which the constraints are not fixed and the evaluation of fitness function is not pre-specified.

2. Network exhibits emergence

We say that a network N of interrelated components in CS exhibit an emergent process E with emergent properties P if and only if, Thompson (2007):

(i) *Nonlinear dynamics*: E is a global process that instantiates P, and arises from the coupling of N's components and the nonlinear dynamics D of their local interactions, Bar-Yam (1997), Hilborn (2003).

(ii) *Global to local influence*: E and P have a global to local (downward) causation or determinative influence on the dynamics D of the components of N, and possibly:

(iii) *Relational Holism*: E and P are not exhaustively determined by the intrinsic properties of the components of N, that is they exhibit relational holism.

Emergence has similarities to the phase transition encountered in physics, where local changes result in a global change in which new properties emerge abruptly. In particular, under emergence, the many degrees of freedom arising due to its component parts collapse into a fewer new ones with a smaller number of globally relevant parameters. That is the properties of emergence cannot be predetermined and the system evolves on a qualitatively new level with respect to its global constraints. In this sense, emergence implies the failure of the reductionist hypothesis which claims that a total system can be analyzed in terms of the intrinsic properties of the constituent parts.

3. Power-Law Scaling or Scale-Invariance property in emergence
It means that the given functional form F(x) remains unchanged to within a multiplicative factor under the rescaling of the independent variable x. That is, such a property results in the equation:

F(ax) = bF(x), namely invariance under multiplicative changes of scale that results in self- similarity. This property can occur in both space and time for the emergent properties of the CS.

4. Adaptation including fault-tolerance and resilience to damage
Adaptation is a condition that reflects flexibility or capacity to change in relation to changing conditions in a viable way. CS can self-modify their past behaviour and adapt to environmental changes, available resources, as well as, tolerate failures or non-cooperation of some of their components, in achieving a specified goal.

Classical (Darwinian) evolutionary theory claims that organisms tend to pass on from generation to generation those properties that are more likely to help them adapt to their environment. The organisms carrying out this selection process most efficiently will survive the longest, and that the organisms will tend to find gaps in the natural world in order to evolve in new ways that are not already exhausted by existing organisms. This means that over a period of time, organisms exhibit emergent properties.

Also a mathematical functional or a recursively computable relationship is not derivable between adaptation, and mutation, selection and their interplay. In evolutionary biology, many random mutations and selections take place over a long period of time to adapt to a new environment resulting in the synthesis of new proteins, or stopping of the synthesis, or modification of existing proteins. These result in newer biological functions, such as: sensitivity to ultraviolet radiation, modification of haemoglobin, production of antifreeze, production of venoms, toxins, and loss of visual apparatus in animals living in darkness or evolution of new organs. It is now well established that all these effects arise due to changing chemistry, through mutation of codons, to adapt to a new environment. The set of all possible chemical compounds that can evolve during a long period of time, their physical, structural properties (nature of bonds, their angles, energy functions and reactions) and biological activity properties, due to environmental changes turn out to be noncomputable or nonalgorithmic. Also, the cause- effect relation among them tuns out to be circular, as to whether the mutation was the cause or it is an effect due to other means. Further, most functional properties cannot be predicted in advance, and can only be analysed after they have happened.

In summary, there are good reasons to believe that the above four properties are interrelated to self-organization, stigmergy and self-assembly, as well as, to systems exhibiting positive metric entropy, self-similarity and the small world phenomena. It is hard to separate these concerns as they form fuzzy boundaries, Camazine (2002), Serugendo (2004), Serugendo et al (2006).

Since CS are often highly nonlinear, it is not possible to obtain analytical expressions and study the behaviour of these systems. Thus we need to resort to simulation-based methods to understand the behaviour of such systems. This simulation- based approach can be viewed as "*soft computation*", since unlike in conventional computation (called "hard computation") where exactness is our goal, we allow for the possibility of hybrid (both analog and digital) computation with error and randomness. Soft computation needs to be supported by a suitable choice of a datastructure and an associated programming paradigm.

It is the object of this chapter to describe a Multi-set of Agents Paradigm (MAP) to model, simulate and animate CS and related computational schemes: for example, Genetic algorithms, Goldberg (1989) Holland et al.(1987), Genetic Programming, Koza (1999), Immunocomputing, Gonchorova et al (2003), Stepney er al (2003), de Castro and Timmis (2002), Ishida (2004), Self-organized criticality, Guerin and Kunkle (2004), and Active Walker models (ants with scent or multiwalker-paradigm, where each walker can influence (repel or attract) the other through a shared landscape based on probabilistic selection, Bonabeau et al. (1999), Chu et al.(2003), Dorigo et al.(2002), Dorigo and Stutzle(2004), Meuleau and Dorigo (2002), Kennedy and Eberhart (2001), Kennedy (2006), Pacino (2002), Bio-inspired robotics and computing, Bar-Cohen and Breazeal (2003),Werfel et al.(2005), Boahen (2005), Dorigo et al. (2006), Brownian Agents, Ebeling and Schweitzer (2003), *in silico* integrative biological modeling, Belloquid and Delitala (2006), Mc Cullogh and Huber (2002), Keele and Wray (2005) and brain and cognition modelling, Ivancevic and Ivancevic(2006).

Three crucial properties of Agents, Woolridge (2002), make them suitable for the simulation of CS:

(i) *Autonomy*: Agents can make decisions on actions they want to do and perform these actions deterministically, nondeterministically or probabilistically, without explicit control from the user.

(ii) *Reactive*: Agents can respond appropriately depending upon the context.

(iii) *Proactive*: Agents can act in anticipation of future goals that can meet the specified objectives.

In order to realize the above computational schemes using the Multiset of Agents Paradigm (MAP) we need to consider the following issues:

1. *What kind of mathematical models are needed to study CS and for simulating within the framework of MAP?*

Two major ingredients are involved in modelling CS:

(i) The individual dynamics of a component and
(ii) Information sharing among the components, represented by an interactive network.

Three interrelated mathematical models that have a direct bearing on these two ingredients are:

(i) Fractal and percolation model
(ii) Stochastic/ Chaotic (nonlinear) dynamical model and
(iii) Topological (network) or graph model.

The above models have been very well studied in recent years and are closely related to the properties such as: self-similarity, scale-free, resilience to damage, self-organization and emergence.

2. *Is the emergence of CS analogous to the critical phenomenon in physics or percolation? What are the suitable parameters to describe this phenomenon?*

We discuss this aspect from the point of view of geometric parameters such as: Lyapunov exponents, strange attractors, metric entropy, as well as, topological indices such as, Cluster coefficient, Average degree distribution and the correlation length of the interacting network.

Section 2 describes the fractal and percolation models and the role of positive metric entropy in CS. Sections 3 and 4 describe the multi-set of agents-based paradigm (MAP). In Section 5 we describe the advantages of agent-based modelling, simulation and animation. Section 6 outlines the statistical properties of interacting networks and their role in CS. Section 7 deals with the simulation of Swarm dynamics; this is illustrated by a few examples-ant colony heuristic based on group swarming, bacterial colonies, human-animal trails, and graph-growth. Section 8 briefly considers the engineering of CS, the role of scales, and our inability to simulate nano and molecular levels due to the limitations arising from quantum mechanics. Section 9 describes in brief the currently available agent-tool kits. Section 10 summarizes the results and the directions for research in this area.

2 Fractal and Percolation Models in CS

A Complex System (CS) has an associated data domain or space. In nature, this space is usually the three dimensional space; it is called the geometric dimension of the CS. When the system is placed in an environment, it communicates through its surface area. Since the amount of communication is proportional to the surface area, a simple way to control the communication rate is to choose a suitable geometrical or topological structure that can spontaneously and easily be modified to vary the surface area, Bunde and Havlin (1994), Grimmett (2004) Hilborn(2003), Moon (1999),Torquato (2002), Wolfram (2002). In biology, chemistry, and materials science, where surface phenomena play a crucial role, nature has preferred the evolution of fractal objects. Such a choice can optimize the interaction with the environment and provides for adaptation and the survival.

In heterogeneous chemistry, the structure and geometry of the environment at which the reaction takes place plays an important role. It alone can dictate whether a reaction will take place at all. The geometric parameter is the fractal dimension. In fact environmental interaction can change geometrical features and conversely, geometrical feature modify the interaction; see Chapter on "Fractal Analysis" by Avnir et al. in Bunde and Havlin (1994), Torquato (2002).

To simulate CS, using computing agents, the agents should share knowledge among its peers by altering the pattern of their communication pathways, that is, the topology and geometry of the network, resulting in the change of a fractal dimension. Examples of such systems abound in nature: lungs, flock of birds, ant-swarms, formation of bacterial colonies, Maini and Othmar (2001),cellular signal propagation, Weiss (1996), and Animal- human trails, Schweitzer (2002), Traffic patterns, Nagel (2003).

However, yet another model is required to provide us a tool based on probabilities to compute the connectivity structure among the components in a network arising from cooperative and competitive interactions. This model is the Percolation model, Grimmett (2004) Torquato (2002). The percolation theory is concerned with the study of the diffusion / penetration of certain materials from an environment into the material of system placed in that environment through the boundary or surface of the system. It has been found experimentally that such a penetration is determined by a single parameter p. We call a particular value $p = p(c)$ the "threshold" for percolation, if the pathways are infinite when $p > p(c)$ and the pathways are nonexistent or have limited access within the material of the system when $p < p(c)$. The value p(c) is called the critical point. Also we say that there is a percolation above p(c). The region above p(c) is called the supercritical phase and the region below p(c) is called the subcritical phase. Also when $p = p(c)$ the infinite open cluster or pathway is called an incipient infinite cluster.

Percolation theory is studied using an internal connected structure of the material of the system, either as a lattice or as a connected graph. Random fractals can be highly multiply connected and the topological structure can change dramatically, when a continuously varying parameter increases through a critical value resulting in a simply connected path, Thus self-organized, unforeseen new paths can emerge between points of interest beyond a critical threshold, e.g., nest to the food when communication takes place among ants in an environment, Kauffman (1993).

Also percolation-like cooperative threshold phenomenon arises in cell biology; for example, the sharpness of response of cells increases (decreases) with an increase in the concentration (number) of effector (inhibitor) molecules binding to a target molecule or extra-cellular signalling (a kind of sigmoidal relation), Alberts et al. (2002). Similar cooperative property yielding the sigmoidal relation arises in the binding of oxygen to haemoglobin. Here the configuration of haemoglobin changes as oxygen is bound or released. The oxygen affinity can be varied through alterations of the binding constants thus providing for evolutionary microadaptation of closely related species living at different altitudes for different oxygen levels.

2.1 Role of Positive Metric Entropy

The properties of CS mentioned in Section 1 and the requirement mentioned earlier in this section, having a readily modifiable connectivity pattern among its components with a fractal structure, belong to those of both the computational and the nonlinear dynamical systems. The fractal communication structure is very sensitive to small parameter changes and can lead to an abrupt change in behaviour, e.g bifurcation. Fractal dimension, metric entropy, and Lyapunov exponent serve as important invariants to quantify the behaviour of such a non-linear system, Falcioni et al. (2003). A system with negative Lyapunov exponents imply stability and a system with the positive Lyapunov exponents imply instability and the presence of positive feedback. Thus positive feedback (cooperativity) in a nonlinear system implies a positive metric entropy. This is analogous to positive thermodynamical entropy export to the environment in a living system leading to far-from equilibrium states, Prigogine (1980), and to self-replication of molecules. Equilibrium Systems have zero entropy and they do not export positive entropy and correspond to nonliving system. In fact, evolution stops if entropy is zero.

Also, a chemical reaction cannot happen spontaneously unless the entropy turns positive. For example, the conversion of ADP(Adenosine Di-Phosphate) to ATP (Adenosine Tri-Phosphate) is not spontaneous. So the conversion requires another reaction to provide energy. This is Glycolysis-conversion of Glucose to the Lactate anion. This is a spontaneous reaction exporting positive entropy, Petrucci et al. (2002). The positive entropy keeps the reaction far away from equilibrium. As far as biological systems are concerned, reactions reaching equilibrium will result only in dead cells, Eigen (1992).

Thus the introduction of positive entropy (due to chaos) or randomness seems essential in any computational and dynamical system to simulate the far-from equilibrium systems, Chan and Tong (2002). Mathematically speaking, such a positive entropy system turns out to be topologically transitive and sensitive to initial conditions, or chaotic. This means, that the system is no longer expressible as a sum of its component parts and the system functions as a whole leading to what is known as "Emergence". In such a case, the algorithmic independence among the subsystems is lost, Chaitin (2003). This reasserts the fact that a complex system lies in between the totally random system and the algorithmic system, Gell-Mann (1994), Crutchfield (2003).

In general, the evolutionary rules of systems with nonpositive entropy (or Lyapunov exponent less than or equal to zero) are predictable (Turing expressible or algorithmic) while such rules are not exactly predictable for positive entropy machines. Systems with a positive metric entropy represent "disordered" machines; their evolutionary rules are in general, Turing non-expressible or nonalgorithmic.

The metric entropy distinguishes the two major classes of machines: ordered (O) and disordered) (D) machines:

1. Ordered or nonpositive Metric Entropy Machines (O)
These are completely structured, Deterministic, Exact behaviour (or Algorithmic) Machines. This class contains, the machines in Chomskian hierarchy:

(i) Finite state machines (regular grammar),
(ii) Push-down stack machine (context-free grammar),
(iii) Linear bounded automata (context sensitive grammar)
(iv) Turing machines that halt
(v) Exactly integrable Hamiltonian flow machines.

These machines are in principle, information loss-less and instruction obeying; their outputs contain all the required information, as dictated by the programs.

2. Positive Metric Entropy or Disordered Machines (D)
Positive entropy machines exhibit various degrees of irregular dynamics, Livi et al. (2003):

(i) Ergodicity: Here the set of points in phase space behave in such a way that the time-average along a trajectory equals the ensemble average over the phase space. The term "Ergodicity" means statistical homogeneity; here, the trajectory starting from any initial state can access all other states in the phase space.
(ii) Mixing: The initial points in the phase space can spread out to cover it in time but at a rate weaker than the exponential (e.g. inverse power of time).
(iii) Bernoullicity, K-flow or Chaos: The trajectories cover the phase space in such a way that the initially neighbouring points separate exponentially and the correlation between two initially neighbouring points decays with time exponentially. It is with this class of irregular motion we define classical chaos. These belong to non-halting or looping Turing computations.
(iv) Far-from--equilibrium systems: Systems exhibiting emergent behaviour - such as Chemical and Biological machines and living systems, Prigogine (1980).

Each of the above properties imply that all the preceding properties hold; that is, within a chaotic region the trajectories are ergodic on the attractor and wander around the desired periodic orbit.

The properties of CS are governed by three important parameters: Metric Entropy, Lyapunov exponents, and Fractal dimension. There is a close relationship between positive Lyapunov exponents (or equivalently the eigenvalues of the interaction matrix) and fractal dimensions. Lyapunov exponents measure how the trajectories move apart and they are measured as bits per second.Thus if the Lyapunov exponent is positive the rate at which information is generated increases exponentially and we need to measure increasing amount of information. If the Lyapunov exponent is negative, we need to measure decreasing amount of information over time as we are moving towards a stable state.

If h(1),h(2),...,h(j) are Lyapunov exponents in ascending order that are greater than zero, and, h(j+1),..h(k) are less than zero in descending order, then we have j stretching and (k-j) contracting directions in the k dimensional trajectories, then the quantity, Lyapunov dimension, Moon (1999), Baker and Blackburn (2005):

$d(L) = j + 1/|h(j+1)| \sum_{i=1}^{j} h(i)$. It is conjectured that d(L) = d(b) the box-counting dimension.

Another dimension that is used is the topological dimension. Here we add up the positive and negative Lyapunov exponents until the sum is just greater than or zero. The largest number j of exponents h(j) added to keep the sum nonnegative is the topological dimension D.

The Kaplan-Yorke dimension D(KY) is then defined by, Sprott (2003):

$$D(KY) = D + 1/|h(j+1)| \sum_{i=1}^{j} h(i),$$

Note that if there are no negative Lyapunov exponents the system is unbounded; if all the exponents sum to zero the system is volume conserving and usually structurally unstable, unless the system is energy conserving. For nondissipative conservative system, the Lyapunov exponents occur as a pair of negative and positive exponents of equal values that sum to zero. The system is then time reversible.

In dissipative systems the sum of Lyapunov exponents is negative with at least one negative Lyapunov exponent; then the state space contracts and collapses onto an attractor.

The fractal dimensions d(L)and D(KY) can change rapidly depending upon h(i) as a function of the network size, the nature of the connectivity structure and the interaction among the agents (e.g. ant and bacterial colony, protein networks Sections 7.1 and 7.2).

Thus, to understand CS quantitatively, we need to know the spatial structure of attractors and the temporal aspects of the trajectories. The former provides information about phase transitions, while the latter tells us whether a trajectory governed by a positive Lyapunov exponent falls in a given attractor. Real systems including biological systems are nonstationary in which the parameters are changing with time. Thus quantitative understanding of a CS is extremely difficult except through simulation. To simulate CS, we need to combine the properties of zero and positive entropy machines; that is we introduce positive entropy through the chaotic (deterministic randomness) or statistical randomness, in appropriate decision or control steps of a deterministic machine. We will discus these aspects in Section 4.2.

3 Agents

We define the following terms:

System: A system is a set of objects together with relationships between objects and their attributes.

Environment: For a given system, the environment is the set of all objects, a change in whose attributes affects the system, and also those objects whose attributes are changed by the behaviour of the system. It is assumed that the system and its environment are separated by a boundary.

Agent: The AOIS (Agent-Oriented System Community) defines an agent as a system that is capable of perceiving events in its environment or representing information about the current state of affairs and of acting in its environment guided by perceptions and stored information, Woolridge (2002). Since the term "Agent" is used in a variety of ways in Artificial intelligence and social sciences, we will restrict ourselves to the definition that the agent is a software module having the transactional programming model, Krishnamurthy and Murthy (1992), with all the important features-such as: atomicity of commitment, short duration transactions, sensing and aborting, and its own thread of control. Further, we will not use psychological terms such as intention and desire, although we will be using the term "revision of belief " for rewriting (updating or deleting) existing information (or knowledge) in its memory state.

Also we define the following terms in the structure of an agent, see Figure 1. This model is similar to the evolutionary biology model for the developmental pathways, Wilkins (2002), where environmental cues and sensory processing lead to morphogenesis (the process of development that lead to the creation of particular shapes and forms):

(1) Worldly states or environment U: All those states which completely describe the universe containing all the agents (called the agent space).

(2) Percept: Depending upon the sensory capabilities (input interface to the universe or environment) an agent can receive from U an input T (a standard set of messages), using a sensory function Perception (PERCEPT): PERCEPT :U $\rightarrow$ T.

PERCEPT can involve various types of perception, e.g., see, read, hear, smell. The messages are assumed to be of standard types based on an interaction language that is interpreted identically by all agents. Since U includes both the environment and other agents the input can be either from the agents directly or from the environment that has been modified by other agents. This assumption permits us to deal with agents that can communicate directly, as well as, indirectly through the environment as in active walker model; this is called "stigmergy" in Ant colony, where one ant can modify its environment and affect the behaviour of another ant (see also EFFECT).

(3) States of Mind M (D,P): The agent has a mind M (essentially a problem domain knowledge consisting of an internal database D for the problem domain data and a set of problem domain rules P) that can be clearly understood by the agent without involving any sensory function. Here, D is a set of beliefs about objects, their attributes and relationships stored as an internal database and P is a set of rules expressed as preconditions and consequences (conditions and actions). When T is input, if the conditions given in the left-hand side of P match T, the elements that correspond to the right-hand side are taken from D, and suitable actions are carried out locally (internally in M) as well as externally on the environment.

In all our discussions here, we use the computer science terminology "*nondeterministic choice*" to mean that the choice is made arbitrarily without any a priori probability assignment, Krishnamurthy (1989).

The nature of internal production rules P, their mode of application and the action set determines whether an agent is deterministic, nondeterministic, probabilistic or fuzzy.

Rule application policy in a production system P can be modified by:

(1) Assigning probabilities/fuzziness for applying the rule
(2) Assigning strength to each rule by using a measure of its past success
(3) Introducing a support for each rule by using a measure of its likely relevance to the current situation.

The above three factors provide for competition and cooperation among the different rules. Such a model is useful for many applications, e.g., Active-walker, Self-organization and swarm models, Chemotaxis. Accordingly, we assume that each agent can carry out other basic computations, such as having memory, simple addition capability, comparison, simple control rules and the generation of random numbers, Pacino (2002). These mechanisms enable us to simulate and animate tumbling, as well as running of organisms for foraging.

(4) Organizational Knowledge (O): Since each agent needs to communicate with the external world or other agents, we assume that O contains all the information about the relationships among the different agents, e.g., the connectivity relationship for communication, the data dependencies between agents, interference among agents with respect to rules and information about the location of different domain rules.

(5) INTRAN: On the receipt of T, the action in the agent M is suitably revised or updated by the function called Internal transaction (INTRAN).

Revision: Revision means acquisition of new information about the environment, that requires a change in the rule system P. This may result in changes in the database D. In a more general sense, revision may be called "mutation' of the agent since the agent exhibits a mutation in behaviour due to change of rules or code.

Example: The inclusion of a new tax-rule in the Tax system.

Update: Update means adding new entries to the database D; the rules P **are not changed**. In a more general sense, the update may be called "Reconfiguration" since the agent's behaviour is altered to accommodate a change in the data.

Example: Inclusion of a new tax-payer in the Tax system. Both revision and update can be denoted in set-theoretic notation by:

INTRAN: $M \times T \rightarrow M(D,P)$

Both mutation and reconfiguration play important roles in Nature-inspired computing. These are achieved in our model by introducing changes in D and P as required. This can be interpreted as updating or revising a set of database instances. Hence, if one or several interaction conditions hold for several non-disjoint subsets of objects in the agent at the same time, the choice made among them can be nondeterministic or probabilistic. This leads to *competitive parallelism*. The actions on the chosen subset are executed atomically and committed. In other words, the chosen subset undergoes an 'asynchronous atomic update'. This ensures that the process of matching and the follow-up actions satisfy the four important ACID properties: Atomicity (indivisibility and either all or no actions or carried out), Consistency (before and after the execution of a transaction), Isolation (no interference among the actions), Durability (no failure). Once all the actions are carried out and committed the next set of conditions are considered.

As a result of the actions followed by commitment, we may revise or update and obtain a new database within each agent; this may satisfy new conditions of the text and the actions are repeated by initiating a new set of transactions. These set of transformations halt when there are no more transactions executable or the agent space does not undergo a change for two consecutive steps indicating a new consistent state.

However, if the interaction condition holds for several disjoint subsets of elements within an agent at the same time, the actions can take place independently and simultaneously. This leads to *cooperative parallelism*; e.g. vector parallelism, pipeline parallelism.

(6) EXTRAN: External action is defined as an external transaction (EXTRAN) that maps a state of mind and a partition from an external state into an action performed by the agent. That is:

EXTRAN: M X T $\rightarrow$ A; it means the current state of mind and a new input activates an external action from the action set A.

(7) EFFECT: The agent also can affect U by performing an action from a set of actions A (ask, tell, hear, read, write, speak, send, smell, taste, receive, silent), or more complex actions. Such actions are carried out according to a particular agent's role and governed by an etiquette called protocols. The effect of these actions is defined by a function EFFECT that modifies the world states through the actions of an agent:

EFFECT: A X U $\rightarrow$ U;

EFFECT can involve additions, deletions and modifications to U. Thus an agent is defined by a set of nine entities, a 9-tuple:

(U,T,M(D,P),O,A,PERCEPT,INTRAN,EXTRAN,EFFECT).

The interpreter repeatedly executes the rules in P until no rule can be fired.

Thus, we can interpret all the abstract machine models (finite state machines, state charts, Turing machines) and concurrent computational models (Petri nets, Transition networks, Deutch et al (2007)) as subclass of the agents, by a suitable choice of the entities in the 9-tuple.

3.1 Interacting Multiagent System

A multi-agent system is realised as a loosely coupled network of single agents shown in Figure 1; they interact among themselves and through the environment to solve a problem. Thus if N agents are involved each of the agents will be

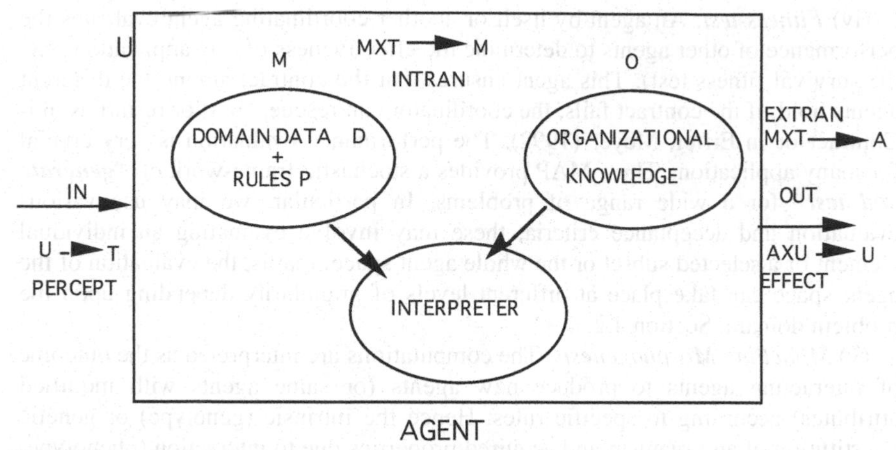

Fig. 1 Structure of an Agent

denoted with a label (i), i = 1,2, , , N. Operationally, the Multiagent system is *"structurally similar"* to the operations occurring within a single agent except that, instead of operating on objects in the object space wthin an agent, now the operations take place among different agents in the agent space through EXTRAN (Section 3). Thus in the simulation of a CS the agents serve as components, while the multiagent system serves as a macroscopic system. This scheme can be iterated as a hierarchy, if required. Thus, we can simulate evolutionary models from micro state to intermediate mesostates and to the macrostates, guided by selection and genetic diversification, exhibiting emergence at different levels of the hierarchy.

The multiset of agents paradigm (MAP) consists of the following features:

(i) *Active datastructure*: A multiset M that contains evolving agents whose information is structured in an appropriate way to suit the problem at hand. A multiset or a bag is a collection of objects in which a member can have multiplc occurrences. Conventionally, the elements of the multiset are passive datastructures. Here each element of the multiset is an agent. Thus apart from the advantages resulting from the use of intelligent agents as elements, we have a strong theoretical background from the abstract rewriting systems on multisets, Suzuki et al. (2001).

(ii) *Context dependence*: A set of interaction rules that prescribes the context for the applicability of the rules to the agents. Each rule consists of a left-hand side (a pattern or property or attribute) describing the conditions under which the agents can communicate and interact, and a right hand side describes the actions to be performed by the agents, if the rule becomes applicable, based on deterministic, nondeterministic, fuzzy or probabilistic criteria.

(iii) *Rule-interference, conflicts*: A control strategy that specifies the manner in which the agents will be chosen and interaction rules will be applied, the kinetics of the rule- interference (inhibition, activation, diffusion, chemotaxis) and a way of resolving conflicts that may arise when several rules match at once.

(iv) *Fitness test*: An agent by itself or another coordinating agent evaluates the performance of other agents to determine the effectiveness of rule application (e.g. the survival fitness test). This agent ensures that the contract among the different agents hold; if the contract fails, the coordinator can rescue, abort or restart as in i-Contract or in Eiffel, Meyer (1992). The performance evaluation is very crucial for many applications Thus, MAP provides a stochastic frame-work of ***"generate and test"*** for a wide range of problems, In particular, we may use various evaluation and acceptance criteria; these may involve evaluating an individual element or a selected subset or the whole agent space; that is, the evaluation of the agent space can take place at different levels of granularity depending upon the problem domain, Section 4.2.

(v) *Modelling Morphogenesis*: The computations are interpreted as the outcome of interacting agents to produce new agents (or same agents with modified attributes) according to specific rules. Hence the intrinsic (genotype) or genetic constitution of an organism and acquired properties due to interaction (phenotype) with the environment can both be incorporated in the agent space model. Thus this model can be used in the evolution of the developmental pathways, Wilkins (2002), where environmental cues and sensory processing lead to morphogenesis.

(vi) *Cooperation and competition*: Since the interaction rules are inherently parallel, any number of actions can be performed ***cooperatively or competitively*** among the subsets of agents, so that the new agents evolve toward an equilibrium or unstable or chaotic state.

(vii) *Content-based activation*: The next set of rules to be invoked is determined solely by the contents of the agent-space, as in the context of chemical reactions.

(viii) *Pattern matching*: Search takes place to bind the variables in such a way to satisfy the left hand side of the rule. It is this characteristic of pattern (or attribute) matching that gives the agent-based paradigm its distinctive capabilities for nature-inspired computing.

(ix) *Choice of objects, topology, and actions*: We can choose several types of objects, as the basic elements of computation to perform suitable actions on them by defining a suitable topology, geometry or a metric space.

In order to use the multi-agent paradigm to realise cooperative and competitive computational tasks, we need to consider how the agents can interfere with each other. These interference rules are similar to those enunciated by Turing (1952) to describe the development of shape, form and pattern in organisms (chemical morphogenesis rules), Wilkins(2002). Such interferences can take place in four ways.

1. Enabling dependence (ED): Agent A(i) and agent A(j) are called enable dependent (or dataflow dependent) if the messages from A (i) creates the required precondition in A(j) to carry out a specific action.

2. Inhibit dependence (ID): Agents A (i) and A (j) are called inhibit dependent, if the actions of A (i) creates the required precondition in A(j) to prevent it from executing a specific action.

3. INTRAN Conflict (IC): Agents A (i) and A (j) are opposition dependent (also called data-output dependent) through A(k)), if the order in which A (i) and

A (j) enable A(k) and update A(k) produce different results in A(k); that is the objects A(i) and A (j) perform operations on A(k) that are not order reversible. That is, local serializability is not ensured in the INTRAN within A(k), if the actions are carried out within an agent in different order.

4. EXTRAN Conflict (EC): Agents A (i) and A(j) are data antidependent through A(k) if the order in which A(i) enables (inhibits) A(k), and A(j) enables (inhibits) A(k) result in different external actions (EXTRAN) by A(k) on the environment. That is the order in which information arrives from the environment and other agents, affects the global serializability of the actions of an agent.

Note that various types of Petri net models, Krishnamurthy (1989), are used currently in Systems Biology, Cardelli (2005), Pinney et al. (2003), Effroni et al (2005), Harel (2003), Aviv and Shapiro (2002). The agent models are more general and provide greater flexibility and also help in animation.

Remark: ED and ID:
The two properties ED and ID are crucial for modelling any life-like system and in modelling the evolution of developmental pathways, Wilkins(2002), where both positive and negative regulation and molecular switches are needed.

These rules permit an agent to enable itself (autocrine signalling for development or for autocatalysis),or inhibit itself, and also an agent A(i) to enable A(j) and A (j) to enable A(i) cyclically.

For example, A(i) can create the required precondition in A(k), so that A(j) can enable A(k). Also, A(i) can inhibit the required precondition in A(k) so that A(j) is prevented from enabling A(k).

Example: In cellular signal processing proteins whose conformation can be altered by the equilibrium binding of a regulatory ligand, switches between active and inactive states when the concentration of the regulator varies. Also situations such as- two ligands whose binding sites in a protein are coupled can reciprocally affect each others binding or proteins that can simultaneously bind two ligands - can be modelled by ED and ID, Graham and Duke (2005).

3.2 Concurrency and Conflicts

In distributed computing and transaction processing: we require that the following two conditions are satisfied for global serialization when concurrent operations take place.

1. At each agent the actions in local actions are performed in the non-conflicting order (**Local serializability**).

2. At each agent the serialization order of the tasks dictated by every other agent is not violated. That is, for each pair of conflicting actions among transactions p and q, an action of p precedes an action of q in any local schedule, if and only if, the preconditions required for p do not conflict with those preconditions required for execution of the action q in the required ordering of **all tasks in all agents (Global serializability)**.

The above two conditions require that the preconditions for actions in different agents A(i) and A(j) do not interfere or cause conflicts. These conditions are necessary for the stabilization of the multi-agent systems so that the computations are locally and globally consistent.

Termination: For the termination of agent –based program, the interaction among the agents must come to a halt. When the entire set of agents halt we have an equilibrium state (or a fixed point) also called stability while dealing with exact computation in a deterministic system.

Non–termination, instability, multiple equilibria and chaos: These cases arise when the agents continue to interact indefinitely as in chemical oscillations, biological reactions, game theoretic problems, and cellular signal processing. It is also possible that the evolution of the agent system is sensitive to initial conditions leading to chaos and self-organization. Then the multiagent-system can have several attractors or equilibrium states.

Conflicts: Resolution or compromise? When agents are used for modelling behaviour, under concurrency, the conflicts arising in INTRAN and EXTRAN may require resolution or to an agreeable compromise. This situation can arise in animation and robotics. For example, the actions, “fold your right arm” or “stretch your right arm” are conflicting concurrent actions. We can achieve a compromise by keeping the arm folded half –way. Alternatively, one can blend the behaviour of actions, e.g., as in the actions “walk” and “run”, if the quantitative parameters can be suitably averaged over. These rules should be based on the context.

4 Multi-Agent Programming Paradigm (MAP)

As mentioned earlier, in all our discussions in this chapter, we use the computer science terminology *“nondeterministic choice”*” to mean that the choice is made arbitrarily without any apriori probability assignment, Krishnamurthy (1989).

The skeletal structure of the Multi-agent Programming paradigm (MAP) is the program specification:

G(R, A)(S) = If there exists elements a, b, c,.. in an agentspace S such that an interaction rule

R (a, b, c,...) involving agents a, b, c is applicable, then

G(R, A) ((S- {a, b, c,.. }) + A(a, b, c,...)), else S

Here S denotes the initial agent space with components of appropriately chosen data type. This is a multiset or a bag in which a member can have multiple occurrences. The operator - denotes the removal or modification(or annihilation) of the interacted elements; it is the multiset difference; the operator + denotes the insertion (or creation) of new elements after the action A; this is multiset union of appropriately typed components. Note that R is a condition text (or interaction condition that is a boolean) that tests when some of the elements of the agent space S can interact. The function A is the action text that describes the result of this interaction.

In the case of deterministic/ nondeterministic programming R and A are deterministic/ nondeterministic. Testing for R involves a deterministic/ nondeterministic search, and evaluation of truth or falsity of Boolean predicates.

The function R can be interpreted as the query evaluation function in a database S and the function A can be interpreted as the updating function for a set of database instances. Hence, if one or several interaction conditions hold for several non-disjoint subsets of the agents at the same time, the choice made among them can be nondeterministic or probabilistic. This leads to *competitive parallelism.* If the interaction condition holds for several disjoint subsets of agents at the same time, the actions can take place independently and simultaneously. This leads to *cooperative parallelism.*

Then as described in Section 3.1, the actions on the chosen subset of agents are executed atomically and committed. Once all the actions are carried out and committed the next set of conditions are considered. As a result of the actions followed by commitment, we derive a new agent space that may satisfy new conditions of the text and the actions are repeated by initiating a new set of transactions. These set of transformations halt when there are no more transactions executable or the agent space does not undergo a change for two consecutive steps indicating a new consistent state of the multiagent system.

4.1 Deterministic and Nondeterministic Iterative Computation

This consists of applications of rules that consume the interacting elements of the agent space and produce new or modified elements in the multiset. This is essentially *Dijkstra's guarded command program.* It is well-known that the *Guarded command* approach serves as a universal distributed programming paradigm for all conventional algorithms with deterministic or nondeterministic components, Krishnamurthy (1989).

Termination: For the termination of rule application, the interaction conditions R have to be designed so that the elements in the object space can interact only if they are in opposition to the required termination condition. When the entire elements meet the termination condition, the rules are not applicable and the computation halts leaving the object space in an equilibrium state (or a fixed point).

Non–termination, instability, chaos: These cases arise when the rules continue to fire indefinitely as in chemical oscillations. Then the object space can be in a non-equilibrium state. It is also possible that the evolution leads to instability, chaos of the deterministic iterative dynamics as well as self-organization, Sprott (2003).

Examples

(i) Chaotic System

Consider the rule-based iterative dynamical system:

For X(0) in the range [-1,1], if X(i) $\geq$ 0 then G(X(i+1)) = -2X(i) +1 else G(x(i+1)) = 2X(i)+1.

The rules $X(i) \geq 0$ and $X(i) < 0$, are deterministic,mutually exclusive and non-competitive; they generate a chaotic dynamical system, unstable, having a dense orbit in [-1,1].

(ii) Bak's Sand Pile

A typical self organizing system is Bak's sand pile. Here the agents are arranged in a square lattice of ((N+2)X(N+2)) cells and a 4-neighbourhood (column and row neighbours) of each cell (i,j) with $1 \leq i,j \leq N$ and include a boundary of cells A(0,j), A(i,0), A(N+1,j), A(i,N+1) that surround NXN cells. The boundary agents are set to value zero.

To start with, each agent has an arbitrarily or randomly assigned value $1 \leq A(i,j) \leq 3$.

At each iterative step k, a randomly or nondeterministically chosen agent A(i,j) is updated asynchronously by adding unity:

That is $A(k+1,i,j) := A(k,i,j) + 1$.

Then for any agent whose value $A(i,j) > 3$, the contents of the nearest neighbours are updated using the assignment rules:

$A(k+1,i,j) := A(k,i,j) - 4$; $A(k+1,i\pm1,j) := A(k,i\pm1,j) + 1$; $A(k+1,i,j\pm1) := A(k,i,j\pm1) + 1$

These actions amount to unloading the sand grains to neighbouring agents whenever the slope gets larger. The above rules are reiterated until no agent contains a value greater than 3.

For a large matrix N one can simulate sand pile using colours and observe the Self-organization.

4.2 Stochastic MAP

In Section 2.1 we mentioned that the introduction of chaos (positive entropy) or randomness is essential in any computational and dynamical system to model the far-from equilibrium systems. While it is still not known whether chaotic or stochastic approach is superior in computational performance, the stochastic methods seems to be more easily amenable for proof techniques and seems to be more robust under dynamic noise. This is a major research area, Chan and Tong (2002).

The introduction of randomness in a rule-based system has several advantages:

(i) It provides ergodicity of search orbits. This property ensures that searching is done through all possible states of the solution space since there is a finite probability that an individual can reach any point in problem space with one jump.

(ii) It provides solution discovery capabilities (as in genetic programming) and enables us to seek a global optimum rather than a local optimum.

(iii) It cuts down the average running time of an otherwise worst–case running time-algorithm. We achieve this gain by producing an output having an error with a small probability.

(iv) Applicable to problems in many discipline.

The stochastic MAP is obtained by introducing probabilities for selection when one or more reaction conditions hold for several non-disjoint subsets at the same time. In this case the choice made among these subsets is determined by a random number generator that selects the ith possible subset with a probability p(i) to perform the required actions - thus providing for probabilistic competition among the different choices. This results in the probabilistic MAP and is defined by the function:

PG (R(p(i), A) (S) = if there exists elements a, b, c,. .. belonging to an agent space S (a multiset) such that R(a, b, c,...) then G(R,A)((S-{a, b, c,.. }) + A(a, b, c,..)) else S

where each of the possible number of subsets i that satisfy the conditions R is chosen with a probability p(i) and the corresponding text of action A is implemented and the components of the multiset are updated appropriately. Note that the sum of p(i) equals 1. Also when p(i) is absent the choice can be deterministic or nondeterministic. Thus a general MAP can contain within itself the deterministic, nondeterministic and probabilistic subprograms.

The implementation of MAP consists of the following four basic steps:

Step 0: Initialization: Initializing the multiset of agents representing the problem domain.

Step 1: Search (sampling): Deterministic or random searching for the candidate agents that satisfy a given rule (interaction condition) exactly or within a probabilistic bound.

Step 2: Rule Application: Carrying out the appropriate actions on these chosen the elements as dictated by the given rule.

Step 3: Stopping (Evaluation and acceptance): It is a common practice in probabilistic method, not to explicitly state a stopping criterion. A key reason for this is that the convergence theory can provide only asymptotic estimates, as the number of iterations goes to infinity. Therefore, we choose a suitable stopping criterion for the given problem by using suitable evaluation and acceptance criteria-otherwise, we may be wasting the resources.

In step 3, we may use various acceptance criteria; these may involve evaluating an individual element or a selected subset or the whole object space; that is, the evaluation of the object space can take place at different levels of granularity depending upon the problem domain. Also, the acceptance criteria may be chosen dependent or independent of the number of previous trials and the choice of probabilities can remain static or can vary dynamically with each trial. Thus depending upon the evaluation granularity, acceptance criteria and the manner in which the probability assignments are made, we can devise various strategies by suitably modifying the skeletal structure of MAP. For example one may choose a reaction rule from a rule-base probabilistically or vary the frequency of application of competing rules. Also one may carry out any operation probabilistically.

The stochastic MAP is useful for simulation of many problems:

1. It can be used to realise algorithms- such as classifier systems, probabilistic, bucket brigade learning, the genetic algorithms (see Belew and Forest (1988);

Booker et al., (1986); Forrest (1991a), (1991b); Goldberg (1989); self-organized criticality and active walker models (ants with scent or multiwalker-paradigm where each walker can influence the other through a shared landscape based on nondeterministic or probabilistic action, Bonabeau et al.(1999), Dorigo et al. (2002), Kennedy and Eberhart (2001), Lam (1998), Spall (2003). Also it is useful in distributed exploration of a space by autonomous robots when the probabilities are based on geometric relationship among the location of the robots, Parker et al.(2000).

2. MAP provides a suitable model for understanding a large class of evolutionary events. Such a model is applicable to very wide areas in biological and social systems which are characterized by different kinds of attractors that belong to four classes, called Wolfram classes, Wolfram (2002), Hofstadter (1995).

a. Evolution to a fixed homogeneous state in living systems (limit points in dynamical systems) corresponding to fixed points in programming.

b. Evolution to simple separated periodic structures in living systems (or limit cycles in dynamical systems) corresponding to competitive cycles of deadlock or livelock in concurrent computation.

c. Evolution to a chaotic behaviour yielding aperiodic patterns in living systems (strange attractors in dynamical systems) that has no correspondence in computer science.

d. Evolution to complex patterns of localized structures in living systems - has no analog in dynamical systems or in computer science.

In addition, phase transitions can arise between the various classes. Thus one proceeds through a complexity hierarchy from simple to complex up to the transition region and beyond that complex to simple dynamics. The phase transition therefore separates the space of computation into an ordered and a disordered regime- which can be thought to be in correspondence to halting and nonhalting computations.

3. The multiset datastructure used in MAP is suitable to describe physical events. It can represent point-like variables in Physics (time, space, velocity or other quantity) or discreteness of events intrinsic to the physical processes (intrinsic point processes) or arising out of observations (observational point processes), Milne, (2001), Scargle and Babu (2003). Also it can represent iterative dynamical systems, including Cellular Automata,Ilachinski (2002), Wolfram (2002), Hofstadter(1995), and Evolving Networks, Dorgovtsev and Mendes (2003) and transition networks, Noe and Smith (2007).

Example

Ising model

The Ising model is simulated in manner analogous to Bak's sand pile we described, except that we now use a different set of rules for update and a probabilistic action in acceptance. Here we take a square grid of cells each containing an agent representing a spin value $\pm$ 1. It is assumed that the periodic boundary conditions correspond to mapping around a torus. Initially the spins are chosen at random and each agent is updated asynchronously. The energy is a minimum when all the agents have contents with the same signs; this can happen only at temperature zero.

Assuming that the total energy is given by the sum of energies of nearest neighbours:

$E= - \Sigma S(i) S(j)/2$ where the division by two compensates for double counting.

When the spins in cell i flips, the change in total energy is $\Delta E = -S(i) \Sigma S(j)$; here the sum over takes place over the 4-neighbouring agents. This sum is an integer in the range $-4 \leq \Delta E \leq 4$.

If $\Delta E \leq 0$, accept the change; however, if $\Delta E > 0$ accept the change with the probability

$P= Exp (-\Delta E/T)$ where T is a normalized temperature.

This is implemented by choosing a uniform random number r in the range $0 \leq r \leq 1$ and the action of flipping of the spin takes place probabilistically, only if $r < P$.

The simulation will show that at lower temperature the spins are all nearly aligned and as temperature increases they tend to orient suitably and at the critical trmperature they form a fractal boundary while as temperature increases further, the organization turns nearly random with no perceptible structure.

4.3 Markov Chain Monte-Carlo

The Markov Chain Monte Carlo (MCMC) methods, Robert and Casella (1999), that include Simulated Annealing, Data Augmentation, and Metropolis Hastings type algorithms are used to construct the *a posteriori* probability density function of a random dynamical system. The MAP can realize such Markov chain Monte Carlo methods as follows: the Production rule (condition text R /pi) prescribes the nature of random variable (i.e, probability distribution function) to be used for the selection of the agents of the agent space and also a set of **deterministic criteria** for the acceptance or rejection of the elements (agents). The corresponding action text (A) implements these conditions and accepts or rejects the elements as and when they are randomly generated (on the fly). Following the rejection of elements, the second production rue (condition text) (say R* /p*i) prescribes a **probabilistic criterion** to accept (or reject) some of the earlier rejected elements; the corresponding action text A* implements these conditions. The condition text R* /p*i is then varied as a function of the current number of trials and a parameter called "temperature". The effect of this variation is such that the probability of accepting a new solution that is worse than the current solution decreases with the degree of the deterioration of the solution, and more significantly with the run time of the method.

The recursive (real-time) Markov Chain Monte Carlo (MCMC) algorithm, called particle filters, Doucet et al. (2001), iteratively updates the currently available set of particles into a new set of particles so that the empirical probability distribution of the particles closely follow the true distribution. That is, the simulated evolution of the particles mimics the real system. These are based on randomized grids for propagating the conditional density of the state of a dynamical system given noisy observations. Then sequential importance sampling together with Bayes rule is used to update the weights of the grid points from a prior distribution to a posterior distribution. In order to avoid degeneracy problems, a random re-sampling method is used to eliminate low probability points in the grid. This step mimics evolution in the sense it eliminates most poorly adapted species (selective extinction or death). These steps can be implemented by the MAP. In MCMC methods, the granularity of the evaluation takes place at the elemental level and hence they permit on-the-fly acceptance of elements thereby providing high concurrency in the implementation.

MAP can realize the Multiple Particle Filter approach described by Yuen and MacDonald (2004). It is computationally advantageous, since it splits a multidimensional problem into several lower dimensional ones provided there is sufficient degree of independence among the agents in the estimation problem.

5 Multi-Agent-Based Modeling and Simulation

Multi-Agent-based modelling requires:

a. The agent space- a multi set of agents
b. An environment- the interaction geometry
c. Interaction matrix

In this modelling when we assign agents to objects, their interaction scale becomes crucial: as to whether it is Molecular, Nano, Micro or Macro. Hence agent-based modeling needs to be carefully designed to reflect the true situation. Further, any simulation process (and a model) should not be treated as an end product. They can only be regarded as tools to gain insight rather than a true representative of the real and experimental validation should be followed up, Deutsch et al.(2007).

5.1 Advantages of Agent Based Simulation (ABS)

(i) Agent-based simulation (ABS) starts with simple rules of interaction among the individual components that drive the system to the complex behaviour observed. It works bottom up by examining what low-level rules and what kind of heterogeneous, autonomous agents are required to synthesize the required higher level behaviour.

(ii) ABS can exhibit the emergence of global behaviour of a complex system from the local interactive behaviour of its components; also it can cause global to local influence on components as required.

(iii) ABS enables us to make predictions about the processes occurring at the intermediate mesoscopic scale due to the interplay between the microscopic dynamics and the macroscopic environment.

(iii) Using ABS, we can study a hierarchy of models where iterated application of simple rules can generate complex forms.

5.2 Special Purpose Agents in Simulation

Agents can be designed to realize special purpose functionalities. These functionalities depend on the problem domain. In building cellular biological models based on agents, it will be more convenient to use special purpose agents – such as sensor agents, goal agents, skill agents and compose these modules to build reusable, composable, blendable behaviour; e.g., in cell system biology, the protein networks organize themselves into special purpose, Alberts et al (2002), Cardelli (2005), Wooley and Lin (2005). Agent-based animation can be helpful to understand specific sequence of reactions and the evolution of developmental pathways, Wilkins(2002).

5.3 Agents and Avataras in Animation

Expansion of simulation to animation can be useful for visualization of the dynamics, Szarowicz et al. (2001), Stith (2004). For example, in cell biology the molecules are involved in various interactions with other cells through cell-signalling cascades. Cell signalling cascades consist of several steps of activation and inhibition, Alberts et al. (2002). These are usually illustrated in static form as a series of steps. Animation can introduce each step individually and in order to emphasize multiple effects of one protein and the cellular location of each effect. The animation emphasises the order of events, the binding of ligands and the accompanying reactions and the movements along with the timing of the events. Ideally, such a realization would require a total manifestation of the real object and the events that are occuring in time and the reactions of the associated objects. This can be achieved through the introduction of "Avatara" which is realised as an extension of the Agent model (Figure 1) by including a body with a desired geometry and related functions. The body provides for the visualization part of the events. The (actions) animation of the body is carried out by a scheduler that is controlled by one or more agents. In this sense Avatara provides a visible part of the environment; however, it cannot perceive or react to events on its own.

In reactive animation, the system reacts to various kinds of events, signals and conditions that are often distributed and concurrent. Also the system can contain components that signal each other and are repeatedly created and destroyed. Further, they can be time critical exhibiting both digital and analog (or hybrid) behaviour, Efroni et al. (2005), Harel (2003).

6 Interactive Connectivity Structure among Agents

We mentioned in the introductory section that two major ingredients are involved in modelling CS:

(i) The individual dynamics of a component and

(ii) Information sharing among the components, represented by an interactive network.

Sections 2 and 4 dealt with agents modelling individual components in a CS and their dynamics and the role of randomness or chaos in introducing positive entropy. The second ingredient, namely, the information sharing among agents, results in a very large disordered interconnection network having a positive metric entropy in the state space of all the agents. These disorderly interconnection patterns play a key role for applications to various modeling and simulation aspects of CS. Three important statistical properties of the networks, namely average degree, characteristic path length and cluster coefficient, to be defined below are used as measures to distinguish the regular networks from the three typical disordered networks:

(i) Random networks (ii) Power-law scaling networks and (iii) Small World Networks.

For a detailed study of networks having various statistical and nonlinear dynamical properties having different fractal dimensions and emergent properties, see, Barabasi et al.(1999), Ben Naim et al (2004), Bornholdt and Schuster (2003), Dorogovtsev and Mendes (2003), Newman (2004), Strogatz (2003), Watts (1999).

Let us consider a finite graph G(V,E) where V is the set of n nodes and E the set of edges. Let us assume that the graph is represented as an adjacency matrix A with A(i,j) =1, if there is an edge from node i to node j; and A(i,j)=0, otherwise. We assume A(i,i) = 0, that is no self loops.

The following local and global parameters are derived from adjacency matrix:

(i) *Average degree*: $K = 1/n \sum_{i} k(i)$, and $k(i) = \sum_{j=1}^{n} A(i,j)$,or

k(i) is the degree of node, $0 \le K \le (n-1)$

(ii) The *Characteristic path length* L measures the global property, namely, the average path length of the network.

Given L(i,j) the shortest distance between nodes i and j, L is defined by:

$$L= 2/n(n-1) \sum_{i=1}^{n-1} \sum_{j=i+1}^{n} L(i, j);\ 1 \le L \le (n-1)$$

This is referred to as "global connectivity" in the context of lattice percolation, if we need to only infer whether they are connected or not. Thus the notion of percolation in lattice grid is closely related to the small-world and scale-free networks.

(iii) *The cluster coefficient* C is the average of C(i), where C(i) = Number of E(i) existing between k(i) neighbours of node i / Total number of possible edges [k(i)(k(i)-1)]/2.

$$\text{or, } C=1/n \sum_{i} C(i) \quad \text{Note that } 0 \le C \le 1$$

The above three properties serve as guidelines to roughly classify the three classes of disordered graphs as below.

6.1 Random, Scale-free and Small World Networks

(i) Random Network: In Random network, Cannings and Penman (2003), the degree distribution is a binomial or Poisson distribution in the limit of a large graph. Most of the nodes have the average degree and few nodes have more than average and few nodes have less than the average degree. Also L and C are small in random graphs.

(ii) Scale-free Network: In this network, many nodes have a low degree (few links) and a few nodes have a high degree (many links). The distribution of the degree of the nodes has an unusual fat-tailed form or a power-law scaling property: i.e., P(k) the degree distribution of a network is given by $P(k) = k^{-g}$ where $2 < g < 3$.

This power-law degree distribution or scale-invariant property arises from two kinds of operations on a random graph:

1. *Creating new nodes*: Growth of the graph by adding new nodes into an initial group of nodes as time progresses and
2. *Preferential attachment of Links*: The new nodes created are linked to old nodes, with a probability based on certain dominant properties the old nodes possess, e.g. the nodes having a higher degree (or attractiveness), chemical or physical interaction strength. In each case, the neighbourhood is appropriately defined as a conceptual graph. As the network grows the ratio of well-connected nodes to the number of nodes in the rest of the network remains nearly a constant.

It has been experimentally observed that the biological networks have a range $1 < g < 2.5$ and the social networks have a range $2<g<3$, Chung and Lu (2004), Barabasi et al (1999). Dorogovtsev et al. (2003) prove that the range $2 < g < 3$ is crucial to have the following properties:

(a) Self organization and (b) Resilience against random damage.

Also power-law graphs provide for faster search through the graphs, Adamic et al.(2003).

(iii) Small-world graphs: A graph is called a small-world graph, by Watts (1999), if it exhibits the following two properties (when compared to a random graph of same number of nodes and average degree):

1. *Higher clustering coefficient C closer to unity:* This implies that two nodes are more likely to be adjacent, if they share a common neighbour and
2. *Smaller average distance L between any two nodes:* L scales logarithmically with the number of nodes. This measures a global property.

It results in a compact network- where the average length of the shortest directed path between two randomly chosen nodes is of the order of the logarithm of the size of the graph. This property ensures that communication and interaction between distant neighbours can take place over long ranges. This is called the small world effect. Therefore, in agent based systems, where a very large number of agents are interconnected, the small-world network can arise so as to permit distant neighbours to communicate and interact. In evolutionary context this is useful to propagate superior characteristic quickly through the entire population.

6.2 Assortative and Disassortative Networks

In some networks, the high degree nodes are connected to high degree nodes. These are called *assortative or homophilic networks.* In *disassortative networks* high degree nodes avoid being connected to high degree nodes. These two types of networks are distinguished by using a degree-correlation coefficient that is positive for assortative networks and negative for disassortative networks, Newman (2004). Assortative mixing seems more prevalent among social networks. However, both assortative and disassortative networks seem to occur in biological networks according to the requirement as to whether percolation or stability to fluctuations is needed.

The *assortative networking* results in larger positive Lyapunov exponents (eigenvalues) of the interacting matrix of the dynamical system. As the network size grows *the real part of the eigenvalues scales as a power of the size of the network.* This means the system can quickly become unstable resulting either in the formation of giant components in graph networks exhibiting the percolation phenomenon or collapsing like a *house of cards*. Such a phenomenon occurs in self-organizing critical systems, such as ion channels. Here the sub units of proteins can alter their conformation to permit an ion to enter or exit through an interior hole connecting the inside of the cell and the exterior solution, Mosekilde and Mosekilde (1991), Palla and Vattay (2006).

In *disassortative networking* high degree nodes avoid being connected to high degree nodes and result in a smaller positive Lyapunov exponent (or positive eigenvalues); as the size of the network grows the *real part of the eigenvalues have only a logarithmic dependence on the size*; hence the dynamical fluctuations are not amplified and the system can reach stability more quickly. Maslov et al (2002), (2003) observe that protein interactions happen in this manner.

6.3 Information Sharing among Interacting Agents

In a finite undirected graph, as the number of edges E increase with respect to the number of nodes N, and E/N >0.5 a giant connected component arises in the graph. As E/N increases beyond a threshold value 1, the graph becomes connected and cycles of all length emerge, indicating the presence of strange attractors of various cycle lengths (due to ergodicity) and the presence of chaos, Kauffman (1993). For a directed graph, all the nodes get connected for E/N >1 and self-loops emerge when E/N>2. As the number of edges increase beyond this threshold, the shorter pathways emerge among the nodes and the mean distance among the nodes become shorter. Also, many nodes have fewer links and few nodes have many links indicating the presence of scale-free distribution. Further, the self-loops and multiple loops interlock themselves so that there are minimal pathways among the nodes of the order of the logarithm of the number of nodes.

If we consider the network of N agents, each node i being associated with a variable x(i), the time evolution of the graph can be represented by the equation: $dX/dt = -X+F(X)$ where F is nonlinear.

It is well known that the eigenvalues of the Jacobian matrix at the fixed points play a crucial role in determining the stability of the fixed points. Lyapunov exponents and Lyapunov numbers are related to the eigenvalues and they play an important role respectively in continuous and discrete systems, Sprott (2003). A bounded dynamical system with a positive Lyapunov exponent is chaotic and it tells us the average rate at which predictability is lost. Also it is useful in determining the sensitive dependence on initial conditions, stability of the fixed points, periodic orbits, and the chaotic behaviour. A few differences between eigenvalues and Lyapunov exponents are worth noting: Eigenvalues are computed, usually at an equilibrium point and provide only local information; Lyapunov exponents are averaged over the trajectory and provide global information. While eigenvalues can be complex, Lyapunov exponents are always real numbers.

The Lyapunov exponent h(i) is the natural logarithm (base e) of the Lyapunov number m(i) (or m(i)= exp (h(i)). It measures the average of the log of the absolute values of the n first order derivatives of the local slopes, as measured over the entire attractor or fixed point.

The equilibrium at a point X* is determined by the Lyapunov exponents of the Jacobian that represents the interaction-this is a function of the size of the graph, nature of connectivity and the interaction strength between the agents. As mentioned earlier in section 2.1, the fractal dimension is a function of the Lyapunov exponents or eigenvalues of the Jacobian at the equilibrium points. These can change rapidly as the connectivity and interaction strength (e.g., attractive, repulsive) change.

For continuous systems, governed by differential equations, the Lyapunov exponents h(i) are defined as the real parts of the eigenvalues λ(i) at the equilibrium point. They indicate the rate of contraction or expansion near the equilibrium point when t tends to infinity (in the limit).

For discrete systems governed by difference equations, we use the Lyapunov numbers m(i). Note that m(i) and h(i) provide the same stability information, that is:

$|m(i)| < 1$, **if and only if** $Re[h(i)] < 0$ and $|m(i)| \geq 1$ **if and only if** $Re[h(i)] \geq 0$.

As mentioned in Section 2.1, positive feedback (due to cooperation) implies a positive metric entropy and it implies $|m(i)| \geq 1$ or $Re[\lambda(i)] \geq 0$. Thus if we have the information that $Re[\lambda(i)] < 0$ for the continuous system, it implies that $|m(i)|$ lies within the unit circle for the corresponding discretized version, as long as the time step for discretization is chosen adequately, Parker and Chua (1980).

For a related study of a dynamical system based on graph of interactions, see Jain and Krishna (2003). They relate the autocatalytic cycle formation, stability, self-organization and other evolutionary aspects to the Perron-Frobenius (PF) eigenvectors and eigenvalues of the adjacency matrix of the graph. Also they show that a necessary condition for the formation of autocatalytic cycle is given by, PF eigenvalue ≥ 1; this implies the presence of positive feedback or positive metric entropy. Such cycles nested within one another in a symmetrical and reciprocal way lead to a sustainable system forming the basis of all living systems.

For a more detailed mathematical discussion on the bounds of eigenvalues of adjacency matrix, and relevant applications, see Chung and Lu (2006), Palla and Vattay (2006).We will now provide some examples to clarify the above concepts.

6.4 Examples

(i) Artificial Neural Networks

A communicating ring of N agents (neurons) can simulate an N dimensional iterated map. Assume that each agent i (i=1,2,..N) receives an input from its k nearest neighbours on either side of the ring and use the functional iteration, with n as the index of the iteration:

$X(n+1,i) = \tanh \Sigma\, a(i)\, [X(n,i-j) + X(n,i+j)]$, where Σ denotes the sum taken over $j = 1$ to k.

Here the weights a(i) are symmetric on either side, but can vary along the ring. Typically a(i) can be chosen randomly or nondeterministically in the range $-4 < a(i) < 4$, Sprott (2003).

(ii) Kaneko's coupled map lattice

To simulate this map, we choose (in the above example) in each cell a logistic map $f(x) = ax(1-x)$, weakly coupled with a strength c to its k nearest neighbours on either side of the ring, and generate the iterated functional with n as the iteration index:

$X(n+1,i) = (1- c)f(X(n,i)) + (c/ 2\,k)\, \Sigma\, [f(X(n,i-j) + f(X(n,i+j)]$, where Σ denotes the sum taken over $j = 1$ to k.

If c=0, then the maps behave independently; if c=1, only the second term is present and each agent i is affected only by its neighbours.

(iiii) Metabolic networks

In metabolic networks the number of nodes is of the order of 800 and edges 3600; thus they contain many cycles and have a mean degree 10, node-to node distance 3; they are disassortative.

(iv) Protein Networks

In protein networks, the number of nodes are of the order of 2000, number of edges of the order 2000, the mean degree is 2 and node to node distance is 6.8, and cluster coefficient is 0.07, Newman (2004). The degree distribution satisfies $P(k) = k^{-g}$ with $g = 2.4$ showing that they are scale-free, self-organizing and resilient against damage. Similar results have been reported by Atilgan et al (2004);

namely, the ratio of cluster coefficient of protein networks to random networks is large and the average path length scales logarithmically with the number of nodes.

(v) Protein folding problem, dynamics and stability

The folding of protein to a pattern known as its native state is a computationally complex optimization problem. Numerous papers and books have been written on this subject over thirty years; yet no "final solution" seems to be in sight. The native state is the thermodynamicaly most stable state out of the possible numerous conformational states 2^N, where N is the number of residues. Levinthal paradox, Finkelstein et al (2003), arises from the observation that to reach a stable state from the exponentially large number of states to a native state through exhaustive sampling is impossible due to the time complexity involved. However, this paradox is resolved by Finkelstein et al (2003), based on cooperative transitions that lead to an all-or none behaviour and thermodynamical arguments. They show that a protein needs only a time $\exp(1\pm0.5)N^{2/3}$ nanoseconds, and most reasonable folding time is of the order of only minutes. This has also been verified by simulation by Steinhofel et al (2007).

Also Finkelstein et al (2003) observe that in multidomain proteins, each domain has a compact globular structure and is an independent folding unit; further all the different domains can fold simultaneously. From the graph theory point of view, the residues within a domain are more densely connected than with the residues in different domains, and different domains are sparsely connected. Thus a protein forms a disassortative network. Hence its adjacency matrix nearly assumes a block diagonal form. The dynamical stability of the total system is essentially governed by the stability of each individual domain, and the small positive Lyapunov exponents in each domain are nearly unaffected by its neighbouring domains. Being disassortative, this network becomes quickly stable reaching the attractor of the system, namely, its native state. This is perhaps the reason, why protein networks self-organize into mutidomains,each domain being an autonomous stable unit with a minimal interference from its neighbouring domains, enabling them to fold independently and simultaneously, all of the trajectories converging to their respective folds or attractors. However, computing Lyapunov exponents for folding and unfolding of proteins is extremely challenging due to their large phase space dimensionality and the rate at which information is generated.

An interesting example is the simulation of chaotic dynamics of Abductin, a rubber protein, consisting of the sequence of ten Amino acids: Phe-Gly-Gly-Met-Gly-Gly-Gly-Asn-Ala-Gly, that has a 10 dimensional phase space, Villani (2002). This system exhibits emergence and hence cannot be described in terms of its simple vibrational and rotational movements. It reaches a self-organizing critical state and is chaotic with 6 positive and 4 negative Lyapunov exponents. Villani (2003) has computed the Lyapunov exponents in forward and backward in time. These are respectively:

(0.4815, 0.3959, 0.3161, 0.2346, 0.1392, 0.0190, -0.1171, -0.3131, -0.6109, -1.2779) and (0.5513, 0.4666, 0.3772, 0.2907, 0.1969, 0.0786, -0.0576, -0.2505, -0.5426, -1.1906).

Since the first of these nine exponents sum to nonnegative values, namely: 0.5452 and 1.1106, the Kaplan-Yorke dimension (Section 2.1) for forward and backward in time are:

9+ (0.5452/1.2779)= 9. 42 and 9 + (0.1.1106/1.1906) = 9.93.

This indicates that in dimension 10 the hypervolume is contracting and the attractor has a dimension between 9.42 to 9.93.

Shakhnovich(2006) observes that long range interactions result in two-state cooperativity, while short range forces interactions accelerate the folding of proteins. This is in agreement with our earlier observation that protein networks can be both assortative or disassortative according to stability requirement and exhibit either percolation like transition or stability to local fluctuations.

(vi) Neuronal Networks

The neuronal network develops in three phases:

1. In the first phase the different parts develop according to their own local programs.
2. In the second phase axons and dendrites grow along specific routes setting up a provisional, but orderly network of connections between different parts of the system.
3. In the final phase, the connections are adjusted and refined thro interactions among far-flung components in a way that depends upon the electrical signals that pass between them.

The central challenge of neural development is to explain how the axons and dendrites grow out find their right partners and synapse with them selectively to create a functional network. This growth is very sensitive to diffusible chemotactic factors (as in ant swarms). The different pathways grow out as in the movement of cars in a rush-hour traffic in highway junctions. At the same time the component failures are tolerated. These indicate that the scale –free and small world networks play a role in the design of nervous system and Hebbian networks, Szirtes et al (2001).

7 Swarm Dynamics

Swarming tactics are widely used in nature by ants, flock of birds and in warfare; Edwards (2000), Bonabeau et al. (1999), Dorigo et al (2002). Swarming requires four basic steps: Locate a target, Converge to the target, Attack the target and Disperse from the target. Swarming entities must be capable of *sustainable pulsing* or a swarm must come rapidly and efficiently on a target, disperse and be able to recombine for a new pulse of attack. These steps require the iterative solution of a nonlinear dynamical problem to obtain the required information on the strange attractors and the trajectories that fall into them.

In using swarming as a foraging or battlefield tactic, we need to consider the following aspects: logistics, command and organization, degree of autonomy of agents, nature (tacit or explicit) and amount of communication, and technology used in communication between the agents. In the military context three factors contribute to the success of swarming - Elusiveness thro mobility or concealment, a longer range fire power-stand-off capability, and superior situational awareness (having more information about the location, activity and intent of the enemy).

A multiset of agents that uses inferences, beliefs and computation can evolve into self-organizing swarms. We can use two different forms of communication to enable (connect) or inhibit (disconnect) agents to form interactive networks, exhibiting the properties of small world graphs or scale-free property or their combined properties.

1. Tacit (Indirect) communication: Agents with simple intelligence (e.g.,ants): *Use of markings similar to a chemical gradient or diffusion mechanism or a communication field.* This provides a common spatial resource, where each agent can leave a mark that can be perceived by other agents. This requires minimal amount of memory and communication overhead.

2. Explicit (Direct) communication: Agents with more complex intelligence: *Use of voice, signals, radio resulting in a positive feed-back or nonlinear response to the information available from knowledge other agents may possess (by connecting or disconnecting with other agents at random).* This involves a greater amount of memory and communication overhead. Also this would require that *each agent knows what other agents know, and how much they know measured in a taxonomic scale* so that each agent can have a score about its neighbours to link, de-link and form clusters. This would result in nonlinear iterative schemes among the agents. Here, individual agents are points in space, and change over time is represented as movement of points, representing particles with different properties and the system dynamics is formulated using the rules:

(1) Stepping (or local coupling) rule: The state of each individual agent is updated or revised in many dimensions, in parallel, so that the new state reflects each agent's previous best success.

(2) Landscaping (or global coupling) rule: Each agent assumes a new best value of its state that depends on its past best value and a suitable function of the best values of its interacting neighbours, with a suitably defined neighbourhood topology and geometry.

All agents in the universe or selected chunks are updated using rules (1) and (2).

The above two rules permit us to model Markovian random walks which is independent of the past history of the walk and non-Markovian random walks, dependent upon past history- such as self-avoiding, self-repelling, communicating, and active random-walker models. This can result in various kinds of attractors having fractal dimensions presenting a swarm-like, flock-like, bacterial colony-like appearances depending upon the Jacobian of the mapping, Wolfram (2002), Ben- Naim et al (2004).

7.1 Agents Simulating Group-Swarms

In order to illustrate the simulation procedure, we now describe a multiset of agents based self-organizing heuristic that uses inferences, beliefs and computation. We assume that the number of agents is very large to exhibit the statistical features. The agents use randomization in the application of the rules and interconnectivity among them that can alter the important neighbourhood properties; these result in different kinds of disordered networks with varying fractal dimensions (and hence Lyapunov exponents).

The simulation of group-swarms for solving problems has the following features:

1. The basic principle in swarm intelligence is cooperation and knowledge sharing. The multiset of agents approach can realise both exploration and exploitation by grouping the particles in different zones of the search space to set up cooperation and competition among members of a same group or different groups. This will permit two types of strategies: (a) competition, self or peer-evaluation through communication among groups and extinction of unfit groups; or (b) competition, external evaluation and exclusion of unfit groups Thus, the creation of different groups permits the use of evolutionary paradigm to annihilate the unfit populations and create fit populations.

2. Using different grouping prevents possible quasi-ergodic behaviour that can arise resulting in the entrapment of the orbit in isolated regions, since the grouping amounts to using several Markov chains, with the initial states reasonably apart, in the search space. Besides, it simulates in parallel the act of sustainable pulsing (repeated attacks).

3. Different types of interaction topologies (such as the wheel, the ring, the star and other lattice structures), Kennedy (2006) can be assigned within and among different groups and their influence can be studied. For example, we can implement Evolutionary / Genetic Algorithms in special type of graph structures (Complete, Petersen,Torus etc)], scale-free graphs and small-world graphs on genetic algorithms.

4. We can optimize several objectives. If many different equally likely independent modes m are to be searched for, it is more efficient to choose a multiset of agents with m sets, each with k agents so that mk = N where N is the total number of agents; i.e., the agents are partitioned into distinct groups each with their own speciality, Coello et al. (2002), Yang et al.(2007).

5. The agent model, permits the simulation of a shared landscape through a blackboard; self-avoiding walks can be simulated using a table that stores the locations visited earlier.

The Ant Heuristics, Dorigo et al (2002),Chu et al. (2003), Blackwell and Branke (2004),Meleau and Dorigo (2002) are based on the model of real ants finding optimal solutions to the shortest path between their nests and the food sources. Each ant leaves a trail of pheromone, thereby creating a new landscape so that other ants are attracted towards the food source by the scent. This problem is analogous to modifying a random graph that grows with time with preferential attachment, Barabasi et al (1999) to those strong pheromone nodes leading to the food source from the nest. The optimal path turns out to be the path with a maximum density of the scent allowing for evaporation of the scent with time and degradation of the scent with distance. The communication among the ants take place through a shared landscape, by establishing links with the other agent nodes having the same belief, so as to form clusters.

To start with, the agents initialize their beliefs, by randomly linking for information. Then with time they update their beliefs through preferential linking to nodes with maximal pheromone intensity and forming local groups of nearly equal fitness, thus forming an assortative network. Then they modify the landscape further, by contacting other groups to obtain collective intelligence, and reach an equilibrium state resulting in an optimal path from the nest to the food source.

Use of multi-swarms rather than a single swarm sets up competition within a swarm and between different swarms. It results in a more efficient heuristic in which preferential attachment happens to high degree nodes, leading to "power-law graphs" with "scale free distribution" of degrees. As shown by Adamic et al. (2003) creating a power-law graph results in faster, more efficient search.

As an illustrative example, we now give a formalized agent-based heuristic for a shortest path is given below:

Step 1:
Generate a multiset of agents each agent representing an ant. Let there be G sets contained in the multiset, each set containing identical agents N(j) for j = 0,1,2,..,(G-1).

Randomly select an initial node r for each agent. The initial pheromone level on every edge of the graph is set to a small positive quantity f(0) in the blackboard.

Step 2: Initialization of beliefs:
Let J (i,j,r) is the set of cities that remain to be visited by the i th agent in the j th group starting from node r; let f (j,r,s) is the pheromone level between node r and node s, for the j th group.

Let v(r,s) = 1/D(r,s) where D(r.s) is the distance between the nodes r and s; let x be a parameter which determines the relative degradation of pheromone level inversely with the distance.

Stepping Rule: Starting from r, calculate the next visited node s, for the i th agent in the j th group as per the rule:

(i) To begin with choose q, a random number in $0 < q < 1$; let q(0) be a constant chosen in the interval between 0 and 1.
(ii) **If** $q \leq q(0)$, **then**

$$s = \text{Arg Max}_{u \in J(i,j,r)} [f(j,r,u)].[v(r,u)]^{x},$$

else P(i,j,r,s) .

Here P(i,j,r,s) is the transition probability from node r to node s for the i th agent in the j th group, given by:

If $s \in J(i,j,r)$, **then** $f(j,r,s).(v(r,u))^{x} / \Sigma_{\{u \in J(k,r)\}} [f(j,r,u)].[v(r,u)]^{x}$,
else 0;

Updating the beliefs from Intra-group knowledge:

Step 3: Local Landscaping Rule:
Update the pheromone level between nodes for each group using:

$$f(j,r,s) = (1-y).f(j,r,s) + y\,\Delta f(r,s) \text{ where } \Delta f(r,s) = f(0) = 1/[nL(n,n)]$$

where f(j,r,s) is the pheromone level between nodes r and s for the agents in the j th group. and L(n,n) is the approximate distance of the route between all nodes. Here n is the number of nodes and $0<y<1$ is the decay parameter of pheromone.

Continue Stepping and local landscaping until each agent in each group completes the route. In this manner, each agent updates its beliefs and communicates the beliefs to concerned agents.

Step 4: Evaluation:
Calculate the total length travelled by each agent in each group.

Step 5: Global Landscaping Rule:
Let L(j) is the shortest length for the agents in the j th group and z ia pheromone parameter.

Update the pheromone level between the nodes for each group using the information on the best available route of the j th group thus:
$f(j,r,s) = (1-z).f(j,r,s) + z. \Delta f(j, r,s)$
where we set
$\Delta f(j, r,s)$ = **If** (r,s) belongs to the best route of the j th group, **then** (1/L(j)
else ZERO.

Step 6: Updating belief from Inter-group Knowledge:
Let L(best) be the best length among the best route of all the groups, i.e., L(best) < L(j), for all j, j = 0,1,2,...,(G-1)

Reassign
$f(j,r,s) = f(j,r,s) + w. \Delta f(best, r,s)$
where: $\Delta f(best, r,s)$= **if** (r,s) belongs to the best route of all groups **then** 1/ L(best),
else ZERO.

Step 7: Stability:
Repeat steps 2 to 6 **until the system reaches stability,** when **almost all** the agents travel the same route.

Simulation Results

The above heuristic was simulated with a multiset of agents consisting of 100 groups each with 500 agents and the performance studied. In comparison to using a single group of 50000 agents, the multi-group performance was markedly better. Also it turns out that the many different equally likely independent modes m are to searched for, and there are N agents, it seems to be more efficient to choose a multiset each with k elements such that mk = N or k =N/m. If there are correlations between different modes, it will be helpful to use this knowledge with advantage. Such a correlation leads to selective linking or assortative or homophilic mixing.

Ant colony optimization algorithm has been successfully applied for finding longest common subsequences, Shyu and Tsai (2009), and to the protein folding problem, Shmygelska et al. (2005). Shmygelska et al (2005) also observe that use of multiple colonies improve the performance of their algorithm.

Since the scale-free networks and small world networks exhibit self similarity (in the sense their highly connected components themselves belong respectively to the same class), the hierarchical decomposition arising due to multiswarming results in each group linking to another group through a short path length.

In this set-up, if some of the individual nodes or ants are randomly destroyed, the heuristic is not affected at all exhibiting the resilience of the scale-free distribution under random failure. However, if some of the hubs of well-connected group of nodes are selectively destroyed, the structure of the swarm turns totally erratic, in a manner analogous to the disruption of air-services that occur in a major airport-hub (e.g., Chicago) due to a severe snow storm, and collapses like a house of cards, being an assortative network. Also if the food source is absent the swarm disperse rapidly; that is, if preys are extinct, the fitness of predators change rapidly.

Also in this heuristic, since the initial tendency is to form local links, this results in higher clustering and if many nodes find the same nodes to link with no degree restriction, we obtain a correlation length smaller than the random network.

Further, if (i) if some agents are assumed to be aging, non-cooperative and they are allowed to die, and (ii) we limit the number of links to each agent by a cost factor, then the scale-free distribution is markedly affected. Our estimate of the box-counting fractal dimension of the swarm lies between 1.9 and 2.1 (resembling Diffusion limited aggregation, Grimmett (2004),Torquato (2002) This value agrees with the expected value the power law exponent for biological networks, Newman (2004).

Simulation shows that the swarm network topology is very sensitive to the nature of interaction and threshold values, cost and aging of nodes. The ant swarms seem to self-organize into assortative or disassortative networks according to the task to be carried out, namely, locating a target, converging to the target, attacking the target and dispersing from the target, Deneubourg, et al. (2005).

The above heuristic can be improved in several ways. We can use different scent types, different scent strengths, use special purpose agents for exploration and trail selection and allow non-cooperative and lazy agents to die. These depend upon the level of intelligence and nature of the agents. Also, as suggested by Vaughan et al. (2000), we can improve the trail laying in the heuristic by using the waggle dance approach of bees that requires a much higher communication and memory overhead. This will have application in cooperative robotics.

The agent- based approach of randomly communicating with other agents to revise belief based on new information available, by preferential linking locally in a group and globally among different groups, form the basis for evolutionary selection and adaptation. The principle of group-swarming applies in foraging, breeding and cell-assembly. It can be extended to simulate adaptive evolution (molecular evolution and phylogenetic tree evolution) by identifying tree hierarchy with agent hierarchy, where each subtree is an agent,Koza (1999).

Group-swarming approach will be useful to understand the pattern formation - (both micro (local) and macro (global)- that result in multifractals; e.g., bacterial colonies, human-animal trails, crowd patterns and biological patterns. In living organisms these patterns arise due to the intrinsic (genotype) or genetic constitution of an organism and the acquired properties (phenotype) due to interaction with the environment, e.g., sex determination in turtles and crocodiles, caste determination in eusocial insects (ants, bees, wasps), and other morphogenetic events, Wilkins (2002).

7.2 Simulating Bacterial Colony

Microorganisms often need to cope with hostile environment. To do this they have developed sophisticated cooperative behaviour and intricate communication capabilities, Ben Jacob et al (1997). In particular, a variety of branching and chiral (handedness) fractal patterns emerge. Invoking the ideas from pattern formation in nonliving systems and using generic modelling, the studies of Ben Jacob et al (1997) reveal novel survival strategies which account for the evolved patterns. Using the communicating walkers model (Discrete model) in which bacteria are represented by discrete, random walking entities consuming nutrients, reproduce, perform random biased movement, and produce and respond to chemicals whose time evolution is described by reaction-diffusion equations they show how communication leads to self-organization via a cooperative behaviour of the cells. In this regard, pattern formation in microorganisms can be viewed as the result of the exchange of information between the microlevel (individual cells) and the macro-level (colony). Bacterial colonies alter their fractal patterns so as to have different dimensions depending upon the growth conditions. Starrus et al (2007) show that swarming pattern in bacteria can arise due to their rod shape and self-propulsion, by using implicit interaction through volume exclusion without any kind of explicit signalling. We can learn a lot from the smart bacterial colony as to how to build agent-controlled robots for special purpose missions.

7.3 Simulating Animal-Human Trails

Active walker models in which animals or humans interact through indirect communication mediated by the environment and leave a purposeful trail either from nest to the food source or between other points of interest, can be interpreted qualitatively as a small world phenomena on graphs, or as percolation phenomena in lattices, by choosing a proper lattice structure (square or hexagonal). These trails result due to agglomeration process of several walkers moving arbitrarily leaving markings (clearing vegetation or leaving chemicals), but eventually producing an attractive effect on another walkers. Clearly, there is a preferential choice among the possible trails and the most frequently used trails combine to become popular. Also rarely used old trails disappear and frequently used trails are reinforced although many new entry points may arise and destinations may branch off. Also fitness is evaluated for each trail as to its cost and utility Schweitzer (2002), Ebeling and Schweitzer (2003).

Similar approach can be useful for the study of traffic patterns and crowd dynamics and the way to control them. The use of nondeterminsm in MAP permits agents to make different possible decision in a traffic situation. For a detailed study of Traffic networks, see Nagel (2003).

7.4 Adaptive Evolution

The group-swarming approach (of randomly communicating with other groups to revise belief based on new information available), by preferential linking locally in a group and globally among different groups, mimics the process of ***epochal evolution and innovation*** advocated by Crutchfield (2003), in which the groups assemble partial solutions that are metastable and eventually reach equilibrium. The individual groups correspond to subbasins of approximately isofitness and are strongly or weakly connected to other subbasins thus mapping genotype-to phenotype, and phenotype to fitness. Adaptation takes place through innovation by linking with other groups (subbasins) to gain better information. Thus Crutchfield's principle seems to hold universally, wherever there is a selection pressure- in foraging, breeding, protein function modification, and cell-assembly.

We can simulate adaptive evolution (molecular evolution and phylogenetic tree evolution) by identifying tree hierarchy with agent hierarchy, where each subtree is an agent or a group swarm. Here, a population of trees is kept in every generation; these trees "breed" to produce trees of the next generation (or group swarm).The algorithm uses operations similar to mutation and recombination to generate new trees from the current ones. Survival of each tree into the next generation depends on its "fitness" which is the optimality criterion based on factors such as: parsimony (minimal changes in tree), maximum likelihood (optimizing branch lengths and related parameters), minimum evolution (sum of tree lengths), Bayesian (posterior probability, calculated by integrating over branch lengths and related parameters). This is a vast area in bioinformatics, Yang (2007), Albert (2005), Bininda-Emonds (2004), Felenstein (2007), Twyman(2003), and is closely related to Genetic programming (GP) where the fitness of the individual program generated locally and globally, Koza (1999).

In GP each program construct is a tree constructed from tasks, functions and terminal symbols. Then we perform crossover and mutation by swapping program sub-trees leading to feasible programs, taking care of the nature and type of the task. These operations resemble Monte-Carlo methods, Newman (2004), to create transitivity in a graph from a given node to a desired attractor node. The GP operations correspond to an ergodic move-set in the space of graphs with a given set of parameters and repeatedly generating the moves and accepting them with probability p or rejecting them with probability (1-p).

Suitable move-sets are: creation of new nodes, aging and annihilation of nodes, Mutation, i.e., movement of edges from one place to another, mating, i.e., swapping edges of the form (s,t), (u,v) to (s,u), (t,v), adding new edges based on a cost function. Such moves can create a phase transition (or percolation) to reach a global goal through successive local goals, when a large number of trials are performed. In GP is the fitness of the individual program generated locally and globally. In self-organization, ideally, one requires that the fitness is a *self-awareness* function, that is, the individual who does the work evaluates itself, ensuring that the global fitness is guaranteed. This is widely prevalent in Nature for activities such as: nest building (stigmergy), food searching (foraging).

7.5 Synchronization among Agent Population

A simplest adaptive system arises in 'Synchronization". Here two oscillatory systems (or repetitive systems) adjust their behaviours relative to each other so as to attain a state where they work in unison. This is a universal phenomenon, Pikovsky et al.(2003), Manrubia et al. (2004), Strogatz (2003), Baker and Blackburn (2005). Agents provided with nonlinear oscillatory capabilities can couple through different choice of interacting functions. This kind of coupling can result in the emergence of dynamical order and the formation of synchronous clusters and swarming. Such systems can function like Chemical systems, Biological systems, Molecular machines and will have applications in designing task specific biomorphic robotics, Bar Cohen et al. (2003), Manrubia et al.(2004). Also reference is made to the several chapters in this book devoted to this topic.

8 Engineering Complex Systems

Since CS is inherently nonlinear and hysteretic, agents made out of Piezoceramics, magnetostrictives, Shape-Memory Alloys (SMA), Electrostrictive polymers, Ferromagnetic SMA, can serve as components for specific applications. However, the central question in designing CS is how to program the components so that the system as a whole self organizes. This is the basic question addressed in the design of Amorphous computers and Spray computers, as well as, in the synthesis of heterogeneous materials with certain specified macro properties from a knowledge of the properties of the microsturctures, Torquato (2002). Emergence is a global behaviour (or a goal) that evolves from the local behaviours (goals) of components, Serugendo et al.(2006). The evolutionary rules for arriving at the global goal is non-computable, since the global goal cannot be expressed as a finite composition of computable deterministic function of local goals for any arbitrary problem domain. Thus we cannot design a general purpose program based on modularity to create a CS with exactly specified properties.

However, for specific applications and a pre-defined interactive topology among the agents, the geometric, dynamical, topological parameters, and statistical properties can be obtained through simulation and tuned to build a specific CS. Recent advances in Cell biology reveals that Nature has always concentrated on devising special purpose computing elements having both the analog and digital features, and exploiting cooperative and competitive schemes using percolation like threshold phenomena, positive feedback leading to self-enforcement, nonlinearity, and short and long range signalling to facilitate interaction using disordered protein networks. For accomplishing any special task, the protein machines are interconnected through softwiring, and are temporally and spatially coordinated through linked processes to achieve maximal efficiency.

8.1 Quantum Mechanics and Complex System

Nature exhibits power law behaviour only in macrosale, e.g. inverse-square law in gravitational, electrical and magnetic forces. On molecular level and below, such simple laws no longer hold giving rise to molecular and nuclear forces resulting in atoms and molecules that are autonomous objects and their rules of interactions vary and not fully understood. It is only in the macro scale, the agent-based simulation plays a realistic role. All our arguments are not valid at nano and molecular scales, where quantum mechanical (QM) effects take over. Hence, the geometry, nature on interaction and other details are not easy to model using classical computational tools because:

1. In QM there is no chaos: Quantum mechanics is a linear theory and no chaos exists. There is no exponential time divergence of trajectories. The quantum mechanical view is time-reversible. However, in classical chaos, the presence of a positive Lyapunov exponent prevents the recovery of initial state and so there is a time-arrow and a complex system will age.

2. Algorithmic complexity: Since there is no chaos, Lypaunov exponents and metric entropy are not definable. Chaotic dynamics results in an evolved pattern that requires a storage space that is far larger than the original space of coarse grained cells. Hence in practice, if we use floating point computation, we need to extend our precision indefinitely to capture the chaotic evolution, since the number of bits required grows exponentially with the positive Lyapunov exponent resulting in an enormous algorithmic cost. However, if indeed one can build quantum registers, we can capture this information without extra cost as quantum dynamic evolution opens up the Hilbert space of states.

For example, the Arnold cat map, although time-reversible, Livi et al. (2003):

$$x(t+1) = x(t)+y(t) \mod 1$$
$$y(t+1)= x(t)+2y(t) \mod 1$$

where x,y are real numbers in [0,1] and index t indicates discrete time intervals t= 0,1,..., can run forwards as well as backwards in Quantum computing recovering the initial states, whereas in classical computing this information is destroyed due to exponential growth of errors.

3. Built-in-ergodicity: Whereas a classical agent will travel precisely through a specific deterministic path, if there is no added randomness, a quantum mechanical agent inherently explores all possible pathways in space time to reach a given final point producing an exponentially growing information. This means the Quantum mechanical agent is endowed with the ability to explore and exploit,without a necessity to add any randomness or nondeterminsm- a kind of self-awareness,!

4. Emergence and Decoherence: Quantum mechanics always works on a holistic view. The quantum system exhibits both kinematic and measurement-induced nonlocality. In a composite system composed of parts and the state of whole is not definable in terms of the states of its parts. There are quantum correlation between different subsystems due to the superposition principle. In fact, the whole is more than its parts, even if we assume there is no interaction among the subsystems (non-separability). Quantum systems already exhibits emergence leading to many different views. Decoherence provides a particular view.

5.Simulating Quantum Systems: This is a grand challenge!

9 Multi-Agent Architecture and Toolkits

Shakshuki et al. (2004), evaluate multiagent tool kits, such as: Java Agent development framework (JADE), Zeus Agent building toolkit and JACK Intelligent Systems. They consider Java support, and performance evaluation.

Gorton et al. (2004) have evaluated agent architectures: Adaptive Agent architecture (AAA), Aglets developed by IBM, and the Java based architecture Cougaar. The paradigm described here is well-suited for implementing in Cougaar, a Java based agent architecture, since Cougaar is based on human reasoning. A Cougaar agent consists of a blackboard that facilitates communication and operational modules called plug-in that communicate with one another through the blackboard and contain the logic for the agent's operations. The use of blackboard and direct communication are useful for simulating the problems described in this chapter. Further developments in agent-based languages and architecture are needed to simulate very large scale complex systems.

10 Summary and Directions for Research

We described some important properties a Complex System (CS) need to possess and the models needed to understand, simulate and animate CS. Multi-agent systems in a network can exhibit the properties of both the computational and dynamical systems at the macro-level and can undergo phase transition and emerge into an CS. However, to understand CS quantitatively, we need to obtain the spatial structure of attractors and the temporal aspects of the trajectories, and topological or graph parameters. The first two aspects provide information about phase transitions and tells us whether a trajectory governed by a positive Lyapunov exponent falls in a given attractor. The topological aspects give us a statistical behaviour of the network of connectivity among the agents. Real systems including biological systems are nonstationary, in which the parameters are changing with time. Thus quantitative understanding of an CS using the above parameters with arbitrary time varying interactive topology seems difficult. These difficulties arise due to the statistical nature of the models available and inability to compute the exact parameters numerically. However, for specific applications and a pre-defined static interactive topology among the agents, the statistical parameters can be obtained through simulation and tuned to build a special purpose macro scale CS. Such a tuning scheme seems to have formed the basis behind anthropic coincidences and intelligent design, Carr and Reeves (1979), Dembski (1999).

Introduction of chaos or randomness in a machine provides a greater power (more complex) than an algorithm by providing an internal mechanism for exploration and discovery. As a result, in such a machine programmability is lost. The evolution of a resilient complex system is linked to its adaptability. Streamlining it through an algorithmic structure interferes with its self-organization. Thus high adaptability is mutually exclusive to programmability; whenever maximum efficiency is needed adaptive systems can do better than programmable systems.

The study of complex networks is in its early stage. The interaction of agents within this framework is very complex and requires the availability of tools for creating a very large number of agents and dealing with them in a suitable environment. This would provide us an insight into the complexity of behaviour of natural systems we see around us.

Protein-based computational networks have an important role to play in the design of CS- particularly, in Genetic regulatory systems, Prosthetics, Systems Biology and understanding the organization of the living cells. However, there are major obstacles in understanding protein networks and their functions. First of all the definition of functions in biology is entirely different from computable or mathematical functions. Three kinds of interrelated functions occur in biology:

(i) The *biochemical function* referring to the chemical activity, binding and catalytic property and conformational changes.

(ii) The *cellular function* that is context dependent with respect to a tissue or an organ.

(iii)The *phenotypic function* that determines the behavioural and physiological properties of an organism in its environment.

In language theory, the above three functional aspects are analogous to the syntax that deals with the structure and grammar of a sentence (the chemical structure, bonds and reaction), the semantics that assigns a meaning to a sentence (its role in cellular action), and thirdly the pragmatics of a language which deals with the usability of the language in successfully meeting its goals (survival of an organism). All these functions are nonunique and context dependent. That is, similar functions can result in different outcomes and different functions can result in similar outcomes due to their many to one, and one to many mapping properties. This is called multiple realizability and leads to the failure of reductionism.

In mathematical logic these functions correspond to higher-order logical functions that deal with the nondenumerable properties of functions of another function of yet another function, and hence turn out to be noncomputable. In addition, proteins are multi domain structures, undergoing extensive conformational changes and rapidly evolve their folds through order-disorder transitions making the complexity of folding problems in proteins computationally very complex.

Another challenging problem is to consider the role of scales and the role of quantum computational systems in the design of CS. This will require understanding and reconciling several fundamental issues of classical nonlinear dynamics and biodynamics with the philosophy of quantum mechanics.

Finally, one wonders how to enlarge CS to include life-like systems. We assumed that CS satisfies some of the basic requirements such as: cooperative and coordinated interaction, emergence, power-law scaling, adaptation, fault tolerance and resilience. These requirements are not adequate to realize life-like systems.

According to Shapiro (2007) a life-like system (LLS) needs to satisfy the five requirements:

1. A boundary to separate life from nonlife. This is essential to increase the entropy of the environment and decrease the entropy of the living system.
2. An energy resource to drive the organization of the network of components.
3. A coupling mechanism must link the release of energy to the organization process that produces and sustains life.
4. A chemical network must be formed that permits adaptation and evolution.
5. The network grows and reproduces; means it gains materials from outside faster than it loses.

CS as defined is abstract and differs from a life-like system, since it need not be a chemical network. It appears that the requirement of being a network of multiset of chemical agents hierarchically organized on the right scale, is essential for life providing the right type of content-based activation, cooperative and competitive interaction rules among them and with their environment, for enabling, inhibiting, autocatalysis, altering conformation, shape matching and binding, self-nonself discrimination, diffusion, chemotaxis, conflict resolution based on energy and other physical quantities, percolation and swarming, Zewail(2008). If this is not, do we have alternative methods of producing life- like systems that are not based on chemistry? Such questions are very basic and are related to anthropic principle and anthropic coincidences proposed by Carr and Rees (1979) and to Intelligent design, Dembski (1999). These basic questions can keep the upcoming generation of computational biologists active for many decades.

References

[1] Adamic, L.A., et al.: Local search in Unstructured Networks. In: Born holdt, S., Schuster, H.G. (eds.) Handbook of Graphs and Networks, pp. 295–317. Wiley-VCH, New York (2003)

[2] Albert, V.A.: Parsimony, Phylogeny and Genomics. Oxford University Press, Oxford (2005)

[3] Alberts, B., et al.: The Molecular Biology of the Cell. Garland Science, New York (2002)

[4] Atilgan, A.R., et al.: A Small-world Communication of Residues and Significance in Protein Dynamics. Biophysical Journal 86(1), 85–91 (2004)
[5] Aviv, R., Shapiro, E.: Cellular Abstractors: Cellular computation. Nature 419, 343 (2002)
[6] Baker, G.L., Blackburn, J.A.: The Pendulum. Oxford University Press, Oxford (2005)
[7] Barabasi, A., et al.: Emergence of Scaling in Random Networks. Science 286, 509–512 (1999)
[8] Bar-Cohen, Y., Breazeal, C.: Biologically-Inspired Intelligent Robotics. S.P.I.E.Press, Bellingham (2003)
[9] Bar-Yam, Y.: Dynamics of Complex Systems. Addison Wesley, Reading (1997)
[10] Belew, R.K., Forrest, S.: Learning and Programming in classifier Systems. Machine Learning 3, 193–223 (1988)
[11] Belloquid, A., Delitala, M.: Mathematical Modelling of Complex Biological Systems. Birkhauser, Boston (2006)
[12] Ben-Jacob, E., et al.: Smart bacterial colonies in Physics of Biological systems: From Molecules to Species. Lecture Notes in Physics, vol. 480, pp. 307–340. Springer, New York (1997)
[13] Ben-Naim, E., et al. (eds.): Complex Networks. Lecture Notes in Physics, vol. 650. Springer, New York (2004)
[14] Bininda-Emonds, O.R.P.: Phyogenetic Super trees: Combining Information to reveal the tree of life. Kluwer Academic Press, Dordrecht (2004)
[15] Blackwell, T., Branke, J.: Multi-swarm Optimization in Dynamic Environments. In: Raidl, G.R., Cagnoni, S., Branke, J., Corne, D.W., Drechsler, R., Jin, Y., Johnson, C.G., Machado, P., Marchiori, E., Rothlauf, F., Smith, G.D., Squillero, G. (eds.) EvoWorkshops 2004. LNCS, vol. 3005, pp. 489–500. Springer, Heidelberg (2004)
[16] Boahen, K.: Neuromorphic microchips. Scientific American 292, 38–41 (2005)
[17] Booker, L.K., et al.: Classifier systems and Genetic Algorithms. Artificial Intelligence 40, 235–282 (1989)
[18] Bonabeau, E., et al.: Swarm Intelligence: From Natural to Artificial Systems. Oxford University Press, London (1999)
[19] Born holdt, S., Schuster, G.H.: Handbook of Graphs and Networks. Wiley-VCH, New York (2003)
[20] Bunde, A., Havlin, S.: Fractals in Science. Springer, New York (1994)
[21] Camazine, S.: Self-Organization in Biological Systems. Princeton University Press, Princeton (2002)
[22] Cannings, C., Penman, D.D.: Models of random Graphs and their Applications. In: Rao, C.R. (ed.) Handbook of Statistics, vol. 21, pp. 51–91. North Holland, Amsterdam (2003)
[23] Cardelli, L.: Abstract Machines in Systems Biology. Springer Transactions on Biological Systems (2005)
[24] Carr, B., Rees, M.: The anthropic principle and the structure of the physical world. Nature 278, 605–612 (1979)
[25] Chaitin, G.: Two Philosophical Applications of Algorithmic Information Theory. In: Calude, C.S., Dinneen, M.J., Vajnovszki, V. (eds.) DMTCS 2003. LNCS, vol. 2731, pp. 1–10. Springer, Heidelberg (2003)
[26] Chan, K.S., Tong, H.: Chaos: A statistical Perspective. Springer, New York (2002)

[27] Chu, S., et al.: Parallel Ant colony Systems. In: Zhong, N., Raś, Z.W., Tsumoto, S., Suzuki, E. (eds.) ISMIS 2003. LNCS (LNAI), vol. 2871, pp. 279–284. Springer, Heidelberg (2003)
[28] Chung, F., Lu, L.: Complex Graphs and Networks, American Mathematical Society. In: CBMS, Providence, Rhode Island, vol. 107 (2006)
[29] Coello, C.A.C., et al.: Evolutionary algorithm for Solving Multi-objective Problem. Kluwer, New York (2002)
[30] Crutchfield, J.P., Schuster, P.: Evolutionary Dynamics. Oxford University Press, Oxford (2003)
[31] de Castro, L.N., Timmis, J.I.: Artificial Immune Systems: A New computational Intelligence Approach. Springer, New York (2002)
[32] Dembski, W.A.: Intelligent Design. InterVarsity Press, Downers Grove, Ill (1999)
[33] Deneubourg, J.L., et al.: Optimality of communication in self-organized behaviour. In: Hemelrijk, C.K. (ed.) Self-Organization and Evolution of Social Systems, ch.2, pp. 25–35. Cambridge University Press, Cambridge (2005)
[34] Deutsch, A., et al.: Mathematical Modeling of Biological Systems, vol. 1 and 2. Birkhauser, Boston (2007)
[35] Dorigo, M., et al.: Ant Algorithms 2002. LNCS, vol. 2463. Springer, Heidelberg (2002)
[36] Dorigo, M., Stutzle, T.: Ant Colony optimization. M.I.T. Press, Cambridge (2004)
[37] Dorigo, M., et al.: Swarm-Bot: design and implementation of colonies of self-assembling robots. In: Yen, G., Fogel, D.B. (eds.) Computational Intelligence, pp. 103–136. IEEE Press, New York (2006)
[38] Dorogovtsev, S.N., Mendes, J.F.F.: Evolution of Networks. Oxford University Press, Oxford (2003)
[39] Doucet, A., Gordon, N., Krishnamurthy, V.: Particle Filters for State Estimation of Jump Markov Linear Systems. IEEE Trans. Signal Processing 49, 613–624 (2001)
[40] Ebeling, W., Schweitzer, F.: Self-organization, Active Brownian dynamics and biological Applications. Nova Acta Leopoldina 88(332), 169–188 (2003)
[41] Edwards, S.J.: Swarming on the Battlefield, National Defence Research Institute, RAND,U.S.A (2000)
[42] Effroni, S., et al.: Reactive animation: Realistic Modeling of Complex Dynamic Systems. IEEE Computer, 33–46 (January 2005)
[43] Eigen, M.: StepTowards Life. Oxford University Press, Oxford (1992)
[44] Falcioni, M., et al.: Kolmogorov's legacy about entropy, Chaos and Complexity. Lecture Notes in Physics, vol. 636, pp. 85–108. Springer, New York (2003)
[45] Felenstein, J.: Inferring Phylogenesis. Sinauer associates, Sunderland (2007)
[46] Finkelstein, A.V., Ptitsyn, O.B.: Protein Physics. Academic Press, New York (2003)
[47] Forrest, S.: Parallelism and Programming in classifier systems. Morgan Kauffman, San Mateo (1991a)
[48] Forrest, S.: Emergent Computation. M.I.T Press, Cambridge (1991b)
[49] Gell-Mann, M.: The Quark and the Jaguar. W.H.Freeman, New York (1994)
[50] Goldberg, D.E.: Genetic algorithms in search, optimisation and machine learning. Addison Wesley, Reading (1989)
[51] Goncharova, L.B., Melnikov, Y., Tarakanov, A.O.: Biomolecular Immunocomputing. In: Timmis, J., Bentley, P.J., Hart, E. (eds.) ICARIS 2003. LNCS, vol. 2787, pp. 102–110. Springer, Heidelberg (2003)

[52] Gorton, I., Haack, J., McGee, D.R., Cowell, A.J., Kuchar, O., Thomson, J.: Evaluating Agent Architectures: Cougaar, Aglets and AAA. In: Lucena, C., Garcia, A., Romanovsky, A., Castro, J., Alencar, P.S.C. (eds.) SELMAS 2003. LNCS, vol. 2940, pp. 264–278. Springer, Heidelberg (2004)
[53] Graham, I., Duke, T.: The logical repertoire of ligand-binding proteins. Physical Biology 2, 159–165 (2005)
[54] Grimmett, G.: Percolation. Springer, New York (2004)
[55] Guerin, S., Kunkle, D.: Emergence of Constraint in Self-organizing Systems, Nonlinear dynamics. Psychology and Life Sciences 8(2), 131–146 (2004)
[56] Harel, D.: A grand challenge for computing: towards full reactive modeling of a multicellular animal, EATCS Bulletin (2003), `http://www.wisdom.weizmann.ac.il/~dharel/papers/grandchallenge.doc`
[57] Hilborn, R.C.: Chaos and Nonlinear Dynamics. Oxford University Press, Oxford (2003)
[58] Hofstadter, D.: Fluid concepts and creative analysis. Basic Books inc., New York (1995)
[59] Holland, J.H., et al.: Induction. M.I.T.Press, Cambridge (1987)
[60] Ilachinski, A.: Cellular Automata. World Scientific, Singapore (2001)
[61] Ishida, Y.: Immmunity Based Systems. Springer, New York (2004)
[62] Ivancevic, V.G., Ivancevic, T.T.: Neuro-Fuzzy Associative Machinery for Comprehensive Brain and Cognition Modelling. Springer, New York (2006)
[63] Jain, S., Krishna, S.: Graph theory and Autocatalytic networks. In: Born holdt, S., Schuster, H.G. (eds.) Handbook of Graphs and Networks, pp. 355–394. Wiley-VCH, New York (2003)
[64] Kauffman, S.A.: The origins of Order. Oxford University Press, Oxford (1993)
[65] Keele, J.W., Wray, J.E.: Software Agents in molecular computational Biology. Briefings in Bioinformatics 6(5), 370–379 (2005)
[66] Kennedy, J., Eberhart, R.C.: Swarm Intelligence. Morgan Kauffman, London (2001)
[67] Kennedy, J.: Swarm Intelligence. In: Zomaya, A. (ed.) Handbook of Nature-Inspired & Innovative Computing, pp. 187–221. Springer, New York (2006)
[68] Koza, J.R.: Genetic programmingIII. Morgan Kaufmann, San Francisco (1999)
[69] Krishnamurthy, E.V.: Parallel Processing. Addison Wesley, Reading (1989)
[70] Krishnamurthy, E.V., Murthy, V.K.: Transaction Processing Systems. Prentice Hall, Sydney (1992)
[71] Lam, L.: Nonlinear Physics for Beginners. World Scientific, Singapore (1998)
[72] Livi, R., et al.: Kolmogorov Pathways from integrability to chaos and beyond. Lecture Notes in Physics, vol. 636, pp. 3–32. Springer, New York (2003)
[73] Maini, P.K., Othmar, H.G. (eds.): Mathematical models for biological pattern formation. Springer, New York (2001)
[74] Manrubia, S.C., et al.: Emergence of dynamical order. World Scientific, Singapore (2004)
[75] Maslov, S., et al.: Specificity and stability in topology of protein networks. Science 296, 910–913 (2002)
[76] Maslov, S., et al.: Correlation profiles and motifs in complex networks. In: Bornholdt, S., Schuster, H.G. (eds.) Handbook of Graphs and Networks, pp. 168–198. Wiley-VCH, New York (2003)

[77] McCullogh, A.D., Huber, G.: Integrative Biological modelling in silico biological processes. In: Bock, G., Goode, A. (eds.) Novatis Foundation symposium, pp. 4–19. John Wliey and Sons, Chichester (2002)
[78] Meuleau, N., Dorigo, M.: Ant colony optimization and Stochastic Gradient Descent. Artificial Life 8, 103–121 (2002)
[79] Meyer, B.: Applying design by contracts. IEEE Computer 25(10), 40–52 (1992)
[80] Milne, R.K.: Point processes and some related Processes. In: Rao, C.R. (ed.) Handbook of Statistics, vol. 19, pp. 599–641. North Holland, Amsterdam (2001)
[81] Moon, F.C.: Chaotic and Fractal Dynamics. John Wiley, New York (1999)
[82] Mosekilde, E., Mosekilde, L. (eds.): Complexity, Chaos and Biological Evolution. Plenum Press, New York (1991)
[83] Nagel, K.: Traffic Networks. In: Bornholdt, S., Schuster, H.G. (eds.) Handbook of Graphs and Networks, pp. 248–272. Wiley-VCH, New York (2003)
[84] Newman, M.E.J.: The Structure and Function of complex Networks, Santa Fe Institute (2004)
[85] Noe, F., Smith, C.: Transition Networks: A unifying theme for Molecular Simulation and Computer Science. In: Deutsch, A., et al. (eds.) Mathematical Modeling of Biological Systems, ch. 11, vol. 1, pp. 121–135. Birkhauser, Boston (2007)
[86] Orengo, C.A., et al.: Bioinformatics. BIOS Scientific Publishers, New York (2003)
[87] Pacino, K.M.: Biomimicry of bacterial foraging for distributed optimisation and control. IEEE Control System Magazine 22(3), 52–68 (2002)
[88] Palla, G., Vattay, G.: Spectral Transitions in Networks. New Journal of Physics 8, 306–314 (2006)
[89] Parker, L.E., et al. (eds.): Distributed Autonomous Robotic Systems. Springer, New York (2000)
[90] Parker, T.S., Chua, L.O.: Practical Numerical algorithms for chaotic systems. Springer, New York (1989)
[91] Petrucci, R.H., et al.: General Chemistry. Prentice-Hall, NJ (2002)
[92] Pikovsky, A., et al.: Synchronization. Cambridge University Press, Cambridge (2003)
[93] Pinney, J.W., et al.: Petri net representations in systems biology. Biochemical Society Transactions 31, Part 6 (2003)
[94] Prigogine, I.: From being to becoming. W.H.Freeman and Co., San Francisco (1980)
[95] Robert, C.P., Casella, G.: Monte Carlo Statistical Methods. Springer, Heidelberg (1999)
[96] Scargle, J.D., Babu, G.J.: Point Processes in Astronomy. In: Rao, C.R. (ed.) Handbook of Statistics, vol. 21, pp. 795–825. North Holland, Amsterdam (2003)
[97] Schweitzer, F.: Brownian Agents and Particles. Springer, Berlin (2002)
[98] Serugendo, G.D.M., et al.: Self Organization: Paradigms and Applications. In: Di Marzo Serugendo, G., Karageorgos, A., Rana, O.F., Zambonelli, F. (eds.) ESOA 2003. LNCS (LNAI), vol. 2977, pp. 1–19. Springer, Heidelberg (2004)
[99] Serugendo, G.D.M., et al.: Self-organization and Emergence in MAS: A overview. Informatica 30, 45–54 (2006)
[100] Shakhnovich, E.: Protein folding Thermodynamics and dynamics: Where Physics, Chemistry and Biology meet. Chemical Reviews 106, 1559–1588 (2006)
[101] Shakshuki, E., Jun, Y.: Multi-agent development toolkits: An Evaluation. In: Orchard, B., Yang, C., Ali, M. (eds.) IEA/AIE 2004. LNCS (LNAI), vol. 3029, pp. 209–218. Springer, Heidelberg (2004)

[102] Shapiro, R.: A simpler origin for life. Scientific American 296, 24–31 (2007)
[103] Shmygelska, A., et al.: An ant colony optimization algorithm for the 2D and 3D hydrophobic polar protein folding problem. BMC Bioinformatics 6, 30–47 (2005)
[104] Shyu, S.J., Tsai, C.-Y.: Finding the longest common subsequences for multiple biological sequences by ant colony optimization. Computers and Operations Research 36, 73–91 (2009)
[105] Spall, J.C.: Introduction to Stochastic Search and Optimization. Wiley Interscience, New York (2003)
[106] Sprott, J.C.: Chaos and Time Series Analysis. Oxford University Press, Oxford (2003)
[107] Starrus, J., et al.: Bacterial Swarming Driven by Rod shape. In: Deutsch, A., et al. (eds.) Mathematical Modeling of Biological Systems, ch. 14, vol.1, pp. 163–174. Birkhauser, Boston (2007)
[108] Steinhofel, K., et al.: Relating time complexity of protein folding simulation to approximations of folding time. Computer Physics Communications 176, 465–470 (2007)
[109] Stepney, S., et al.: Artificial Immune System and the grand challenges for non-classical computation. In: Timmis, J., Bentley, P.J., Hart, E. (eds.) ICARIS 2003. LNCS, vol. 2787, pp. 204–216. Springer, Heidelberg (2003)
[110] Stith, B.J.: Use of animation in teaching cell biology. Cell. Biology Education 3(3), 181–188 (Fall 2004)
[111] Strogatz, S.H.: Sync:The emerging science of spontaneous Order. Hyperion Press, New York (2003)
[112] Suzuki, Y., et al.: Artificial Life applications of a class of P systems: Abstract rewriting systems on Multisets. In: Calude, C.S., Pun, G., Rozenberg, G., Salomaa, A. (eds.) Multiset Processing. LNCS, vol. 2235, pp. 299–346. Springer, Heidelberg (2001)
[113] Szarowicz, A., et al.: The application of AI to automatically generated animation. In: Stumptner, M., Corbett, D.R., Brooks, M. (eds.) Canadian AI 2001. LNCS (LNAI), vol. 2256, pp. 487–494. Springer, Heidelberg (2001)
[114] Szirtes, G., et al.: Emergence of Scale-free Properties in Hebbian Networks. International Journal of Neural Systems (2001)
[115] Thompson, E.: Mind in Life. Harvard University Press, Cambridge (2007)
[116] Torquato, S.: Random Heterogeneous Materials. Springer, New York (2002)
[117] Turing, A.M.: The Chemical Basis for Morphogenesis. Phil.Trans. Roy. Soc. London 237, 37–79 (1952)
[118] Twyman, R.M.: Principles of Proteomics. BIOS Scientific Publishers, New York (2003)
[119] Vaughan, R.T., et al.: Blazing a trail: Insect - inspired reource transportation by a robot team. In: Parker, L.E., et al. (eds.) Distributed Autonomous Robotic Systems, pp. 112–120. Springer, New York (2000)
[120] Villani, V.: Complexity of polypeptide dynamics:Chaos, Brownian motion and elastcity in aqueous solution. Journal of Molecular Structure:THEOCHEM 621, 127–139 (2003)
[121] Watts, D.: Small Worlds. Princeton University Press, Princeton (1999)
[122] Weiss, T.F.: Cellular Biophysics, vol. 1and 2. MIT Press, Cambridge (1996)
[123] Werfel, J., et al.: Construction by robot swarms using extended stigmergy,Technical Report, AI Memo AIM -2005-011, MIT,C S and AI Lab (2005)

[124] Wilkins, A.S.: The Evolution of Developmental Pathways. Sinauer Associates, Inc., Sunderland (2002)
[125] Wolfram, S.: A New Kind of Science. Wolfram Media Inc., Champaign, Ill (2002)
[126] Wooley, J.C., Lin, H.C. (eds.): Catalyzing inquiry at the interface of computing and biology. National Research council of the National Academies, National Academies Press, Washington, DC (2005)
[127] Woolridge, M.: Introduction to Multi-Agent Systems. John Wiley, New York (2002)
[128] Yang, S., et al.: Evolutionary computation in Dynamic and Uncertain Environments. Springer, New York (2007)
[129] Yang, Z.: Computational Molecular Evolution. Oxford University Press, Oxford (2007)
[130] Yuen, D.C.K., MacDonald, B.: Theoretical considerations of Multiple particle Filters for simultaneous Localization and Map-Building. In: Negoita, M.G., Howlett, R.J., Jain, L.C. (eds.) KES 2004. LNCS (LNAI), vol. 3213, pp. 203–209. Springer, Heidelberg (2004)
[131] Zewail, A.H.: Physical Biology-From Atoms to Medicine. Imperial College, London (2008)

Simulation of Nonlinear Dynamics and Synchronization for Structural Control at Seismic Excitations

Svetla Radeva

Abstract. A new branch of dynamic theory of earthquake engineering was developed during last decades where analysis and calculation of dynamic behaviour of structures is simulated with synthetic and real registrated accelerograms. The purpose of structural response simulation is to analyze behavior of structures during vibrations caused by earthquakes. Their direct measurement are not always possible, especially when under investigation is behavior of structures at strong motion seismic excitations. The structural control devices and systems for seismic protection are very expensive and individual for each structure. That's why they are tested not only with computer simulation but as well experimentally via simulation prototype models under simulated seismic excitations, maximal close to the real ones.

1 Active and Semi-active Forcing Systems for Structural Control

Structures have been designed to resists natural hazards through a combination of strength, deformability and energy absorption. The purpuse of structural control systems is to protect structures at earthquakes and strong winds via absorbing some part of seismic energy. The absorbed seismic energy can be presented as a sum of:

(1) absolute kinetic energy and regenerated energy of elastic sagging of absorbed from construction energy;
(2) dissipated from the construction seismic energy at non-elastic or other deformations;
(3) energy, absorbed from the *system of seismic protection* of structure.

In this way the energy absorbed from the structure at non-elastic deformations can be reduced from the system of seismic protection of structure. This leads to many innovation approaches, divided into three main groups – passive, active

Svetla Radeva
University of Architecture, Civil Engineering and Geodesy, Bulgaria
e-mail: svetla_fce@abv.bg

K. Kyamakya (Eds.): Recent Adv. in Nonlinear Dynamics and Synchr., SCI 254, pp. 201 – 234.
springerlink.com

and semi-active structural control [10, 16, 18]. The ***passive control*** systems are connected with seismic base isolation, which reduce the level of energy which can be transmitted to the structure [3]. The most important requirements for an isolation system are its flexibility to lengthen the natural period of the structure and the seismic excitation. Typical seismic isolation devices may include elastomeric bearings, lead rubber bearings, high damping rubber bearings, sliding friction pendulum bearings etc.

Another approach at passive control systems is connected with increasing the energy dissipation capacity of the structure. This is realized with dynamic vibration absorbers, such as tuned mass dampers, tuned liquid dampers etc.

1.1 Active Forcing Systems

Active forcing systems have the ability to adapt the structure to different loading conditions at seismic excitations and to control different vibration modes of the structure for it synchronization. An active forcing system is a part of structural control [2, 7],which operates by using external energy supplied by actuators to impart forces on the structure. On the base of measured structural response is determined the appropriate control action.

The basic configuration of an active forcing system is presented on Fig.1.

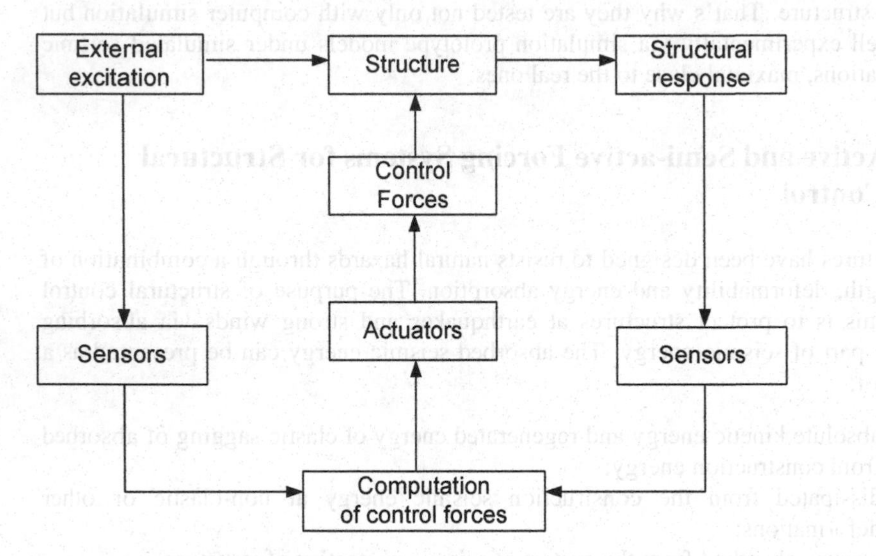

Fig. 1 Diagram of active forcing system

The active forcing system consists of sensors located on different places on the structure to measure external excitations and structural response, devices to process the measured information and to compute necessary control forces for given control algorithm and actuators to produce the required forces. The objective of ***active structural control*** is to reduce structural motion during external excitations like earthquakes and winds [5, 17].

There were developed different active control mechanisms, like active tendon system, active bracing system, active tuned mass damper/driver, active aerodynamic appendage mechanism etc. For each system are developed various control algorithms, like:

(1) Open-loop control algorithms;
(2) Closed-loop control algorithms;
(3) Control algorithms which account the force and stroke limitations of control actuators;
(4) Nonlinear control algorithms;
(5) Neural based control algorithms;
(6) Fuzzy control algorithms;

The most fundamental disadvantage of active control is that it in general requires an external power supply, which presents questions about both reliability and practicality. In civil engineering applications an active control system strongly depends on its power sources, which can be destabilized and blackout during earthquake. From the other side, the power demands of active control systems for large buildings are typically too large to be met with local supplies.

The second disadvantage of active control systems is that by designing an actuator which accepts power from an external source, the system is in certain way a bounded-energy system. For such system arise questions about it stability and robustness. Significant efforts were done for solving this problem with application of robust control theory, but the nature of model uncertainty in civil engineering structures is such that significant concerns remain unresolved.

Another important emerging technology in structural control based on active control systems are ***hybrid systems*** where to active forcing systems is added passive structural control. In hybrid systems a combination of active and passive systems are working in tandem. Hybrid systems are more reliable, because the passive part of the actuator will still work in the absence of electric power. Operating both systems together enhances the robustness of the passive system and reduces the energy requirements of the active system.

There are two main approaches for the implementation of hybrid systems: the hybrid mass damper and the hybrid seismic isolation system. A hybrid mass damper combines a tuned mass damper with an active actuator to enhance its robustness to reduce structural vibrations under different loading conditions. Usually the energy required by a hybrid mass damper is far less than that required by an active mass driver with comparable performance. Hybrid seismic isolation systems, through the installation of additional active devices into seismic isolation systems, can achieve the isolation effect while keeping the base displacement at low levels.

1.2 Semi-active Forcing Systems

The semi-active forcing systems are based on low-power, high-performance force actuators called semi-active devices [4, 8, 6]. These devices are dissipative like passive devices, but prossess fast-acting variable properties which may be “tuned” in real time to optimize the dynamic response of the structure. Despite their limitations, such devices exhibit some of appealing traits of active systems, in that they are capable of real-time force control, although in a limited range.

A semi-active control device has properties that can be adjusted in real time but can not inject energy direct into the controlled system and can't destabilize it. The parameters of a semi-active device situated in one location in the structure may be controlled based on the dynamic response of the entire structure. The only power necessary for operation of semi-active forcing systems is that which is needed for the sensors and control intelligence, and to tune the device parameters.

A theoretical example of semi-active device is presented on Fig. 2. The actuator accepts an input parameter γ, and for different values of γ, the relationship between the actuator velocity $\dot{x}$ and the actuator force f has a different shape.

The input parameter γ is formulated through some control system to be such that the resultant force f will be of a desired magnitude. This parameter corresponds to some property of the device which may be modified to yield the different curves in Fig. 1.2.

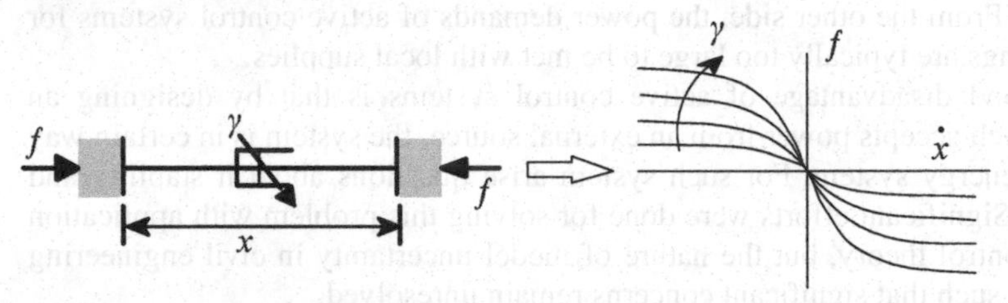

Fig. 2 Idealized semi-active damper

The simplest example of a real semi-active device is the variable-orifice damper, where the parameter γ represents the size of the orifice through which the viscous fluid in the damper is forced to flow, due to the motion of the piston. Theoretically, it requires no work to change the size of the orifice, which simply controls the amount of energy dissipated by the surrounding mechanical system. Therefore, if the device is ideal, there isn't power flow, associated with parameter γ. In a real system there is a small amount of work associated with γ, because there will be losses in the electromechanical system which changes the orifice size, but these losses are extremely small, in comparison to the power flow of the actuator. Variable orifice dampers can achieve variable damping by changing the hydraulic fluid flow resistance using an electromechanical variable orifice. This type of dampers was worked out to control the motion of bridges experiencing seismic motion. Because of its time-varying damping coefficient, which makes it difficult to apply standard control methods, bang-bang control and instantaneous optimal control methods were developed for this system.

The control parameters of semi-active actuators don't have significant power flow associated with them. Most semi-active devices are capable of operating for extended periods of time only on battery for power supply, even those which have force capabilities of over 10 tons.

Semi-active devices and control systems received a fair amount of attention in recent years, because of its potential for reliable, low power structural control. There are two distinct advantages of semi-active control:

(1) For a structure which is open-loop stable (stable without control), the implementation of a control system with semi-active actuators physically cannot destabilize the structure.
(2) Semi-active actuators are unaffected by external power supply failures, which make them more reliable.

Recent works has indicated that semi-active control systems, when appropriately implemented, achieve significantly better results than passive control systems. They may even outperform fully-active control systems, demonstrating significant potential for controlling structural responses to a wide variety of dynamic loading conditions.

The control strategy of a semi-active control system is based on the feedback of structural motions. Different control algorithms can be adopted directly from active control systems. However, semi-active control systems are typically nonlinear due to the intrinsic nonlinearities of semi-active devices. The development of efficient control strategies is still an open research topic.

2 Regenerative Force Actuation

The *Regenerative Force Actuation* is realized with a separate class of actuators for structural control. Such devices are capable of two-way power flow, like active and semi-active devices. The regenerative actuators have external power supply demands which are orders of magnitude below their power flow capabilities. They have two main characteristics:

(a) *Power storage and reuse:* Semi-active device must always remove energy from the mechanical system. The regenerative force actuation differs from semi-active devices with the fact that have the capability of storing at least a fraction of the energy they remove from the mechanical system, and of re-injecting that energy back into the mechanical system at a later time.
(b) *Power coupling in actuation networks:* When multiple regenerative actuators are distributed throughout a structure, they may be capable of "sharing" power with each other. For example, one device may remove energy from a mechanical system from one location, while another device simultaneously re-inject that energy back into the mechanical system.

The concept of regenerative actuation is implemented in numerous areas of structural control and there exist some differences as to exactly what qualifies an actuation system as "regenerative". These discrepancies arise from how the energy storage system is modeled and if into the regenerative force actuation are implemented multiple actuators, are they sharing the common power supply.

2.1 Electromechanical Dissipation

The actuators dissipate absorbed seismic energy through mechanical means. For example, the variable-orifice and the controllable fluid dampers dissipate energy as liquid passed through an orifice. The variable friction damper dissipates energy at the contact surface. However, energy may also be dissipated through electrical means. For example, in the control of flexible structures with piezoelectric materials, considerable reduction in vibration can be achieved by application of electrical RL shunt impedances across the terminals of the piezoelectronic actuator. Such impedances store and dissipate electrical energy, transducted by the piezoelectric material from the vibrating structure were expanded to a semiactive framework by allowing the resistor to be real time controllable.

Because of the physical scale of civil structures, the piezoelectric materials used in lightweight, flexible aerospace structures cannot be employed. Instead an electric motor, by operating as a generator, could be used to convert mechanical to electrical energy. Then this energy could be dissipated in an electrical network. Semi-active control through electrical dissipation has an advantage over mechanical methods in that such devices may also be operated as active devices if electrical power is available.

A semi-active device which uses an electric motor to facilitate energy dissipation may be driven as an active device by a simple switch in the circuitry to which the machine is connected. This would give to actuator versatility not attainable with most other semi-active devices. One of the best electromechanical actuators for civil structures is the permanent –magnet brushless DC machine, presented on Fig. 3.

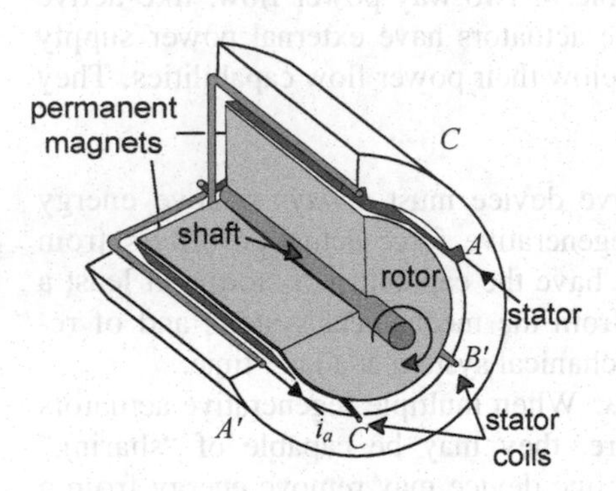

Fig. 3 A permanent magnet brushless electromechanical actuator

The permanent magnets are mounted to the rotor which provides a rotating magnetic field as the rotor spins. The interaction of this magnetic field with the stator windings provides the avenue by which mechanical power on the machine shaft is converted to electrical power in the stator coils, and vice versa.

These machines are used in the design of semi-active electromechanical devices, where schematic diagram of such device is shown on Fig. 4.

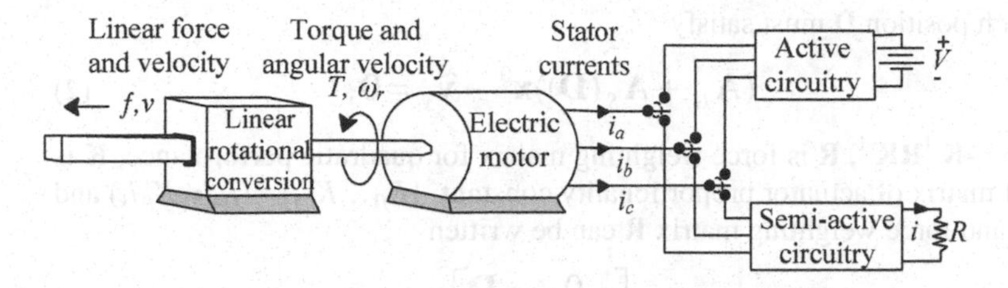

Fig. 4 Diagram of device with active/semi-active capability

This diagram shows a rotational electric motor being used as a linear force actuator, by means of a linear to rotational converter consisting of a gear reduction and screw mechanism. Two electrical circuits are connected to the motor, depending on the positions of three switches. The upper circuitry is used for active control and is connected to an external power source and is designed to actively drive the force actuator. When the switches are in down position, the motor is connected to semi-active circuitry, consisting of a network of passive components, like resistors, inductors, etc., and transistors. The active circuitry on Fig. 4 is used when external power supply voltage V is available – otherwise is realized switch to semi-active circuitry.

In semi-active mode of operation, the device uses the motor as a generator which converts mechanical energy to electrical energy, which is dissipated in the network. The energy dissipation is regulated by controlling the transistors, which are used as electronic switches in the network. By power selection of the voltage, the transistor resembles a mechanical switch in parallel with a diode. This mode of operation consumes very little power, even if the electric power flowing through the network is very high. For a semi-active system the only demand for an external power source are connected with these transistor switches together with the sensors and control intelligence.

2.2 Switching Equilibrium

At regenerative force actuation the description of the system dynamics does not explicitly state which operating points (i.e. which electromechanical forces $\mathbf{f}_e$ at a velocity $\mathbf{v}$) the system is capable of producing. Clearly, there exist some operating points that are possible, and other that are not possible. Even if the regenerative force actuation network is ideal, its limitations would constrain the forces it could achieve.

The switch position $\{D_1 \dots D_m\}$ can be either 1 or -1, and $\mathbf{D}$ lie in a set of 2^{m+1} possible vectors. For a switch position $\mathbf{D} = [D_1 \dots D_m D_R]^T$, where D_R is the switch position for dissipative interface, to bring $\mathbf{x}$ to equilibrium about $\mathbf{x}^0$ at constant velocity $\mathbf{v}$, in the traditional sense, it would have to satisfy

$$\dot{\mathbf{x}}(\mathbf{x}^0, \mathbf{v}, \mathbf{D}) = 0 . \tag{1}$$

For a switch position **D** must satisfy

$$(\mathbf{A}_H + \mathbf{A}_S(\mathbf{D}))\mathbf{x}^0 - \tilde{\mathbf{v}}_x = 0, \quad (2)$$

where $\mathbf{A}_H = -\mathbf{K}^{-1}\mathbf{R}\mathbf{K}^{-1}$, **R** is force weighting matrix for quadratic performance, **K** is a diagonal matrix of actuator proportionality constant $\{K_{f1} \ldots K_{fm}\}$ - $(f_{ek} = K_{fk} i_k)$ and for $\mathbf{A}_s(\mathbf{D})$ and force weighting matrix **R** can be written

$$\mathbf{A}_S(\mathbf{D}) = \mathbf{K}^{-1} \begin{bmatrix} 0 & \mathbf{D} \\ -\mathbf{D}^T & 0 \end{bmatrix} \mathbf{K}^{-1}, \quad (3)$$

$$\mathbf{R} = \begin{bmatrix} \mathbf{R}_f & & \\ & R_R & \\ & & 1/R_S \end{bmatrix}. \quad (4)$$

For each **D** (2) has a unique solution for the state $\mathbf{x}^0$, because $\mathbf{A}_H$ + $\mathbf{A}_S(\mathbf{D})$ is nonsingular. So in general (1) is only possible for 2^{m+1} values of $\mathbf{x}^0$. Thus, due to the switching nature of the control input for regenerative force actuation is used a weaker conception of equilibrium. This issue is common to all dynamic switching systems.

There exist two common approaches:

(1) Duty cycle-based switching;
(2) Lyapunov-based switching.

For consideration the first one assume that each switch position goes through a cycle with period T_S. During each cycle the switch starts at position one, toggles to position 2, and at the end of the cycle returns to position 1. Let **d** is the average of **D** over one cycle. Assuming that the value of **d** changes very little from one cycle to the next, it may be viewed as a continuous function of time $\mathbf{d}(t)$.

The components of $\mathbf{d}(t)$ are called duty cycles or duty ratios of the switches. Assuming that $\mathbf{d}(t)$ varies slowly over any duration T_S, the dynamics of the system may be approximated by a “switch-averaged” model, where for equilibrium the state $\mathbf{x}^0$ must be such that there exists an admissible **d** such that satisfy

$$(\mathbf{A}_H + \mathbf{A}_S(\mathbf{d}))\mathbf{x}^0 - \tilde{\mathbf{v}}_x = 0. \quad (5)$$

This approach allows for control input to be viewed as a continuous signal. This in turn allows linearization of non-linear system about an operating point. In this way the control algorithm for **d** can be designed through linear control system theory.

In practice duty cycle-based switching approach isn’t often implemented, because most regenerative force actuation networks implement *sliding-mode control* which doesn’t involve linear system analysis.

In Lyapunov-based switching approach to the analysis of equilibrium for switching systems the goal is to determine the set of $\mathbf{x}^0$ values for which there exists a switching control algorithm $\mathbf{D} = \mathbf{F}(\mathbf{x})$ such that in the closed-loop system **x** is stable about $\mathbf{x}^0$ in the Lyapunov sense. It is assumed that the switches in the network have negligible switching transition times and that they may alternate infinitely rapidly.

Define the state vector $\mathbf{x}$ and switch position vector $\mathbf{D}$ as

$$\mathbf{x} = [\mathbf{f}_e^T \quad i_R \quad V_S]^T \quad \mathbf{D} = [D_1 ... D_m D_R]^T. \tag{6}$$

Then the differential equation for description of the entire electrical network is

$$\dot{\mathbf{x}} = \Delta(\mathbf{A}_H \mathbf{x} + \mathbf{A}_S(\mathbf{D}) - \tilde{\mathbf{v}}_x). \tag{7}$$

with

$$\mathbf{K}_f = diag(K_{f_1} ... K_{f_m}), \quad \Delta_f = \mathbf{K}_f^2 diag(1/2L_1 ... 1/2L_m), \quad R_f = diag(2R_1 ... 2R_m)$$

$$\mathbf{K} = \begin{bmatrix} \mathbf{K}_f & \\ & \mathbf{I} \end{bmatrix}, \quad \Delta = \begin{bmatrix} \Delta_f & & \\ & 1/L_R & \\ & & 1/C_S \end{bmatrix}, \quad \mathbf{R} = \begin{bmatrix} \mathbf{R}_f & & \\ & R_R & \\ & & 1/R_S \end{bmatrix}.$$

where C_S is the capacitance of DC bus, K_{fk} is actuator proportionality constant: $f_{ek}=K_{fk}i_k$, R_R is resistance of dissipative interface, R_S is the resistance at state S, $\mathbf{v}$ is actuation velocity vector and m is number of actuators.

Definition: Let $\mathbf{v}$ is assumed to be constant and $\mathbf{D}$ is determined by a memoryless feedback law $\mathbf{D} = \mathbf{F}(\mathbf{x})$. Then the system is in *switching equilibrium* about a point $\mathbf{x}^0$ if there exists a function U, such that $U(\mathbf{x})$ is a Lyapunov function, i.e.

$$U(\mathbf{x}) \geq 0, \quad \frac{dU}{dt} \leq 0, \quad \frac{dU}{dt} = 0 \Leftrightarrow \mathbf{x} = \mathbf{x}^0. \tag{8}$$

A state $\mathbf{x}^0$ is called a switching equilibrium point.

2.3 Stability of Operating Points

By definition the switching equilibrium points are always stable and characterized by an ever-shrinking neighborhood, inside which lie $\mathbf{x} - \mathbf{x}^0$. In general, it is nontrivial to prove stabilityfor feedback systems because there may be sliding modes in the system dynamics on which $\mathbf{D}$ is undefined [11].

In switching equilibrium energy should be conserved, i.e. the aggregate amount of electrical power generated by the machines should equal the amount dissipated in the network

$$\mathbf{x}^{0T} \mathbf{A}_H \mathbf{x}^0 - \mathbf{x}^{0T} \tilde{\mathbf{v}}_x = 0, \tag{9}$$

where it is necessary and sufficient condition for $\mathbf{x}^0$ to be a switching equilibrium point.

3 Simulation of Nonlinear Dynamics of Structures at Seismic Excitation

The process of simulation of nonlinear dynamics of structures at seizmic excitations is considered as an *three component elasto-mechanical model* of interaction between *seismic waves*, *soil layers* and *structure*. During this simulation under investigation are relative displacements and absolute accelerations of each floor and the basement of the structure. The simulation experiments are provided with and without control to compare the received results after implementation of chosen type of structural control. Such simulations are realized with strong motion real or synthetic seismic signals. Their spectral characteristics are strong dependent on the structure and parameters of the soil layers. That's why it is necessary to investigate the soil layers under structure which leads to additional modeling and simulation.

3.1 Strong Motion Seismic Waves Simulation

The simulation experiments provided with and without structural control at seismic excitations have to be realized with great many strong motion seismic records with different spectral characteristics. In real practice there exists only few real register seismic records for considered region. The testing of structural control systems have to be produced with variety of strong motion seismic waves with different spectral characteristics. This leads to developing of approaches for generating of sinthetic strong motion seismic records (acelerograms) [13, 15].

The characteristics of accelerograms depend on the distance from the source of seismic excitation, it magnitude and geological characteristics of soil layers. Most often separately are modeled three main phases of seismic wave: primary (P- waves), transversal or secondary (S- waves), and converted and guided waves (C-/G- waves). The destructive are S- waves and their behavior is under investigation. For evolutionary power spectrum estimation is used the time dependent stochastic principal axes method. According to this method earthquake records (accelerograms) are delivered as representations of the three-dimensional acceleration vector in a Cartesian coordinate system, generally with axes parallel to east –west, north-south and vertical direction.

Strong motion seismic waves simulation is realized with space multiple signals processing, where are estimated the energy spectra of one-dimensional and multiple-dimensional seismic signals. Such procedure includes the spectral correlation matrix and generation of sets of synthetic one-dimensional signals. The spectral correlation matrix is estimated via combinations of spectral estimations for all of the phase differences of surface wave delays at the half-space, received for given frequency.

In a homogeneous elastic half-space with no damping simulation of surface seismic waves is realized at a constant phase velocity, i.e. different frequencies propagate with the same velocity. In a layered or vertically heterogeneous half-space, such as an ideally layered soil profile, different frequency surface waves propagate and are modelled with different phase velocities and different wavelengths. The relationship between phase velocity, frequency and wave number is called *dispersion relation*.

The simulation is done for a sequence of small windows, where the nonlinear parameters are modelled like linear. The lower frequency waves tend to propagate and are modelled with higher velocities, where several modes may exist in a single soil profile. Higher modes propagate with longer wavelengths and sample the soil profile to larger depths.

The main simulated parameters are dispersion relation and attenuation, which is modelled as a nonlinear regression. To these parameters are added several physical and environmental phenomena like possible presence of multiple modes, interfering waves and interfering background or ambient noise energy.

The phase velocity as a function of frequency, or dispersion curve is one of the primary functions of interest at simulation. Traditionally, the phase velocity is estimated over a range of two-point spatial lags, and the two-point estimates are averaged to estimate a single wave number as a function of frequency. Using spatial arrays, the phase velocity can be estimated with several sensors simultaneously, rather than utilizing only two point estimates of phase change.

Two traditional attenuation estimation methods can be implemented at simulation:

(1) Linear regression technique, which relies on a simple model design;
(2) Non-linear methods which attempts to more accurately model of modes superposition and geometric spreading.

The traditional linear regression estimator uses the $\sqrt{r}$ decay approximation, where r is the distance from the earthquake source, and estimates a single attenuation coefficient as a function of frequency $\alpha(\omega)$.

The traditional non-linear regression technique uses the theoretical multiple mode geometric spreading function $G(\omega,r)$, obtained using the method of reflection and transmission of coefficients, and fits a single multimodal attenuation coefficients as a function of frequency $\alpha(\omega)$.

Spherical wave propagation follows the following wave equation

$$s(x,t)=\frac{A}{x}e^{j(\omega t-kx)}, \tag{10}$$

where $s(x,t)$ is displacement at spatial position x at time t and A is initial amplitude of the wave. Spherical wave propagation is symmetric, so the wave front depends only on the scalar distance x from the source and the scalar wave number k.

Cylindrical waves follow a Henkel type solution, given by

$$s(x,t)=A\mathrm{H}_0(kx)e^{-j\omega t}, \tag{11}$$

where H_0 is the Henkel function. The exponential term gives the motion harmonic time dependence, where the Henkel function is determined as

$$\mathrm{H}_0(kx) = \mathrm{J}_0(kx) + j\,\mathrm{Y}_0(kx), \tag{12}$$

where J_0 is the Bessel function. At fixed time $t = 0$, the real valued motion due to the earthquake source motion with amplitude A and frequency ω equals the real part of (11) or it can be written

$$s(x,0) = A\mathrm{J}_0(kx), \tag{13}$$

Traditional surface seismic wave analysis uses the far-field Henkel function approximations for both phase velocity estimation and energy attenuation. The Henkel approximation ca be given in several different forms as

$$\mathrm{H}_0(kx) \approx \sqrt{\frac{2}{\pi kx}} \exp\left[i\left(kx - \frac{\pi}{4} \right) \right], \tag{14}$$

The real valued Rayleigh surface wave vertical motions, which are destructive are given as

$$\mathrm{J}_0(kx) \approx \sqrt{\frac{2}{\pi kx}} \cos\left(kx - \frac{\pi}{4} \right), \tag{15}$$

the horizontal motion associated with Rayleigh surface wave are $\pi/2$ radians out of phase with the vertical motions in elastic media.

The asymptotic expansion of the Henkel function (14) clearly shows the $1/\sqrt{x}$ decay and plane wave nature of the cylindrical wave equation in the far-field.

3.2 Soil Layers Simulation

The simulation experiments provided with and without structural control at seismic excitations have to be realized with a model of soil layers under the structure, where for each layer the shear modulus and shear damping ration have to be determined [14].

Soil exhibit complex material response characteristics, depending on factors including strain level, state of effective stress, and loading. In general, soils are non-linear, inelastic materials, but at very small strain levels, the material response can be assumed linear viscoelastic. The primary dynamic material properties, *shear modulus* and *shear damping ratio*, can be determined from several different types of tests, each offering different advantages and disadvantages. Their simulation is very important because the shear modulus controls the *velocity* of shear wave propagation, and the shear damping ratio controls *energy dissipation*.

The material properties of layered or vertically heterogeneous soil profiles strongly affect the magnitude of ground surface shaking due to energy propagation along the surface of the earth. The model consists of n homogeneous soil layers above a homogeneous half-space. For each layer, the density ρ and the Lame parameters λ and μ, specify the material response to dynamic excitation.

Energy dissipation is realized with the following three mechanisms:

(1) Geometric spreading;
(2) Apparent attenuation;
(3) Material attenuation.

Geometric spreading refers to the spreading of a fixed amount of energy over a large area during propagation away from the earthquake source, e.g. cylindrically and spherically spreading waves. Apparent attenuation includes scattering, reflection, and mode conversion energy losses due to obstacles and material boundaries. Material attenuation is an intrinsic property of the material being modeled and simulated.

Particle displacement magnitudes are simulated as a function of distance from the seismic source. Traditional attenuation coefficients are fit to the experimental data using the following general wave-field model:

$$A(\omega,r) = A_0(\omega)Ge^{-\alpha(\omega)r}, \tag{16}$$

where $A(\omega,r)$ is the magnitude of the particle displacement as a function of frequency and distance r from the active point source of earthquake, $A_0(\omega)$ is the magnitude of the source as a function of frequency, G is a function accounting for geometric spreading and $\alpha(\omega)$ is attenuation as a function of frequency.

Since the material attenuation enters into the model as an exponential power, changing to the natural logarithm domain allows the attenuation coefficient to become linear with spatial offset. Taking the natural logarithm of both sides of (16) yields

$$Ln(A(\omega,r)) = Ln(A_0(\omega)) + Ln(G) - \alpha(\omega)r\,. \tag{17}$$

Geometric spreading the energy from a point source complicates simulation of active surface wave material attenuation. In the far-field, point source surface waves decay at a rate proportional to $\sqrt{r}$, and the far-field model is

$$A(\omega,r) = \frac{A_0(\omega)}{r^{0.5}} e^{-\alpha(\omega)r} e^{j(\omega t - kr)}, \tag{18}$$

where $A(\omega,r)$ is spectral displacement magnitude at spatial offset r and frequency ω, and in this case $G = r^{-0.5}$ is constant for all frequencies. Two parameters are necessary to fit the model – the attenuation coefficients $\alpha(\omega)$ and the initial signal amplitude $A_0(\omega)$.

Numerical models of multiple mode wave-fields indicated that the $\sqrt{r}$ decay model is inadequate to describe geometric spreading in different vertically heterogeneous profiles. For this purpose is introduces the Rayleigh geometric spreading function

$$A(\omega,r) = A_0(\omega)G(\omega,r)e^{-\alpha(\omega)r} e^{j(\omega t - \psi(\omega,r))}, \tag{19}$$

where $A_0(\omega,r)$ is summation of all modal Rayleigh wave magnitudes for a given frequency, $G(\omega,r)$ is the geometrical spreading function as a function of frequency and spatial offset, and $\psi(r, \omega)$ is the integrated phase argument from the source position to r due to the superposition of all modes of propagation in soil layers.

For power spectra estimation of spreading and attenuation of surface wave in different soil layers under construction is used *linear prediction*. This approach is included into software part of structural control devices as a prognoses module for estimation of possible appearance of strong motion seismic waves. Linear prediction is a parametric-based spectrum estimation technique, which models the outputs of selected sensor measured displacements as the weighted linear combination of all other sensor outputs. The output of the reference sensor is given by

$$S_{S_0}(\omega) = \sum_{S \neq S_0} w_S^* S_S(\omega), \tag{20}$$

where s_0 is the reference sensor, w_S is the complex-valued linear predictive weights and S_S is the Fourier spectrum of a data vector from sensor s. introducing a column vector δ_{S0} of length S, which is a zero vector except for a one in the s_0 position, the optimum linear predictive weight vector equals

$$w(\omega)_{LP} = \frac{R(\omega)^{-1}\delta_{S_0}}{\delta_{S_0}^H R(\omega)^{-1}\delta_{S_0}}, \tag{21}$$

where $R(\omega)$ is the spatio-spectral correlation matrix for each frequency ω.

An example of power spectrum estimation with linear prediction approach as a function of temporal frequency and wave number received during the simulation of seismic excitation for a model consisted of three soil layers and bedrock is shown on Fig. 5.

The linear prediction method yielded good results for standard three soil layers models, but for models with five layers produce biased dispersion curve estimates.

3.3 Structural Control Simulation

Structural control simulation is realized with *earthquake simulator*, which includes strong motion seismic wave simulation, soil layers simulation and experimental setup of regenerative force actuation network with:

(1) Physical model of investigated structure on shaking table;
(2) Controller;
(3). Control actuators;
(4) Sensor network for measurement relative displacements and absolute accelerations of each floor of the experimental model.

The earthquake simulator produces signals as a result of seismic wave simulation and soil layers simulation. These signals are measured with physical sensors, mounted on the model of investigated structure on shaking table and sent

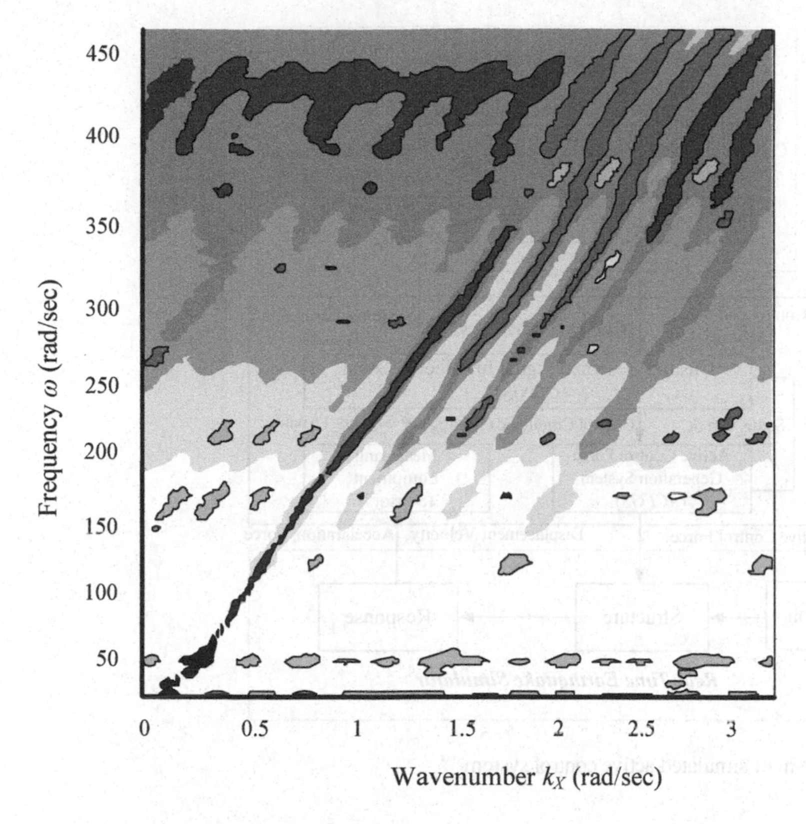

Fig. 5 Power spectuum estimation

to controller. The controller calculates appropriate control signals on the base of measured structural responses. These control signals have to be sent to actuators for active structural control. The control actuators are a function of responses of the system. The basic task for an active control system designer is to determine a control strategy that uses the measured structural responses to calculate appropriate control signals to be sent to actuators.

The objective of active/semi-active structural control is to reduce structural motion during severe external excitation. The block diagram in Fig. 6 shows the detailed hardware function of an active control system. As is seen from the diagram, from *sensors* are measured:

(a) the response of the structure under seismic excitation;
(b) the active device status;
(c) the remote control status;
(d) the fail-safe monitoring status of Active Control Force Generation System.

The results from these measurements are sent to a Custom Designed Signal Interface System. The purpose of this device is to condition and filter signals

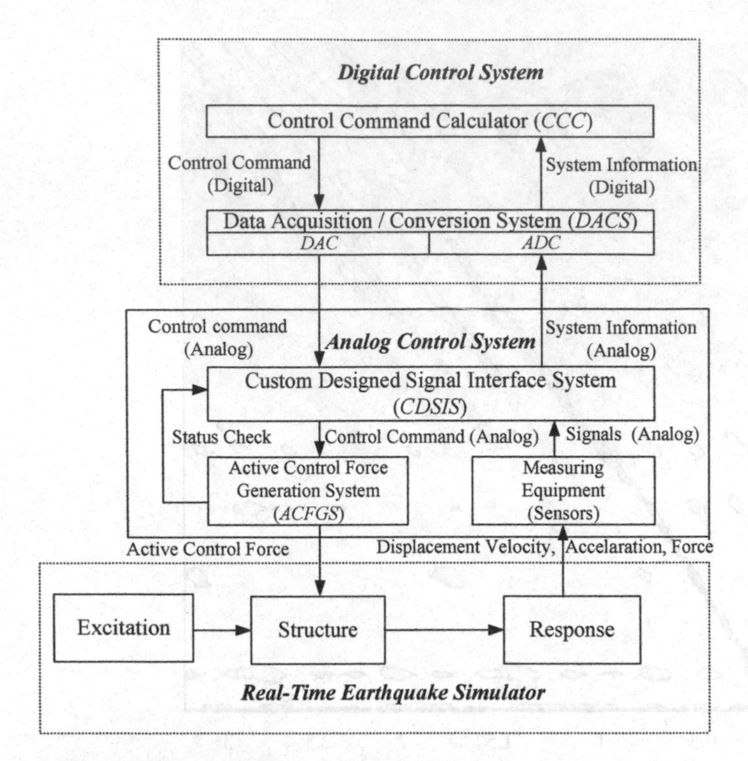

Fig. 6 Diagram of simulated active control system

produced by the sensors. In this interface system are implemented also some customized fail-safe functions, the remote control functions of *Analog Control System*, and specified signal communication functions. The required information is filtered and then sent to the *Digital Control System*. To perform the real-time control the analog signals of the controlled system are measured by sensors and sent through analog-to-digital converter (*ADC*) as corresponding digital numbers with a resolution determined by the *ADC*.

The required control forces are calculated by Control Command Calculator and send to the force generator at Active Control Force Generation System, through the interface Custom Designed Signal Interface System.

The active control force is calculated by a personal computer and Digital Signal Processing (DSP) controller. This DSP controller not only performs the control force calculation, but also conducts large amounts of signal manipulation processes, such as monitoring all the control hardware status, trigging the fail-safe protection function, recording the required measurements information and further more, integrating acceleration measurements to their velocities and displacements. The real-time integration feature overcomes the limitation on sensors or measurement.

As an example, consider a simulation of a controlled single story structure subjected to one-dimensional earthquake excitation $\ddot{x}_g$. The structure is controlled

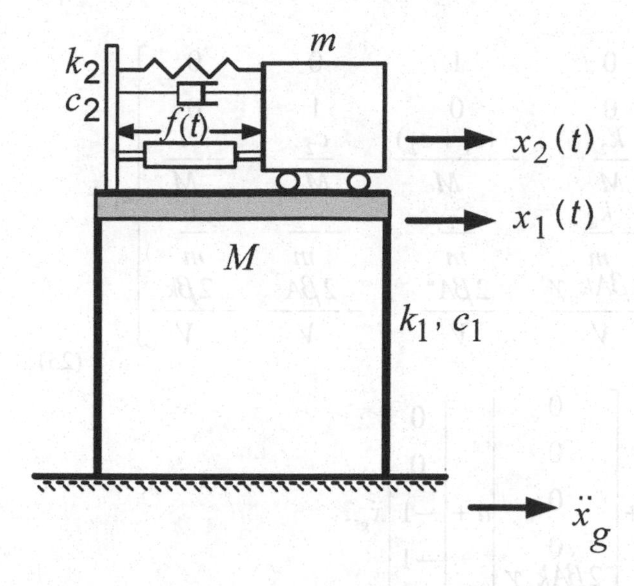

Fig. 7 Single story building with active mass driver

with active mass driver, as shown on Fig. 7, where M is the mass of the building and m is the mass of active mass driver, including the mass of the hydraulic actuator.

The equations of motion are

$$M\ddot{x}_1 + (c_1 + c_2)\dot{x}_1 - c_2\dot{x}_2 + (k_1 + k_2)x_1 - k_2 x_2 = -f - M\ddot{x}_g, \quad (22)$$

$$m\ddot{x}_2 + c_2(\dot{x}_2 - \dot{x}_1) + k_2(x_2 - x_1) = f - m\ddot{x}_g, \quad (23)$$

where x_1 and x_2 are the displacement relative to the ground of the building and the moving mass, respectively, c_1 and k_1 are the damping and stiffness coefficients of the building, c_2 and k_2 are the damping and stiffness coefficients of the active mass driver system and f is the control force applied by the hydraulic actuator.

The equation of dynamics of hydraulic actuator system under unity-gain feedback of the actuator displacement (i.e., x_2 - x_1) can be written as

$$\dot{f} = \frac{2\beta}{V}(Ak_q\gamma\{u - (x_2 - x_1)\} - k_c f - A^2(\dot{x}_2 - \dot{x}_1)), \quad (24)$$

where f is the control force generated by the actuator, β is the bulk modulus of the fluid, V is the characteristic hydraulic fluid volume for the actuator and γ is the proportional feedback gain stabilizing the actuator.

Defining the state vector of the system as $\mathbf{z}_1 = [x_1\ x_2\ \dot{x}_1\ \dot{x}_2\ f]'$, the equation of motion can be written in matrix form as

$$\dot{\mathbf{z}}_1 = \begin{bmatrix} 0 & 0 & 1 & 0 & 0 \\ 0 & 0 & 0 & 1 & 0 \\ -\frac{(k_1+k_2)}{M} & \frac{k_2}{M} & -\frac{(c_1+c_2)}{M} & \frac{c_2}{M} & -\frac{1}{M} \\ \frac{k_2}{m} & -\frac{k_2}{m} & \frac{c_2}{m} & -\frac{c_2}{m} & \frac{1}{m} \\ \frac{2\beta A k_q \gamma}{V} & -\frac{2\beta A k_q \gamma}{V} & \frac{2\beta A^2}{V} & -\frac{2\beta A^2}{V} & -\frac{2\beta k_c}{V} \end{bmatrix} \mathbf{z}_1 + \begin{bmatrix} 0 \\ 0 \\ 0 \\ 0 \\ \frac{2\beta A k_q \gamma}{V} \end{bmatrix} u + \begin{bmatrix} 0 \\ 0 \\ -1 \\ -1 \\ 0 \end{bmatrix} \ddot{x}_g . \tag{25}$$

The measurements are chosen to include the displacement of the first floor mass relative to the ground, the displacement of the active mass driver relative to the first floor mass, the absolute accelerations of both masses and the force applied by the actuator, i.e., $\mathbf{y} = [x_1 \; (x_2 - x_1) \; \ddot{x}_{a1} \; \ddot{x}_{a2} \; f]'$. Thus the measurement equation is

$$\mathbf{y} = \begin{bmatrix} 1 & 0 & 0 & 0 & 0 \\ -1 & 1 & 0 & 0 & 0 \\ -\frac{(k_1+k_2)}{M} & \frac{k_2}{M} & -\frac{(c_1+c_2)}{M} & \frac{c_2}{M} & -\frac{1}{M} \\ \frac{k_2}{M} & -\frac{k_2}{m} & \frac{c_2}{M} & -\frac{c_2}{M} & \frac{1}{M} \\ 0 & 0 & 0 & 0 & 1 \end{bmatrix} \mathbf{z}_1 + \begin{bmatrix} v_1 \\ v_2 \\ v_3 \\ v_4 \\ v_5 \end{bmatrix} , \tag{26}$$

where v_i is the noise in the *i*th measurement.

The next step is to determine the transfer functions from the ground acceleration to each of the measured responses. This can be obtained by exciting the structure with band-limited white noise ground acceleration (0-100 *Hz*) with the control actuator in place and the actuator command set to zero. Similarly, the experimental transfer functions from the actuator command signal to each of the measured outputs have to be determined by applying a band-limited white noise (0-100 *Hz*) to the actuator command, while the ground is held fixed.

Methods for experimental determination of transfer function break down into two fundamental types: swept-sine and the broadband approaches using fast Fourier transforms. Both methods can produce accurate transfer functions estimates. The swept-sine approach is rather time-consuming, because it analyzes the system one frequency at a time. The broadband approach estimates the transfer function simultaneously over a band of frequencies. The first step is to independently excite each of the system inputs over the frequency range of interest. Exciting the system at frequencies outside this range is typically counter productive; thus the excitation should be bound limited (e.g., pseudo-random). Assuming the two continuous signals (input $u(t)$ and output $y(t)$) are stationary, the transfer function is determined by dividing the cross spectral density of the two signals S_{uy} by the auto-spectral density of the of the input signal S_{uu} as

$$H_{yu}(j\omega)=\frac{S_{uy}(j\omega)}{S_{uu}(\omega)}. \tag{27}$$

However, experimental transfer functions are usually determined from discrete-time data. The continuous time records of the specified system input and the resulted responses are sampled at N discrete-time intervals with an A/D converter yielding a finite duration, discrete-time representation of each signal $u(nt)$ and $y(nt)$, where T is the sampling period and n=1,2,…,N. For the discrete case the transfer function (27) can be presented as

$$H_{yu}(jk\Omega)=\frac{S_{uy}(jk\Omega)}{S_{uu}(k\Omega)}, \tag{28}$$

where $\Omega=\omega_S/N$, ω_S is the sampling frequency, k=0,1,..N-1. The discrete spectral density functions are obtained via standard digital signal processing methods. This frequency transfer function can be thought of as frequency sampled version of the continuous transfer function in (27).

In practice, one collection of samples of length N does not produce very accurate results. Better results are obtained by averaging the spectral densities of a number of collections of samples of the same length. Given that M collection of samples are taken the equations for averaged functions are

$$\bar{S}_{uu}(k\Omega)=\frac{1}{M}\sum_{i=1}^{M}S_{uu}^{i}(k\Omega);\quad \bar{S}_{uy}(j\Omega)=\frac{1}{M}\sum_{i=1}^{M}S_{uy}^{i}(jk\Omega);\quad \bar{H}_{yu}(jk\Omega)=\frac{\bar{S}_{uy}(jk\Omega)}{\bar{S}_{uu}(k\Omega)} \tag{29}$$

where S^i denotes the spectral density of the i-th collections samples and the over bar represents the ensemble average. Note, that increasing of number of samples N, increases the frequency resolution, but does not increase the accuracy of the transfer functions.

On Fig. 8 is shown one experimental obtained transfer function.

The experimental transfer function obtained in the system identification procedure is modeled as a ratio of two polynomials in the Laplace variable s. In this step the effects of control-structure interaction should be consistently incorporated into the identification process. Thus, the zeros of the transfer function from the actuator command to the applied force are the poles of the structure (when the actuator is not attached). As is seen from Fig. 8, the resulting experimental and analytical models of transfer function are almost overlaid for

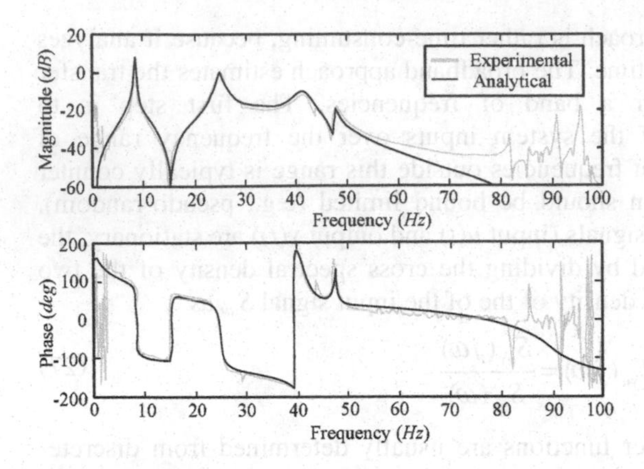

Fig. 8 Experimental obtained transfer function

frequency values between 0-100 *Hz*. The quality of the mathematical models for the remaining transfer functions is similar to depicted here.

Next, the model has to be assembled in state space form. At least ten states have to be used to model the system corresponding to the ground acceleration input

$$\dot{\mathbf{x}}_1 = \mathbf{A}_1\mathbf{x}_1 + \mathbf{B}_1\ddot{x}_g,$$
$$\mathbf{y} = \mathbf{C}_1\mathbf{x}_1 + \mathbf{D}_1\ddot{x}_g, \tag{30}$$

over the frequency range of interest, where $\mathbf{A}_1$, $\mathbf{B}_1$, $\mathbf{C}_1$ and $\mathbf{D}_1$ are in controller canonical form, $\mathbf{x}_1$ is the state vector, and $\mathbf{y}$ is the vector of measured structural responses.

At least twelve states have to be used to form the second state equations for modelling the input/output relationship between the command to each actuator u_i for i=1,2,…,l and the responses $\mathbf{y}$ given by

$$\dot{\mathbf{x}}_{21} = \mathbf{A}_{2i}\mathbf{x}_{2i} + \mathbf{B}_{2i}u_i,$$
$$\mathbf{y} = \mathbf{C}_{2i}\mathbf{x}_{2i} + \mathbf{D}_{2i}u_i, \tag{31}$$

where $\mathbf{A}_{2i}$, $\mathbf{B}_{2i}$, $\mathbf{C}_{2i}$ and $\mathbf{D}_{2i}$ are in controller canonical form, $\mathbf{x}_{2i}$ is the state vector for system corresponding to each input u_i.

These systems typically contain all of the poles corresponding to the building modes, plus additional dynamics required to model the actuators. Because the transfer function characteristics from the ground to the building response are dominated by the dynamics of the building, the system in (30) typically requires fewer states, corresponding to the dominant modes of the building, to accurately model the experimental transfer functions over the frequency range of interest.

Once each of the two state space systems is assembled a model reduction is performed, all of the transfer functions of the reduced – order model are compared to the experimental data. The quality of the transfer functions corresponding to the actuator command input are of higher quality than those corresponding to the ground acceleration input.

For the design of effective and robust controllers, the model should accurately predict the response of the system to the control signal. Therefore, in the system identification procedure, efforts are focused on obtaining an accurate model from the actuator command to the structural responses in the frequency range of interest.

The identified model of the system for structural control is used for control design and experimental verification of acceleration feedback strategies for structural control.

4 Switching Control and Control Algorithms

Let for a given *actuation velocity* $\mathbf{v}$, the region of feasible forces for a regenerative force actuation is $S(\mathbf{v})$. This region has boundaries with approximately elliptical shape and arises from losses in the electrical system. It is assumed that $\mathbf{v}$ is a slowly-varying parameter, compared to the electrical dynamics. The switching control is realized with a feedback controller which facilitates zero-error tracking for a force command $\mathbf{f}^*_e$, assuming that $\mathbf{f}^*_e \in S(\mathbf{v})$. This controller relates the switch position vector $\mathbf{D}$ to the electrical state vector $\mathbf{x}$.

The closed-loop performance of the system must be robust to parametric uncertainty in the electrical network, because it is assumed that the system parameters during a seismic excitation are not known precisely. This problem is made more complicated by the fact that because of this system uncertainty, the boundary $\partial S(\mathbf{v})$ is also uncertain. Thus, for a force command $\mathbf{f}^*_e$ which is in the nominal (i.e., modelled) system, the region of feasible forces $S(\mathbf{v})$ may be outside the physical $S(\mathbf{v})$ and for this situation the force tracking is physically impossible.

4.1 Control Design

The primary goal of the control design is to minimize the accelerations of each floor of the structure. The controller should be designed to be robust, such that in these circumstances the system equilibrates to a value of the vector of electromechanical forces $\mathbf{f}_e$ which is on the boundary of the actual region of feasible forces $S(\mathbf{v})$, and "close" to the $\mathbf{f}^*_e$ command connected with implementation of the optimal value of the vector of electromechanical forces.

An additional complication to the control system design is the fact that the switches have a maximum frequency at which they can operate. It is possible to design switching controllers without considering this constraint. Then, the controller design can be modified to accommodate switching frequency limitations.

Consider a state $\mathbf{x}^0 = [\mathbf{f}_e^{0T}\, i_R^0 V_S^0]^T$ to be a switching equilibrium point

$$V_S^0 \geq \max\{\| \mathbf{K}_f \mathbf{v} \|_\infty, \| \mathbf{K}_f \tilde{\mathbf{v}} + \mathbf{K}_f^{-1} \mathbf{R} \mathbf{f}_e^0 \|_\infty\} \tag{32}$$

$$0 \le i_R^0 \le \max\{0, (V_S^0 - V_{swR}) / R_R\} \tag{33}$$

$$\mathbf{f}_e^{0^T}(\mathbf{C}_c^{-1}\mathbf{f}_e^0 + \tilde{\mathbf{v}}_f) + V_S^{0^2} / R_S + i_R^0(V_{swR} + R_R i_R^0) = 0 . \tag{34}$$

Let the vector $\mathbf{x}^* = [\mathbf{f}_e^{*T} i_R^* V_S^*]^T$ be a desired set point for the state of the system. If **x*** satisfies (22)-(34) than it is a feasible operating point for the system. The controller for a closed-loop system will be of the form

$$\mathbf{D} = \mathbf{F}(\mathbf{f}_e, V_S; \mathbf{f}_e^*, \mathbf{v}) , \tag{35}$$

where **F** is a memoryless function of $\mathbf{f}_e$ and V_S, but not of i_R (current in resistor R_R). Furthermore, **F** also depends on the command $\mathbf{f}_e^*$ and **v**. The closed-loop performance of the controller has the following requirements:

(1) In closed loop, $\mathbf{f}_e^0 = \mathbf{f}_e^*$ whenever $\mathbf{f}_e^* \in S_{Kv}(\mathbf{v})$;

(2) In closed loop, $V_S^0 \ge V_S^*$ whenever $\mathbf{f}_e^* \in S_{Kv}(\mathbf{v})$.

The feedback law that satisfy both requirements can we written as

$$\mathbf{F}(\delta \mathbf{f}_e, \delta V_S^*) = Dsat\left(\begin{bmatrix} -\mathbf{1} & \mathbf{f}_e^* u(-\delta V_S) \\ & V_S^* \\ \mathbf{0} & 1 \end{bmatrix} \begin{bmatrix} \delta \mathbf{f}_e \\ \delta V_S \end{bmatrix} \right) \tag{36}$$

Consider an arbitrary n-degree of freedom base excited shear structure equipped with a regenerative force actuation network. The response of such structure at seismic excitation can be determine as

$$\mathbf{M}_S \ddot{\mathbf{q}} + \mathbf{C}_S \dot{\mathbf{q}} + \mathbf{K}_S \mathbf{q} = -\mathbf{M}_S \mathbf{G} a_g + \mathbf{N}\mathbf{f} . \tag{37}$$

where **q** is the structural displacement vector relative to the base, a_g is the base acceleration, **f** is the m-vector of actuator forces, m is the number of actuators included in RFAN for protection of the structure against destructive influences of strong motion seismic waves, $\mathbf{C}_S$ is the capacitance of DC bus, **N** is the structural force control input matrix and **G** is the matrix of input seismic excitations on the structure.

The forces **f** are degenerate if the matrix **N** has rank less than m (i.e., if **N** has a nontrivial null space). It is assumed that the actuation system is non-degenerate.

The structural force control input matrix **N** relates the actuation velocities **v** to the structural velocities $\dot{\mathbf{q}}$ through

$$\mathbf{v} = \mathbf{N}^T \dot{\mathbf{q}} . \tag{38}$$

The relationship between the vector of actuator forces **f** and the vector of regenerative actuation electromechanical forces $\mathbf{f}_e$ determines the structural dynamics as

$$\mathbf{f} = \mathbf{H}(\mathbf{f},\mathbf{v})[\mathbf{f}_e - \mathbf{L}^{-2}\mathbf{B}_A\mathbf{N}^T\dot{\mathbf{q}} - \mathbf{L}^{-2}\mathbf{J}_A\mathbf{N}^T\ddot{\mathbf{q}}]. \tag{39}$$

where $\mathbf{H}(\mathbf{f},\mathbf{v})=\mathrm{diag}\{h_1(f_1v_1),...,h_m(f_mv_m)\}$, $\mathbf{L}=\mathrm{diag}\{l_1,\ldots,l_m\}$,$\mathbf{B}_A=\mathrm{diag}\{B_1,\ldots,B_m\}$ is the acceleration input matrix for nominal system model, $\mathbf{J}_A=\mathrm{diag}\{J_1,\ldots,J_m\}$ is the structural control system performance measure. In this way the response of the structural system can be expressed as

$$\mathbf{M}_S + \mathbf{N}\mathbf{H}(\mathbf{f},\mathbf{v})\mathbf{L}^{-2}\mathbf{J}_A\mathbf{N}^T]\ddot{\mathbf{q}} + [\mathbf{C}_S + \mathbf{N}\mathbf{H}(\mathbf{f},\mathbf{v})\mathbf{L}^{-2}\mathbf{B}_A\mathbf{N}^T]\dot{\mathbf{q}} + \mathbf{K}_S\mathbf{q} = -\mathbf{M}_S\mathbf{G}a_g + \mathbf{N}\mathbf{H}(\mathbf{f},\mathbf{v})\mathbf{f}_e. \tag{40}$$

Equation (40) together with the electrical dynamics of $\mathbf{f}_e$ fully describes the *physical model* of the structure – actuator system. This model is suitable for the purposes of system analysis and for simulation purposes. It contains the *attenuation* of low-frequency dynamics of the structure and as well the *switching dynamics* of the electronics. The model also *includes nonlinearities*, some of which make dynamic system analysis very difficult. That's why were made some simplifications in this analytical model for practical purposes and as a result was received so called *Nominal System Model* (NSM). It contains the most fundamental traits of the electromechanical control system with the following approximations:

(1) *Lossless electronic switches*: this assumption is equivalent to assuming that the semiconductor switches in the electrical system are ideal. The influence of semiconductor losses on $S(\mathbf{v})$ is only significant for small $\|\mathbf{v}\|$. If these effects are neglected, the boundary $\partial S(\mathbf{v})$ is described by a simple quadratic equation.
(2) *Instantaneous electrical dynamics*: this assumption is equivalent to assuming that the electric control system can instantaneously realize any force, i.e. $\mathbf{f}^*_e=\mathbf{f}_e$. It is a reasonable assumption because the electric time constants are designed to be much smaller than the mechanical time constants.
(3) *Lossless screw conversion*: For high-efficiency conversions $\mathbf{H}(\mathbf{f},\mathbf{v}) \approx \mathbf{I}$. This approximation allows for the mechanical system to be represented as a linear differential equation. If this approximation is realized than (39) has a closed-form solution for $\mathbf{f}$, because $\mathbf{f}$ contributes to the accelerations $\ddot{\mathbf{q}}$ in (39).
(4) *No electrical parameter uncertainty or time delay*: The switching controller is designed to be robust to such uncertainties.

All these assumptions are used for the purposes of system analysis and controller design. The true physical model is a more accurate description of the system behavior. The simplified model resulting from these assumptions is a linear mechanical system. The viscosity and inertia of the rotor may be reflected into the structural damping and inertia to obtain the following system description

$$\mathbf{M}_{SA}\ddot{\mathbf{q}}+\mathbf{C}_{SA}\dot{\mathbf{q}}+\mathbf{K}_S\mathbf{q}=-\mathbf{M}_S\mathbf{G}a_g+\mathbf{N}\mathbf{f}_e, \tag{41}$$

where

$$\mathbf{M}_{SA}=\mathbf{M}_S+\mathbf{N}\mathbf{L}^{-2}\mathbf{J}_A\mathbf{N}^T \quad \mathbf{C}_{SA}=\mathbf{C}_S+\mathbf{N}\mathbf{L}^{-2}\mathbf{B}_A\mathbf{N}^T. \tag{42}$$

The suggested system incorporates the mechanical dynamics of the rotor shafts with the structural dynamics during earthquakes when $\mathbf{f}_e$ is the control input of the system. The combined mass and damping matrixes $\mathbf{M}_{SA}$ and $\mathbf{C}_{SA}$ retain symmetry with the inclusion of the rotor inertia and viscosity, because $\mathbf{L}$, $\mathbf{J}_A$ and $\mathbf{B}_A$ are diagonal. The mass matrix $\mathbf{M}_{SA}$ in general doesn't retain diagonal.

Consider a feedback control regime in which $\mathbf{u}$ is related to $\mathbf{w}$ through a $m\times m$ feedback matrix $\mathbf{Z}$ as

$$\mathbf{u}=-\mathbf{Z}\mathbf{B}_u^T\mathbf{w}, \tag{43}$$

where $\mathbf{w}$ is normalized structural system state vector and $\mathbf{B}_u$ is the force input matrix for the nominal system model. The value of the feedback matrix $\mathbf{Z}$ may be controlled.

If it is assumed that $\mathbf{w}$ is sufficiently small, or feedback matrix $\mathbf{Z}$ is additionally constrained such that $\mathbf{u}$ doesn't violate $|\mathbf{u}| \le \mathbf{u}_{max}$, then this velocity feedback approach fully characterizes the regenerative force actuation capability for the nominal system model. If feedback matrix $\mathbf{Z}$ is constant, then the closed-loop system is linear. This fact expedites an analysis of the structural response.

The characteristics of structural response will vary with the choice of $\mathbf{Z}$. Some responses will be more attractive than others, if for example, these responses will have smaller maximum inter-story drifts, smaller absolute accelerations, structural energy dissipation etc. Let the functional J be a *measure of performance*, with the convention that lower values of J denote better performance.

4.2 Control Algorithms

Control algorithms are developed for controllers, which are programmable devices for manipulating the actuators for the purposes of structural control. The goal of control algorithms put into the controllers is to allow close to optimal structural control realized with controllers which have minimal computation and data storage for implementation, but have good performance characteristics.

Consider a family of nonlinear feedback controllers, parameterized by some finite-dimensional vector $\boldsymbol{\theta}$. Suppose that the subset of feasible controller parameters in this family is characterized by the condition $\boldsymbol{\theta} \in \Theta$, where Θ is some convex set. Then consider the parametric optimization problem

$$\boldsymbol{\theta}_{opt}=\arg\min_{\boldsymbol{\theta}\in\Theta}\{J(\mathbf{w}_0,a_g;\boldsymbol{\theta})\}. \tag{44}$$

This optimization problem is difficult to be solved. It is nontrivial to prove a convex relationship between $\boldsymbol{\theta}$ and J. Even more, the nonlinear system description makes it difficult to derive analytical expressions for $J(\mathbf{w}_0,a_g;\boldsymbol{\theta})$.

This problem can be solved with the use of inequalities [1]. Instead of seeking $\boldsymbol{\theta}\in\Theta$ which minimizes $J(\mathbf{w}_S,a_g;\boldsymbol{\theta})$, under investigation will be

$$J(\mathbf{w}_0,a_g;\boldsymbol{\theta})\le J_U(\mathbf{w}_0,a_g;\boldsymbol{\theta})\,. \tag{45}$$

where J_U is a closed-form expression of the arguments. Then the optimization problem of minimizing J_U over Θ becomes more realistic.

Algorithms for structural control are developed for the following *control approaches for seismic control synchronization*:

(1) ***Free vibration***: *Clipped Linear* controllers with feedback laws, *Damping Reference* controllers etc. Their parameters can be optimized to minimize J_U with design of sub-optimal, nonlinear quadratic regulator.
(2) ***Forced-vibration*** (reducing the forced response of structures): *Damping Reference* controllers implemented at two different disturbance models. In the *first model* is assumed that a_g is white noise and the system is under stationary excitation. Then Damping Reference controllers yield an upper bound on the expectation of a quadratic function of the states and control. This is a specific application of sub-optimal, nonlinear stochastic control. The *second model* is deterministic and Damping Reference controllers are designed to receive a simple expression for J_U which depends only on the mean-square value of a_g and a quadratic function of $\mathbf{w}_0$. This procedure is a specific application of sub-optimal, nonlinear control.

Different algorithms are developed for each kind of controllers where each controller operates according to certain of following *control laws*:

(1) ***Clipped Linear Control*** is implemented for free vibration with quadratic performance measures. For the deterministic performance $J(\mathbf{w}_0)$ is

$$J(\mathbf{w}_0)=\int_0^{\infty}\begin{bmatrix}\mathbf{w}^T(t) & \mathbf{u}^T(t)\end{bmatrix}\begin{bmatrix}\mathbf{Q} & \mathbf{S}\\ \mathbf{S}^T & \mathbf{R}\end{bmatrix}\begin{bmatrix}\mathbf{w}(t)\\ \mathbf{u}(t)\end{bmatrix}dt\,. \tag{46}$$

(2) ***Generalized Clipped Linear Control*** where $\mathbf{u}$ is related to $\mathbf{w}$ through two consecutive operations. The first step called active feedback signal, consists of the linear feedback function $\mathbf{u}_a$, while the second step called clipping action consists of "clipping" $\mathbf{u}_a$ to accommodate the constraints of the feasible force region. The simplified feedback function $\mathbf{u}(t)$ is

$$\mathbf{u}(t)=\arg\min_{\overline{\mathbf{u}}\in\upsilon(\mathbf{w}(t))}\left\|\tilde{\mathbf{u}}-\mathbf{K}_{CL}(\mathbf{P})\mathbf{w}(t)\right\|_{\mathbf{R}}\,. \tag{47}$$

(3) ***Clipped Optimal Control*** is implemented for energy-constraint controllers. By replacing as closed as possible the optimal active feedback control law at all times, subject to the forcing constraint $\mathbf{u}(t)\in\upsilon(\mathbf{w}(t))$, the performance of

the resultant constrained control system should resemble the active controller's performance. The difficulty with clipped optimal control is the fact that it doesn't process a useful upper bound for J. The controller attempts to approximate the active lower bound of J, through the point-by-point minimization of the integrand of in

$$J(\mathbf{w}_0) = \mathbf{w}_0^T \mathbf{P}_o \mathbf{w}_0 + \int_0^\infty \left\| \tilde{\mathbf{u}}(t) - \mathbf{K}_o \mathbf{w}(\tau) \right\|_R^2 d\tau \,. \tag{48}$$

(4) ***Steepest Gradient Control*** is realized with damping reference controllers in similar was as damping reference control. For semi-active control devices the feasible region analogous to $\upsilon(\mathbf{w})$ is more restrictive. Consequently the $\mathbf{Z}$ domain, over which the upper performance bound is optimized, is constrained to diagonal $\mathbf{Z}$ matrices.

(5) ***Free Reference Control*** reduces to maximum energy dissipation controller design with certain choices for $\mathbf{Q}$, $\mathbf{S}$ and $\mathbf{R}$. For free reference control can be used as well damping reference controllers with feedback matrix $\mathbf{Z} = 0$.

(6) ***Damping Reference Control*** is implemented when for a given choice of $\mathbf{K}$, as parameterized by the feedback matrix $\mathbf{Z}$, the upper bound of performance equal to the performance with linear damping feedback matrix $\mathbf{Z}$. The next step is to find the $\mathbf{Z}$ which minimizes the upper bound for a particular initial condition $\mathbf{w}_0$. For this purpose is used the gradient of the performance

$$\frac{\partial}{\partial \mathbf{Z}} (\mathbf{B}_a^T \mathbf{P}_z \mathbf{B}_a) = 2(-\mathbf{B}_u^T \mathbf{P}_z - \mathbf{S}^T + \mathbf{R}\mathbf{Z}\mathbf{B}_u^T)\mathbf{\Phi}\mathbf{B}_u \,. \tag{49}$$

(7) ***Lyapunov Control*** is implemented with Lyapunov based controllers where the performance of closed-loop system for initial conditions $\mathbf{w}(t)$ is realized with Lyapunov function

$$V(\mathbf{w}(t),t) = \int_t^\infty \begin{bmatrix} \mathbf{w}^T(\tau) & \mathbf{u}^T(\tau) \end{bmatrix} \begin{bmatrix} \mathbf{Q} & \mathbf{S} \\ \mathbf{S}^T & \mathbf{R} \end{bmatrix} \begin{bmatrix} \mathbf{w}(\tau) \\ \mathbf{u}(\tau) \end{bmatrix} d\tau \,. \tag{50}$$

Many control algorithms have been proposed for *free vibration* and for *reducing the forced response* of structures using active or semi-active devices.

In free vibration case the performance measure for structural response was J. In the forced response case, the measure of performance is treated in two different ways, depending on the nature of the input disturbance. Different performance measures are implemented for stationary white noise excitation and for deterministic excitation. Algorithms in controllers for deterministic excitation at forced response case are connected with implementation of performance measures for structural response close to the free vibration case, and performance is

$$J(\mathbf{w}_0, a_g) = \int_0^\infty \phi(t) dt \,. \tag{51}$$

For this case damping reference algorithms will be designed to yield a worst-case upper bound on performance of the form

$$J(\mathbf{w}_0, a_g) \le \mathbf{w}_0^T P_U \mathbf{w}_0 + \nu_a \int_0^\infty a_g^2(t)dt \,. \tag{52}$$

Because this problem is a nonlinear for it solving are implemented damping reference nonlinear controllers. They provide a simple quadratic upper bound on the worst-case performance in forces excitation. Earthquake excitations are almost invariably broadband signals with resemble filtered noise, with time-varying filter parameters. Thus, as with linear systems, nonlinear based controllers design may produce overly-conservative controllers. Consequently, they may yield less-favorable performance on average, as compared with stochastic controllers.

Algorithms for *stochastic excitation* consider the case where a_g is white noise with intensity Φ_a. For this case controllers are sought which yield a low expectation for ϕ and good performance based on the metric

$$J = E[\bar{\phi}] = E\left\{ \begin{bmatrix} \mathbf{w}^T & \mathbf{u}^T \end{bmatrix} \begin{bmatrix} \mathbf{Q} & \mathbf{S} \\ \mathbf{S}^T & \mathbf{R} \end{bmatrix} \begin{bmatrix} \mathbf{w} \\ \mathbf{u} \end{bmatrix} \right\}. \tag{53}$$

The damping reference stochastic controllers yield an upper bound on performance of the form J ≤ $\nu_a \Phi_a$, where ν_a is a function of the controller parameters. Stochastic controllers design has a problem, connected with the fact of implementation only the expected value of ϕ in stationary response, but there is no explicit attention given to the size of the tails of the distribution of ϕ. Thus, some controllers which yield favorable values for $E[\phi]$ may yield unfavorable higher moments in the response distribution. In the design of controllers for structural engineering, where the focus is on reducing the probability of failure, the tail of the distribution for ϕ may be of great importance.

5 Control Synchronization at Frequency Characteristics of the Earthquake, Resonances and Anti-Resonances

Seismic control synchronization is realized via programmable structural control at seismic excitations, with sensor technologies and synthesis of feedback control loads in regenerative force actuation network for protection of structures.

The control synchronization aims to return a structure with n-degree-of-freedom back to the equilibrium with dynamic switching commutation of actuator devices engaged in regenerative force actuation network. The network consists of a set of electromechanical devices positioned on different places into the structure. The synchronization is realized after activation when these devices absorb and dissipate a part of seismic energy. The actuator devices are connected with each other and their electronic help to share common electrical energy.

The seismic control synchronization aims to realize:

(1) Equilibrium of the system for active / semi-active structural control;
(2) Dynamic of commutation of devices for active / semi-active structural control at leaving the equilibrium of the system;
(3) Equilibrium of the system for active / semi-active structural control with feedback and non-determined commutation velocity.

At control synchronization the sensory measurements are compared to analytically computed values of the respective variable. Such computations use present and/or previous measurements of other variables. The mathematical model describes from one side the relationships between the system inputs and states and from the other the measured parameters. The idea can be extended to the comparison of two analytically generated quantities, obtained from different sets of variables. In either case, the resulting differences, called residuals, are indicative whether the system's behaviour coincides with the tested model. Another class of model-based methods relies directly on parameter estimation.

The generation of residuals needs to be followed by residual evaluation. Because of the presence of noise, disturbances and model errors, the residuals are never zero, even if the systems operating conditions are precise the same as modelled ones. One way of solving this problem is by testing the residuals against predefined thresholds, obtained empirically or by theoretical considerations.

Usually, for residual evaluation are used methods in state space, which can remove the negative effect from noise and limited number of disturbances as:

(1) *Kalman filter* can be used as a residual. Its mean is zero if the model is correct one (and no disturbances are present) and becomes nonzero if the model changes. Since the investigated sequence is a white noise, statistical tests are relatively easy to construct. One way to solve this problem of choosing the right model is by the usage of a bank of "matched filters", one of each possible earthquake frequency and for each possible arrival time, and check which filter output can be matched with the actual observations.
(2) *Diagnostic observers.* This method gives the freedom in design of the observer. "Unknown input" design techniques may be used to decouple the residuals from (a limited number of) disturbances. The residual sequence is coloured which makes statistical testing somewhat complicated.
(3) *Parity (consistency) relations.* Parity relations are rearranged direct input-output model equations, subject to a linear dynamic transformation. The residual sequence is coloured, just like in the case of observers. The design freedom provided by the transformation can be used for disturbance decoupling.

The residual evaluation is performed by using standard quadratic programming optimization procedure. As initial model probabilities are used the probabilities from the previous iteration. For the first iteration it is assumed that there is no earthquake. The minimization is based on criteria where the difference between the systems output and calculated values from the models is expected to be minimal. In fact this is minimization of the square error. This solution is chosen because only the absolute value of the error is important. To the smaller values of error corresponds greater probability.

The principle of the optimization remains the same when are used states instead of outputs. One of the problem here comes from the fact that the noise presence in real life systems. In case of strong noise the residuals for the correct model may become equal or even bigger than other residual(s).

The control synchronization can be realized on the base of frequency characteristics, such as resonance frequencies and anti-resonance frequencies. For controller design purposes it is proposed to be used the following quality criterion: *maximum distance between basic natural frequencies of the structure and resonance basic frequencies of the seismic signal*. This criterion is applied on the base of information received from strong motion seismic waves simulation and soil layers simulation. For the spectral composition of the seismic signals or at least his resonance frequencies, some effect can be obtain by the controller if it is tuned in such a way that anti-resonance of the structure neutralize some of the main resonances of the bedrock. In this case the anti-resonance frequency range overlaps whole possible frequency ranges of the seismic signal.

6 Simulation Experiments with Active and Semi-active Control

Consider the received experimental results for active / semi-active control of an open loop and closed loop system. The simulator used for this investigation consists of a hydraulic actuator servo/valve assembly that drives a 122cm × 122 cm aluminium slip table mounted on high-precision, low-friction linear bearings. As the hydraulic actuators are inherently open loop unstable, position feedback is employed to stabilize the control actuator.

The capabilities of simulator are: maximum displacement ±5 cm, maximum velocity ±90 cm/sec, and maximum acceleration ±4 g/s with a 450 kg test load. The operational frequency range of the simulator is nominally 0-100 *Hz*.

The test structure is a model of a three-storey single-bay scale model building. The building frame was constructed on steel with a height of 160 cm. the floor masses of the model weighted a total of 230 kg, distributed evenly between the three floors. The time scale factor was 0,2 making the natural frequencies of the model approximately five times those of the prototype.

The accelerometers positioned on the each floor of the structure measured the absolute accelerations of the model, and an accelerometer located on the base measured the ground excitation. To develop a high quality, control-oriented model, an eight channel data acquisition system consisted of eight Syminex XFM82 3 decade programmable anti-aliasing filters were employed. The data acquisition system consists as well of an Analogical CTRTM-05 counter-timer board and the Snap-Master software package. The XFM82 offer programmable pre-filter gains to amplify the signal into the filter, programmable post-filter gains to adjust the signal so that it falls in the correct range for the A/D converter, and analogue anti-aliasing filters which are programmable up to 25*kHz*. The schematic model of experimental setup is shown on Fig. 9.

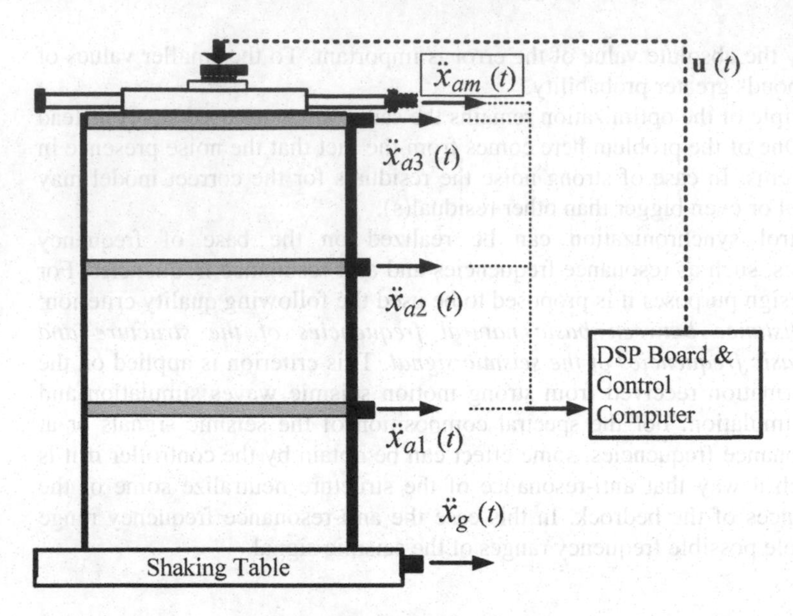

Fig. 9 Schematic model of experimental setup

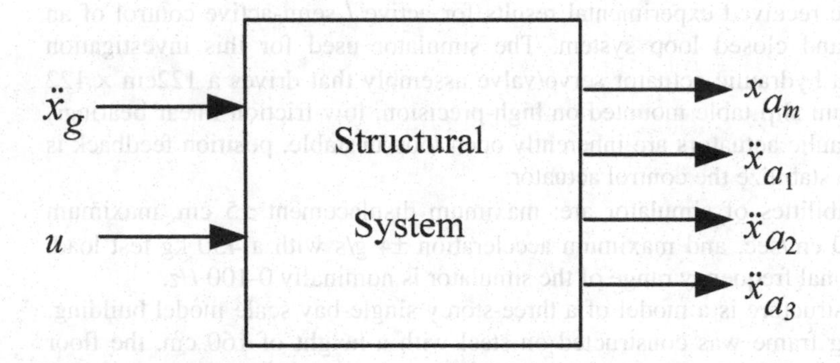

Fig. 10 Block diagram of structural system

A block diagram of structural system to be identified is shown in Fig. 10. The two inputs are the ground excitation $\ddot{x}_g$ and the command signal to the actuator u. The four measured system outputs include the absolute acceleration of the moving mass $\ddot{x}_{am}$, and the absolute accelerations $\ddot{x}_{a1}$, $\ddot{x}_{a2}$, $\ddot{x}_{a3}$, of the floors of the test structure, where $\mathbf{y}=[\ddot{x}_{am}\ \ddot{x}_{a1}\ \ddot{x}_{a2}\ \ddot{x}_{a3}]'$. Thus, a 4×2 transfer function matrix (i.e., eight input / output relations have to be identified to describe the characteristics of the system.

The transfer function from the ground acceleration to each of the four measured responses is obtained by exciting the structure with band-limited signals (0-100Hz) received as a result of seismic wave simulation and soil layers simulation, when the actuator command is set to zero. Similarly, the experimental transfer function from the actuator command signal to each of the measured outputs is determined by applying band-limited signals (0-100*Hz*), received as a result of seismic wave simulation and soil layers simulation to the actuator command while the ground is held fixed.

The experimental transfer functions are determined according to (29) on the base of twenty averages. On Fig. 11 is shown the transfer function from the actuator command signal to the absolute acceleration of the third floor, where the experimental values are in gray and the analytical in black colour.

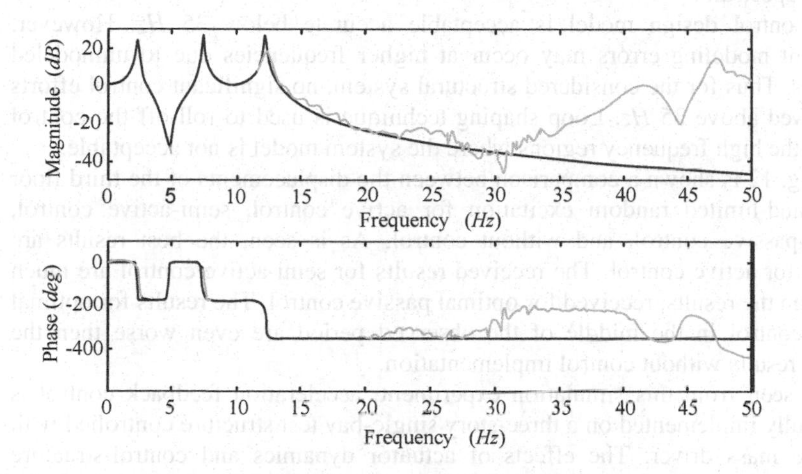

Fig. 11 Transfer function from the actuator command signal to the absolute acceleration of the third floor

Once the experimental transfer functions have been obtained, the next step in the system identification procedure is to model the transfer functions as a ratio of two polynomials in the Laplace variable *s*.

Next the model was assembled in state space form. The transfer functions from the ground to the structural responses were dominated by the dynamics of the structure, so the system related to the ground acceleration input (30) required ten states corresponding to the first five modes of the test structure. The second state equation, modeling the relationship between the actuator command and the structural responses (31), required fourteen states corresponding to the poles identified previously.

Once each of the two state space system is assembled, the states of the individual systems are stacked to form a combined system. The model reduction procedure is applied and the twenty-four state system is reduced to a fourteen-order system. The ten eliminated states corresponded to ten redundant states corresponding to the building dynamics.

To ensure that information is not lost in the model reduction, the transfer functions of the reduced order system are compared to the transfer functions of the original model. All of the eight input / output relationships matched the original model well.

To offer a basis for comparison a number of candidate controllers are designed, each employing a different performance objective. Design which minimize the three relative displacements, inter-story displacements and various combinations of three absolute accelerations of the structure and the absolute acceleration of the active mass driver are considered. In all of the considered controller design the weighting function of the regulated output **W** is a constant matrix (i.e., independent of frequency) and the earthquake filter **F** is modeled based on Kanai – Tajimi spectrum.

The control design model is acceptable accurate below 35 *Hz*. However, significant modeling errors may occur at higher frequencies due to unmodeled dynamics. Thus for the considered structural system, no significant control efforts are allowed above 35 *Hz*. Loop shaping technique is used to roll-off the control effort in the high frequency regions where the system model is not acceptable.

On Fig. 12 is shown a comparison between the displacements of the third floor under band-limited random excitation for active control, semi-active control, optimal passive control, and without control. As is seen, the best results are received for active control. The received results for semi-active control are much better than the results, received for optimal passive control. The results for optimal passive control in the middle of the observed period are even worse then the received results without control implementation.

As is seen from this simulation experiment, acceleration feedback control is successfully implemented on a three-story single-bay test structure controlled with an active mass driver. The effects of actuator dynamics and control-structure interaction were incorporated into the system identification procedure for structural control and synchronization at seismic excitation.

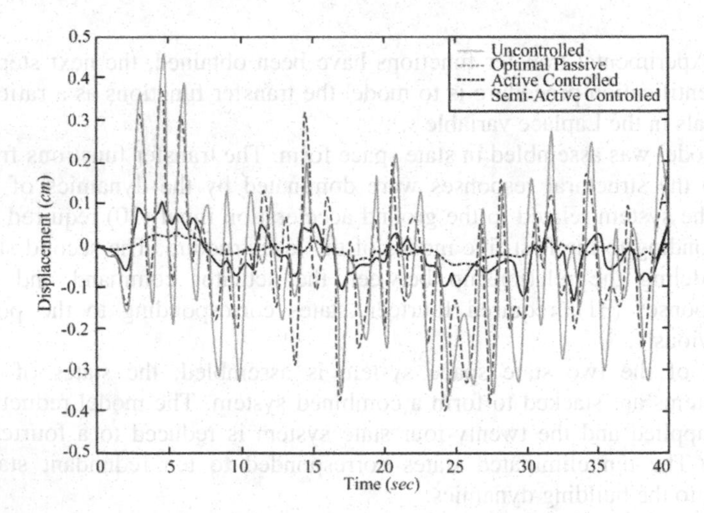

Fig. 12 Comparison between the displacements of the third floor under band-limited random excitation for active control, semi-active control, optimal passive control, and without control

The received results of implementation active structural control show 68% reduction in the peak response of the third floor absolute acceleration. Based on these results, the suggested acceleration feedback control strategies should be regarded as viable and effective for mitigation of structural responses due to seismic excitations.

References

[1] Camino, J., de Oliveira, M., Skelton, R.E.: Convexifying Linear Matrix Inequality Methods for Integrating Structure and Control Design. Journal of Structural Engineering 129(7), 978–988 (2003)

[2] Gluck, J., Ribakov, Y., Dancygier, A.: Predictive Active Control of MDOF Structures. Earthquake Engineering and Structural Dynamics 29(1), 109–125 (2000)

[3] Gibbens, R., Kelly, F.: Distributed Connection Acceptance Control for a Connectionless Network. In: Proc. of 16th International Teletraffic Congress, Edinburgh, UK, pp. 48–56 (1999)

[4] He, W., Agarwal, A., Yang, J.: Novel Semiactive Friction Controller for Linear Structures Against Earthquakes. ASCE Journal of Structural Engineering 129(7), 931–941 (2003)

[5] Ichtev, A., Radeva, S.: Multiple-model seismic structural control. In: Proc. of INDS 2008, International workshop on Nonlinear Dynamics and Synchronization, pp. 97–103 (2008)

[6] Ichtev, A., Scherer, R.J., Radeva, S.: Multiple-model structural control and simulation of seismic response of structures. In: Proc. of the European Conference on Process and Product Modelling (ECPPM), pp. 397–406 (2008)

[7] Nasu, T., Kobori, T., Takahashi, M., Niwa, N., Ogasawara, K.: Active Vibrate Stiffness System with Non-Resonant Control. Earthquake Engineering and Structural Dynamics 30(7), 1594–1614 (2001)

[8] Nishitani, A., Nitta, Y., Ikeda, Y.: Semiactive Structural Control Based on Variable Slip-Force Level Dampers. ASCE Journal of Structural Engineering 129(7), 933–940 (2003)

[9] Radeva, S.: Synthesis Nonstationary Seismic Response Structures. Acta Geodaetica et Geophysica Hungarica 43(2-3), 349–358 (2008)

[10] Radeva, S.: Structural control at seismic excitations. Colbis. Sofia, 178 p. (2008) (in Bulgarian)

[11] Radeva, S., Lockshina, I.: Modeling and simulation of feedback seismic structural control. In: Proceedings of Industrial Simulation Conference ISC 2008, Lyon, France, pp. 371–376 (2008)

[12] Radeva, S.: Simulation and Analysis of Seismic Response of Structures for Sofia Region With Synthetic Accelerograms. Acta Geodaetica et Geophysica Hungarica 43(2-3), 359–367 (2008)

[13] Radeva, S., Scherer, R., Radev, D.: Strong Motion Waves Estimation for Seismic Control of Nuclear Power Plant. Nuclear Engineering and Design, NED 235(17-19), 1977–1988 (2005)

[14] Radeva, S., Arnaudov, S.: Modeling and simulation of soil behavior for stationary and non-stationary load processes. In: Proceedings of the Conf. on Automatics and Informatics 2005, Sofia, pp. 107–110 (2005)

[15] Radeva, S., Scherer, R., Radev, D., Yakov, V.: Real-time estimation of strong motion seismic waves. Acta Geodaetica et Geophysica Hungarica 39(2-3), 297–308 (2004)
[16] Scruggs, J.T., Iwan, W.D.: Control of a Civil Structure Using an Electric Machine with Semiactive Capability. ASCE Journal of Structural Engineering 129(7), 951–959 (2003)
[17] Yuen, K., Beck, J.: Reliability-Based Control of Uncertain Dynamical Systems Using Feedback of Incomplete Noisy Response Measurements. Earthquake Engineering and Structural Dynamics 32(5), 751–770 (2003)
[18] Zhang, Y., Iwan, W.: Active Interaction Control of Civil Structures. Part 2: MDOF Systems, Earthquake Engineering and Structural Dynamics 31(1), 179–194 (2002)

Part III
Applications in Transportation

Emergence of Synchronization in Transportation Networks with Biologically Inspired Decentralized Control

Reik Donner

Summary. The efficient and reliable operation of material flows in transportation networks is a subject of broad economic interest. Important applications include the control of signalized intersections in urban road systems and the planning and scheduling of logistic processes. Traditional approaches to operating material flow networks are however known to have severe disadvantages: centralized controllers suffer from their high computational demands that make an on-line control hardly possible in larger networks, whereas a decentralized control using clearing policies leads under rather general conditions to instabilities.

A particular solution for optimizing flows on a network that has attracted considerable interest in the case of urban traffic networks over the last 50 years is synchronizing sequences of traffic lights so that "green waves" emerge. However, the practical applicability of this very basic synchronization strategy is unfortunately restricted to only few special cases. More recent synchronization approaches to optimizing transportation networks suggest a mutual adjustment of the service periods of neighboring facilities by exclusively local information transfer, yielding self-organized phase synchronization. Traditionally, these concepts have been motivated by theoretical results on the dynamics of coupled phase oscillators, which model the switching cycle of traffic lights at signalized intersections. The corresponding results are thoroughly reviewed in this chapter.

As an alternative approach that may help to overcome the problems of standard central as well as decentralized controllers and make an adaptive and purely demand-driven traffic light control practically applicable, a self-organization mechanism of conflicting flows is proposed that is inspired by oscillatory phenomena of pedestrian and animal flows at intersections or

Reik Donner
Institute for Transport and Economics, Dresden University of Technology,
Andreas-Schubert-Str. 23, 01062 Dresden, Germany
e-mail: donner@vwi.tu-dresden.de

K. Kyamakya (Eds.): Recent Adv. in Nonlinear Dynamics and Synchr., SCI 254, pp. 237–275.
springerlink.com

bottlenecks. For this purpose, a permeability function is introduced that allows to sequentially serve the different possible flow directions at an intersection in a fully demand-dependent way. The self-organized optimization achieved by the presented approach is demonstrated to be closely linked to synchronization of the oscillatory service dynamics at the different intersections in the network. For regular grid topologies, different synchronization regimes are present depending on the inertia of the switching from one service state to the next one. The dependence of this observation on the regularity of the considered network is tested. The reported results contribute to an improved understanding of the conditions that have to be present for efficiently operating material flow networks by a decentralized control, which is of major importance for future implementations in real-world traffic or production systems.

1 Introduction

Many real-world complex systems have (among others) the function of transportation of material and/or information from one place to another. Examples include systems in technology (including vehicular traffic [1, 2, 3], production, logistics, supply networks [4], or telecommunication) as well as biology (for example, the nervous and cardio-vascular system, intracellular transport using the cytoskeleton [5, 6, 7, 8], and nutrient transport in amoeboid organisms [9] or fungal mycelia [10]). One may distinguish continuous-flow systems (for example, power grids, water supply networks, nutrient or blood transport systems in organisms) from such systems which are characterized by a large number of individual and mutually interacting transportation units (which is the case for most information flow networks, road, railway, pedestrian or animal traffic, production and logistics systems).

In the case of systems characterized by discrete flows, the aim of an efficient organization is to minimize the time required for all individual transportation processes. Typically, this optimization is difficult and demanding, since the topology of the underlying networks is composed of a potentially large number of merges and intersections at which there are conflicts between the flows on different routes. To avoid physical collisions, these flows have to be controlled by suitable devices such as traffic lights. The operation strategy of these devices is decisive for the optimization of the system performance.

Whereas in the case of a low network load, the individual service of transportation units is beneficial, due to the necessary safety headways between individual services, it becomes inefficient if the traffic volume in a material flow network exceeds a certain threshold. Hence, in the presence of substantially high traffic volumes, a coordinated operation of the conflicting flows leads to better results [11]. Such a coordinated operation is achieved by bundling material or vehicles into platoons, which is performed in urban road networks by the action of traffic lights, or in logistics by transporting heavy loads on railways instead of roads.

In a coordinated service of conflicting material flows, the switching between flows from and/or in different directions leads to an accumulation of material (like vehicles or products) on the links which are currently not served. The corresponding effects are mathematically described in terms of queueing theory [12, 13]. In switched queueing systems, every intersection of conflicting material flows is characterized by the amount of delayed material on all of the incoming links, which is determined by the lengths $N(t)$ of the associated queues. The arrival and departure rates of material, $A(t)$ and $O(t)$, are bounded by the maximum capacity $\hat{Q}$ that is an intrinsic property of the transportation route and the used devices. For a particular queue, one may distinguish different states of the controller: a "no service" state and a service period, which itself is composed of a "setup" state, a "clearing" state, and a possible "extension" state with free-flow conditions on the served routes (see Fig. 1). In the context of vehicular traffic control, a "setup" state of duration T_{cl} is essential for a safe operation making sure that all vehicles have left the conflict area before the considered traffic stream enters, whereas "clearing" and "extension" states combine to the total green time g during a service period.

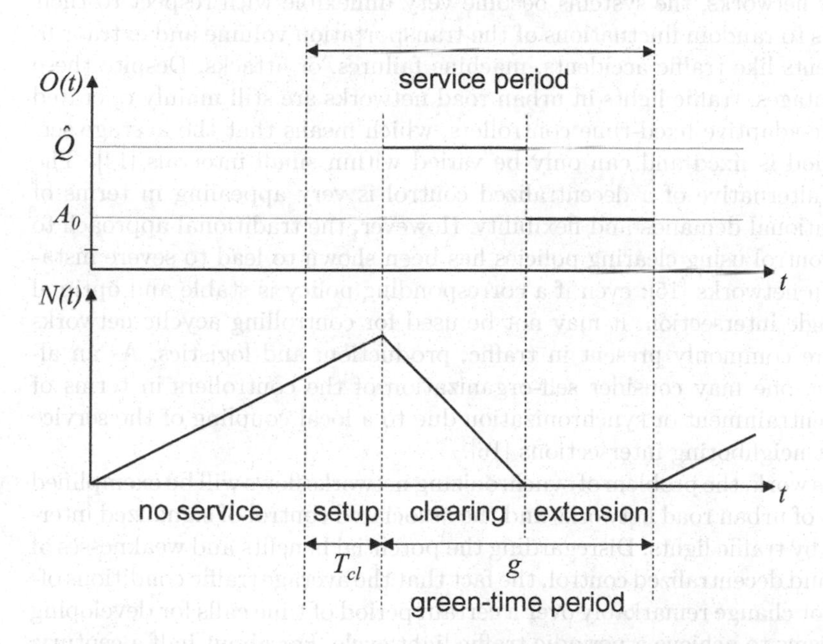

Fig. 1 Evolution of the departure flow (outflow) $O(t)$ and the amount of queued material (vehicles) $N(t)$ for one incoming link with a constant inflow rate $A(t) \equiv A_0 < \hat{Q}$. The different states of the queue are schematically shown.

The queue length $N(t)$ and the respective arrival and departure flows $A(t)$ and $O(t)$ are connected by a simple conservation equation:

$$\frac{dN}{dt} = A(t) - O(t). \tag{1}$$

Each of the states mentioned above is associated with different dynamical regimes of the queueing process, which are characterized by the relationship between the arrival flow and the change in the amount of queued material:

$$\frac{dN}{dt} = \begin{cases} A(t), & \text{“no service” and “setup” states,} \\ A(t) - \hat{Q}, & \text{“clearing” state,} \\ 0, & \text{“extension” state.} \end{cases} \tag{2}$$

Then, it directly follows that one may express the departure flow as:

$$O(t) = \begin{cases} 0, & \text{“no service” and “setup” states,} \\ \hat{Q}, & \text{“clearing” state,} \\ A(t), & \text{“extension” state.} \end{cases} \tag{3}$$

Present day, material flow networks are typically subjected to a global control. Beside the high computational demands of such a control in the case of larger networks, the systems become very unflexible with respect to their reactions to random fluctuations of the transportation volume and extraordinary events like traffic accidents, machine failures, or attacks. Despite these disadvantages, traffic lights in urban road networks are still mainly operated by traffic-adaptive fixed-time controllers, which means that the average service period is fixed and can only be varied within small intervals [14]. The natural alternative of a decentralized control is very appealing in terms of computational demands and flexibility. However, the traditional approach to such a control using clearing policies has been shown to lead to severe instabilities in networks [15]: even if a corresponding policy is stable and optimal for a single intersection, it may not be used for controlling acyclic networks which are commonly present in traffic, production, and logistics. As an alternative, one may consider self-organization of the controllers in terms of mutual entrainment or synchronization due to a local coupling of the service cycles at neighboring intersections [16].

In this work, the problem of synchronizing networks flows will be exemplified in terms of urban road networks and the associated control of signalized intersections by traffic lights. Disregarding the potential benefits and weaknesses of central and decentralized control, the fact that the average traffic conditions often do not change remarkably over a certain period of time calls for developing mechanisms to achieve a periodic traffic light cycle. For about half a century, synchronization of the corresponding signals has been proposed as a promising candidate for such a mechanism in terms of the “green wave” paradigm. Whereas the traditional literature on synchronization control of traffic lights has mainly restricted itself to central control architectures, the substantial advances in our understanding of the process of synchronization [17, 18, 19, 20] and its emergence in complex networks [21, 22] have recently motivated decentralized approaches as well, which are based on a semi-autonomous adjustment of frequency and phase of a traffic light cycle that take only information from neighboring intersections into account.

Section 2 of this chapter gives a detailed review about the history of traffic light control and summarizes in some detail early synchronization approaches based on the green wave idea, modern conceptual models for decentralized controllers based on coupled phase oscillators, and fully self-organized approaches that have emerged in the very recent years. Among these approaches, a model motivated by empirical findings from pedestrian dynamics [23] and studies of social insects is introduced in Section 3. Some details about its derivation and the resulting dynamics of a traffic light operating an isolated intersection are presented. The application to a network structure is presented in Section 4, including a thorough review on the evaluation of different types of synchronization in networks that can be attributed to this kind of system. Finally, the importance of synchronization phenomena in traditional as well as future self-organized traffic light control approaches is discussed, including a list of problems to be solved in the next years.

2 Synchronization Approaches to Traffic Light Control

2.1 History of Traffic Light Control

The idea of traffic lights has been developed almost in parallel with the technology of vehicles. The first signalized intersection was operated in London between 1868 and 1872 as a mechano-optical system [24], however, the application of this first device was not very successful. The first electric traffic lights were developed and operated at the beginning of the 20th century [25], for example, in Salt Lake City (1912) and Cleveland (1914). In the following decades, the technology was fastly progressing, which resulted in the development of combinations of electric controllers with traffic detectors in the 1950s. In contrast to this technological progress, the theoretical understanding of the dynamics of traffic flows necessary for constructing efficient controllers was significantly lagging behind for a considerable period of time. Pioneering work in this field has been done since the 1960s by Little and coworkers [26, 27, 28], who systematically studied the concept of "green waves" (see below) as the simplest example of synchronization in traffic networks and developed a corresponding graphical solution method.

With the development of digital computing technologies, the offline treatment of more complex traffic networks became possible in the late 1960s. The probably most famous approach of this kind is TRANSYT [29], which is gradually extended even these days. The idea of TRANSYT is a systematic search for an optimum switching program utilizing model-based simulations of the traffic dynamics, which are evaluated by means of total waiting times and number of stops. However, as the actual traffic conditions are continuously changing, the use of this strategy calls for a continuous re-evaluation and update of the particular programs, which is computationally demanding.

The graphical and offline optimization solvers can be considered as controllers of the first generation [30]. With the availability of the first computers, it became possible to develop new concepts that are able to reconfigure their switching programs on-line. Nowadays, there is a variety of such adaptive control strategies, including the incremental parameter variation (e.g. SCOOT as an adaptive variant of TRANSYT [31] with the same cycle time for every intersection), prognosis-based strategies (e.g. OPAC [32], PRODYN [33], and CRONOS [34], where the cycle times may differ between the different intersections), hierarchical-decentral decomposition (e.g. UTOPIA [35], RHODES [36], BALANCE [37], and MOTION [38]), and the store-and-forward approach (e.g. TUC [39])).

These days, the majority of traffic lights is controlled by so-called cycle-based strategies, which means that the traffic dynamics across an intersection shows a roughly periodic pattern. A network-wide coordination is achieved by fixing the cycle time for all intersections at a suitable value. The distribution of the different service states over one cycle may be varied according to the average traffic conditions, for example, using one of the traffic-adaptive methods mentioned in the previous paragraph. The on-line optimization of a whole traffic network with such centralized approaches (which are referred to as control structures of the second generation [30]) however requires vast amounts of data to be collected and processed. The minimum number of phases with compatible flows, setup times necessary for clearing an intersection before starting a new phase, and delays due to finite acceleration and passing times have to be taken into account. As a consequence of the complexity of the corresponding optimization problem, a corresponding control can hardly be operated on-line and is thus very unflexible in the case of exceptional events, accidents, temporal building sites, failures of the technical infrastructure, or natural disasters or terrorist attacks, but also in reacting on the short-term fluctuations in the traffic volume during "normal" conditions.

The above stated reasons have called for the development of a third generation of control methods, which are based on decentralized (distributed) controllers which are placed at the individual intersections and may decide autonomously about the operation of the associated traffic light. Obviously, such controllers are much more flexible and computationally by far less demanding [40, 41, 42]. This results in a better adaptivity to the current local traffic conditions, a potentially higher robustness with respect to failures, and the possibility of an on-line operation due to the low complexity of the corresponding optimization approaches. In contrast to this, there is however the disadvantage that there is no guarantee that the sum of locally optimized strategies yields also a globally optimal service of all intersections. In particular, the capability of decentralized approaches to deal with highly congested networks is a benchmark for every corresponding method.

2.2 The "Green Wave" Paradigm

The use of the term "synchronization" for describing an optimal mutual phasing of a sequence of traffic lights can be traced back at least to about 1960 in terms of the idea of green bands along main arterials in road networks ("green wave"). A paper of Gordon F. Newell [43] studying the flow of highway traffic through a sequence of traffic signals is maybe the first internationally recognized publication using the term synchronization in the framework of traffic control. Newell first studied the very simple case of low-density traffic (where interactions between individual vehicles can be neglected for the overall traffic dynamics) and a fixed desired speed. For equally spaced traffic lights, he was able to derive approximate expressions for the mutual phasing of the signals that yield a minimum average delay in the limits of very short as well as very long distances between the individual intersections. Surprisingly, it turned out that in the latter case, a "random synchronization" often performed better than the traditional progressive timing of the lights typically used at that time [43]. In contrast to this, for high traffic flow, the interactions between individual vehicles become important, and platoons emerge. In this case, the best pairwise synchronization also yields the best overall synchronization (in terms of delay times) for one-directional traffic, whereas for bi-directional traffic, the best synchronization corresponds to the solution which is optimal for the direction with the lower flow [44].

The first systematic investigation of the "green wave" scenario was presented by Morgan and Little in 1964 [26]. The authors developed a graphical solver for the switching programs for sequences of signalized intersections, which was - in contrast to the investigations of Newell that tried to optimize the average delays - based on the idea of a maximum green-time window (bandwidth) for the traffic flows. The corresponding mathematical optimization problem was later formulated and numerically solved by Little *et al.* [27, 28]. Bavarez and Newell [45] in turn showed that for any given common cycle time of all traffic lights along a one-way street that may be arbitrarily spaced, there is a choice of offsets between the individual switching cycles that simultaneously minimizes the total delay and the number of stops, but does not necessarily produces a green wave of maximum bandwidth. They also observed that there are signal settings for which some signals operate on half or a third of the cycle times of other lights, which may be used to minimize the delays of vehicles coming from the side streets. In modern language, such settings correspond to a 2 : 1 and 3 : 1 synchronization, respectively. Hillier and Rothery [46] used empirical data from different sites in the road network of London (England) in order to determine the phasing of traffic lights that yielded the minimum delay. As one might intuitively assume, it turned out that the optimal offset is a linear function of the distance between the considered intersections. Newell [47] presented a generalization of the problem of determining optimal offsets on arterials to two-way streets.

It has to be noted that in all mentioned publications, the term synchronization has been used in a rather pragmatic sense as a synonyme for finding an optimal mutual phasing of neighboring traffic lights. The physical meaning of synchronization as a mutual adjustment of oscillatory (traffic) signals by a sophisticated coupling [17] emerged only somewhat later. However, from a practical perspective, one may consider the process of (manually or automatically) adjusting the phasing of different traffic lights according to some optimization principle as a physical (phase) synchronization of the overall system.

Whereas the first investigations mentioned above restricted themselves to one-dimensional configurations, i.e. chains of intersections, it has been a natural generalization to apply the idea of synchronization of traffic signals to two-dimensional network configurations as well. For example, Chang [48] started studying the synchronization of traffic lights in grid networks and applied his results to parts of the urban road systems of Cleveland and San Jose. Gartner [49] defined circuit constraints on the phasing of fixed-time synchronized signal networks using a graph-theoretic approach. Basing on the mixed-integer linear programming method introduced by Little [27], Gartner, Little, and Gabbay later presented a detailed analysis of the problems associated with the coordination and synchronization of a traffic network [50, 51].

2.3 Synchronization in Microscopic Traffic Simulations

The early works on synchronization of traffic lights presented above restricted their attention to the application of fixed-time controllers, i.e. traffic lights with a fixed cycle length. Obviously, this setting yields a fundamental restriction to the ability of reacting to changing traffic conditions. In 1993, Faieta and Huberman [52] presented a study using microscopic simulations of vehicular traffic in a network with external in- and outflows for a comparative evaluation of four different basic strategies for traffic signal synchronization:

(i) In the *random offsets strategy,* all traffic signals have the same green-time periods g, but different phases which are chosen at random and kept fixed during the whole simulation.

(ii) In the *alternating system strategy,* all traffic signals switch the right of way at the same time (i.e. are perfectly synchronized without any mutual phase shifts). The green-time period g is chosen to be equal to the time a vehicle needs to travel between successive intersections on average. If in this case, subsequent traffic signals alternate the right of way, this setup may be used to generate a green wave along the main arteries.

(iii) The *vehicle-actuated strategy* uses a linear combination of the queue lengths (i.e. the numbers of vehicles waiting to be served in the different flow directions) to decide whether the served direction should be switched or not. In this respect, this strategy is a prototype of a decentralized controller. In order to avoid long waiting times on side roads, if the actual green-time period g exceeds a maximally tolerable value, the service state has to be immediately switched as well.

(iv) The *synchronizing strategy* is based on the idea of synchronization in coupled oscillator networks. Initially, the traffic signals in the system are initialized as in the random offsets strategy. Every time a controller switches the right of way in its intersection, a small amount of time gets added to the current green-time period in such a way that the service cycle becomes closer to an alternating system strategy with respect to the subsequent intersection. Faieta and Huberman called this approach the *firefly* strategy, pointing out that it has been motivated by observations of synchronized flashing in groups of fireflies [53, 54, 55, 56].

Using extensive simulations, Faieta and Huberman demonstrated that the synchronizing strategy allowed to enhance the average speed of vehicles in the network as well as the overall flow with respect to the three other strategies.

In the last 15 years, several studies have followed the line of research opened by Faieta and Huberman. For example, Brockfeld et al. [40] used the Chowdhury-Schadschneider cellular automaton model [57] to compare different strategies for optimizing a regular grid-like traffic network. In opposite to the *synchronizing strategy* of Faieta and Huberman, they studied a *synchronized strategy* (where all traffic lights switch at the same time) and demonstrated that this setting behaves under general conditions worse than a two-dimensional green wave scenario (i.e. a (partially) phase synchronized system) or even a random offset strategy. In particular, the latter one turned out to be beneficial if one requires a control strategy that is not very sensitive to the adjustment of the cycle times. In a similar way, Huang and Huang [42, 58] studied a simple cellular automaton model and demonstrated that a phase synchronization strategy (i.e. traffic light cycles with fixed periods and non-zero mutual offsets) is beneficial in one-dimensional sequences of traffic lights.

2.4 Phase Oscillator Models and Self-organized Traffic Lights

Microscopic simulations of traffic networks allow a detailed investigation of various dynamical effects associated with fluctuations of the traffic flow. However, as such detailed information is typically not available in the real systems, one may seek for alternative conceptual approaches of synchronization-based traffic light control. A promising alternative is to use simple models of networks of coupled oscillators (where each oscillator represents a signalized intersection) that are adapted to the specific features and requirements of traffic lights. An ideal starting point for such investigations is the phase oscillator model introduced 1975 by Yoshiki Kuramoto [59, 60], which later became famous as the Kuramoto model with numerous applications in various scientific disciplines [61, 62]. In particular, this model system has motivated several studies on possible decentralized approaches to traffic light control:

Model 1: In 2001, Sekiyama *et al.* [63] presented a first conceptual study explicitly using the Kuramoto model for the development of a decentralized control strategy for traffic lights. Starting from the standard setting of the Kuramoto model as a system of mutually coupled phase oscillators described by simple linear interactions,

$$\frac{d\phi_i}{dt} = \dot{\phi}_i(t) = \omega_i + \sum_{j=1}^{N} K_{ij} \sin(\phi_j - \phi_i) \tag{4}$$

with $\phi_i(t)$ being the phases of the individual oscillators, ω_i their natural frequencies, and $\mathbf{K} = (K_{ij})$ the matrix of mutual coupling strengths, the authors specified the couplings according to the actual utilizations of the individual roads, i.e.

$$K_{ij} = \frac{K}{k_i} A_{ij} \epsilon_{ij} \tag{5}$$

where K is a global coupling constant, k_i is the degree of the node (intersection) i (the number of incoming and outgoing road segments), $\mathbf{A} = (A_{ij})$ is the adjaciency matrix of the network, and

$$\epsilon_{ij} = \frac{1}{\hat{Q}_{i\leftarrow j} T_i} \int_t^{t+T_i} A_{i\leftarrow j}(s) ds \tag{6}$$

is the average traffic flow $\langle A(t) \rangle$ on the road from a node j to node i, which is normalized by the associated road capacity $\hat{Q}_{i\leftarrow j}$. In the latter equation, T_i is the cycle period of intersection i, which may vary with time.

Sekiyama *et al.* described the demand-dependent self-organized adaptation of the cycle frequency and timing for the case of traffic lights at simple four-armed intersections assuming that these can be operated by only two consecutive service phases. Neglecting pedestrian traffic and related effects in the real-world system, the full service of such an intersection however typically requires more than two phases (see Fig. 2, since there is always one direction of turning traffic that would collide with the straight-on traffic in the opposite direction (in particular, this is the case for left-hand turning in case of a right-driving policy like in Central Europe, as dispalyed in Fig. 2, and right-hand turning in case of a left-driving policy used in the UK or Japan, respectively). Hence, both flows are not compatible and should not be served in a common phase s at least in case of a high traffic flow. For the sake of simplicity, the model of Sekiyama *et al.* neglects these separate

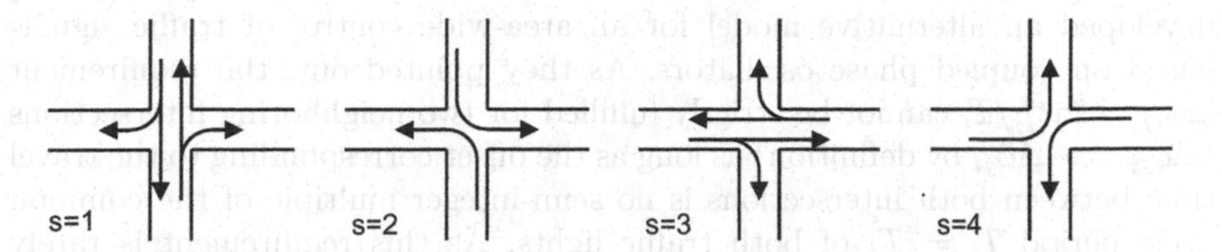

Fig. 2 Schematic representation of compatible flows on a four-armed intersection with right-driving policy (modified after [64])

turning phases, which can be considered as a network where the corresponding turning modes are forbidden. Note that in regular grid-like networks, the corresponding turning can be indirectly realized by three successive turns in the opposite direction. There are indeed some road networks within which this solution has been practically realized.

In order to describe the model for traffic light control in mathematical terms, it is convenient to introduce the abbreviation $\mathcal{L}_s$, $s = 1, \ldots, S$, for the different sets of flows (which are specified by the used pair of incoming and outgoing road) at an intersection that can be served simultaneously, i.e. that are mutually compatible with each other. The adaptation of the traffic light cycle in the two-phase model of Sekiyama *et al.* consists of three consecutive steps:

First, the green time $g = g_{i,1} + g_{i,2} = T_i - 2T_{cl}$ (with T_i being the period of a full cycle and T_{cl} representing the time required for completely clearing the intersection, which corresponds to the setup or amber light phase) is splitted on the two service phases according to the respective demands as

$$g_{i,s}(\tau_{i+1}) = (1-\gamma)g_{i,s}(\tau_i) + \gamma \frac{\sum_{j\in\mathcal{L}_s} \epsilon_{ij}}{\sum_{s=1}^{S}\sum_{j\in\mathcal{L}_s} \epsilon_{ij}} (T_i(\tau_i) - 2T_{cl}) \tag{7}$$

where γ is an updating constant. One may easily convince oneself that this updating rule does not change the full cycle period T_i.

Second, the offset of the switching cycles of neighboring intersections are mutually re-adjusted according to the modified green-time periods. The desired offset is given by

$$\Delta\phi_{ij}^* = 2\pi \frac{t_{ij}^* + (g_{i-1,1} - g_{i,1})}{T_i} \tag{8}$$

where t_{ij}^* denotes the desired travel time between the two successive intersections $i-1$ and i, which is determined by the spatial distance and the maximum possible velocity of the corresponding road section. Due to this update, it is necessary to modify the switching periods themselves in a third step as well, which however has to be constrained by predefining a desired average frequency in order to avoid an unlimited drift of the cycle frequencies.

Model 2: Following a similar line of ideas, Nishikawa *et al.* [65, 66, 67] developed an alternative model for an area-wide control of traffic signals based on coupled phase oscillators. As they pointed out, the requirement $\Delta\phi_{ij} = 2\pi t^*_{ij}/T_i$ cannot be strictly fulfilled for two neighboring intersections ($\Delta\phi_{ij} = -\Delta\phi_{ji}$ by definition) as long as the offset corresponding to the travel time between both intersections is no semi-integer multiple of the common cycle period $T_i = T_j$ of both traffic lights. As this requirement is rarely fulfilled in reality, one may use the options known from practical realizations of green waves, i.e. either giving priority to the direction with the higher flow (leading to a green wave only in this direction) or distributing the remaining phase shift equally among both directions in case of approximately symmetric flow conditions. The latter variant leads to a well-known generalization of the Kuramoto model,

$$\frac{d\phi_i}{dt} = \omega_i + \sum_{j=1}^{N} K_{ij} \sin(\phi_j - \phi_i - \delta_{ij}), \tag{9}$$

where $\delta_{ij} = \delta_{ji}$ is a positive phase difference. With a slightly generalized coupling matrix

$$K_{ij} = \frac{K}{k_i} A_{ij} f(A_{i\leftarrow j}) \tag{10}$$

where $f(\cdot)$ is a monotonously increasing function of the flow, their version of the generalized Kuramoto model for traffic light control can be easily formulated and analytically studied. In addition to the simplified case investigated by [63], they presented a detailed examination of the behavior of a corresponding traffic light including higher-order intersections ($k_i > 4$) as well as a full treatment of separate turning phases, which lead to additional phase terms in the arguments of the sine functions.

As an alternative to using the Kuramoto phase oscillator formalism, Nishikawa *et al.* further demonstrated that a corresponding traffic light control can be alternatively realized using a network of complex-valued Hopfield neurons [68, 69, 70, 71, 72]. Such neuronal networks can be mathematically described as

$$\frac{du_i}{dt} = \left(-\frac{1}{\tau_i} + i\omega_i\right) u_i(t) + \sum_{j=1}^{N} w_{ij} x_j(t) + \theta_i(t), \quad x_i = f(u_i), \tag{11}$$

where x_i, u_i and θ_i are the output, internal state, and external input variables, respectively. Representing the internal state and coupling weights by complex-valued quantities such as $u_i(t) = r_i(t)\exp(i\phi_i(t))$ and $w_{ij} = K_{ij}\exp(i\delta_{ij})$, one may derive separate equations for the amplitudes and phases

$$\frac{dr_i}{dt} = -r_i(t) + \sum_{j=1}^{N} K_{ij} f_R(r_j) \cos(\phi_j - \phi_i - \delta_{ij}) \tag{12}$$

$$\frac{d\phi_i}{dt} = \omega_i + \frac{1}{r_i}\sum_{j=1}^{N} K_{ij} f_R(r_j) \sin(\phi_j - \phi_i - \delta_{ij}) \tag{13}$$

which may again be used for traffic light control. It is obvious that the second equation is structurally equivalent to that obtained from the generalized Kuramoto model.

Model 3: In 2006, Lämmer *et al.* [16] studied the emergence of phase synchronization in another Kuramoto-type model for a traffic light. In his formalism, all characteristic periods are firstly expressed in terms of phase angles, i.e. $\phi(t) = 2\pi t/T \mod 2\pi$. In particular, the green times and setup times are expressed in terms of phases as $\Delta\phi_s = 2\pi g_s/T$ and $\Delta\phi_{cl} = 2\pi T_{cl}/T$, respectively. Trivially, all green and setup times during one traffic light cycle have to add up as $\sum_{s=1}^{S}(\Delta\phi_{cl} + \Delta\phi_s) = 2\pi$.

In order to completely serve all arrival flows that enter during on period T of the traffic light, one may easily demonstrate that the corresponding green time has to fulfill

$$\Delta\phi_s \geq 2\pi \max_{i\in\mathcal{L}_s} \frac{\langle A_i\rangle}{\hat{Q}_i} \quad \text{with} \quad \langle A_i\rangle = \frac{1}{T}\int_0^T A_i(t)\, dt = \frac{1}{2\pi}\int_0^{2\pi} A_i(\phi)\, d\phi. \tag{14}$$

If one defines the load of a whole intersection as

$$u = \sum_{s=1}^{S} \max_{i\in\mathcal{L}_s} \frac{\langle A_i\rangle}{\hat{Q}_i}, \tag{15}$$

it follows that it is only possible to serve all arriving flows if the cycle frequency of the traffic light does not exceed a critical value ω_{max} (i.e. the cycle period has at least a duration $T_{min} = 2\pi/\omega_{max}$) which is given by

$$\omega_{max} = \frac{2\pi}{ST_{cl}}(1-u). \tag{16}$$

In turn, this means that it is only possible to serve all flows if the intersection load does not become too large, in particular, $u < 1 - ST_{cl}/T$.

The above results suggest that if one wants to control a traffic light in terms of a phase oscillator model, the associated instantaneous frequencies $d\phi_i/dt$ must be bounded by a maximum value $\omega_{i,max}$ that may be specific for each individual intersection. A corresponding generalization of the Kuramoto model may then read

$$\frac{d\phi_i}{dt} = \omega_i(t) = \min\left\{\omega_{i,max},\ \Omega_i(t) + \frac{1}{\tau_\phi}\sum_{j=1}^{N} A_{ij}\sin\left(\phi_j(t) - \phi_i(t)\right)\right\}, \tag{17}$$

where τ_ϕ is an intrinsic time-scale for the adaptation of the phase $\phi_i(t)$ of a traffic light i to those of the connected (neighboring) traffic lights, which determines the dynamics unless the maximum frequency is reached. In order to allow for the corresponding frequency adaptation, the inherent frequency $\Omega_i(t)$ must be able to adjust itself as well. This may be realized by considering a second coupling of the form

$$\frac{d\Omega_i}{dt} = \frac{1}{\tau_\Omega} \left(\min_j \{\omega_j(t) | A_{ij} = 1\} - \Omega_i(t) + \Delta\Omega \right), \tag{18}$$

where $\Delta\Omega > 0$ assures a drift towards higher frequencies to reduce the resulting waiting times.

If a network of traffic lights is operated by the control strategy sketched above, one may observe a behavior consisting of two phases: As long as the switching frequencies have not yet approached their maximum values, one may observe a completely synchronized system where the individual traffic light cycles follow their designated inherent frequencies. However, due to the desired adaptation to the maximally possible frequencies, these inherent frequencies show a joint linear drift towards larger values with a constant rate $\Delta\Omega/T_\Omega$ until the maximum frequencies are obtained. Then, the cycle period of each intersection becomes locked to the lowest maximum frequency of its neighbors. In this frequency-locked state, the full synchronization of all cycles is lost, however, one finds that for long times, the phases of the different cycles mutually adjust to each other in order to compensate the joint frequency drift $\Delta\Omega$. Lämmer *et al.* [16] demonstrated that such a compensation is possible if $(N-1)T_\Omega\Delta\Omega < 1$.

3 Bio-Inspired Model for Self-organized Traffic Lights

The previous section has summarized the history of synchronization-based approaches to traffic light control. One may recognize two clear tendencies: On the one hand, the historical development started from approaches incorporating a complete synchronization of the complete network towards concepts yielding phase synchronization of different traffic lights. On the other hand, the traditional global control (for example, in terms of a "green wave") is more and more replaced by decentralized approaches that yield phase synchronization of the individual cycles by exclusively local coupling. Recently, several sophisticated strategies have been suggested that actually allow an efficient self-organized control of intersecting or merging traffic flows under rather general and quite realistic situations [11, 16, 73]. In the following, one of these concepts is discussed and studied in some detail as an example for the complex synchronization phenomena that may arise in transportation networks with decentralized control.

3.1 *Empirical Findings: Dynamics of Pedestrian Groups and Social Insects*

The model behind the decentralized control strategy used in this work is motivated by empirical findings from the field of pedestrian dynamics. Suppose there are two flows of pedestrians in opposite directions which have to pass a bottleneck. Using simulations with the social force model that assumes interactions between individuals due to physical and "social" pressure terms, Helbing *et al.* [23, 74] found a clear tendency towards the formation of groups of individuals that pass the bottleneck together in the same direction (see Fig. 3). As a consequence, some sort of oscillatory service is established by means of self-organization, which resembles the action of a traffic light. The theoretical predictions have been later verified by means of sociological experiments [75]. According to both theoretical and analytical results, the varying "net pressure" between the groups of inividuals waiting at both sides for passing the bottleneck seems to be the most important ingredient for the emergence of an oscillatory switching between the different flow directions.

In a similar way, the formation of groups of individuals sharing the same direction of motion has also been observed in social insects, for example, for the traffic of ants at a bottleneck [76]. In particular, experimental studied supported by model simulations revealed that the cross section of the bottleneck has a crucial influence on the formation of groups: the narrower the available route, the stronger the tendency towards the formation of larger groups of individuals. The similarity of the findings for ants (that are known to often act

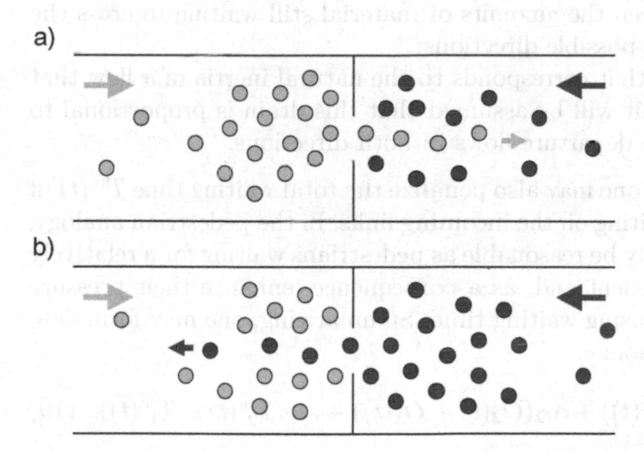

Fig. 3 Schematic representation of the behavior of two flows of pedestrians (indicated by different colors) that want to pass a bottleneck at different directions. For a certain time interval (a), there will always be one group of individuals that jointly passes the bottleneck in the same direction, before this direction switches more or less abruptly to that of the respective counter-flow.

as a "macro-organism") and pedestrians (that may be considered as selfish individuals which often act competetively rather than cooperatively) suggests that the underlying mechanism of self-organization is fundamental and not restricted to specific situations. Indeed, it is known that self-organization phenomena are rather typical at least in the case of pedestrian dynamics. For example, effects that are similar to that occuring for counter-flows at bottlenecks have been found both empirically and by means of simulations in terms of self-organized lane formation in intersecting pedestrian flows [75] and pedestrian counter-flows [77] as well as for intersecting and bottleneck flows of pedestrian and vehicular traffic [78, 79, 80].

3.2 Minimum Model for Self-organized Network Flows

With respect to the dynamics of pedestrian crowds, the above mentioned results suggest that the oscillatory switching between traffic in different directions is induced by the variations of the mutual (physical as well as social) pressure of the waiting and moving pedestrians. Mathematically, this phenomenon can be formulated by a cost function $C(t)$ which controls the permeability of the bottleneck in both directions. For the sake of simplicity, this function will be firstly specified for the case of two merging or intersecting unidirectional flows. In the presented model, there are two main factors entering the cost function [73]:

1. a net "pressure" force of the waiting material, which is proportional to the difference between the amounts of material still waiting to cross the intersection in both possible directions;
2. a net "drain" force that corresponds to the natural inertia of a flow that is currently served; it will be assumed that this drain is proportional to the difference of the departure flows in both directions.

As an additional factor, one may also penalize the total waiting time $T^w(t)$ of the delayed material waiting on the incoming links. In the pedestrian analogy, this additional factor may be reasonable as pedestrians waiting for a relatively long time become impatient and, as a consequence, enhance their pressure on the crowd with increasing waiting time. Summarizing, one may formulate the following cost function:

$$C(t) = \alpha_N(N_2(t) - N_1(t)) + \alpha_O(O_2(t) - O_1(t)) + \alpha_T(T_2^w(t) - T_1^w(t)), \quad (19)$$

where the indices $i = 1, 2$ correspond to the incoming links, and $\alpha_{N,O,T}$ are proper weights. For simplicity, in the following the specific choice $\alpha_T = 0$, $\alpha_N = 1$ and $\alpha_O = \alpha$ will be considered.

Having specified the cost function, the next step is to define a permeability function $\gamma_i(t)$ for the whole set of incoming links i at a given intersection point. This permeability function has to be understood as a theoretical dimensionless quantity that described whether or not (or, more under more general conditions, to which fraction) an intersection may be passed by a flow in the considered direction at a given time t. It should be noted that this quantity has to be clearly distinguished from the instantaneous ratio of the outflows $O_i(t)/\sum_{i=1}^{2} O_i(t)$, which one may understand as the empirical permeability of the intersection. In contrast to this, the permeability function $\gamma_i(t)$ is directly responsible for the outflow $O_i(t)$, whereas the outflow itself influences $\gamma_i(t)$ only indirectly via the cost function $C(t)$. In particular, a vanishing permeability $\gamma_i(t) = 0$ should imply a vanishing outflow $O_i(t) = 0$, but not necessarily vice versa.

In the formalism used in this work [73], the possible choice of this function will be restricted by some general properties: First, $\gamma_i(t)$ is a multiplicative factor entering the dynamic equation for the outflows $O_i(t)$, i.e. there is no outflow from link i if $\gamma_i(t) = 0$. Second, $\gamma_i(t)$ has to be normalized, that is, $\gamma_i(t) \leq 1$. Third, $\gamma_i(t)$ should be a function of the current value of the cost function $C(t)$ exclusively. Fourth, $\gamma_2(t) = 1.0 - \gamma_1(t)$, i.e. in the case of a simple merge or intersection of two flows, the permeability for both incoming directions have to add to one. With this setting, it is possible to fully describe a self-organized oscillatory switching between flows in different directions.

In the case of pedestrian motion where simultaneous flows in both directions (and, hence, collisions between different pedestrians) are still possible, a logistic function [73]

$$\gamma_{1,2}(t) = \frac{1}{1 + \beta e^{\pm \eta C(t)}} \tag{20}$$

can be used as a specification for $\gamma_i(t)$ (see Fig. 4). A similar continuous parametrization may be used for describing transport processes in biological systems like the cardio-vascular system or related biological transportation networks, where a simultaneous service of different directions is possible under the action of pressure gradients or incorportating diffusive processes between the different flows. In contrast to such situations, in technological systems such as vehicular traffic, material transport in production systems by automatically guided vehicles or conveyors, or baggage handling networks at airports, a continuous parametrization with $0 \leq \gamma_i(t) \leq 1$ would allow for potential collisions between objects transported in different directions. However, in the mentioned systems, such collisions have to be avoided to reduce the occurrence of accidents and the resulting damages of the transported objects. For this purpose, it is beneficial to replace the logistic function by a piecewise constant function that can only approach binary values (corresponding to "stop" and "go" commands for the respective flows) [64, 81, 82]. An example for such a function is shown in Fig. 4. As a particular advantage, this choice allows a sharp switching between both states. In order to avoid losses of efficiency due to finite times required for accelerating and decelerating, an additional hysteresis effect may be introduced into the model, which causes the service state to remain the same if the traffic conditions in

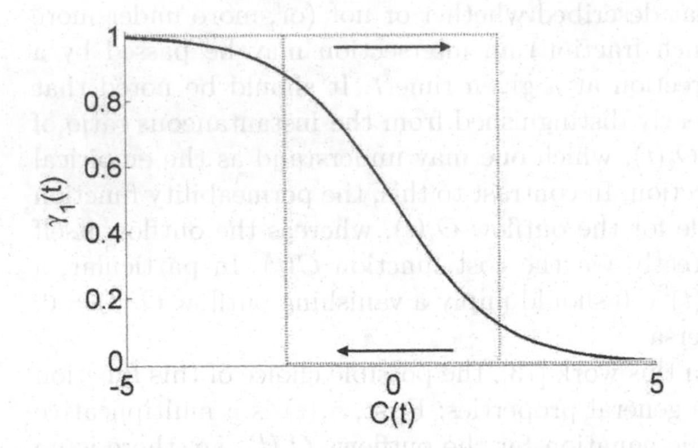

Fig. 4 Two specific parametrizations for the permeability function in dependence on the value of the cost $C(t)$. The solid line shows a logistic function (see Eq. (20)) that may be used for describing intersecting pedestrian or biological flows, whereas the dotted line corresponds to the hysterethic piecewise constant model used for controlling material flows in technological systems within the framework of this study.

the operated and stopped directions are comparable and thus introduces a preferred switching interval at given traffic conditions.

In situations where the material flows can be operated in a two-phase mode (e.g. merges or intersections without turning conflicts), the resulting dynamics of the system is described by a small set of equations relating the amounts of waiting material $N_i(t)$, the arrival flows $A_i(t)$, the departure flows $O_i(t)$, and the permeability $\gamma_i(t)$ of the different directions (service phases) at the "intersection". In addition to the above considerations, one has to specify the relationship between the change of the queue lengths and the associated arrival and departure flows (which is rather trivial) and an explicit expression for the departure flows. Following the considerations from Sec. 1, in the case of a complete permeability of the intersection in direction i (i.e. $\gamma_i(t) = 1$), the departure flow $O_i(t)$ from the respective queue occurs with maximum rate $\hat{Q}_i$ if there is material left in the queue, or otherwise with the arrival rate $A_i(t)$ of new material which is instantaneously processed. Summarizing, one finds the following set of equations:

$$\frac{d}{dt}N_i(t) = A_i(t) - O_i(t) \tag{21}$$

$$O_i(t) = \gamma_i(t) \times \begin{cases} A_i(t), & N_i(t) = 0 \\ \hat{Q}_i, & N_i(t) > 0 \end{cases} \tag{22}$$

$$\gamma_1(t) = \begin{cases} 1, & C(t) < -\Delta C/2, \\ 0, & C(t) > \Delta C/2 \end{cases} \tag{23}$$

$$\gamma_2(t) = 1 - \gamma_1(t) \tag{24}$$

$$C(t) = \sum_{i \in \mathcal{L}_2} (N_i(t) + \alpha O_i(t)) - \sum_{i \in \mathcal{L}_1} (N_i(t) + \alpha O_i(t)), \tag{25}$$

where the permeability functions show a bistable behavior in the interval $[-\Delta C/2, \Delta C/2]$ of the cost function. In order to give different priorities to different queues, it is also possible to use an asymmetric window $[C_{min}, C_{max}]$ (or, under more general conditions, route-specific thresholds C_i) for defining the switching thresholds of the permeability function. Note that the above set of equations neglects finite setup times T_{cl} required for a safe operation of real-world material flows. However, such setup times may be easily incorporated into the model, such that it is general enough to describe network flows in a variety of situations, including urban road traffic, transportation of goods in factories (e.g. by automatically guided vehicles), logistics, biological systems, or even the routing of data packages in certain kinds of information networks.

If the operation of a node in a material flow network requires more than two disjoint service phases (for example, in case of intersections in urban road networks where left-hand turning needs to be taken into account), the above formalism can be generalized in a straightforward way [64], yielding a control of the dynamics by assigning a priority index

$$P_s(t) := \sum_{i \in \mathcal{L}_s} (N_i(t) + \alpha O_i(t)) \tag{26}$$

to each service phase s, operating the phase with the highest priority index until the cost function

$$C_s(t) := P_s(t) - \frac{1}{S-1} \sum_{s' \neq s} P_{s'}(t). \tag{27}$$

falls below a certain threshold, and then switching to the next phase. Again, in order to avoid extraordinarily high waiting times on connections with a low load, it is possible to additionally penalize the total waiting time $T_i(t)$ of the material stored on each link by an additional term proportional to $T_i(t)$ in the definition of the priority index.

3.3 Dynamics of Isolated Intersections with Constant Inflows

In the case of an isolated merge or turning-free intersection of two conflicting material flows, the main properties of the switching dynamics can be analyzed analytically. In order to have stable flow conditions without a successive congestion of any of the links, the average outflow $\overline{O}_i$ over one service period must equal the average inflow $\overline{A}_i$ on every link. Assuming the link capacity being the same on every incoming link ($\hat{Q}_i = \hat{Q}$), there are only two dynamically relevant parameters remaining: the total intersection load u defined in Eq. (15) and the inflow ratio $r := A_2/A_1$ [64].

If the corresponding arrival flows $A_i(t)$ are assumed to be constant with sub-critical values (that is, the sum of all inflows is beyond the capacity limit of the intersection[1], it is possible to derive exact expressions for the switching periods, the duration of eventual extension periods, and the minimum and maximum amounts of material stored on the different transportation routes.

Using a piecewise constant permeability function as described in Sec. 3 with equal switching thresholds $\Delta C/2$ for all possible service states, one may easily convince oneself that for an initial queue length N_i^0, a complete clearing requires a time

$$T_{C,i} = \frac{N_i^0}{\hat{Q}_i - A_i}. \tag{28}$$

However, this complete clearing takes only place if the choice of the switching threshold $\Delta C/2$ is large enough, in particular,

$$\Delta C \geq 2\left[\frac{A_j N_i^0}{\hat{Q}_i - A_i} + N_j^0 - \alpha \hat{Q}_i\right] = \Delta C_1. \tag{29}$$

Otherwise, the switching to a service of the remaining link j occurs already after a time

$$T_{I,i} = \frac{\alpha \hat{Q}_i + N_i^0 - N_j^0 + \Delta C/2}{\hat{Q}_i - (A_i - A_j)}. \tag{30}$$

If this threshold is sufficiently large to allow a complete clearing of the queue i, there may be eventually another extension phase with a duration of

$$T_{E,i} = \frac{\alpha A_i - N_j(T_{C,i}) + \Delta C/2}{A_j} \tag{31}$$

with $N_j(T_{C_i}) = N_j^0 + A_j T_{C,i}$. One may easily establish that such an extension phase may only occur (i.e. $T_{E,i} > 0$) iff

$$\Delta C \geq 2\left[\frac{A_j N_i^0}{\hat{Q}_i - A_i} + N_j^0 - \alpha A_i\right] = \Delta C_2 \equiv \Delta C_1 + 2\alpha\left(\hat{Q}_i - A_i\right). \tag{32}$$

Following these arguments, in all cases the duration of the service period for link i, T_i, increases linearly with the switching threshold $\Delta C/2$. Hence, ΔC may be thought of as a characteristic scale determining the cycle time T of the traffic light for fixed traffic volumes described by the incoming flows A_1 and A_2. Summarizing, the green time g_i may be expressed as follows:

[1] As previously argued, $u < 1 - 2T_{cl}/T$ in the considered case. However, as the setup times T_{cl} are neglected here for the sake of simplicity, the capacity limit of an intersection may be assumed to correspond to $u = 1$.

$$g_i = \begin{cases} T_{I,i}, & \Delta C < \Delta C_1, \\ T_{C,i}, & \Delta C_1 \leq \Delta C \leq \Delta C_2, \\ T_{C,i} + T_{E,i}, & \Delta C > \Delta C_2 \end{cases} . \tag{33}$$

Fig. 5 illustrates the above analytical findings for two symmetric inflows with a road utilization of 0.5. In the dependence of the green time on the switching threshold, two regimes can be distinguished: complete clearing and extension (lowest line with a constant slope) and incomplete clearing (shifted lines with smaller slope). Both regimes are separated by a region where a complete clearing of the queue without an extension phase takes place. The width of this transitional regime is determined by the value of α, and the corresponding time increases linearly with the initial queue length N_i^0. For small values of N_i^0 (here: $N_i^0 < 4$), there is always a complete clearing of the queue within one service period due to the drain force realized by the parameter α. In contrast to this, for larger initial queues, the green time required for complete clearing increases linearly with both switching threshold ΔC and initial queue length N_i^0 as expected.

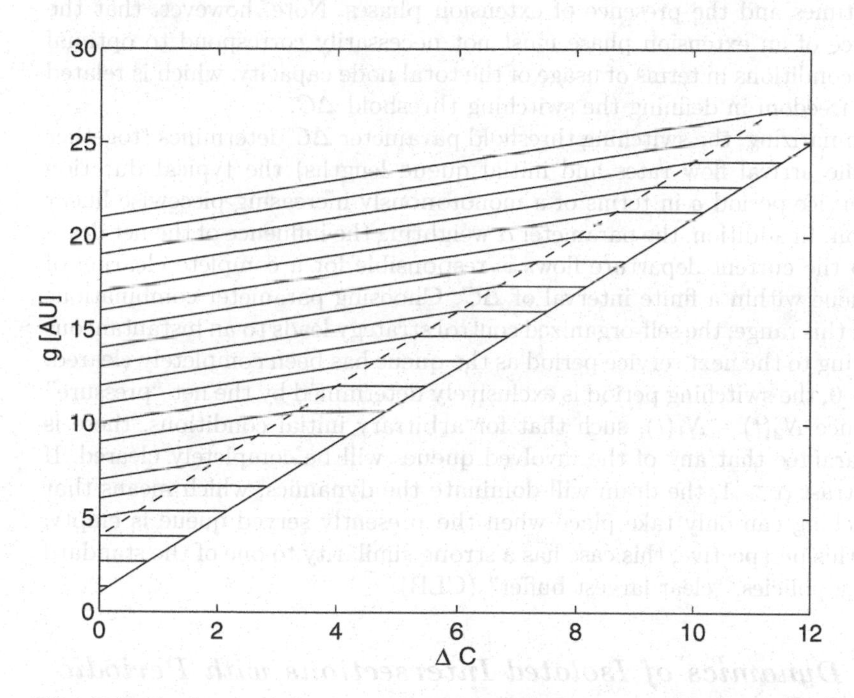

Fig. 5 Dependence of the green time g (in arbitrary units (AU)) on the switching threshold ΔC for $A_1 = A_2 = 0.25$ and $\alpha = 1.0$ in the regimes of incomplete clearing, complete clearing, and complete clearing with extension phase. While the second queue j is assumed to be completely cleared at the beginning of the service period, the different lines correspond to initial queue lengths of $N_i = 20, 19, 18, \dots$ (from top to bottom). The dashed line indicated the transition between incomplete and complete clearing.

The above results suggest that for sufficiently large ΔC, extension phases can be found on both incoming links for a wide range of arrival rates (see [64]). In general, the switching dynamics of an isolated intersection can be qualitatively classified by considering whether the service period of one or both queues leads to an incomplete (I) or complete (C) clearing and even an eventual extension (E) phase. The decision which of the possible combinations (II, IC, IE, CC, CE, or EE) is realized is determined by the inflows A_1 and A_2 (or, alternatively, the quantities u and r), the switching threshold $\Delta C/2$ of the permeability function, and the weight α quantifying the impact of the net drain. Previous investigations [64] have revealed that concerning the qualitative switching dynamics, there are two striking features: First, the presence of an incomplete clearing requires a rather low traffic volume on the considered link, whereas there has to be much more traffic on the second one (i.e. for low values of r and high values of u). Such a phase can only affect the less frequently used link. Second, extension states can be found at almost all traffic conditions, except of such with a sufficiently high intersection load u. Moreover, as one would expect, there is a strong correlation between long green times and the presence of extension phases. Note, however, that the presence of an extension phase must not necessarily correspond to optimal traffic conditions in terms of usage of the total node capacity, which is related to the freedom in defining the switching threshold ΔC.

Summarizing, the switching threshold parameter ΔC determines (together with the arrival flow rates and initial queue lengths) the typical duration of a service period g in terms of a monotonously increasing piecewise linear function. In addition, the parameter α weighting the influence of the net drain due to the current departure flows is responsible for a complete clearing of the queue within a finite interval of ΔC. Choosing parameter combinations within this range, the self-organized control strategy leads to an instantaneous switching to the next service period as the queue has been completely cleared. If $\alpha \to 0$, the switching period is exclusively determined by the net "pressure" difference $N_2(t) - N_1(t)$, such that for arbitrary initial conditions, there is no guarantee that any of the involved queues will be completely cleared. If in contrast $\alpha \gg 1$, the drain will dominate the dynamics, which means that a switching can only take place when the presently served queue is empty. From this perspective, this case has a strong similarity to one of the standard clearing policies, "clear largest buffer" (CLB).

3.4 Dynamics of Isolated Intersections with Periodic Inflows

The case of constant arrival flows of material discussed above is rather artificial. For example, in an urban road network, vehicles are usually bundled to platoons by the action of traffic lights. Since these traffic lights operate in a periodic way, the outflow on a link i and, hence, the inflow to the queue at the downstream end of this link is described by a periodic function, too. If the arrival flows are determined by a periodic function with period T_m, for some distinct interval of these periods (which depends on ΔC), the switching period T of the self-organized control locks to this external demand period [81]. If the amplitude of the periodic arrival flow increases, the width of this locking window increases as well. Moreover, there are other windows of higher-order frequency locking with a similar behavior, which are indicated in Fig. 6. The detailed position and width of the corresponding Arnold tongues is determined by the parameters α and ΔC of the permeability function $\gamma_i(t)$ and the average inflows $\langle A_1 \rangle$ and $\langle A_2 \rangle$. In general, the choice of these parameters again yields a naturally preferred switching frequency in the case of constant inflows, whose existence gives rise to the non-trivial locking intervals.

It has to be mentioned that the existence of various phase-locked states can be observed independently of the shape of the periodic functions. In

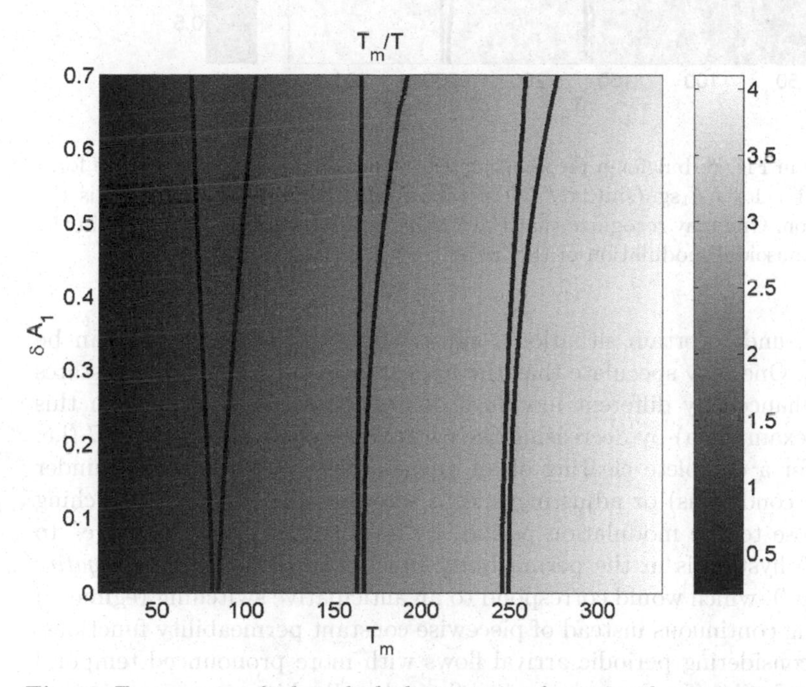

Fig. 6 Emergence of phase-locked states in the case of a periodic arrival flow $A_1(t) = \langle A_1 \rangle (1 + \delta A_1 \sin(2\pi t/T_m))$ on link 1 and a constant arrival flow on link 2. The results are shown for arrival flows with $\langle A_1 \rangle = A_2 = 0.25$ (all values are normalized with respect to the link capacity $\hat{Q}$), $\Delta C = 10$ and $\alpha = 1.0$. Gray-scale colors correspond to the ratio between arrival flow period T_m and switching period T, the black lines indicate parameter regions where the deviation from a perfect 1:1 to 3:1 frequency locking (from left to right) is smaller than 2.5 time units.

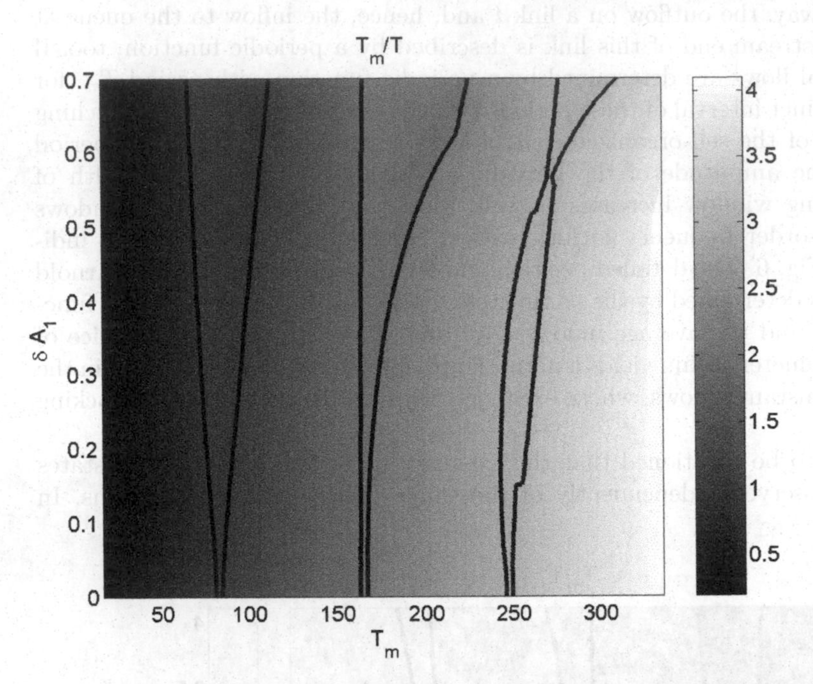

Fig. 7 As in Fig. 6, but for a piecewise constant periodic arrival flow of the form $A_1(t) = \langle A_1 \rangle (1 + \delta A_1 \mathrm{sgn}(\sin(2\pi t/T_m)))$, where $\mathrm{sgn}(x) = x/|x|$ (for $x \neq 0$) is the sign function. One may recognize that the locking windows are broader than in the case of a sinusoidal modulation of the inflow.

particular, under certain situations, non-trivial $m{:}n$ locking states can be found [81]. One may speculate that the appearance and width of such states can be enhanced by different modifications of the scenario studied in this work, for example, a) by decreasing the critical switching threshold ΔC (i.e. allowing for a complete clearing of all queues with extension phases under almost all conditions) or adjusting it in a way that the preferred switching time is close to the modulation period, b) by changing from "positive" to "negative" hysteresis in the permeability function (i.e. choosing a *negative* value of ΔC), which would correspond to an anticipative switching regime, c) by choosing continuous instead of piecewise constant permeability functions, or d) by considering periodic arrival flows with more pronounced temporal variability profiles (for example, arrival flows described by piecewise constant (see Fig. 7) or telegraph (on-off) signals instead of harmonic functions). It should be noted that options b) and c) have been used in a recent study [81], revealing a much larger variety of locking windows than in the scenario shown in Fig. 6.

4 Self-organized Dynamics of Coupled Intersections

Whereas the dynamics of isolated intersections can be (under certain assumptions) treated analytically, the practically more relevant case of controlling networks of intersecting flows is more challenging. In the following, this problem is hence addressed by means of simulations. As the main focus of this contribution is on the emergence of synchronization phenomena in such systems, first a brief review on different methods for synchronization analysis is presented. In particular, in the case of transportation networks, two types of synchronization may be of considerable interest: phase synchronization (if traffic lights are considered as self-sustained oscillators) and event synchronization (only the joint temporal distribution of events of switching from one service phase to another is studied).

4.1 Phase Coherence/Synchronization Analysis

The mutual phase synchronization of two coupled oscillators j and k can be tested using a variety of different statistical quantities. If the corresponding phase difference at a time t is denoted as $\Delta\phi_{jk}(t) = \phi_k(t) - \phi_j(t)$, a widely used characteristic is the Rayleigh coefficient or mean resultant length [17, 83],

$$r_{jk} = \left| \left\langle e^{i\Delta\phi_{jk}(t)} \right\rangle_t \right| = \frac{1}{T} \left| \sum_{t=1}^{T} e^{i\Delta\phi_{jk}(t)} \right| \tag{34}$$

which is a traditional measure of directional statistics that describes the clustering of the distribution of phase differences by considering their mean projection on the unit circle. In case of phase synchronization, this mean projection vector has a large modulus (i.e. the "center of gravity" is well separated from the origin of the coordinate system), which leads to values of r_{jk} near 1.

Alternative, but sometimes less reliable statistics are the standard deviation [17] and the normalized Shannon entropy [84] of the phase differences. Whereas the standard deviation may be stronger affected by outliers than the mean resultant length (as larger differences from the mean phase difference contribute quadratically instead of linearly) and is not normalized to values between 0 and 1, the major disadvantage of the Shannon entropy is its dependence on a particularly chosen discretization of the range $[0, 2\pi]$, which is known to influence the estimated value [85, 86]. Further alternative approaches to quantify bivariate phase synchronization include the consideration of the cross-mutual information between the individual phases [87], and the application of statistics based on recurrence plots [88, 89] of the original time series [90, 91]. Note that in contrast to all other mentioned methods, the latter one does not require the explicit definition of a phase variable (for example, by means of Poincaré sections in phase space, Hilbert, or wavelet transforms [92]), which may also be a critical point in phase synchronization analysis of real-world systems [93, 94].

According to the above mentioned results, in the following, only measures based on the mean resultant length will be further considered. Whereas the measure itself is a characteristic for mutual phase synchronization (or phase coherence if the examined signals are not originated in self-sustained socillatory systems) of only two systems, the corresponding quantification in the case of complex networks of more than two oscillators requires a more sophisticated statistical treatment. In particular, for quantifying phase coherence in a multivariate way, different approaches based on the mean resultant length may be considered [82]: a) global or neighbor-based averages of the pairwise index values, b) the average mean resultant length of all oscillators with respect to their mean-field, c) the synchronization cluster strength computed using the synchronization cluster analysis algorithm of Allefeld and Kurths [95], d) eigenvalue statistics obtained from the matrix of pairwise indices in terms of the generalized synchronization cluster analysis [96], and e) the LVD dimension density method [81, 82] which describes the average exponential scaling of the residual variances obtained from the aforementioned eigenvalues.

For simple model systems like a network of Kuramoto phase oscillators with long-range interactions, all mentioned measures have been found to yield comparable results [82]. However, this may not necessarily remain true for complex systems like transportation networks. The main reason for this is the different sensitivity of the individual measures with respect to the spatial heterogeneity of such networks. In particular, the global and neighbor-based averages and the average coherence with the mean-field quantify only the mean degree of phase coherence, but are not sensitive to detect effects of spatial heterogeneity. The synchronization cluster strength assumes implicitly the presence of a unique synchronization cluster, which is under general conditions a too strong assumption, especially during the transition from non-coherent to phase-coherent dynamics. The number and average strength of synchronization clusters based on the generalized synchronization cluster analysis allow a better characterization of heterogeneity effects, but yield rather coarse measures. Finally, the phase coherence parameter based on the LVD dimension density approach is well suited to quantify heterogeneities, but is only a coarse and uncertain measure for the strength of phase synchronization. In summary, a combined consideration of different approaches may be desirable to distinguish information about the average phase coherence and its heterogeneity.

4.2 Event Synchronization Analysis

If one wishes to apply the notation of phase synchronization to the problem of traffic light control, one has to be aware of the fact that in the decentralized control framework investigated in this work, traffic lights are no self-sustained oscillators, as the switching from one service state to another is triggered by the emergence of nontrivial arrival flows which are caused by the action of the neighboring intersections. Moreover, the phase of a traffic light cycle is not uniquely defined for all times t, but depends on whether $\phi(t)$ is chosen to be, for example, a linear or piecewise linear function of time. In order to improve the interpretability of the results of synchronization analysis, it may be beneficial to use alternative concepts instead that refer to the isochronicity (or fixed-lag synchronicity) of switching events exclusively instead of taking the whole cycle into account. Similar approaches are already used for quantifying synchronization of events, for example, in biological systems in terms of event synchronization of neurophysiological signals, heartbeats, etc. [97].

Consider the switching dynamics of two individual intersections in the network, which consist of two distinct phases under the assumptions described above. In the case of a phase coherent dynamics, the cycle length at all intersections must be the same. Then, for every pair of oscillations, the following quantities are constructed:

- For the second last switching event at intersection j, the switching event at intersection k is identified which has a minimum time lag with respect to this event. This time lag is computed. After this, the sequences of switching events are traced backwards in time for both intersections event-by-event, leading to a series of time differences $\Delta t_{jk}^{(0)}$.
- In a similar way, a sequence of time differences $\Delta t_{jk}^{(1)}$ is constructed where the reference event at intersection k is replaced by its predecessor.

In the case of (almost) isochronous switching, the values of $\Delta t_{jk}^{(0)}$ will be centered around zero with a narrow distribution. If the switching is synchronous with a certain lag, $\Delta t_{jk}^{(0)}$ and/or $\Delta t_{jk}^{(1)}$ are characterized by a small standard deviation with non-zero mean (depending on the symmetry of the switching dynamics). Following this idea, the mean value of $\Delta t_{jk}^{(0)}$ can be used as a measure of isochronicity, whereas the minimum of the standard deviations of both sequences yields a measure for the presence of (lagged) event synchronicity, and the corresponding mean value determines the average lag. The concept of event synchronization analysis can be further refined, however, this is outside of the scope of the presented work. For example, in the case of delays covering more than one switching cycle, an optimal lag could be inferred by considering the standard deviations of all event sequences that are shifted with respect to each other by $\pm k$ events and identifying the minimum of the resulting function.

4.3 Specification of the Scenario

In the following, it shall be studied whether the presence of the biologically inspired decentralized control strategy introduced above actually leads to a self-organized emergence of synchronization of the traffic light cycles in a network. In order to keep the number of variables as low as possible, some simplifications will be made for the following investigations:

1. For avoiding an unlimited congestion of the incoming links, which would finally affect the whole network, it is required that the sum of the maximum inflows for all necessary service phases of a given intersection is sufficiently smaller than the maximum link capacity $\hat{Q}$, which is assumed to be equal for all links.
2. The studied networks consist only of nodes of degree $k = 4$ (i.e. with four pairs of ingoing and outcoming links which connect neighboring pairs of nodes). A consideration of nodes with $k = 3$ (i.e. a merge or diverge) is also possible, however, the case of $k > 4$ is not considered here. All links are assumed to allow a bi-directional traffic.
3. Motivated by the problem of vehicular traffic in networks with right-hand driving policy, for every intersection and every incoming link, only right-hand turning is permitted with a given probability $p \in [0, 1]$. In contrast to this, direct left-hand turning will remain forbidden, as the corresponding possibility would call for two additional turning phases at least if the arrival flows are sufficiently large as already discussed above. Moreover, in the case of road networks with bi-directional traffic, left-hand turning may be effectively achieved by a sequence of right-hand turnings.

Under these assumptions, the cycle of successive service periods consists of only two phases. The dynamic coupling of different intersection points in a material flow network requires to represent the inflows at a given node by the outflows from neighboring nodes at an earlier time (see Fig. 8). In a zeroth-order approximation, the corresponding time delay is assumed to equal the free-flow travel time t^*_{ij} on the respective link,

$$A_i(t) = \sum_{j \neq i} \alpha_{ji} O_j(t - t^*_{ij}), \tag{35}$$

where α_{ji} is the fraction of the flow on link j which is turning to link i ($\alpha_{ji} = p$ for right-hand turning, $\alpha_{ji} = 0$ for left-hand-turning, and $\alpha_{ji} = 1 - p$ otherwise). As a necessary condition of material conservation, $\sum_i \alpha_{ji} = 1$ for all links j. To approach more realistic conditions, the free-flow travel time t^*_{ij} has to be replaced by a load-dependent travel time $t_{ij}(t)$, i.e. a time-dependent travel time which is determined by the amount of material waiting on the link. A very simple way for doing this would be setting

$$t_{ij}(t) = \frac{l_i - N_i(t)\Delta l}{v^*}, \tag{36}$$

where l_i is the total length of the road between two neighboring intersections, Δl the space occupied by one of the queued objects, and v^* the desired velocity under free flow conditions.

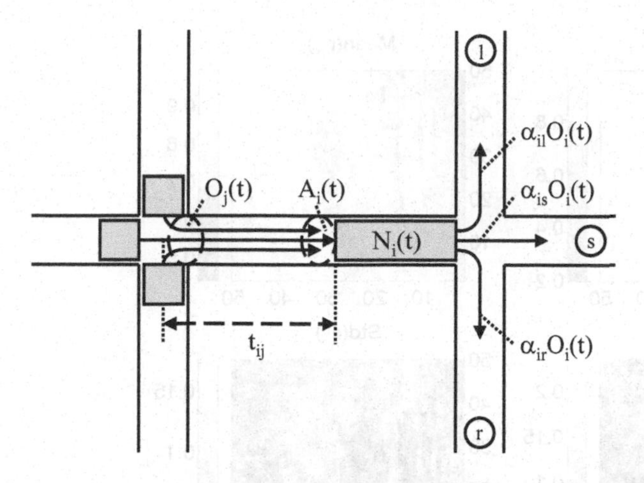

Fig. 8 Basic variables of the material flow model for two coupled intersections (after [82]).

Using the described coupling between neighboring nodes, the switching dynamics has been studied for regular grid networks with 25 nodes where the material flows are operated the same way (i.e. with the same parameters of the permeability function) at all intersections. For the sake of simplicity, it shall be further assumed that the travel times on the individual roads do not depend on the loads (i.e. the states of the queue on these roads). Under these conditions, one may show that the network-wide self-organization of the flows leads to a certain minimization of the total amount of waiting material [82]. As one increases the preferred switching threshold determined by the parameter ΔC, the total capacity of the network, but also the amount of delayed material successively increase.

4.4 Emergence of Synchronization in Regular Grid Networks

In order to better understand the dynamics of the self-organization process due to the proposed decentralized control, a detailed phase coherence analysis[2] has been performed for the case of a fully symmetric grid network consiting of 25 intersections.

[2] Note again that the corresponding measures for quantifying phase coherence have been introduced in the context of phase synchronization analysis, while the notation of phase synchronization is doubtful in the context considered here since the intersecting material flows subjected to decentralized control do not represent self-sustained oscillators, which are a prerequisite for phase synchronization in the standard definition [17].

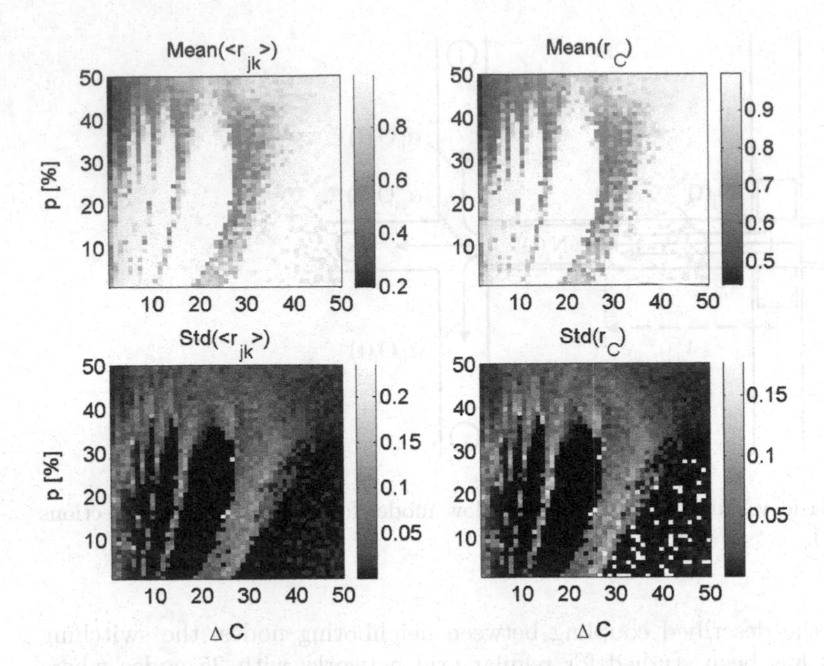

Fig. 9 Mean values (upper panels) and standard deviations (lower panels) of the mean pairwise mean resultant length $\langle r_{jk} \rangle$ (left) and synchronization cluster strength r_C (right), obtained from different parts of a set of long simulation runs for a regular 5×5 grid network with constant travel times $t_{ij} = 30$ time units, fixed randomly chosen inflows A_i on the links entering from outside the network, and $\alpha = 0.1$. Whereas in the phase coherent regimes, this coherence is found to be stable in time (low standard deviations), in the incoherent parts, the considered measures fluctuate also significantly in time (high standard deviations). Note that the results concerning the synchronization cluster strength r_C may be less reliable, as the corresponding algorithm does not necessarily converge correctly in all cases.

For evaluating the presence of phase coherence in this context, the appropriate definition of a monotonously increasing phase variable is necessary. Without loss of generality, the initial phase of node j has been defined in a way that $\phi_j = 0$ corresponds to the time of the first switching of its permeability function. In a similar way, $\phi_j = (n-1)\pi$ then corresponds to the time of the n-th switching at this node. Between these switching times, the phase variable $\phi(t)$ is defined by linear interpolation. Although this definition leads to an increase of the phase which may be periodically modulated if the "on" and "off" times for one specific direction are not symmetric, in the long-term limit, these variables may be used for a phase coherence analysis.

In a two-parameter study of the behavior of the traffic network in dependence on both p and the switching threshold ΔC, multivariate phase coherence analysis reveals pronounced Arnold tongues that correspond to different

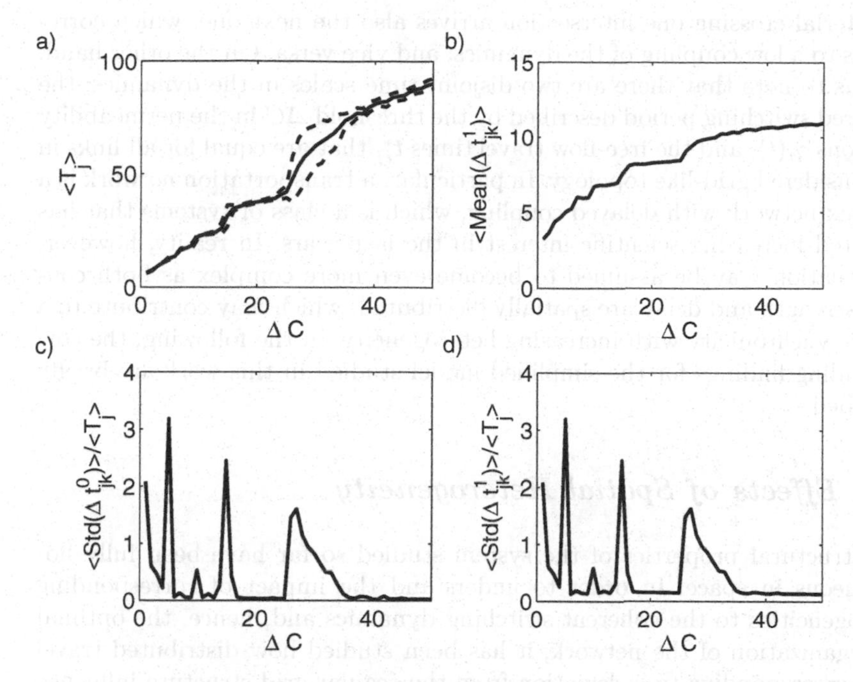

Fig. 10 a) Average of the cycle periods T_j of all intersections (solid line) and $\pm 1.96\sigma_{T_j}$ confidence levels (corresponding to 95% coverage probability in case of Gaussian distributed cycle periods). b) Mean time delay between two successive events at two intersections. c,d) Averaged standard deviations of the switching-time differences $\Delta_{jk}^{(0)}$ and $\Delta_{jk}^{(1)}$ at different intersections as a function of the switching threshold parameter ΔC for $\alpha = 0.1$ and $p = 0.05$. In order to approach a dimensionless parameter, the latter quantities have been normalized by the average cycle period $\langle T_j \rangle$.

phase coherent regimes in the system (see Fig. 9) [82]. These tongues are separated by parameter regions that correspond to an incoherent switching at the different nodes. The corresponding findings are supported by the results of event synchronization analysis (see Fig. 10), which are in good qualitative agreement. In particular, for a fixed turning probability p, there are distinct values of the switching threshold for which the standard deviations of the mutual event-time differences between different intersections approach values near zero, indicating a possibly lagged synchronization of the switching.

A more detailed understanding of the emergence of multiple synchronization regimes is provided by a closer look into the structural properties of the considered system. On the one hand, the probability $\alpha_{ji} = p$ of right-hand turning may be understood as a generalized coupling parameter between neighboring intersections (at least as long as possible sources or sinks along the links of the network are neglected): If p is large, only a lower amount of material crossing one intersection arrives also the next one, which corresponds to a low coupling of the dynamics, and vice versa. On the other hand, one has to note that there are two disjoint time scales in the dynamics: the preferred switching period described by the threshold ΔC in the permeability functions $\gamma_i(t)$, and the free-flow travel times t_{ij}^* that are equal for all links in the considered grid-like topology. In particular, a transportation network is a complex network with delayed coupling, which is a class of systems that has attracted increasing scientific interest in the last years. In reality, however, the situation may be assumed to become even more complex as both coupling strength and delay are spatially distributed, which may contribute to a loss of synchronicity with increasing heterogeneity. In the following, the corresponding findings for the simplified model studied in this work are briefly described.

4.5 Effects of Spatial Heterogeneity

The structural properties of the system studied so far have been fully homogeneous in space. In order to understand the impact of corresponding heterogeneities to the coherent switching dynamics and, hence, the optimal self-organization of the network, it has been studied how distributed travel times corresponding to a deviation from the regular grid structure influence the degree of synchronization in the switching dynamics of the network. In a recent paper [82], it has been demonstrated that the measures of phase coherence discussed above decay only slowly as the disorder of the system increases. In general, this disorder is reflected in an increasing amount of delayed material and a systematic tendency towards shorter service intervals. From this, one may argue that the presence of heterogeneities limits the capability of material flow systems to optimally self-organize in terms of the local control strategy used in this work.

It is likely that also an uneven spatial distribution of other parameters (in particular, turning probabilities, switching thresholds, or the mutual weights of the different force terms entering the cost function $C(t)$) will lead to similar consequences. Of more fundamental interest, however, is the question how a consideration of dynamically adjusted travel times incorporating queued road sections as well as acceleration and deceleration effects may affect the resulting material flow dynamics. This question will be further addressed in future studies.

5 Conclusions

In many socio-economic systems, in particular in the fields of production and logistics, there has already been a paradigm shift from a centralized control towards a decentralized self-organization of material flows [98]. In the special case of urban traffic networks, the situation is however still quite different, as local control strategies have so far not been very successful in practical implementations. A particular reason for this is that "traditional" decentralized control approaches based on clearing policies may lead to unstable traffic conditions due to the presence of dynamic feedbacks. The results reported in this contribution may be an important step in overcoming these problems. However, for specific practical applications, there may be a need for incorporating additional mechanisms for stabilizing the large-scale dynamics, for example, by including demand anticipation over a finite time horizon [14, 99] or by implementing additional heuristic strategies for avoiding local deadlocks [14, 100].

In the presented work, a general concept has been discussed for serving conflicting material flows in general networks in a fully demand-dependent way. Apart from the potential applicability of this approach to controlling material flows in real-world networks, it may also be used for modelling and understanding flows in a variety of other (in particular, biological) systems. The key ingredient of a permeability function which is determined by gradient forces like net pressures or drains can be specified in such systems for describing discrete as well as continuous flows.

It has been shown that an appropriate specification of the permeability function leads to a fully self-organized and synchronized dynamics of the traffic at intersections, which is significantly more flexible than traditional approaches using a centrally enforced cyclic traffic light control. Hence, using the presented approach for implementing a decentralized control strategy at all intersections of a material flow network, the time-delayed local coupling between the flows at neighboring nodes leads to the emergence of a network-wide phase coherent switching dynamics that can be understood in the context of event synchronization. As a particularly interesting feature, the presence of multiple disjoint synchronization regimes in networks has been revealed, which is a so far not yet fully understood dynamical phenomenon possibly related to the presence of two intrinsic time scales of the dynamics, corresponding to the preferred switching intervals prescribed by the parameter ΔC of the considered model and the typical travel times t_{ij} on the different links.

Although in the presented considerations, a continuous flow approximation has been used for describing all material flows, theoretical investigations [11] as well as simulations not discussed here in detail suggest that the resulting dynamics on the network is very similar if the individual transportation units or agents are explicitly considered. Moreover, in this study, a variety of simplifications have been made, which are necessary for a detailed analytical treatment of the dynamical properties of the permeability model. Additional effects have to be explicitly taken into account in future studies, including the influence of spatial heterogeneity and distributed parameters, the dynamical feedback between queue lengths and travel times, the definition of setup phases, acceleration effects leading to a delayed clearing of queues, etc. It will be of particular interest how these effects influence the emergence of a synchronized switching and, as a consequence, the efficiency of a self-organized traffic light control within the network.

The question how the observed self-organization processes are related to an optimization of network flows has not been discussed in much detail in this paper. From the perspective of temporal variability, synchronization introducing regularity of material flows on large spatial scales can be considered as a particular optimization goal. However, regarding the minimization of throughput (or, alternatively, waiting) times in the network, one has to recall that the duration of service periods increases monotonously with the switching threshold ΔC. However, when tuning this parameter in a way that corresponds to a transition from a non-synchronous to a synchronous switching, the corresponding increase of delayed material is significantly reduced. These preliminary results have so far only been verified for switching intervals without setup times. Recent results in the field of urban traffic networks [100] however suggest that the transition to a synchronized, not necessarily phase-coherent service could lead to a significant decrease in the total waiting times. The further validation of this hypothesis within a more general framework of arbitrary material flow networks will be an important topic of future research.

Acknowledgement

His work has been financially supported by the German Research Foundation (DFG projects He 2789/8-1), the Volkswagen foundation, and the Daimler-Benz foundation for their partial support of this study. Discussions with Stefan Lämmer, Dirk Helbing, Aude Hofleitner, Kathrin Padberg, Thomas Seidel, Mamen Romano, Marco Thiel and Tetsuo Yanagita are gratefully acknowledged.

References

1. Chowdhury, D., Santen, L., Schadschneider, A.: Statistical physics of vehicular traffic and some related systems. Phys. Rep. 329, 199–329 (2000)
2. Helbing, D.: Traffic and related self-driven many-particle systems. Rev. Mod. Phys. 73, 1067–1141 (2001)
3. Kerner, B.S.: The Physics of Traffic: Empirical Freeway Pattern Features, Engineering Applications, and Theory. Springer, Berlin (2004)
4. Helbing, D.: Modelling supply networks and business cycles as unstable transport phenomena. New J. Phys. 5, 90.1–90.28 (2003)
5. Howard, J.: Mechanics of Motor Proteins and the Cytoskeleton. Sinauer Associates, Sunderland (2001)
6. Parmeggiani, A., Franosch, T., Frey, E.: Phase Coexistence in Driven One-Dimensional Transport. Phys. Rev. Lett. 90, 086601 (2003)
7. Alberts, B., Bray, D., Lewis, J., Raff, M., Roberts, K., Watson, J.D.: Molecular Biology of the Cell, 3rd edn. Garland, New York (2004)

8. Chowdhury, D., Basu, A., Garai, A., Greulich, P., Nishinari, K., Schadschneider, A., Tripathi, T.: Intra-cellular traffic: bio-molecular motors on filamentary tracks. Europ. Phys. J. B 64, 593–600 (2008)
9. Nakagaki, T., Kobayashi, R., Nishiura, Y., Ueda, T.: Obtaining multiple separate food sources: behavioural intelligence in the Physarum plasmodium. Proc. R. Soc. Lond. B 271, 2305–2310 (2004)
10. Bebber, D.P., Hynes, J., Darrah, P.R., Boddy, L., Fricker, M.D.: Biological solutions to transport network design. Proc. R. Soc. B 274, 2307–2315 (2007)
11. Helbing, D., Lämmer, S., Lebacque, J.-P.: Self-organized control of irregular or perturbed network traffic. In: Deissenberg, C., Hartl, R.F. (eds.) Optimal Control and Dynamic Games, pp. 239–274. Springer, Dordrecht (2005)
12. Newell, G.F.: Applications of Queueing Theory. Chapman and Hall, London (1971)
13. Bose, S.K.: An Introduction to Queueing Systems. Springer, New York (2001)
14. Lämmer, S.: Reglerentwurf zur dezentralen Online-Steuerung von Lichtsignalanlagen in Straßennetzwerken, PhD thesis, Dresden University of Technology (2007)
15. Kumar, P.R., Seidman, T.I.: Dynamic instabilities and stabilization methods in distributed real-time scheduling of manufacturing systems. IEEE Trans. Automat. Contr. 35, 289–298 (1990)
16. Lämmer, S., Kori, H., Peters, K., et al.: Decentralised control of material or traffic flows in networks using phase synchronisation. Physica A 363, 39–47 (2006)
17. Pikovsky, A., Rosenblum, M., Kurths, J.: Synchronization - A Universal Concept in Nonlinear Sciences. Cambridge University Press, Cambridge (2001)
18. Boccaletti, S., Kurths, J., Osipov, G., et al.: The synchronization of chaotic systems. Phys. Rep. 366, 1–101 (2002)
19. Pikovsky, A., Maistrenko, Y. (eds.): Synchronization: Theory and Application. Kluwer, Dordrecht (2003)
20. Strogatz, S.: Sync: The Emerging Science of Spontaneous Order. Hyperion, New York (2003)
21. Osipov, G.V., Kurths, J., Zhou, C.S.: Synchronization in Oscillatory Networks. Springer, Berlin (2007)
22. Arenas, A., Diaz-Guilera, A., Kurths, J., et al.: Synchronization in Complex Networks. Phys. Rep. 469, 93–153 (2008)
23. Helbing, D., Molnár, P.: Social force model for pedestrian dynamics. Phys. Rev. E 51, 4282–4286 (1995)
24. Mueller, E.A.: Aspects of the History of Traffic Signals. IEEE Trans. Vehic. Technol. 19, 6–17 (1970)
25. Marsh, B.W.: Traffic Control. Ann. Amer. Acad. Polit. Soc. Sci. 133, 90–113 (1927)
26. Morgan, J.T., Little, J.D.C.: Synchronizing traffic signals for maximal bandwidth. Operat. Res. 12, 896–912 (1964)
27. Little, J.D.C.: The Synchronization of Traffic Signals by Mixed-Integer Linear Programming. Operat. Res. 14, 568–594 (1966)
28. Little, J.D.C., Kelson, M.D., Gartner, N.H.: MAXBAND: A program for setting signals on arteries and triangular networks. Transp. Res. Rec. 795, 40–46 (1981)
29. Robertson, D.I.: TRANSYT method for area traffic control. Traffic Engin. Contr. 10, 276–281 (1969)

30. Bazzan, A.L.C.: A distributed approach for the coordination of traffic signal agents, Autonom. Agents Multi-Agent Syst. 10, 131–164 (2005)
31. Robertson, D.I., Bretherton, R.D.: Optimizing networks of traffic signals in real time – the SCOOT method. IEEE Trans. Vehic. Technol. 40, 11–15 (1991)
32. Gartner, H.N.: OPAC: A demand-responsive strategy for traffic signal control. Transp. Res. Rec. 906, 75–84 (1983)
33. Henry, J.J., Farges, J.L., Tufal, J.: The PRODYN Real Time Traffic Algorithm. In: 4th IFAC/IFIP/IFORS Int. Conf. Control in Transportation Systems, Baden-Baden, pp. 307–311 (1983)
34. Boillot, F., Blosseville, J., Lesort, J., et al.: Optimal signal control of urban traffic networks. In: 6th Int. Conf. Road Traffic Monitoring and Control, vol. 355, pp. 75–79 (1992)
35. Mauro, V., Di Taranto, C.: UTOPIA. In: Perrin, J.-P. (ed.) Control, Computers, Communications in Transportation, pp. 245–252. Pergamon, Oxford (1990)
36. Mirchandani, P., Head, L.: A real-time traffic signal control system: architecture, algorithms, and analysis. Transp. Res. C 9, 415–432 (2001)
37. Friedrich, B.: Ein verkehrsadaptives Verfahren zur Steuerung von Lichtsignalanlagen. PhD Thesis, Technical University of Munich (1997)
38. Bielefeldt, C., Busch, F.: MOTION – a new on-line traffic signal network control system. In: 7th Int. Conf. Road Traffic Monitoring and Control, London, pp. 55–59 (1994)
39. Dinopoulou, V., Diakaki, C., Papageorgiou, M.: Simulation investigations of the traffic-responsive urban control strategy TUC. Intell. Transp. Syst., 458–463 (2000)
40. Brockfeld, E., Barlovic, R., Schadschneider, A., et al.: Optimizing traffic lights in a cellular automaton model for city traffic. Phys. Rev. E 64, 056132 (2001)
41. Fouladvand, M.E., Nematollahi, M.: Optimization of green-times at an isolated urban crossroads. Eur. Phys. J. B 22, 395–401 (2001)
42. Huang, D.-W., Huang, W.-N.: Traffic signal synchronization. Phys. Rev. E 67, 056124 (2003)
43. Newell, G.F.: The flow of highway traffic through a sequence of synchronized traffic signals. Operat. Res. 8, 390–405 (1960)
44. Newell, G.F.: Synchronization of traffic lights for high flow. Quart. Appl. Math. 21, 315–324 (1964)
45. Baverez, E., Newell, G.F.: Traffic Signal Synchronization on a One-Way Street. Transp. Sci. 1, 55–73 (1967)
46. Hillier, J.A., Rothery, R.: The Synchronization of Traffic Signals for Minimum Delays. Transp. Sci. 1, 81–94 (1967)
47. Newell, G.F.: Traffic Signal Synchronization for High Flows on a Two-Way Street. In: Leutzbach, W., Baron, P. (eds.) Beiträge zur Theorie des Verkehrsflusses: Referate anläßlich des IV. Internationalen Symposiums über die Theorie des Verkehrsflusses in Karlsruhe im Juni, Proc. Fourth Intern. Symp. on the Theory of Traffic Flow, University of Karlsruhe, pp. 87–92. Bundesministerium für Verkehr, Bonn (1968)
48. Chang, A.: Synchronization of Traffic Signals in Grid Networks. IBM Journal 11, 436–441 (1967)
49. Gartner, N.: Constraining Relations Among Offsets In Synchronized Signal Networks. Transp. Sci. 6, 88–93 (1972)

50. Gartner, N., Little, J.D.C., Gabbay, H.: Optimization of Traffic Signal Settings by Mixed-Integer Linear Programming. Part I: The Network Coordination Problem. Transp. Sci. 9, 321–343 (1975)
51. Gartner, N., Little, J.D.C., Gabbay, H.: Optimization of Traffic Signal Settings by Mixed-Integer Linear Programming. Part II: The Network Synchronization Problem. Transp. Sci. 9, 321–343 (1975)
52. Faieta, B., Huberman, B.A.: Firefly: A Synchronization Strategy for Urban Traffic Control. Techn. Rep., Xerox Palo Alto Research Center (1993)
53. Carlson, A.D., Copeland, J.: Flash Communication in Fireflies. Quart. Rev. Biol. 60, 415–433 (1985)
54. Buck, J.: Synchronous Rhythmic Flashing of Fireflies. II. Quart. Rev. Biol. 63, 265–286 (1988)
55. Mirollo, R.E., Strogatz, S.H.: Synchronization of Pulse-Coupled Biological Oscillators. SIAM J. Appl. Math. 50, 1645–1662 (1990)
56. Ermentrout, B.: An adaptive model for synchrony in the firefly Pteroptyx malaccae. J. Math. Biol. 29, 571–585 (1991)
57. Chowdhury, D., Schadschneider, A.: Self-organization of traffic jams in cities: Effects of stochastic dynamics and signal periods. Phys. Rev. E 59, R1311–R1314 (1999)
58. Huang, D.-W., Huang, W.-N.: Optimization of traffic lights at crossroads. Int. J. Mod. Phys. C 14, 539–548 (2003)
59. Kuramoto, Y.: Self-entrainment of a population of coupled non-linear oscillators. In: Araki, H. (ed.) International Symposium on Mathematical Problems in Theoretical Physics. Lecture Notes in Physics, vol. 39, pp. 420–422. Springer, New York (1975)
60. Kuramoto, Y.: Chemical Oscillations, Waves, and Turbulence. Springer, New York (1984)
61. Strogatz, S.H.: From Kuramoto to Crawford: exploring the onset of synchronization in populations of coupled oscillators. Physica D 143, 1–20 (2000)
62. Acebrón, J.A., Bonilla, L.L., Pérez Vicente, C.J., et al.: The Kuramoto model: A simple paradigm for synchronization phenomena. Rev. Mod. Phys. 77, 137–185 (2005)
63. Sekiyama, K., Nakanishi, J., Takagawa, I., et al.: Self-Organizing Control of Urban Traffic Signal Networks. In: Proc. IEEE Int. Conf. on Systems, Man, and Cybernetics, vol. 4, pp. 2481–2486 (2001)
64. Donner, R., Lämmer, S., Hofleitner, A., Helbing, D.: Decentralised control of traffic flows in urban networks; Schneider, J.J. (ed.), Lectures on Socio- and Econophysics. Springer, Berlin (in preparation) (accepted for publication)
65. Nishikawa, I., Nakazawa, S., Kita, H.: Area-wide Control of Traffic Signals by a Phase Model. Trans. Soc. Instr. Contr. Engin. 39, 199–208 (2003) (in Japanese)
66. Nishikawa, I., Kita, H.: Information Flow Control by Autonomous Adaptive Nodes and Its Application to an On-line Traffic Control. In: Proc. 7th Joint Conf. on Information Sciences, pp. 1369–1374. Association for Intelligent Machinery, Durham (2003)
67. Nishikawa, I.: Dynamics of Oscillator Network and Its Application to Offset Control of Traffic Signals. In: Proc. Int. Joint Cof. Neural Networks, pp. 1273–1277 (2004)

68. Nishikawa, I., Kuroe, Y.: Dynamics of complex-valued neural networks and its relation to a phase oscillator system. In: Pal, N.R., Kasabov, N., Mudi, R.K., Pal, S., Parui, S.K. (eds.) ICONIP 2004. LNCS, vol. 3316, pp. 122–129. Springer, Heidelberg (2004)
69. Nishikawa, I., Iritani, T., Sakakibara, K., Kuroe, Y.: Phase dynamics of complex-valued neural networks and its application to traffic signal control. Int. J. Neur. Syst. 15, 111–120 (2005)
70. Nishikawa, I., Iritani, T., Sakakibara, K., Kuroe, Y.: 2 Types of Complex-Valued Hopfield Networks and the Application to a Traffic Signal Control. In: Proc. Int. Joint Conf. Neural Networks, pp. 782–787 (2005)
71. Nishikawa, I., Iritani, T., Sakakibara, K., Kuroe, Y.: Phase Synchronization in Phase Oscillators and a Complex-Valued Neural Network and Its Application to Traffic Flow Control. Prog. Theor. Phys. Suppl. 161, 302–305 (2006)
72. Nishikawa, I., Iritani, T., Sakakibara, K.: Improvements of the Traffic Signal Control by Complex-Valued Hopfield Networks. In: Proc. Int. Joint Conf. Neural Networks, pp. 459–464 (2006)
73. Helbing, D., Siegmeier, J., Lämmer, S.: Self-organized network flows. Netw. Heterog. Med. 2, 193–210 (2007)
74. Helbing, D., Johansson, A., Mathiesen, J., Jensen, M.H., Hansen, A.: Analytical approach to continuous and intermittent bottleneck flows. Phys. Rev. Lett. 97, 168001 (2006)
75. Helbing, D., Buzna, L., Johansson, A., Werner, T.: Self-Organized Pedestrian Crowd Dynamics: Experiments, Simulations, and Design Solutions. Transp. Sci. 39, 1–24 (2005)
76. Dussutour, A., Deneubourg, J.-L., Fourcassié, V.: Temporal organization of bi-directional traffic in the ant Lasius niger (L). J. Exp. Biol. 208, 2903–2912 (2005)
77. Burstedde, C., Klauck, K., Schadschneider, A., Zittartz, J.: Simulation of pedestrian dynamics using a two-dimensional cellular automaton. Physica A 295, 507–525 (2001)
78. Helbing, D., Jiang, R., Treiber, M.: Analytical investigation of oscillations in intersecting flows of pedestrian and vehicle traffic. Phys. Rev. E 72, 046130 (2005)
79. Jiang, R., Wang, Q.-S.: Interaction between vehicle and pedestrians in a narrow channel. Physica A 368, 239–246 (2006)
80. Jiang, R., Helbing, D., Kumar Shukla, P., Wang, Q.-S.: Inefficient emergent oscillations in intersecting driven many-particle flows. Physica A 368, 567–574 (2006)
81. Donner, R., Hofleitner, A., Höfener, J., Lämmer, S., Helbing, D.: Dynamic stabilization and control of material flows in networks and its relationship to phase synchronization. In: Proc. 3rd Int. Conf. Physics and Control, PhysCon 2007, p. 1188 (2007)
82. Donner, R.: Multivariate analysis of spatially heterogeneous phase synchronisation in complex systems: Application to self-organised control of material flows in networks. Eur. Phys. J. B 63, 349–361 (2008)
83. Rayleigh, L.: On the resultant of a large number of vibrations of the same pitch and of arbitrary phase. Phil. Mag. Ser. 5(10), 73–78 (1880)
84. Tass, P., Rosenblum, M.G., Weule, J., et al.: Detection of n:m Phase Locking from Noisy Data: Application to Magnetoencephalography. Phys. Rev. Lett. 81, 3291–3294 (1998)

85. Schürmann, T., Grassberger, P.: Entropy estimation of symbol sequences. Chaos 142, 414–427 (1996)
86. Finn, J.M., Goettee, J.D., Toroczkai, Z., et al.: Estimation of entropies and dimensions by nonlinear symbolic time series analysis. Chaos 13, 444–456 (2003)
87. Paluš, M.: Detecting phase synchronization in noisy systems. Phys. Lett. A 235, 341–351 (1997)
88. Eckmann, J.-P., Oliffson Kamphorst, S., Ruelle, D.: Recurrence plots of dynamic systems. Europhys. Lett. 4, 973–977 (1987)
89. Marwan, N., Romano, M.C., Thiel, M., Kurths, J.: Recurrence plots for the analysis of complex systems. Phys. Rep. 438, 237–329 (2007)
90. Romano, M.C., Thiel, M., Kurths, J., et al.: Detection of synchronization for non-phase-coherent and non-stationary data. Europhys. Lett. 71, 466–472 (2005)
91. Romano, M.C., Thiel, M., Kurths, J., et al.: Synchronization Analysis and Recurrence in Complex Systems. In: Schelter, B., Winterhalder, M., Timmer, J. (eds.) Handbook of Time Series Analysis, pp. 231–264. Wiley-VCH, Weinheim (2006)
92. Donner, R.: Phase Coherence Analysis of Decadal-Scale Sunspot Activity of Both Solar Hemispheres. In: Donner, R.V., Barbosa, S.M. (eds.) Nonlinear Time Series Analysis in the Geosciences. Springer Lecture Notes in Earth Sciences, vol. 112, pp. 355–385. Springer, Berlin (2008)
93. Kralemann, B., Cimponeriu, L., Rosenblum, M., et al.: Uncovering interaction of coupled oscillators from data. Phys. Rev. E 76, 055201 (2007)
94. Kralemann, B., Cimponeriu, L., Rosenblum, M., et al.: Phase dynamics of coupled oscillators reconstructed from data. Phys. Rev. E 77, 066205 (2008)
95. Allefeld, C., Kurths, J.: An approach to multivariate phase synchronization analysis and its application to event-related potentials. Int. J. Bifurcation Chaos 14, 417–426 (2004)
96. Allefeld, C., Müller, M., Kurths, J.: Eigenvalue decomposition as a generalized synchronization cluster analysis. Int. J. Bifurcation Chaos 17, 3493–3497 (2007)
97. Quian Quiroga, R., Kreuz, T., Grassberger, P.: Event synchronization: A simple and fast method to measure synchronicity and time delay patterns. Phys. Rev. E 66, 041904 (2002)
98. Hülsmann, M., Windt, K.: Understanding Autonomous Cooperation and Control in Logistics. Springer, Berlin (2008)
99. Lämmer, S., Donner, R., Helbing, D.: Anticipative control of switched queueing systems. Eur. Phys. J. B 63, 341–347 (2008)
100. Lämmer, S., Helbing, D.: Self-Control of Traffic Lights and Vehicle Flows in Urban Road Networks. J. Stat. Mech. Theor. Exp., P04019 (2008)

Synchronization of Movement for a Large-Scale Crowd

Boris Goldengorin, Dmitry Krushinsky, and Alexander Makarenko

Abstract. Real world models of large-scale crowd movement lead to computationally intractable problems implied by various classes of non-linear stochastic differential equations. Recently, cellular automata (CA) have been successfully applied to model the dynamics of vehicular traffic, ants and pedestrians' crowd movement and evacuation without taking into account mental properties. In this paper we study a large-scale crowd movement based on a CA approach and evaluated by the following three criteria: the minimization of evacuation time, maximization of instantaneous flow of pedestrians, and maximization of mentality-based synchronization of a crowd. Our computational experiments show that there exist interdependencies between the three criteria.

1 Introduction

Recently synchronization phenomena became one of the most interesting topics of investigations on nonlinear dynamics [1-7]. Analysis of the development of synchronization theory, just as of the theory of nonlinear dynamical systems, leads to a conclusion that there are two main sources of the development of such theories: the first one is an application of new mathematical methods and concepts to the systems with synchronization while the second source derives from the expanding fields of investigations with development of new mathematical objects that can serve as a background for further investigations and interpretations. Usually nonlinear dynamical systems of neutral or delay type had been the classical objects for synchronization investigations (e.g. ordinary differential equations, partial differential equations, chains of coupled maps, discrete equations with delay etc.).

Boris Goldengorin and Dmitry Krushinsky
University of Groningen, Operations Department, P.O. Box 9700 AV, Groningen, The Netherlands
e-mail: b.goldengorin@rug.nl

Dmitry Krushinsky and Alexander Makarenko
Institute for Applied System Analysis, National Technical University of Ukraine "KPI", 37 Pobedy Avenue, 03056, Kyiv-56, Ukraine
e-mail: makalex@i.com.ua

K. Kyamakya (Eds.): Recent Adv. in Nonlinear Dynamics and Synchr., SCI 254, pp. 277–303.
springerlink.com

Recently some new less known mathematical objects had been found - dynamical systems with an anticipation property. It should be mentioned that a natural source of systems with anticipation are social systems which take into account mental properties of individuals (see e.g. [8, 9]). Another example of nonlinear dynamic systems with synchronization can be found in [10, 11].

In this paper we describe nonlinear models with anticipation for problems of pedestrian movements modelled by means of cellular automata (CA). This choice is enforced by the fact that CA is a well known, applicable and useful type of tools [12-15]. Some examples of cellular automata with anticipation applied to traffic and crowd movement modelling have been studied in [8, 9, 16-24].

In this chapter we study a synchronization phenomenon in pedestrian traffic, taking into account different anticipation models of pedestrians with respect to their speed, radius and shape of observed neighbourhood, including local stochastic predictions for neighbouring cells to be occupied. We consider two types of situations: evacuation of finite number of pedestrians from premises with static geometry and (endless) continuous flow of pedestrians through a corridor with simplified geometry that allows quantitative analysis on a strict background. We estimate performance of the crowd in these two situations by evacuation time (that must be minimized) and flow through the exit (that must be maximized), respectively. In fact, these two characteristics of performance are closely related, as increase in the flow at the exit leads to a faster evacuation and to decrease in the evacuation time. Further, we introduce a measure of synchronization as a measure of collaborating behaviour of (anticipating) pedestrians and claim that in a perfectly synchronized crowd the minimum value of evacuation time and the maximum value of instantaneous flow are achieved.

In section 2 we give a description of the basic CA model that was further used for introduction of the anticipation property. In section 3 we show four different implementations of anticipation property, provide the results of numerical experiments with variations of our anticipating model and describe a framework for generalization of our model for the case of arbitrary size of the extended neighbourhood and arbitrary horizon of prediction (a number of steps for which a prediction is made). In section 4 we introduce a notion of synchronization in terms of pedestrian crowd movement and discuss a relation between anticipation and synchronisation. Section 5 is devoted to conclusions and prospective of future research.

2 CA Models of Pedestrian Traffic

In this paper we study special cases of CA models with discrete space and time as follows [15-21]:

- microscopic: every pedestrian is simulated by a separate single cell;
- stochastic: local rules contain random values;
- space- and time-discrete.

The basic assumptions behind the models are:

- dynamics of pedestrian motion can be represented by a CA;
- global route is pre-determined;

- irrational behaviour is rare;
- persons are not strongly competitive, i.e. they do not hurt each other;
- individual differences can be represented by parameters determining the behaviour.

A CA has two layers (Fig. 1). The first one – data layer – embeds the information about the geometry of the scene, i.e. placement of pedestrians and obstacles. Every cell in this layer has 3 possible states: "empty", "obstacle", "pedestrian".

The second layer embeds a vector field of directions and stores the information about the global route. This field of directions is constructed so that to minimize evacuation time of a sole pedestrian. If there are several possibilities at a particular point, they are considered to be equally probable. Every pedestrian receives information about the global route via interaction with this so-called static floor field [21].

At every time step, for every pedestrian probabilities of shift for all the directions are being computed according to the following principles:

- if a target cell is occupied (by obstacle or other pedestrian), the corresponding probability is set to 0;
- pedestrians try to follow the optimal global route.

In fact, primary "forces" driving the pedestrians' behaviour in our basic model are somewhat different from those used by other authors (see e.g. [21]) and include:

1. determination: striving for the global route (similarly to [21]);
2. inertia: attempt to preserve current direction of movement (relaxed in [21]);
3. randomization: tendency to random changes in direction of movement (in contrast to [21] we consider a mix of uniformly distributed movements with the properties 1 and 2).

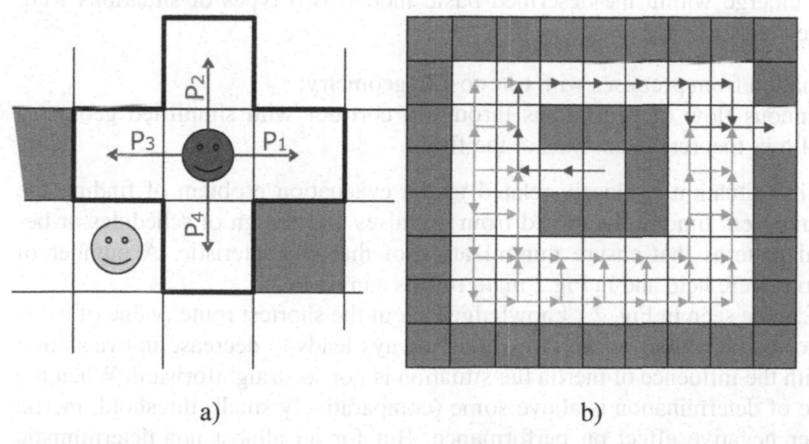

Fig. 1 Structure of a general CA model of pedestrian traffic: data layer (a) that embeds a description of placement of obstacles and pedestrians and auxiliary layer (b) that embeds a description of the global route in a form of vectors of direction for every cell.

The influence of the mentioned factors is determined by three parameters p_d, p_i, p_r, reflecting the primary forces 1, 2, and 3, respectively. If only the first of them is nonzero then we have a completely deterministic crowd with each member trying to follow the shortest path to an exit. However, in real life even in the most deterministic group of pedestrians there is a place for randomness.

So, what is the origin of randomness in a well organized crowd? Let us consider a situation when a pedestrian cannot follow the shortest path at a particular time step because some other pedestrian is standing at his way. Then the first one has two major options:

a) to stand still and wait while his way is free;
b) to move in any possible direction so as to pass this jam around.

In real crowds persons tend to apply a mixture of the mentioned behavioural patterns and in our model we have one more parameter ranged between 0 and 1 that expresses probability of type b) behaviour.

In order to implement probabilistic approach to resolve conflicts, arising when any two or more pedestrians attempt to move to the same target cell, at every step a sequence of pedestrians' shifts is randomly chosen.

Persons also differ in their maximum speed. These differences are implemented via division of every time step into v_{max} sub-steps $\tau_1,\ldots,\tau_{V\max}$; v_{max} is a maximum possible speed over the crowd. An *i*-th person tries to move at a sub-step *k* only if $v_i<k$, where v_i – his maximum speed, and both values of v_i and *k* are numerical parameters independent on their informal interpretations.

2.1 *Evacuation and Max Throughput: Major Behavioural Patterns and Characteristics*

In the current subsection we describe major behavioural patterns of pedestrians, that can emerge within the described basic model, two types of situations were considered:

A. Evacuation from premises with true-to-life geometry;
B. Continuous flow of pedestrians through a corridor with simplified geometry that allows theoretical analysis of the flow.

The first situation is closely related to the evacuation problem of finding the minimum egress time of the crowd from premises and design of schedules or behavioural patterns that ensure minimization of that characteristic. A number of simulations were held and in Fig.2 main results can be found.

As it can be seen in Fig. 2, "knowledge" about the shortest route (value of p_d) is critical to the behaviour of the crowd and always leads to decrease in evacuation time. With the influence of inertia the situation is not so straightforward. When the influence of determination is above some (comparatively small) threshold, inertia has rather negative effect on performance. But for an almost non-deterministic crowd ($p_d \rightarrow 0$) there exists some optimal nonzero ratio p_i/p_r that ensures the best possible performance in case of fixed other parameters (in the given example this ratio is about 10).

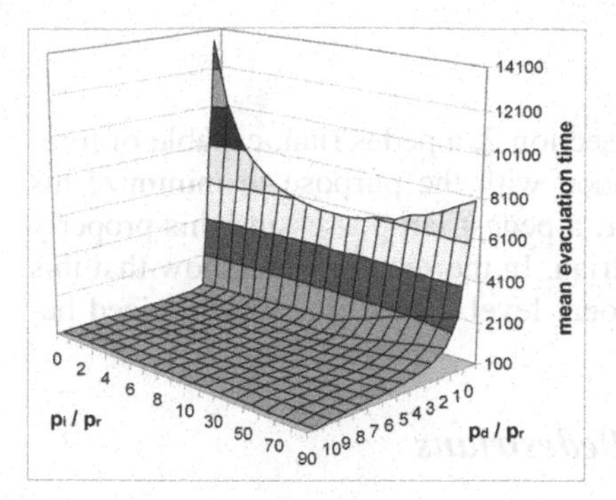

Fig. 2 Mean evacuation time for different values of parameters of the model.

Behaviour of a continuous flow of pedestrians is also of interest as it represents somewhat simpler problem and is more amenable to theoretic methods of analysis. Unlike evacuation that represents itself a transient process, continuous flow tends to a behavioural stationarity that can take different forms: the virtually constant flow, periodic oscillations. Chaotic oscillations, though being rather uncommon, are also possible but lead to difficulties in modelling and analysis. Of particular interest is a phenomenon of emergence and propagation of shock waves.

Results demonstrating the model performance in a case of continuous flow can be found in Fig.3. According to the results of simulations, in a case of continuous flow the situation is somewhat different from the case of evacuation. Now both determination and inertia have positive impact on performance; however, some peculiarities should be mentioned. First of all, for the values of p_I beyond some threshold there exists optimal finite value of p_d that maximizes the flow. Another peculiarity is that at high values of p_d impact of inertia becomes less substantial, which is illustrated by the flat right part of the surface in Fig.3a.

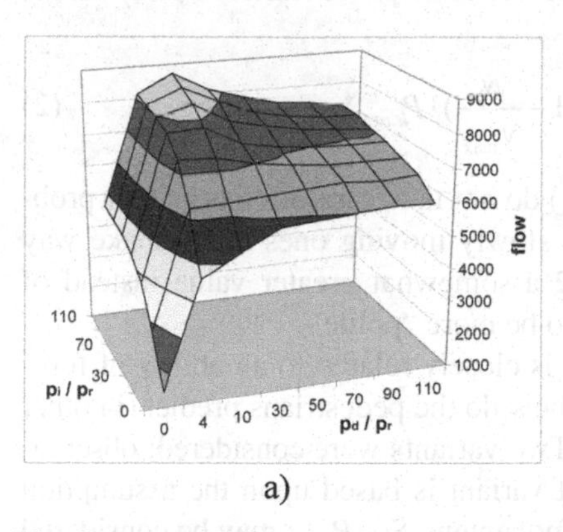

a)

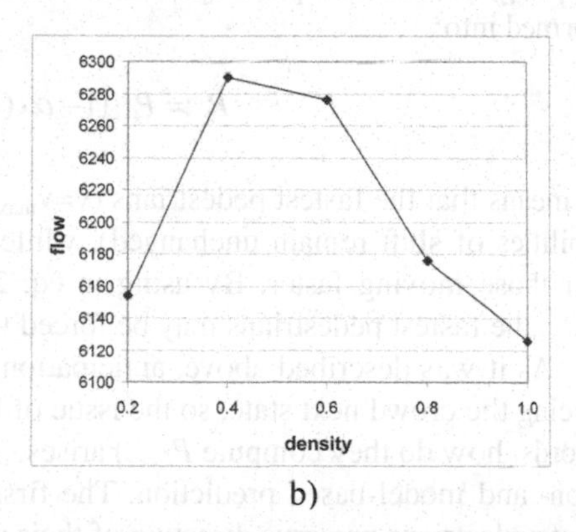

b)

Fig. 3 Average flow for different values of parameters of the model (a) and a typical density-flow diagram (b) that illustrates a phenomenon specific to crowd movement: increase in density over a certain threshold leads to a decrease in flow because pedestrians start blocking each other.

3 Anticipating Pedestrians

Starting from the basic model described in section 2, a pedestrian, capable of foreseeing the situation within his neighbourhood with the purpose to minimize his evacuation time, may be generated. Further, a pedestrian possessing this property will be referred to as an anticipating pedestrian. In the section 4 we show that this property, while being introduced at a personal level, leads to a synchronized behaviour of the whole crowd.

3.1 Basic Models of Anticipating Pedestrians

As it was mentioned at the beginning of the section 2, at every step a person determines probabilities of shift (P_k, k=1,...,4). These are these values that may be subjected to influence of an anticipation. Let us assume, that pedestrians try to avoid collisions, i.e. a person tries not to move into a particular cell of his neighbourhood if (as he predicts) it will be occupied by another person at the next time step. This may be achieved by changing the probabilities in the following manner:

$$P_k := P_k \cdot (1 - \alpha \cdot P_{k,occ}), \tag{1}$$

where $\alpha \in [0,1]$ is a free parameter expressing influence of anticipation, $P_{k,occ}$ – the probability of occupation of k-th cell in a neighbourhood by one of neighbours. It is quite natural, that values P_k have to be normalized, so that their sum is equal to one (if at least one of them is nonnegative). It should be noted, that in this case all the pedestrians are assumed to have equal rights. If α is set to 1, a situation, when two pedestrians attempt let each other move and stand still, may occur. Such deadlocks can be completely excluded only by selecting the value of $\alpha < 1$. However, the number of these deadlocks can also be reduced by granting certain (e.g. fast-moving) pedestrians privileges. In this case the shift probabilities will be transformed into:

$$P_k := P_k \cdot (1 - \alpha \cdot (1 - \frac{v}{v_{max}}) \cdot P_{k,occ}) \tag{2}$$

It means that the fastest pedestrians ($v=v_{max}$) do not take care of others (their probabilities of shift remain unchanged), while slowly moving ones try to make way for those moving faster. By using in eq. 2 a somewhat greater value instead of v_{max}, the fastest pedestrians may be forced to be more "polite".

As it was described above, anticipation is closely related to an ability of foreseeing the crowd next state, so the issue of how do the pedestrians predict (in other words, how do they compute $P_{k,occ}$) arises. Two variants were considered: observation- and model-based prediction. The first variant is based upon the assumption that pedestrians preserve direction of their movement. So, $P_{k,occ}$ may be considered to be a linear function of the number of pedestrians "facing" the k-th cell (the direction of their sight is defined by the direction of their previous shift):

$$P_{k,occ} = \frac{m}{M}. \tag{3}$$

where m – number of pedestrians "facing" the k-th cell; M = n-1, in our case M = 3, and n is a number of cells in the neighbourhood.

Such an approach, though being the most simple and natural, is, at the same time, the least accurate. Thus, for the sake of comparison, the second approach was considered, according to which a target pedestrian for every cell of his neighbourhood computes P_k of its neighbours (excluding himself) and the resulting probability is defined as follows:

$$P_{k,occ} = \sum_{i=1}^{3} P_i - \sum_{i \neq j} P_i P_j + \sum_{i \neq j, j \neq k} P_i P_j P_k \tag{4}$$

It is quite evident that this approach allows a more accurate evaluation of $P_{k,occ}$, while being somewhat unnatural, as every pedestrian must know behavioural models of the others.

A number of computational experiments were held and typical performance of all the mentioned configurations of the model (corresponding to two ways of introducing anticipation given by eqs. 1 and 2, and two ways of prediction given by eqs. 3 and 4) is presented in Fig.4.

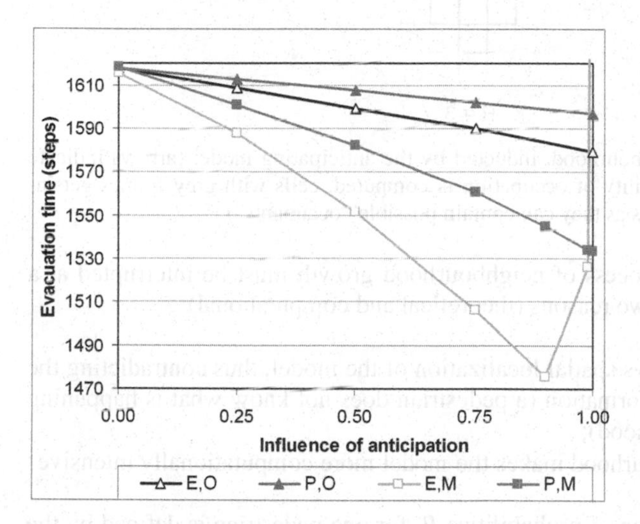

Fig. 4 Performance of CA models with different types of anticipation. (E/P – equality/priority of fast-moving pedestrians (eqs. 1 and 2, respectively); O/M – observation-/model-based prediction (eqs. 3 and 4, respectively)).

The results of simulation reveal the fact, that granting fast-moving pedestrians a priority results in greater overall evacuation time, thus making little sense. On the other hand, the more accurately $P_{k,occ}$ are computed, the better the performance. This proves the consistency of the proposed method of anticipation accounting (given by eq. 1).

3.2 *Spatial "de-localization"*

In subsection 3.1 an anticipating pedestrian was generating his prediction based on non-anticipating model of his neighbours. Thus, it is quite straightforward to make his prediction more accurate by involving an anticipating model of neighbours. For that, every pedestrian (within the neighbourhood of radius 2) is subjected to the procedure described in subsection 3.1 for the target pedestrian: calculation of P_k, calculation of $P_{k,occ}$ (eq. 4) and correction of P_k (eq. 1). It is evident, that in this case cells lying at a distance of 3 cells from the target pedestrian (centre of the neighbourhood) are involved in evaluation of $P_{k,occ}$. At the same time, pedestrians standing 2 cells apart from the centre use non-anticipating model of their neighbours (standing 3 cells apart from the centre). If they have used anticipating model instead, pedestrians standing 4 cells apart from the centre of the neighbourhood would have become involved. Thus, a neighbourhood is growing until it "covers" the entire scene (Fig. 5).

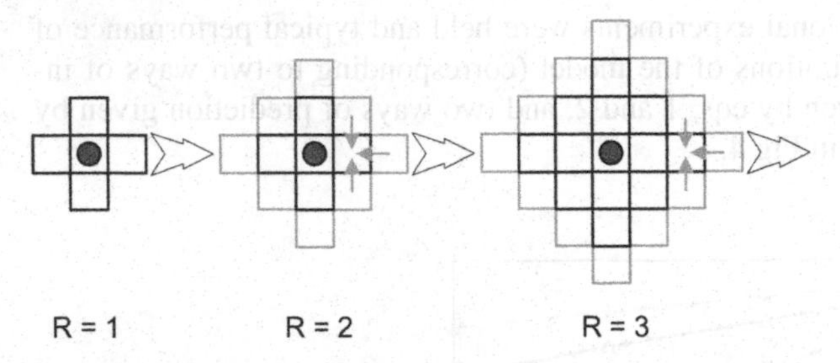

Fig. 5 Growth of the neighbourhood, inducted by the anticipating model (arrows indicate the cell for which a probability of occupation is computed, cells with grey frames get involved into the computations as they can contain possible "occupants")

It is clear that this process of neighbourhood growth must be interrupted at a certain step, because of two reasons (theoretical and computational):

- every next step destroys spatial localization of the model, thus contradicting the hypothesis of local information (a pedestrian does not know what is happening beyond his neighbourhood);
- growth of the neighbourhood makes the model more computationally intensive.

Time-cost of calculation of probabilities P_k for one pedestrian is defined by the number of cells in his (extended) neighbourhood. In our case (4-cell elementary neighbourhood) this number makes up $(R+1)^2+R^2-1 \sim R^2$, where R – radius of an extended neighbourhood. So, this radius should be limited by a certain value, through which different extent of information distribution may be simulated. From a point of view of the target pedestrian, this may be given the following interpretation: all the neighbours inside the extended neighbourhood are considered to be anticipating, unlike those standing on a border. On the other hand, pedestrians on a border may be also considered to be anticipating under an assumption that there are no pedestrians beyond the neighbourhood (in this case for these pedestrians

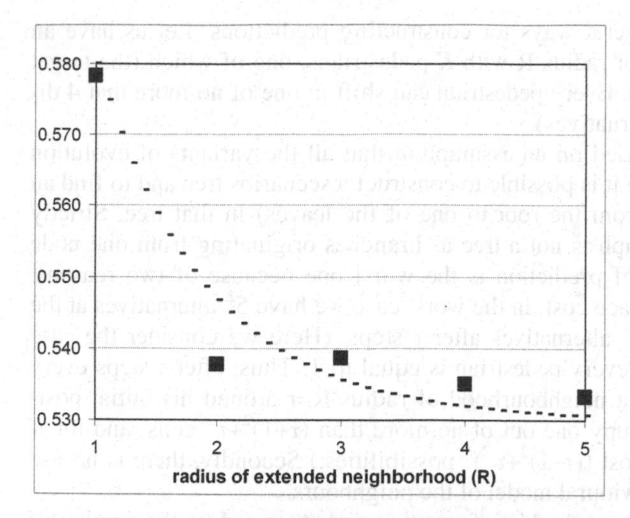

Fig. 6 Impact of neighbourhood extension on evacuation time (evacuation time is given in relative scale with 1.0 corresponding to 150 time steps).

holds $P_{k,occ} = 0$). The algorithm for implementation of the described above scheme can be found in [4].

As it becomes clear from the simulations (see Fig.6), extension of the neighbourhood has positive effect on performance of the crowd leading to shorter evacuation times. Decreasing slope of the curve in the picture is quite natural, as distant pedestrians have little influence on each other. Therefore, within the described model for practical reasons it makes sense to limit the value of R by 4 or 5.

According to the results of simulations, increasing extent of awareness of pedestrians (extending neighbourhood) leads to a decrease in evacuation time, thus certifying adequateness of the proposed approach. However, from a quantitative point of view improvements are not as high, while information about extended neighbourhood allows further optimization of behaviour via multi-step anticipation (anticipation based on multi-step prediction).

3.3 *Temporal "de-localization"*

The implemented models of anticipation (see subsections 3.1, 3.2) allow pedestrians to utilize additional information about their nearest environment within a neighbourhood that is somewhat broader than elementary one. Thus, comparatively to the basic model (see section 2) every pedestrian is provided with two types of additional information:

- information about placement of more distant neighbours;
- information about behavioural model of other pedestrians.

It is quite straightforward that this information is not used to its full extent, as only one-step forecasts are computed while the former suffice for multi-step prediction.

Naturally, there are several ways for constructing predictions. Let us have an extended neighbourhood of radius R with K pedestrians, one of which (the target pedestrian) is in the centre. Every pedestrian can shift in one of no more that 4 directions or stay still (5 alternatives).

The simplest way is based on an assumption that all the variants of evolution are equally probable. Then it is possible to construct a scenarios tree and to find an optimal trajectory (path from the root to one of the leaves) in that tree. Strictly speaking, the obtained graph is not a tree as branches originating from one node may intersect. This way of prediction is the worst one because of two reasons. Firstly, it has high time-space cost: in the worst case we have 5^K alternatives at the first step and $((\tau+1)^2+\tau^2)^K$ alternatives after τ steps. (Here we consider the case when maximum speed of every pedestrian is equal to 1. Thus, after τ steps every pedestrian will be within a neighbourhood of radius R=τ around his initial position, therefore he will occupy one out of no more than $(\tau+1)^2+\tau^2$ cells, and for K pedestrians there are at most $((\tau+1)^2+\tau^2)^K$ possibilities.) Secondly, there is no use of information about behavioural model of the neighbours.

At the next stage we can apply this information and try to reduce the number of alternatives (branches in the tree) at every step taking into account probabilities of shift of pedestrians. Let us have a set of pedestrians (of capacity K), a set of cells in the neighbourhood (of capacity $(R+1)^2+R^2$) and a list of probabilities for every pedestrian to shift to every cell. Now we have to find such an alternative or a set of pairs <pedestrian, cell> for which the sum of probabilities reaches the maximum. It is now clear that selection of the most probable alternative may be found as a solution to the assignment problem [25] (see Fig.7). For the sake of simplicity we are able to apply the necessary and sufficient conditions of uniqueness of an optimal assignment problem solution which can be stated as follows. The set of optimal assignment problem solutions contains a single solution if and only if all corresponding upper tolerance values are strictly positive [26]. Based on the upper tolerances we are able to adjust the number of optimal assignment problem solutions such that our computational efforts will be reasonable. The advantage of this approach is based on the fact that assignment problem has been extensively investigated and there exist efficient algorithms for solving it [27].

However, this approach needs some revision. As a matter of fact, obtained in such a way predictions are based on an assumption that probabilities for target pedestrian are known, while all these computations are held in order to adjust them. So, it makes sense to consider all possible variants for the target pedestrian. As a result, even in a case of unique solution of an assignment problem at a particular step up to 4 leaves (this value is bounded by the number of cells in an elementary neighbourhood) will be added to the tree. Thus, in order to construct a one-step prediction the following steps are necessary:

1. to construct up to 4 (number of non-zero P_k, k=1,...,4) graphs, similar to that in the Fig.7; in each such graph for the target pedestrian holds $P_i=\delta_{ik}$ (δ – Kronecker symbol);
2. to solve an assignment problem for every graph (obtained at the previous step) that will give one or several most probable variants of movement of pedestrians;

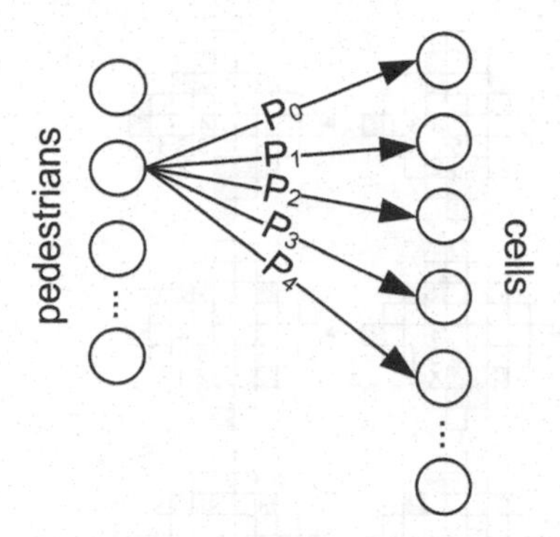

Fig. 7 One-step prediction as a solution to the assignment problem (connections are shown only for one pedestrian).

3. to construct the alternatives by shifting pedestrians in the directions corresponding to solutions obtained at the previous step and to add them to the scenarios tree.

In order to construct a T-step prediction (for an arbitrary T) one has to repeat the above mentioned procedure T times. This will result in a tree, each node of which corresponds to some state of the extended neighbourhood and every simple path (sequence of nodes from the root to one of the leaves) – to one of the most probable scenarios for the nearest T steps (see Fig.8). Now we have to find the optimal one among these scenarios (strict formulation of the optimality criterion is provided in subsection 3.4). Knowledge about the optimal scenario will allow the pedestrian to make his step in a direction that will enforce this scenario and thus he will make his positive impact (in terms of minimization of evacuation time and maximization of the flow) on the overall behaviour of the crowd. In the following subsection 3.4 we describe a general framework for finding an optimal path in a scenarios tree and two approaches to solving this problem in order to demonstrate its solvability, however numerical experiments with these approaches is a matter of future research.

3.4 Finding Optimal Trajectories in a Scenarios Tree

As it was mentioned above, a scenarios tree is strictly speaking not a tree but a directed graph $G(T) = (V, A)$ (see Fig.8b), T is a horizon of prediction (a number of steps for which a prediction is made), $V=\{v_j\}$ – a set of vertices (each of them corresponds to one of the future states of the extended neighbourhood that are present in the scenarios tree), $A \subset V \times V$ – set of arcs (each arc corresponds to a transition between two states of the extended neighbourhood). Any two vertices i and j from V are connected by an arc (i, j) if and only if there is a transition in the scenarios tree between the states of the extended neighbourhood that correspond to these two vertices. Let us denote by s (source, or root of the scenarios tree) the vertex of G(T), that corresponds to the current state of an extended neighbourhood, by t_i – vertices that correspond to final states of an extended neighbourhood (leaves of

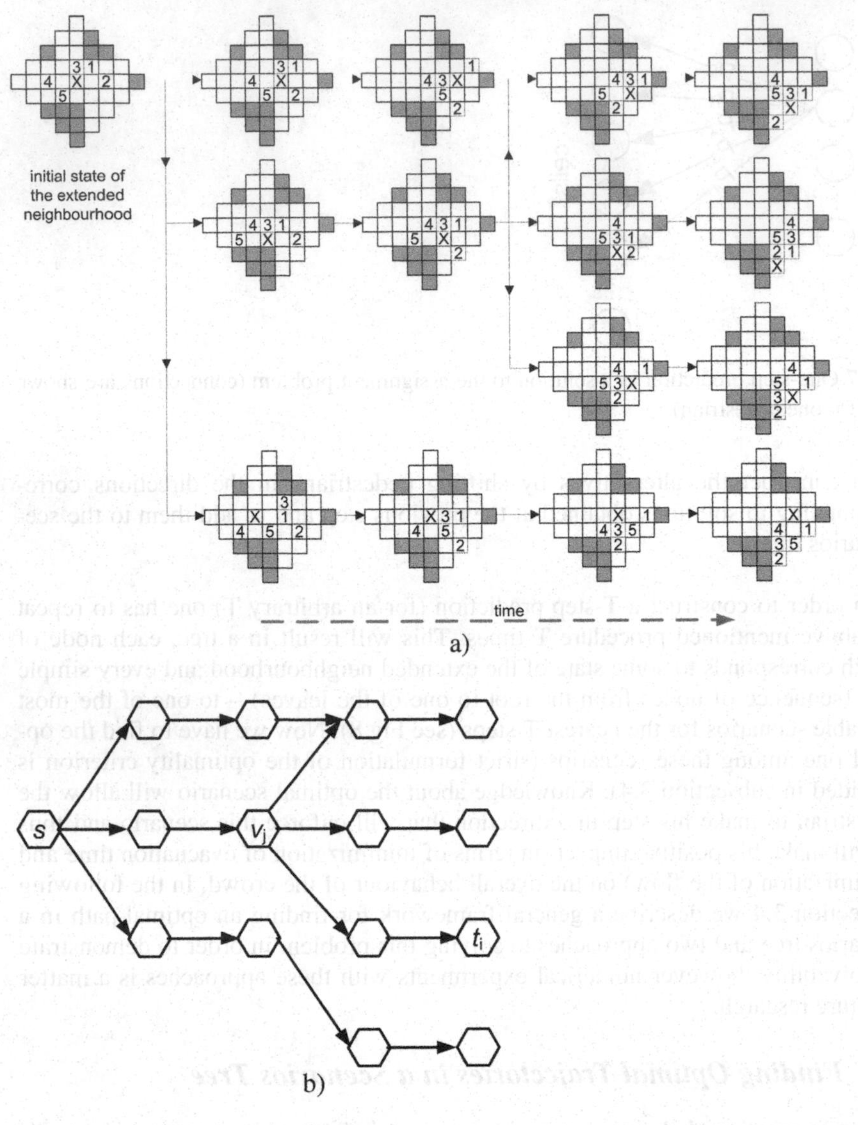

Fig. 8 An example scenarios tree (a) (X - target pedestrian, other pedestrians are numbered 1,…,5, grey cells indicate obstacles, arrows indicate one-step transitions between states of the extended neighbourhood) and a corresponding graph G(T) (b). Any chain of states of the extended neighbourhood linked by arrows reflects a possible evolution of the CA (restricted to the neighbourhood under consideration). In this case a horizon of prediction T=4, a radius of the extended neighbourhood R=4, a number of pedestrians within the neighbourhood *K*=6.

the scenarios tree), *K* is a number of pedestrians in the extended neighbourhood. Under introduced terms we can formulate the following specific properties of a graph G(T):

1. G(T) has one vertex s that has no incoming arcs (s has in-degree equal to 0);
2. G(T) has no more than $((T+1)^2+T^2)^K$ vertices t_i, that have no outgoing arcs (t_i have out-degree equal to 0);
3. G(T) has a layered structure, vertices from a layer are not connected to each other and are connected only to vertices of the next layer; there are T+1 layers;
4. every vertex (except s) has at least one and at most 5^K ingoing arcs;
5. every vertex (except t_i) has at least one and at most 5^K outgoing arcs.

Over the set of vertices (states of the extended neighbourhood) V we can define a quality function $q(v)$ ($v\in$ V), that will be used as a criterion of optimality (quality). Naturally, $q(v)$ can be defined in different ways, below is an example of how it can be done.

Let us introduce an auxiliary function C(.) (C: $V\rightarrow Z^2$) that returns coordinates of the target pedestrian. Now the optimality criterion can be defined as follows:

$$q(v) = -\min_i[\mu(C(v), X^i_{exit})],\ v\in \mathrm{V}, \tag{5}$$

where X^i_{exit} – coordinates of i-th exit, $X^i_{exit}\in Z^2$. In this case a value of the optimality criterion depends on a distance from the target pedestrian to the nearest exit, minus sign provides growth of the value of the criterion as a pedestrian approaches one of the exits. Thus, according to this criterion, scenarios that will lead the target pedestrian closer to the exit will be preferred.

Alternatively, $q(v)$ can be defined as:

$$q(v) = \mu(C(v), C(s)),\ v, s\in \mathrm{V}, \tag{6}$$

where $\mu(.)$ is a Manhattan metrics. In this case optimality of a vertex v depends on a distance between predicted position of the target pedestrian (in the state of the extended neighbourhood that corresponds to the vertex v of G(T)) and his current position (in the initial state of the extended neighbourhood that corresponds to the vertex s of G(T)). Such definition of the criterion ensures that scenarios that allow more movements of the target pedestrian will be preferred.

An advantage of the latter definition of the optimality criterion (see eq. 6) is that pedestrians do not need to know positions of exits (distances to exits). This allows to simulate situations when pedestrians either unaware of where the exits are or can hardly orientate themselves in dark or filled with smoke premises.

Optimality criterion of an arbitrary path $p=\{s,...,t_i\}$ (a sequence of vertices from the root s to one of the leaves t_i, any two consecutive ones of which are connected by an arc in G(T)) in the most general case can be defined as a linear function of optimality values of its vertices, e.g.:

$$Q(p) = \alpha_0 q(s) + \alpha_1 q(v_1) + ... + \alpha_{T-1} q(v_{T-1}) + \alpha_T q(t_i), \tag{7}$$

where $v_1,\ldots,v_{T-1}\in p$, values of constants $\alpha_j\in$ IR (j=0,...,T) are up to the user's choice.

3.4.1 Network Flow Approach

Let us construct a graph $G_k(T) = (V', A')$ from $G(T) = (V, A)$ (introduced in subsection 3.4) in the following way (in this subsection we consider optimality criterion $q(.)$ defined in eq. 5):

1. $V' \supset V$;
2. $A' \supset A$;
3. every arc $(v_i,v_j)\in A'$ has the following capacity $c(v_i,v_j)=q(v_j)-q(v_i)+2$ (2 ensures positive sign of $c(v_i,v_j)$ as $|q(v_j)-q(v_i)|\le 1$ holds)
4. arcs of type (s,v_j) have zero capacity if $j\neq k$ and infinite (large enough) capacity if $j=k$;
5. let us add one more vertex t (sink) to V' and arcs of type (t_i,t) to A' such that $c(t_i,t)=q(t_i)-q(s)+T$

In graphs $G_k(T)$ preferred transitions (arcs) have larger capacity (see Fig.9). As each of these graphs has one vertex s that can be treated as a source and one vertex t that can be treated as a sink, it is possible to solve a max-flow/min-cut problem (e.g. using the Ford-Fulkerson algorithm) [28]. It is clear that arcs that belong to optimal paths have the largest capacities and, therefore, will make up the maximum flow.

However, a graph $G_k(T)$ has one negative property: the number of arcs tends to increase from layer to layer while capacities have the same order ($c(v_i,v_j)\in\{1,2,3\}$ for all $(v_i,v_j)\in A'$). So, the closer an arc to the source s is, the faster it saturates, thus, the value of max-flow will be defined mainly by the arcs that are close to the source and capacities of distant arcs will have little influence. In order to avoid this, capacities $c(v_i,v_j)$ were adjusted by addition of the following fixed value

$$\sum_l c(t_l,t), \qquad v_i,v_j \neq t_l, \forall l:(t_l,t)\in A'. \tag{8}$$

Thus, the resulting capacities of arcs in $G_k(T)$ are as follows:

$$c(v_i,v_j) = q(v_j) - q(v_i) + 2 + \sum_l c(t_l,t), \ \forall l:(t_l,t)\in A' \tag{9}$$

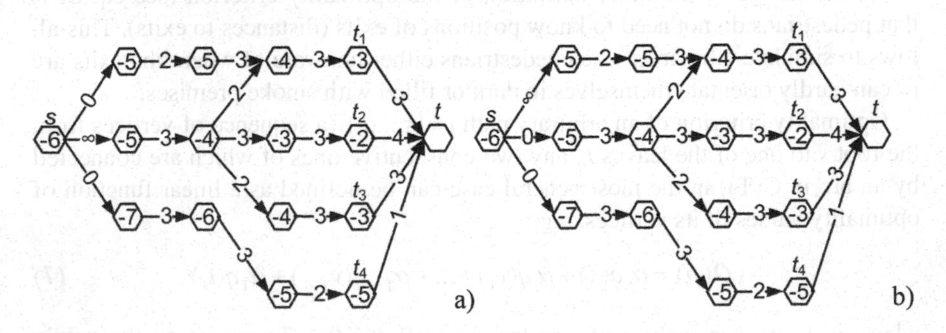

Fig. 9 Construction of $G_k(T)$: numbers in vertices correspond to $q(v_j)$, capacities of edges are determined according to eq. 9, T=4. $G_1(T)$ and $G_2(T)$ are shown.

It is quite straightforward that the value of max-flow $f_{\max}$ in this graph will satisfy

$$\min_i c(t_i,t) \le f_{\max} \le \sum_i c(t_i,t),\ \forall i:(t_i,t)\in A' \tag{10}$$

- it will not be smaller than the capacity of thinnest arc incident to the sink and will not exceed total capacity of the arcs incident to the sink) and

$$\min_i(q(t_i)-q(s))+T \le f_{\max} \le \sum_i (q(t_i)-q(s)+T),\ \forall i:(t_i,t)\in A' . \tag{11}$$

Now $f_{\max}$ can be defined as $f_{\max} = \alpha_0 q(s) + \sum_i \alpha_i q(t_i)$ where $\alpha_0,\alpha_i \in$ IR and sum is taken over all the leaves t_i.

Comparing the latter expression and (3.7), having $\alpha_1,\ldots,\alpha_{T-1}$ in (3.7) equal to 0, value of max-flow can be expressed as

$$f_{\max} = \sum_i Q(p_i), \tag{12}$$

where p_i – simple paths (sequences of vertices, each two consecutive ones of which are connected by an arc in $G_k(T)$) of type $(s,..,t)$ in $G_k(T)$.

Thus, the value of max-flow can serve as a consistent criterion of joint optimality of simple paths in a graph $G_k(T)$.

After having computed values of maximum flow for all $G_k(T)$ (no more that 4 – the number of arcs incident to the vertex s that is equal to the number of possible directions of shift that is bounded by the number of cells in an elementary neighbourhood) - $f_{k,\max}$, one can obtain a quantitative measure of optimality of shift in k-th direction for the target pedestrian: the higher the value of $f_{k,\max}$ is, the more opportunities he has in following the optimal trajectory (optimal for at least next T steps).

Let us introduce an auxiliary value $P_{k,f}\in[0,1)$, that will be used for adjustment of probabilities of shift P_k:

$$P_{k,f} = 1 - \frac{f_{k,\max} - \min_i(f_{i,\max}) + 1}{\max_i(f_{i,\max}) - \min_i(f_{i,\max}) + 1},\ i=1,\ldots,4. \tag{13}$$

Now we can adjust the probabilities of shift P_k according to the mentioned above scheme (compare with eq. 1):

$$P_k := P_k \cdot (1 - \alpha \cdot P_{k,f}),\ i=1,\ldots,4. \tag{14}$$

3.4.2 Neural Network Approach

Layered structure of a scenarios tree and, therefore, of the corresponding to it graph G(T) = (V, A) (V – set of vertices that correspond to states of the extended neighbourhood in a scenarios tree, A – set of arcs that correspond to one-step transitions between states of the extended neighbourhood, that are present in a scenarios tree) allows to treat the latter as a multilayer perceptron and to apply corresponding algorithms.

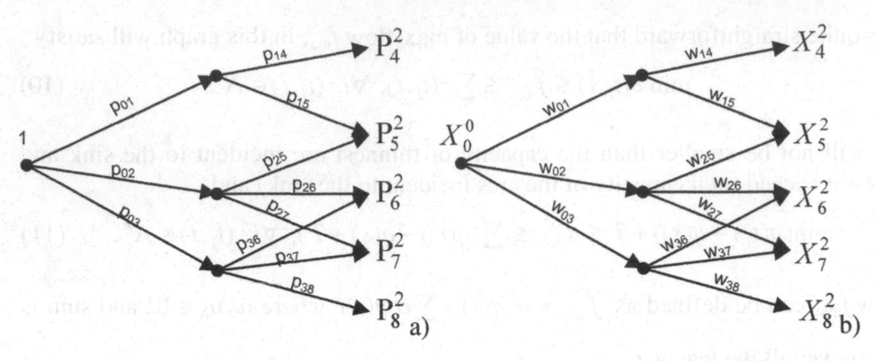

Fig. 10 An example scenarios tree (a) and a corresponding neural network (b)

Let us consider an example of a scenarios tree and a corresponding graph G(2) (see Fig.10a). Here by p_{ij} we denote a probability of transition between states *i* and *j* of the extended neighbourhood, by P_i^τ – probability of the extended neighbourhood to appear in state *i* after τ steps (τ =0,…,T), for the case in Fig.10 T=2, n_i – a number of descendants of *i*-th state of the extended neighbourhood (a number of outgoing arcs in the corresponding graph G(T)), $n_i = \text{card}\{(i,j)| \ \forall j : (i, j) \in A\}$, m_i – a number of predecessors of *i*-th state of the extended neighbourhood (a number of ingoing arcs in the corresponding graph G(T)), $m_i = \text{card}\{(j,i)| \ \forall j : (j,i) \in A\}$. Under the described terms, the probabilities p_{ij} of transitions are subject to:

$$\sum_j p_{ij} = 1, \ j=1,\ldots,n_i \tag{15}$$

At the same time the values P_i^τ subject to $\sum_i P_i^\tau = 1$ (*i* enumerates all final states of the extended neighbourhood). Moreover, the following holds:

$$P_i^\tau = \sum_j p_{ji} P_j^{\tau-1}, \ \tau=1,\ldots,T, \ j=1,\ldots,m_i \tag{16}$$

$P_1^0 = 1$ – the initial state is definite and known.

Taking into account that $P_i^\tau \in [0,1]$, (3.16) implies that:

$$0 \le \sum_j p_{ji} P_j^{\tau-1} \le 1, \ \tau=1,\ldots,T, \ j=1,\ldots,m_i \tag{17}$$

Let us now consider a multilayer perceptron that corresponds to the scenarios tree (the number of vertices and neurons, as well as a pattern of their connection, coexist, see Fig.10b). After having assigned connections in the neural network some weights w_{ij} and denoting output of *i*-th neuron in τ-th layer as X_i^τ, functioning of the T-layer perceptron can be expressed as:

X_1^0 - arbitrary input of the network,

$$X_i^\tau = \sigma\left(\sum_j w_{ji} X_j^{\tau-1}\right), \tau=1,\ldots,T,\ j=1,\ldots,m_i \tag{18}$$

where σ(.) is a sigmoid function [29] that bounds the output of a neuron by 0 and 1.

If one defines sigmoid function σ(.) as

$$\sigma(x) = \begin{cases} x, x \in [0,1] \\ 0, x < 0 \\ 1, x > 1 \end{cases} \tag{19}$$

then it becomes straightforward that (3.16) and (3.18) are equivalent. Thus, given appropriate input, the output of the neural network can be interpreted as a vector of probabilities of final states.

On the other hand, we can consider an inverse problem: some desired final probability distribution P^T over the set of final states of extended neighbourhood is known and we have to find such a set of probabilities at the first step p_{0j} that will ensure the given final distribution. Within the framework of neural networks the following problem emerges: to find such weights w_{0j} that will transform a particular input of the network ($X_1^0 = P_1^0 = 1$) into a particular output ($X^T = P^T$). In other words we have a problem of learning of a perceptron that can be resolved by a variety of algorithms, one of which is BackPropagation [29].

Remark

BackPropagation algorithm requires existence of the first derivative of σ(.), which is not true in our case (see eq. 19). There are two ways to deal with this problem:

1. to define the derivative of piecewise-linear σ(.) in 0.0 and 1.0;
2. to approximate piecewise-linear σ(.) by exponential sigmoid [29] with some accuracy level ε.

Let us now return to the problem of finding the optimal path in a scenarios tree G(T). In order to implement the described above neural network approach, arcs have to be assigned weights corresponding to probabilities of relevant transitions. The probability of transition from state i to state j is:

$$p_{ij} = \prod_{l=1}^{K} \sum_{k=0}^{4} P_{k,l} \delta_{k,d(l)}, \tag{20}$$

where K – the number of pedestrians in a neighbourhood, $P_{k,l}$ – the probability of shift of l-th pedestrian in k-th direction, $d(l)$ – the direction of shift of l-th pedestrian while the neighbourhood turns from state i into state j. As the scenarios tree contains not all the transitions but the most possible ones (see subsection 3.3), in case of $w_{ij}=p_{ij}$ eq. 15 does not hold, thus weights of arcs are to be normalized:

$$w_{ij} = \frac{p_{ij}}{\sum_l p_{il}},\ l=1,\ldots,n_i \tag{21}$$

Input of the network is 1 ($X_1^0=1$). Output represents distribution of probabilities of different final states of the neighbourhood. Naturally, the more the value of optimality criterion $q(t_i)$ (see eqs. 5, 6) of some final state t_i, the higher the corresponding probability should be. Thus, the desired output of the perceptron can be defined as:

$$X_i^{\mathrm{T}} = \frac{q(t_i) - \min_j(q(t_j))}{\sum_l \left[q(t_l) - \min_j(q(t_j)) \right]}, \tag{22}$$

where j and l enumerate all neurons in the last layer of the neural network (of graph G(T)), T is a number of the output (last) layer of the neural network (of graph G(T)). After the process of learning is completed, we have desired probabilities of shift at the current step $\mathrm{p}_{0i}=w_{0i}$ (i=1,…,n_0). Now the values of P_k can be adjusted either according to the mentioned above scheme (see eq. 1)

$$P_k := P_k \cdot (1 - \alpha \cdot (1 - \mathrm{p}_{0k})),\ k\text{=1,…,4} \tag{23}$$

with subsequent normalization ($\sum_{k=1}^{4} P_k = 1$) or

$$P_k := (1-\alpha) P_k + \alpha \cdot \mathrm{p}_{0k},\ k\text{=1,…,4}. \tag{24}$$

3.5 Asymptotic Estimates

There is no doubt that performance of a crowd drastically depends on the behavioural characteristics of individuals (parameters of the model). It was shown above that a thorough selection of values of the parameters allows to improve the overall performance (to reduce evacuation time or to increase flow) and endowing pedestrians with such a feature as anticipation allows further improvement. But to what extent are these improvements substantial? In order to be able to answer this question we need to assess the best (theoretically) possible performance.

Let us consider the situation with continuous flow of pedestrians. In case of stationary geometry of the scene and constant boundary conditions the problem of finding the maximum flow is rather simple. According to the Ford-Fulkerson theorem,

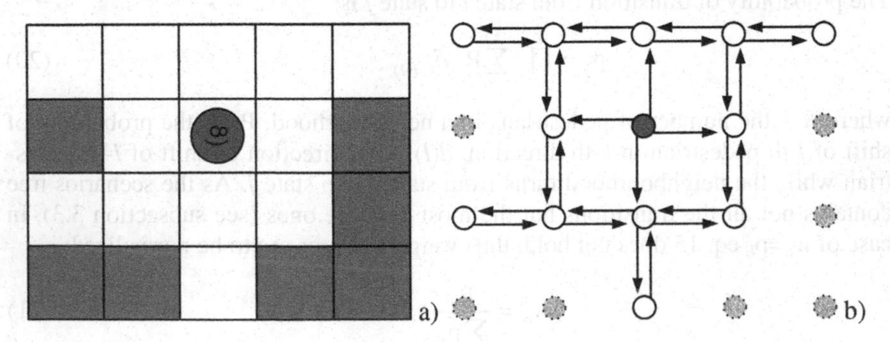

Fig. 11 An example of a scene of CA and corresponding graph (all arcs have unit capacity)

the value of maximum flow is equal to the capacity of minimum cut. Moreover, application of one of the class of so-called augmenting path algorithms will provide a bunch of paths that pedestrians have to follow in order to saturate the minimum cut (and ensure the largest possible flow).

The situation is completely different and somewhat more complex in case of evacuation. As a matter of the fact, evacuation represents a transient process that can enter stationary mode for only very short periods of time (e.g. when the whole crowd runs through a long corridor). Thus, it is not possible to apply the simple formula <time>=<number of pedestrians>/<capacity of min. cut> in this case, mainly because the width of a minimal cut varies in time (see Fig.12).

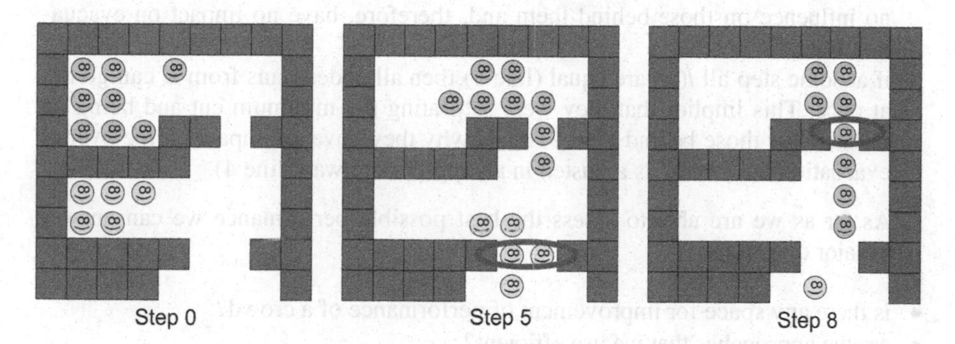

Fig. 12 Variable minimum cut emerging at different time steps of evacuation.

So, a need for some dynamic procedure occurs. Below we give an algorithm that simulates optimal evacuation for the case when all pedestrians have the same speed (equal to 1) and finds minimum egress time based on dynamically updated max-flow/min-cut problem. While solving the latter, pedestrians are treated as sources, exits - as sinks and arcs in a graph are assumed to have unit capacity.

In the algorithm the following definitions are used: by |I| we denote number of the elements in the set I, $l(p_i)$ – length of the corresponding path;

I – a set of the indices of occupied (by pedestrians) cells of an automaton;
K – subset of I that contains indices of the cells occupied by pedestrians for which paths were build at one step of the algorithm, $K \subset$ I;
n – step number.

At the output we have T – minimum evacuation time.

The Evacuation Algorithm (max flow of pedestrians and min evacuation time)

0 T:=0 //initialization
1 while |I|>0 { //while there are some pedestrians
2 build max flow => max flow F^n_{max}, saturated paths p_i, $i \in K \subset$ I, $|K|=F^n_{max}$
3 if max($l(p_i)$)=min($l(p_i)$) // if all $l(p_i)$ are equal
4 T+=max(min(($l(p_i)$)-T , 1)
5 for all $i \in K$
6 if $l(p_i)$=min($l(p_i)$) //first pedestrians reaching the exit…

```
7          I:=I \ i                    //...go out
8 }
```

The main ideas behind the given algorithm are as follows:

1. at every step a maximum set of pedestrians K that are able to reach the exit at the same time is found (number of elements in this set is bounded only by capacities of corridors);
2. as pedestrians from K can initially be at different distances from the exit (differ in values of their $l(p_i)$), only those having minimum value of $l(p_i)$ can go out at this step (lines 6,7 in the Evacuation Algorithm). At the same time, they have no influence on those behind them and, therefore, have no impact on evacuation time;
3. if at some step all $l(p_i)$ are equal (line 3) then all pedestrians from K can go out at once. This implies that they were saturating the minimum cut and being an obstacle for those behind them. This is why they have an impact on the overall evacuation time and T is adjusted in an appropriate way (line 4).

As far as we are able to assess the best possible performance we can answer two major questions:

- is there any space for improvement of performance of a crowd?
- are the approaches that we use efficient?

Let us classify the models of pedestrians according to the extent of information delocalization. Thus, by the term MP(R,T) we denote a model of pedestrian that has information about his neighbourhood of radius R for a time period of the next T steps. It should be mentioned that within our framework information about future steps is not provided to the pedestrians explicitly but is produced by them via anticipation and is not perfectly precise.

To answer the two posed questions a number of experiments were held and a summary of the results obtained can be found in the given graphs (see Fig.13).

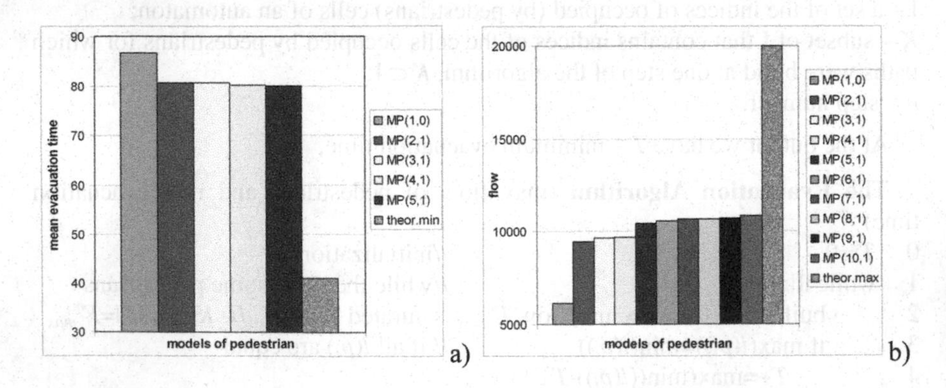

Fig. 13 Performance of the tested models of pedestrians and theoretically best performance

As it can be seen from the picture, all the tricks that we have implemented by this time have positive but rather small impact. It is a quite natural result as those models are very restricted in the amount of information that is available for pedestrians, while asymptotic estimates utilize all available information to the full extent. However, the described above framework allows to build a model MP(R,T) for arbitrary values of R and T, thus making it possible to consider an ultimate model MP(∞,∞). For practical purposes instead of infinity there must stand values that are large enough to cover the entire scene of an automaton and entire period of evacuation.

Now the most interesting question is whether the ultimate model will approach the best possible performance. And the most probable answer to this question is negative because of the following reason. The peculiarity of the model crowd (real crowds usually also have this feature) is that every pedestrian tries to improve his own performance, while, asymptotic estimates are obtained for the perfectly coordinated pedestrians.

4 Anticipation, Synchronization and Complexity

As it was shown in the previous paragraphs, anticipation in its simplest form has positive (see Fig.6) but tiny (see Fig.13) impact on the performance of the crowd. The major cause of this is that anticipating pedestrians are somewhat "egoistic", i.e. they are trying to improve their own egress time and take little care about the others. At the same time, as we know from the game theory [30], cooperating individuals can usually do better than competing ones. This fact allows coming to an idea of introduction of cooperative behaviour into the model in an explicit way. Within the described above framework this can be done by changing the criterion of optimality of predicted states of the neighbourhood $q(.)$ (see eqs. 5 and 6) in such a way that it depends not only on the position of the target pedestrian but also on the positions of his neighbours. In this case a target pedestrian will try to optimize not only his own behaviour but also that of the others (within his extended neighbourhood) and in the limit case (MP(∞,∞)) of all the crowd, thus allowing for the best possible performance. At the same time, the pedestrians in the crowd will behave in a perfectly organized and synchronized way.

Thus, under the term synchronization within the framework of pedestrian crowd movement we mean a phenomenon of cooperating behaviour of individuals that makes their movements correlated and more efficient with regard to maximization of flow or minimization of evacuation time. Thus, we can introduce a measure of synchronization as an "amount" of cooperative behaviour and claim that if for a particular crowd the value of this measure reaches its maximum, then the crowd will have the minimum possible evacuation time and produce the maximum possible flow at the exit (during its evacuation). According to this informal definition of synchronization and its measure we are able to introduce a formalized quantitative approximation to this measure. In particular, synchronization in a crowd can be measured as a percentage of conflicts (situations when two or more pedestrians try to move to the same cell) resolved in a collaborative way relatively to the total amount of conflicts. Here we can also define notions of snap

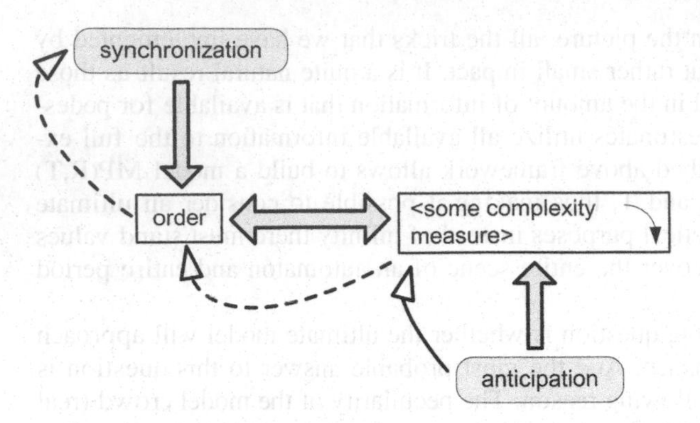

Fig. 14 Implicit relation between anticipation and synchronization via their impact on the ordered behaviour of the crowd.

synchronization and integral synchronization. According to the former we consider behaviour of pedestrians during two consecutive time steps: at first of them we can detect possible conflicts and at the second we can check how they were resolved and to calculate the value of the introduced measure of synchronization. Naturally, integral synchronization is a mean value of snap synchronization over some arbitrary long time period.

However, when collaboration is not introduced explicitly, is there a place for synchronization? In order to answer this question, let us check the relation between one-step anticipation and synchronized behaviour of pedestrians. In Fig.14 it is shown that both these phenomena are implicitly biased via an effect that they have on behaviour and some quantitative characteristic of the crowd – a measure of complexity or disorder. But how can we measure complexity of behaviour of the crowd. The most straightforward measure of complexity is dispersion of some characteristic of the crowd, e.g. evacuation time (that varies from one simulation to other) or flow (that varies from one time step to another). However, such statistical approach demands a substantial number of simulations to be held in order to obtain consistent estimates of dispersion. That is why we proposed another, entropy based, measure that allows estimating the complexity given only a snapshot of the scene.

From the common intuition we assume that behavioural complexity of the crowd is determined by behavioural complexities of its members. This fact makes it possible to define the complexity of the crowd as a sum of behavioural complexities of pedestrians (in fact it is better to use a normalized sum, i.e. divided by the number of pedestrians). As it was mentioned at the very beginning of the chapter (section 2), the behaviour of every pedestrian is driven by the probabilities of shift P_k (k=1,…,4) that depend on parameters of the model and situation in the neighbourhood of that pedestrian. Thus, we can define a snap complexity of the behaviour as Shannon's entropy: $H=-\Sigma P_k \ln(P_k)$ (k=1,…,4).

In order to define relation between this measure and anticipation a number of simulations were held and results are reflected in Fig.15. Curves in the figure correspond to different snapshots of the scene, that were obtained in the following

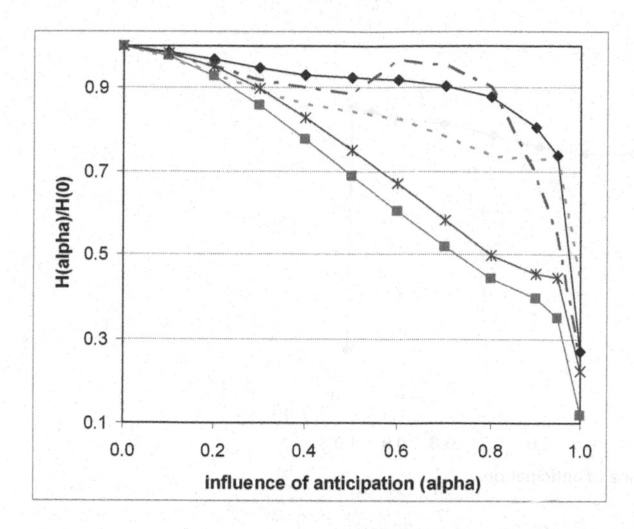

Fig. 15 Decrease of behavioural complexity with increasing influence of anticipation. Curves correspond to different snapshots of the scene (configuration of obstacles and number of pedestrians are the same in all case, only their relative positions are different).

way. The initial conditions (geometry of obstacles, initial placement of pedestrians and their individual parameters) were the same for all trials. Then the model was run for 20 time steps and for each obtained in this way snapshot we varied the value of the parameter α that expresses influence of an anticipation and measured behavioural complexity. As the models that we consider in this paper are stochastic, the obtained snapshots differ in relative positions of pedestrians and depending on how they blocked each other different curves were produced. According to the obtained numerical data, anticipation leads to a decrease in complexity, however this decline is not always monotonous and allows local peaks that do not exceed complexity of non-anticipating crowd.

As far as we have an implicit estimate of the relation between anticipation and synchronization, it is natural to perform some explicit procedures in order to check consistency of the proposed approach. In further simulations we were trying to estimate an extent of coherent behaviour in the crowd. Under the coherent behaviour we mean that pedestrians that started from one point should have same trajectories, thus leading to a more organized and synchronized crowd. In order to obtain numerical data on the issue, we have considered a dense (density ~ 75%) continuous flow of pedestrians through a straight corridor with one L-shaped bend. As a measure of coherency we used a correlation coefficient between vectors of direction of pedestrians' movement at specially chosen points that are critical to the overall performance (in this particular case we considered points at the edge of the bend). The results of simulations are given in the picture below (see Fig.16).

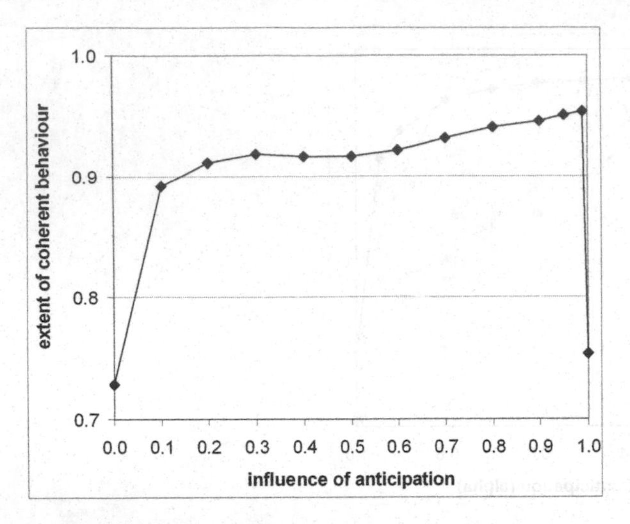

Fig. 16 Coherency of behaviour vs. influence of anticipation.

5 Summary and Future Research Directions

Up-to-date literature suggests to study a synchronization phenomenon by application of mathematical models for nonlinear stochastic dynamical systems of either neutral or delay type, based for example on stochastic differential equations, partial differential equations, chains of coupled maps, discrete equations with delay, etc. (see [1-7]). Each of the above mentioned models leads to intractable computational difficulties in finding the solutions within them. One of the inherent obstacles in finding such solutions is the so called "mentality" property induced by multi-valuedness of the corresponding solutions as functions of complex time. In some simple cases these solutions can be expressed as the inversion of a single hyper-elliptic integral. The associated Riemann surfaces are known to be infinitely sheeted coverings of the complex time plane, ramified at an infinite set of points whose projection in the complex time plane is dense [31]. Note that another way for considering such systems is provided by the theory of attractors [32]. We resolve such difficulties by using cellular automata tools with incorporated adjusted classical max-flow-min-cut models [25]. They allow studying global properties of pedestrian flows in terms similar to asymptotic behaviour of solutions, periodic orbits and their stability or sensitive dependence on initial conditions and on input parameters.

Our experiments show that by increasing the probability (see eq. 1), characterizing a personal anticipation, the maximization of flow and minimization of evacuation time of a crowd are achieved. If this probability is close to 1, then the indicated performance becomes essentially worse. The synchronization of a crowd measured by the number of egoistic pedestrians' behaviour will follow the same dependency as that probability. Note that a crowd is well synchronized even if it is not moving at all, but the above mentioned criteria will be infinitely dropped. In other words our measure of synchronization is applicable only for a crowd with at least a single movement. The following management rule might be extracted from our experiments for the minimization of crowd's evacuation time: all fast moving pedestrians should provide a preference for occupation of nearest free cells to slowly moving ones.

In this chapter we have considered a rather novel class of nonlinear models with anticipation, provided results of simulations of these models as well as interpretations of the obtained solutions. One of the most important results is the emergence of models that combine intrinsic multi-valuedness (which makes the models naturally suitable for studying the processes related to decision-making) with tractability (of their parameters, structure and results of simulations) and computational feasibility. Moreover, we proposed a framework for modelling optimal decision-making of the anticipating pedestrian based on a scenarios tree, and developed two approaches (based on the max-flow/min-cut problem and the problem of multilayer perceptron learning, correspondingly), that can be implemented within this framework. However, computational experiments with both of them remain a matter of future research. Though the proposed models in general case are not amenable to standard mathematical methods of analysis, we managed to estimate their performance in extreme cases.

We have also considered such at first sight different phenomena as anticipation, synchronization and complexity of behaviour and proposed a scheme of their implicit relations. So, another major result is an attempt to bridge the gap between the three: more anticipating crowd has lesser behavioural complexity and is more synchronized.

The described investigations open new possibilities for research on nonlinear dynamics and synchronization. May be the most interesting and important is the transition of concepts from classical theory: synchronization, self-organization, bifurcation and chaos, attractors, etc. to the case of multi-valued dynamical systems. Such transition will require the transformations of the notion and definitions of different types of synchronization, changes in self-organization theory, new aspects of bifurcation theory, definition of chaos etc. Moreover, interpretations of new solutions and properties of these novel mathematical models will lead to advances in the field of different practically important systems, for example in supply chain management [33].

Acknowledgments. The authors would like to thank the editors, professors Jean Chamberlain, Koteswara Anne, Kyandoghere Kyamakya for granting us with an extra week of pleasant work on this paper.

References

1. Picovsky, A., Rosenblum, M., Kurths, J.: Synchronization. A universal Concept in Nonlinear Sciences. Cambridge University Press, Cambridge (2001)
2. Blekhman, L.: Synchronization in Science and Technology. ASME Press, New York (1988)

3. Pikovsky, A., Maystrenko, Y. (eds.): Synchronization: Theory and Application. NATO Science Ser. II. Mathematics, Physics and Chemistry, vol. 109, p. 258. Kluwer AP, Dordrecht (2002)
4. Makarenko, A., Krushinsky, D., Goldengorin, B.: Anticipation and Delocalization in Cellular Models of Pedestrian Traffic. In: Kyamakya, K. (ed.) Proc. INDS 2008, pp. 61–64. Shaker Verlag, Aachen (2008)
5. Kurth, J.: Synchronization in oscillatory networks. In: Kyamakya, K. (ed.) Proc. INDS 2008, p. 9. Shaker Verlag, Aachen (2008)
6. Restrepo, J., Ott, E., Hunt, R.: Emergence of synchronization in complex networks of interacting dynamical systems. Physica D 224, 114–122 (2006)
7. Kreuz, T., Mormann, F., Andrzejak, R.G., Kraskov, A., Lehnertz, K., Grassberger, P.: Measuring synchronization in coupled model systems: A comparison of different approaches. Physica D 225, 29–42 (2007)
8. Makarenko, A.: Anticipating in modeling of large social systems neuronets with internal structure and multivaluedness. Int. J. of Computing Anticipatory Systems 13, 77–92 (2002)
9. Makarenko, A.: Anticipatory agents, scenarios approach in decision-making and some quantum-mechanical analogies. Int. J. of Computing Anticipatory Systems 15, 217–225 (2004)
10. Dubois, D.: Theory of incursive synchronization and application to the anticipation of a chaotic epidemic. Int. J. of Computing Anticipatory Systems 10, 3–30 (2001)
11. Thong, M.H., Nakagava, M.: A secure communication system using projective – log and/or projective – anticipative synchronizations of coupled multi-delay feedback systems. Chaos, Solitons and Fractals 38, 1423–1438 (2008)
12. Illachinski, A.: Cellular Automata. A Discrete Universe, p. 842. World Scientific Publishing, Singapore (2001)
13. Toffoli, T., Margolis, N.: Cellular automata computation. Mir, Moscow (1991)
14. Wolfram, S.: New kind of science. Wolfram Media Inc., USA (2002)
15. Chua, L.: A nonlinear dynamics perspective of Wolfram's new kind of science. In: Kyamakya, K. (ed.) Proc. INDS 2008, p. 6. Shaker Verlag, Aachen (2008)
16. Helbing, D.: Traffic and related self-driven many-particle systems. Rev. Modern Physics 73, 1067–1141 (2001)
17. Helbing, D.: From emergent crowd behavior to self-organized traffic light. In: Kyamakya, K. (ed.) Proc. INDS 2008, p. 8. Shaker Verlag, Aachen (2008)
18. Stepanov, A., Smith, J.: Multi-objective evacuation routine in transportation networks. European Journal of Operational Research (2008) (accepted), doi: 10.1016/j.ejor2008.08.025
19. Klupfel, H.A.: A Cellular Automaton Model for Crowd Movement and Egress Simulation. Ph.D. Thesis, Univ. Duisburg, Essen (2003), http://www.ub.uni-duisburg.de/ETD-db/theses/available/duett-08012003-092540/unrestricted/Disskluepfel.pdf
20. Nagel, K., Schreckenberg, M.: A cellular automation model for freeway traffic. Journal of Physics I France 2, 2221–2229 (1992)
21. Kirchner, A., Schadschneider, A.: Simulation of evacuation processes using a bionics-inspired cellular automation model for pedestrian dynamics. Physica A 312, 260–276 (2002)
22. Goldengorin, B., Krushinsky, D., Makarenko, A., Smilianec, N.: Toward the management of large-scale crowds: current state and prospects. In: Proc. 3rd Int. Conf. Human-Centered Process, HCP 2008, Delft (2008)

23. Goldengorin, B., Makarenko, A., Smilianec, N.: Some applications and prospects of cellular automata in traffic problems. In: El Yacoubi, S., Chopard, B., Bandini, S. (eds.) ACRI 2006. LNCS, vol. 4173, pp. 532–537. Springer, Heidelberg (2006)
24. Makarenko, A., Goldengorin, B., Krushinsky, D.: Game 'Life' with Anticipatory Property. In: Umeo, H., Morishita, S., Nishinari, K., Komatsuzaki, T., Bandini, S. (eds.) ACRI 2008. LNCS, vol. 5191, pp. 77–82. Springer, Heidelberg (2008)
25. Hillier, F.S., Lieberman, G.J.: Introduction to Operations Research, 8th edn. The McGraw-Hill Companies, Inc., New York (2005)
26. Goldengorin, B., Jager, G., Molitor, P.: Tolerances Applied in Combinatorial Optimization. Journal of Computer Science 2(9), 716–734 (2006)
27. Volgenant, A.: In: Korte, B., Vygen, J. (eds.) Combinatorial optimization theory and algorithms, 2nd edn. Algorithms and Combinatorics, vol. 21, Springer, Berlin (2002)
28. Cormen, T.H., Leiserson, C.E., Rivest, R.L., Stein, C.: Introduction to algorithms, 2nd edn. MIT Press and McGraw-Hill Companies, Inc. (2001)
29. Rumelhart, D.E., Hinton, G.E., Williams, R.J.: Learning internal representations by error propagation. In: Rumelhart, D.E., McClelland, J.L. (eds.) Parallel Distributed Processing, vol. 1. MIT Press, Cambridge (1986)
30. Peters, H.: Game theory: a multi-leveled approach. Springer, Heidelberg (2008)
31. Fedorov, Y., Gomes-Ullata, D.: Dynamical systems on infinitely sheeted Riemann surfaces. Physica D 227, 120–134 (2007)
32. Kapustyan, O., Melnik, V., Valero, J., Yasinsky, V.: Global attractors of multi-valued dynamical systems and evolution equations without uniqueness. Naukova Dumka, Kyiv (2008)
33. Anne, K., Chedjon, J., Bhagavatula, S., Kyamakya, K.: Modeling of the three-echelon chain: stability analysis and synchronization issues. In: Kyamakya, K. (ed.) Proc. INDS 2008, pp. 65–71. Shaker Verlag, Aachen (2008)

Part IV
Applications in Security Related System Concepts

Adaptive Synchronization of Chaotic Systems and Its Uses in Cryptanalysis

Ying Liu and Wallace Kit-Sang Tang

Abstract. This chapter discusses adaptive synchronization of a class of chaotic systems based on a new design of adaptive observer. The design is relied on a linear feedback control scheme and the dynamical minimization algorithm. A linear feedback signal is designed to drive the state estimation errors to zero, while the parametric updating rules are obtained with the dynamical minimization algorithm so that parameter estimation can be achieved at the same time. The success of this approach is illustrated by tackling with some typical synchronization problems found in chaotic systems, where its stability is justified by conditional Lyapunov exponents and local Lyapunov function method. The same design also serves as an effective attack, performing the cryptanalysis for chaos-based communication systems. As demonstrated with our simulation results, the securities of many proposed chaos-based cryptosystems are in fact questionable as the information of their transmitters, including the states and/or system parameters, as well as the transmitting messages, may be revealed by simply observing the transmitted signal.

1 Introduction

Chaos synchronization has become an active research topic in nonlinear sciences and received a lot of interests since the work of Pecora and Carroll [1, 2]. To achieve chaos synchronization, it is usually assumed that both the structure and the parameters of a targeted chaotic system are available while the sole unknown is its initial conditions. Obviously, this assumption is impractical as the exact model of the system may not be known and parametric uncertainties exist in common. Therefore, synchronization of chaotic system with unknown parameters, also called *adaptive synchronization*, becomes more important and more attractive.

Ying Liu and Wallace Kit-Sang Tang
Department of Electronic Engineering,
City University of Hong Kong, Hong Kong, China
e-mail: yingliu2@student.cityu.edu.hk,kstang@ee.cityu.edu.hk

K. Kyamakya (Eds.): Recent Adv. in Nonlinear Dynamics and Synchr., SCI 254, pp. 307–346.
springerlink.com

From the viewpoint of system control theory, synchronization problem can be regarded as designing an observer to achieve state estimation while adaptive synchronization demands a design of adaptive observer for the simultaneous state-parameter estimation of a system. In the last two decades, various kinds of strategies for chaos synchronization and adaptive synchronization have been proposed, including adaptive feedback control [3]–[8], auto-synchronization based scheme [9, 10], minimization algorithm [11, 12, 13], adaptive sliding mode [14], geometrical control [15, 16, 17], etc. Their potential usages in different applications, such as secure communications, analysis of time series, system diagnosis, and so on, have also been widely explored [4, 5],[18]–[24].

Amongst all, the most popular potential application must be related to secure communications. The basic working principle can be briefly described as follows. Messages are encrypted by the use of a chaotic system embedded in the transmitter of a communication system, based on different encryption schemes such as chaotic masking [5, 19, 20, 21], chaotic modulation [4, 5, 22, 41], chaotic shift keying [5, 23], etc. The cipher-text is then transmitted to the receiver via an open channel. At the receiver end, original message is obtained when another copy of the same chaotic system is driven to be synchronized by the received signal. To reduce the communication cost and improve the security of the system, only a partial system information, for example, one of the system states, is commonly transmitted. To further relax the stringent requirement of having identical chaotic models at both the transmitter and receiver ends, the concept of adaptive synchronization have also recently been employed [5, 8],[20]–[24].

For any cryptosystem, the most important design issue is the security it provides. This is largely dependent on how resistive to different kinds of attacks the system is. As demonstrated in [25], adaptive observer can be considered as an attack for revealign the information modulated onto a chaotic system. Therefore, as pinpointed in [26], the in-use chaotic system in a cryptosystem must be *anti-adaptive* so as to prevent intruder from retrieving the system states and/or the parameters based on the observability of the transmitted output. A typical example is the Lorenz system, which has been widely used in chaos-based cryptosystem. Its popularity is due to the fact that it is very difficult, if not impossible, to achieve adaptive synchronization if multiple unknown parameters reside in all its three dynamical equations.

This synchronization problem is considered to be typical but challenging, for which the existing global adaptive observer design methods fail. That is because the Lyapunov stability theory or the Lasalle's invariance principle commonly applied in adaptive observer design usually impose some restrictions. For example, the unknown parameters must only reside in the dynamical equation of the observable state, or the system can be transformed in some particular forms, such as state-affine form [27, 28]. Unfortunately, such requirements cannot be fulfilled in many cases, while the synchronization problem of Lorenz system with multiple unknown parameters is one of the examples.

To resolve these limitations, many attempts related to synchronization-based parameter estimation techniques have been suggested [9, 11, 12, 25, 29, 30, 31]. In [9], it is to design an adaptive rule for parameter estimation driven by the synchronization error. However, in order to estimate the Lorenz model's parameters via a scalar time series, a predefined ansatz for the parameter control loop is required and the time for synchronization will be very long. Similarly, a local adaptive Lyapunov function has been suggested in [25], so as to design the parameter adaptive control loops for the synchronization of a given system as well as estimating the unknown parameters. However, a systematic and general approach to obtain this local Lyapunov function is in vain.

Another school of thought is to consider the problem as an optimization problem. In [11], a dynamical minimization algorithm, formulated by adaptive feedback control and dynamical minimization principle, has been suggested. It is demonstrated that the synchronization errors and the parameter estimation errors converge to zero asymptotically, by achieving a negative largest conditional Lyapunov exponent. However, as stated in [11], it still fails to estimate multiple unknown parameters if they appear in the dynamical evolution of the variables other than observable one.

In the light of above developments, an improved design of adaptive observer based on feedback control and dynamical minimization is to be suggested in this chapter, which is applicable for a large class of chaotic systems including the Lorenz system. The organization of this chapter is as follows. In Sec. 2, the design of the new adaptive observer is explained in details, while illustrative simulations are also given to demonstrate its effectiveness. In Sec. 3, this adaptive observer is proposed as an attack for various chaos-based cryptosystems to testify their securities. Chaos-based secure communication systems under various schemes, such as chaotic masking, chaotic modulation and chaotic shift key protocols, are investigated. Finally, conclusions are given in Sec. 4.

2 A New Adaptive Observer Design

In this section, a new design of adaptive observer is proposed to achieve adaptive synchronization of a large class of chaotic systems even if multiple unknown parameters exist. The design is based on the feedback control scheme and dynamical minimization, which have been shown to be effective for the synchronization problem of Lorenz system [29] described in Sec. 1.

2.1 Design Concepts

Consider a class of nonlinear systems expressed as:

$$\begin{cases} \dot{\mathbf{x}} = \mathbf{A}(\mathbf{p})\mathbf{x} + \varphi(\mathbf{x}) \equiv \mathbf{F}(\mathbf{x}, \mathbf{p}) \\ y = \mathbf{C}\mathbf{x} \end{cases} \tag{1}$$

where $\mathbf{x} \in \Re^n$ represents the state vector; y is the system output; $\mathbf{A}(\mathbf{p})$ is a fixed matrix independent of $\mathbf{x}$, contributing the linear part of the system with $\mathbf{p}$ depends linearly on $\mathbf{x}$; $\varphi(\mathbf{x})$ is the only nonlinear function, satisfying the Lipschitz condition in $\mathbf{F}$, i.e. there exists a positive constant $\gamma_\varphi < \infty$, such that $\| \varphi(\mathbf{x}) - \varphi(\hat{\mathbf{x}}) \| \leq \gamma_\varphi \| \mathbf{x} - \hat{\mathbf{x}} \|$, $\forall \mathbf{x}, \hat{\mathbf{x}} \in \Re^n$; $\mathbf{C}$ is a constant matrix.

Assuming that the following conditions hold:

1. $(\mathbf{A}, \mathbf{C})$ is a detectable pair and there exists an observer gain $\mathbf{K}$, such that $\mathbf{A}_c \equiv (\mathbf{A}(\mathbf{p}) - \mathbf{KC})$ is a strictly stable matrix, i.e. the eigenvalues of $\mathbf{A}_c$ lie in the left-half plane.
2. $\mathbf{S}$ is a symmetric positive definite matrix, satisfying the following Lyapunov equation
$$\mathbf{A}_c^T \mathbf{S} + \mathbf{S}\mathbf{A}_c = -\mathbf{T} \tag{2}$$
where $\mathbf{T} = \mathbf{T}^T > 0$ is a positive definite diagonal matrix.

As proved in [14], if Thau condition $\gamma_\varphi < \frac{\lambda_{min}(\mathbf{T})}{2\lambda_{max}(\mathbf{S})}$ is satisfied, then the following observer can be designed to synchronize with (1):

$$\begin{cases} \dot{\hat{\mathbf{x}}} = \mathbf{A}(\mathbf{p})\hat{\mathbf{x}} + \varphi(\hat{\mathbf{x}}) + \mathbf{u}(y, \hat{y}) \equiv \mathbf{F}(\hat{\mathbf{x}}, \mathbf{p}) + \mathbf{u}(y, \hat{y}) \\ \hat{y} = \mathbf{C}\hat{\mathbf{x}} \end{cases} \tag{3}$$

where $\mathbf{u}(y, \hat{y}) = \mathbf{K}e_y = \mathbf{K}(y - \hat{y})$ is a given feedback control.

Following the Proposition 2.1 stated in [6], if system (1) is *persistently exciting* and there exists an observer w.r.t (3) such that $\hat{\mathbf{x}} \to \mathbf{x}$ when $t \to \infty$, then an asymptotical adaptive observer can be designed even if $\mathbf{p}$ is unknown but appears in the observable states. It is also shown in [6] that the original Lyapunov stability proof can be obtained based on some simple parameter updating dynamics so that adaptive synchronization is achieved.

For our proposed adaptive observer, similar concept is adopted while Condition (1) provides a general guideline of our feedback design. It should also be remarked that the importance of the availability of $\mathbf{K}$ has been ignored in many existing designs, and hence their performance may be degraded.

However, due to the existence of multiple unknown parameters and the limited number of measurable states in (1), the parameter updating rules need to be carefully designed. Our approach is based on a modification of dynamical minimization principle [11], which originally mimicking the dynamics of a damping particle for the minimization of synchronization errors. Yet, the original algorithm in [11] works fine with a single unknown, it does encounter difficulties when there are multiple unknown parameters. Based on our studies in its error dynamics, it is observed that the convergence rates of states and unknown parameters are significant for the stability of the adaptive observer design. All the convergence rates must be similar so that synchronization errors as well as parameter estimation errors can reach the minimal simultaneously [29, 40, 44, 46]. Otherwise, these errors may eventually diverge, causing failure in using dynamical minimization algorithm for the case with multiple unknown parameters.

2.2 Conjecture and Design Procedures

Our design proposal is enunciated with the following conjecture.

Conjecture 1

Assuming that system (1) can be synchronized by another system (3) using some feedback control and the dynamical functions of system (1) is *linearly independent*[1], it is possible to design an adaptive observer expressed as follows:

$$\begin{cases} \dot{\hat{\mathbf{x}}} &= \mathbf{A}(\mathbf{q})\hat{\mathbf{x}} + \varphi(\hat{\mathbf{x}}) + \mathbf{K}e_y \\ \hat{y} &= \mathbf{C}\hat{\mathbf{x}} \\ \dot{q}_{ij} &= g_{ij}(\hat{\mathbf{x}}, e_y) = \delta_{ij} h_{ij}(\hat{\mathbf{x}}, e_y)\mu_{ij}(\hat{\mathbf{x}}) \qquad i = 1, \cdots, n; j = 1, \cdots, m. \end{cases} \tag{4}$$

where h_{ij} is a dynamical function that minimizes synchronization errors; μ_{ij} is an auxiliary function, moderating the convergence rate of each parameter so that they become similar; $\mathbf{q}$ is the estimator corresponding to the unknown parameter $\mathbf{p}$; $e_y = y - \hat{y}$ is the observed error and $\delta_{ij} > 0$ are some stiffness constants. Then, adaptive synchronization of (1) with unknown $\mathbf{p}$ can be accomplished by (4), so that $\hat{\mathbf{x}} \rightarrow \mathbf{x}$ and $\mathbf{q} \rightarrow \mathbf{p}$ when $t \rightarrow \infty$, for all $\hat{\mathbf{x}} \in \Re^n$ and $\mathbf{q} = \{q_{ij}\} \in \Re^{nm\times 1}$.

Design Procedures

In practice, the adaptive observer (4) can be designed with the following procedures:

1. **Determine K to stabilize the linear part:**
   Based on the design shown in [14], $\mathbf{K}$ should stabilize the linear part, i.e. $\mathbf{A}(\mathbf{p})$, of the targeted system (1). However, $\mathbf{A}(\mathbf{p})$ is now unavailable, and hence a more practical design criterion is to stabilize the linear part of the observer, i.e. $\mathbf{A}(\mathbf{q})$, instead, so that all the eigenvalues of $(\mathbf{A}(\mathbf{q}^{(0)}) - \mathbf{KC})$ are being negative. Obviously, a better initial guess of $\mathbf{q}$ (i.e. a better knowledge on $\mathbf{p}$) will improve the choices of $\mathbf{K}$.
2. **Design the functions h_{ij}, $i = 1, \cdots, n; j = 1, \cdots m$ by the dynamical minimization algorithm:**
   The updating equations for unknown parameters are designed by applying the dynamical minimization algorithm over the synchronization error e_y.

[1] **Definition 1** [32]: The functions $f_{ij}(\mathbf{x}(t)), i = 1, \cdots, n;\ j = 1, \cdots, m$, are said to be *linearly independent* if $r_{ij} = 0$ is the only solution for the following equation:

$$\sum_{j=1}^{m} r_{ij} f_{ij}(\mathbf{x}(t)) = 0, \qquad i = 1, 2, \cdots, n,$$

Note that Definition 1 is similar to the condition of *persistency exciting* (*P.E.*), which is commonly required in all identification processes. Generally speaking, it is always satisfied by system exhibited in its chaotic mode.

Without loss of generality, x_1 is assumed to be the measurable output (i.e. $y = x_1$), and the square synchronization error can be defined as:

$$E(q_{ij}, t) = e_1^2 = (x_1 - \hat{x}_1)^2. \tag{5}$$

It is noted that if q_{ij} takes precisely the value of p_{ij}, the observer will synchronize with the targeted system, resulting a zero error in (5) which is minimized. Therefore, it is desirable to construct the dynamical equations of $\mathbf{q}$ so that $E(q_{ij}, t)$ is driven to minimal. Following the guidelines given in [11], the below relationship is obtained:

$$\dot{q}_{ij} \propto \frac{\partial E(q_{ij}, t)}{\partial q_{ij}} \tag{6}$$

According to the dependence of the unknown parameter and the observable state, three different cases are classified:

a. For parameters q_{1j} (appears in function $F_1(\hat{\mathbf{x}}, \mathbf{q})$), whose corresponding state x_1 in the targeted system is available, h_{1j} is obtained straightforwardly as:

$$h_{1j} \propto \frac{\partial F_1(\hat{\mathbf{x}}, \mathbf{q})}{\partial q_{1j}} e_1 \tag{7}$$

b. if q_{ij} appears in $\hat{x}_i$, $i \neq 1$ and the evolution of $\hat{x}_1$ depends on state $\hat{x}_i$ directly, then

$$h_{ij} \propto \frac{\partial F_1(\hat{\mathbf{x}}, \mathbf{q})}{\partial \hat{x}_i} \frac{\partial F_i(\hat{\mathbf{x}}, \mathbf{q})}{\partial q_{ij}} e_1, \tag{8}$$

c. if q_{ij} appears in $\hat{x}_i$, $i \neq 1$ but the dynamical function F_1 does not depend on state $\hat{x}_i$ explicitly, then a further dependence according to the dynamical evolution of the system should be considered as given below:

$$h_{ij} \propto \left\{ \sum_k \left[\frac{\partial F_1(\hat{\mathbf{x}}, \mathbf{q})}{\partial \hat{x}_k} \frac{\partial F_k(\hat{\mathbf{x}}, \mathbf{q})}{\partial \hat{x}_i} \right] \frac{\partial F_i(\hat{\mathbf{x}}, \mathbf{q})}{\partial q_{ij}} \right\} e_1 \tag{9}$$

3. **Design $\mu_{ij}, i = 1, \cdots, n; j = 1, \cdots m$ to achieve similar convergence rate of estimation errors:**
Considering the case of having n variables and total nm unknown parameters in (1), the error dynamics can be formulated as an $(n + nm)$ high-dimensional system. Defining the error

$$\begin{aligned} \mathbf{z} &= (e_1, \cdots, e_n, r_{11}, \cdots, r_{nm}) \\ &\equiv (x_1 - \hat{x}_1, \cdots, x_n - \hat{x}_n, p_{11} - q_{11}, \cdots, p_{nm} - q_{nm}), \end{aligned}$$

and by linearizing the error dynamical equations around $\mathbf{z} = 0$, the Jacobian matrix J can be obtained as follows:

$$J = \begin{bmatrix} J_{11} & J_{12} \\ J_{21} & J_{22} \end{bmatrix}$$

where

$$J_{11} = \begin{bmatrix} \left(\frac{\partial F_1}{\partial e_1} - k_1\right) & \frac{\partial F_1}{\partial e_2} & \cdots & \frac{\partial F_1}{\partial e_n} \\ \vdots & \vdots & \cdots & \vdots \\ \left(\frac{\partial F_n}{\partial e_1} - k_n\right) & \frac{\partial F_n}{\partial e_2} & \cdots & \frac{\partial F_n}{\partial e_n} \end{bmatrix} \in \Re^{n \times n},$$

$$J_{12} = \begin{bmatrix} \frac{\partial F_1}{\partial r_{11}} & \cdots & \frac{\partial F_1}{\partial r_{nm}} \\ \vdots & \cdots & \vdots \\ \frac{\partial F_n}{\partial r_{11}} & \cdots & \frac{\partial F_n}{\partial r_{nm}} \end{bmatrix}$$

$$= \begin{bmatrix} \frac{\partial F_1}{\partial r_{11}} \cdots \frac{\partial F_1}{\partial r_{1m}} & 0 & \cdots & \cdots & \cdots & \cdots & 0 \\ 0 \quad \cdots \quad 0 & \frac{\partial F_2}{\partial r_{21}} & \cdots & \frac{\partial F_2}{\partial r_{2m}} & 0 & & 0 \\ \vdots \quad \cdots & & & & & & \vdots \\ 0 \quad \cdots \quad \cdots & \cdots & \cdots & 0 & \frac{\partial F_n}{\partial r_{n1}} & \cdots & \frac{\partial F_n}{\partial r_{nm}} \end{bmatrix} \in \Re^{n \times nm}$$

$$J_{21} = \begin{bmatrix} -\frac{\partial g_{11}}{\partial e_1} & 0 \cdots 0 \\ \vdots & \vdots \cdots \vdots \\ -\frac{\partial g_{nm}}{\partial e_1} & 0 \cdots 0 \end{bmatrix} \in \Re^{nm \times n}$$

$$J_{22} = \mathbf{0} \in \Re^{nm \times nm}.$$

The high dimensionality of J further explains why adaptive synchronization of a chaotic system with multiple unknown parameters is a difficult task. For example, the rank of J is $(n+1)$ if only x_1 is available. Hence, the eigenvalue of J cannot be determined freely. Moreover, the convergence rates for any design are difficult to determine under the same limitation. Therefore, in our approach, it is proposed to consider the case with each single unknown parameter separately in order to approximate the convergency rates of the states and parameter estimation errors.

Now assuming that there is only one unknown parameter, say p_{ij}, one obtains the below time-varying dynamical equations:

$$\begin{cases} \dot{\mathbf{e}} &= J_{11_{(ij)}} \mathbf{e} + J_{12_{(ij)}} r_{ij} \\ \dot{r}_{ij} &= -\dot{q}_{ij} = J_{21_{(ij)}} \mathbf{e} \end{cases} \tag{10}$$

where $J_{11_{(ij)}}$ is the same as the above sub-block of Jacobian matrix J_{11} except that parameter p_{ij} is taken place by its estimator q_{ij}; $J_{12_{(ij)}} = \begin{bmatrix} 0 \cdots 0 \ \frac{\partial F_i}{\partial r_{ij}} \ 0 \cdots 0 \end{bmatrix}^T$ and $J_{21_{(ij)}} = \begin{bmatrix} -\frac{\partial g_{ij}}{\partial e_1} \ 0 \cdots 0 \end{bmatrix}$.

With the system (10), it is derived that

$$\begin{aligned} \Gamma_{ij}(t) &= -J_{21_{(ij)}} J_{11_{(ij)}}^{-1} J_{12_{(ij)}} \\ &= \frac{D_{i1}}{|J_{11_{(ij)}}|} \frac{\partial F_i}{\partial r_{ij}} \frac{\partial g_{ij}}{\partial e_1} \\ &= \delta_{ij} \frac{D_{i1}}{|J_{11_{(ij)}}|} \frac{\partial F_i}{\partial r_{ij}} \frac{\partial h_{ij}}{\partial e_1} \mu_{ij}(\hat{\mathbf{x}}) \end{aligned} \tag{11}$$

where $|J_{11_{(ij)}}|$ is the determinant of $J_{11_{(ij)}}$; D_{i1} is the matrix cofactor; h_{ij} is the dynamical equations derived from (7)–(9). From (11), it is also noticed that $\Gamma_{ij}(t)$ is a scalar when there's only one unknown parameter. Provided that $\dot{\zeta} = -J_{21_{(ij)}} J_{11_{(ij)}}^{-1} J_{12_{(ij)}} \zeta$ is asymptotically stable based on our choices of feedback gain $\mathbf{K}$ and functions h_{ij} determined in the previous steps, it can be proved that the state and parameter estimation errors converge to zero with a rate governed by $\Gamma_{ij}(t)$ [33, 34]. Therefore, by comparing $\Gamma_{ij}(t)$ $\forall i, j$, $\mu_{ij}(\hat{\mathbf{x}})$ can be determined for rectifying the convergence rate in each case.

2.3 Simulation Example I: Lorenz System

The well-known Lorenz system is presented as our first example, which can be formulated as:

$$\dot{\mathbf{x}} = \mathbf{A}(\mathbf{p})\mathbf{x} + \varphi(\mathbf{x}) \equiv \mathbf{F}(\mathbf{x}, \mathbf{p}) \tag{12}$$

where $\mathbf{x} = [x_1 \quad x_2 \quad x_3]^T$ is the state vector; $\mathbf{A}(\mathbf{p})$ is a fixed matrix, given as

$$\mathbf{A}(\mathbf{p}) = \begin{bmatrix} -p_1 & p_1 & 0 \\ p_2 & -1 & 0 \\ 0 & 0 & -p_3 \end{bmatrix}$$

with $\mathbf{p} = [p_1 \quad p_2 \quad p_3]$; $\mathbf{F} \equiv \{F_i\}_{i=1,2,3}$ and $\varphi(\mathbf{x})$ is the only nonlinearity expressed as

$$\varphi(\mathbf{x}) = \begin{bmatrix} 0 \\ -x_1 x_3 \\ x_1 x_2 \end{bmatrix}.$$

Design of Adaptive Observer

Assuming that the state x_1 in (12) is measurable (i.e. $\mathbf{C} = \begin{bmatrix} 1 & 0 & 0 \end{bmatrix}$ as defined in (1) or $y = x_1$) and $\mathbf{p}$ is unknown, one can construct the following adaptive observer,

$$\dot{\hat{\mathbf{x}}} = \mathbf{A}(\mathbf{q})\hat{\mathbf{x}} + \varphi(\hat{\mathbf{x}}) + \mathbf{K}e_y \equiv \mathbf{F}(\hat{\mathbf{x}}, \mathbf{q}) + \mathbf{K}e_1 \tag{13}$$

where $\mathbf{q} = \{q_i\}$ is the estimator of $\mathbf{p}$ governed by the following dynamical equations:

$$\dot{q}_i = \delta_i h_i(\hat{\mathbf{x}}, e_1)\mu_i(\hat{\mathbf{x}}); \quad i = 1, 2, 3. \tag{14}$$

with $e_y = e_1 = x_1 - \hat{x}_1$.

Following the design procedures given in Sec. 2.2,

1. $\mathbf{K}$ is determined as follows:
Since $\mathbf{A}(\mathbf{p})$ is unknown, $\mathbf{K}$ is selected to stabilize $\mathbf{A}(\mathbf{q})$. For example, with an initial values of $\mathbf{q}$, $\mathbf{q}^{(0)} = [13, \quad 25, \quad 2.4]^T$, one can obtain $\mathbf{K} = [30 \quad 30 \quad 0]^T$ such that the eigenvalues $\lambda_{1,2,3} = -2.4000, -2.6093$ and -41.3907 are all negative. It should be remarked that this value of $\mathbf{K}$ should also stabilize the error dynamics for the case with only one single unknown parameter.
2. The functions h_i, $i = 1, 2, 3$, are designed as follows:
Applying the minimization algorithm given in Sec. 2.2 to the Lorenz system (12), one obtains:

$$
\begin{aligned}
\frac{\partial F_1(\hat{\mathbf{x}},\mathbf{q})}{\partial q_1} &= \hat{x}_2 - \hat{x}_1 \\
\frac{\partial F_1(\hat{\mathbf{x}},\mathbf{q})}{\partial \hat{x}_2}\frac{\partial F_2(\hat{\mathbf{x}},\mathbf{q})}{\partial q_2} &= q_1\hat{x}_1 \\
\left\{\sum_i \frac{\partial F_1(\hat{\mathbf{x}},\mathbf{q})}{\partial \hat{x}_i}\frac{\partial F_i(\hat{\mathbf{x}},\mathbf{q})}{\partial \hat{x}_3}\right\}\frac{\partial F_3(\hat{\mathbf{x}},\mathbf{q})}{\partial q_3} &= q_1\hat{x}_1\hat{x}_3
\end{aligned}
\tag{15}
$$

Hence, it can be obtained that

$$
\begin{aligned}
h_1 &= \mathrm{sgn}(\hat{x}_2 - \hat{x}_1)e_1 \\
h_2 &= \mathrm{sgn}(\hat{x}_1)e_1 \\
h_3 &= \mathrm{sgn}(\hat{x}_1\hat{x}_3)e_1
\end{aligned}
\tag{16}
$$

where

$$
\mathrm{sgn}(x) = \begin{cases} 1 & x > 0 \\ 0 & x = 0 \\ -1 & x < 0 \end{cases}
$$

It should be noticed that a sign function is introduced in (16), although (15) can be directly applied. This modification not only provides a simpler structure for realization and maintains the same updating trend, its performance is also found to be better. The dependence of q_1 in the functions $h_{2,3}$ is further omitted as it is assumed to be always positive.
3. The functions μ_i are obtained as follows:
Firstly, let's assume that there is only one single unknown parameter, say p_1. By linearizing the error dynamics of (12) and (13) evaluated on a typical trajectory, the following linear time-varying system can be obtained:

$$\begin{cases} \dot{\mathbf{e}} = J_{11_{(1)}}\mathbf{e} + J_{12_{(1)}} r_1 \\ \dot{r}_1 = J_{21_{(1)}}\mathbf{e} \end{cases} \tag{17}$$

where $\mathbf{e} = \mathbf{x} - \hat{\mathbf{x}}$, $r_1 = p_1 - q_1$, and

$$J_{11_{(1)}} = \begin{bmatrix} -q_1 - k_1 & q_1 & 0 \\ p_2 - \hat{x}_3 - k_2 & -1 & -\hat{x}_1 \\ \hat{x}_2 - k_3 & \hat{x}_1 & -p_3 \end{bmatrix}$$

$$\begin{aligned} J_{12_{(1)}} &= [\hat{x}_2 - \hat{x}_1 \quad 0 \quad 0]^T \\ J_{21_{(1)}} &= [-\delta_1 \mu_1(\hat{\mathbf{x}})\text{sgn}(\hat{x}_2 - \hat{x}_1) \quad 0 \quad 0] \end{aligned} \tag{18}$$

with $\mathbf{K} = \begin{bmatrix} k_1 \ k_2 \ k_3 \end{bmatrix}^T$ as obtained in the first step.
The convergence rate of (17) is then governed by:

$$\begin{aligned} \Gamma_1 &= -J_{21_{(1)}} J_{11_{(1)}}^{-1} J_{12_{(1)}} \\ &= -\frac{\delta_1}{|J_{11_{(1)}}|}\mu_1(p_3 + \hat{x}_1^2)(\hat{x}_2 - \hat{x}_1)\text{sgn}(\hat{x}_2 - \hat{x}_1) \\ &= -\frac{\delta_1}{|J_{11_{(1)}}|}\mu_1(p_3 + \hat{x}_1^2)|\hat{x}_2 - \hat{x}_1| \end{aligned} \tag{19}$$

where $|J_{11_{(1)}}| = (q_1 + k_1)(p_3 + \hat{x}_1^2) - q_1(p_2 p_3 - p_3\hat{x}_3 - \hat{x}_1\hat{x}_2 - p_3 k_2 + k_3\hat{x}_1)$. Similarly, if p_2 or p_3 is the sole unknown parameter, one obtains

$$\Gamma_2 = -\frac{\delta_2}{|J_{11_{(2)}}|}\mu_2 p_1 p_3 |\hat{x}_1| \tag{20}$$

and

$$\Gamma_3 = -\frac{\delta_3}{|J_{11_{(3)}}|}\mu_3 p_1 |\hat{x}_1 \hat{x}_3| \tag{21}$$

respectively, where $|J_{11_{(2)}}| = (p_1 + k_1)(p_3 + \hat{x}_1^2) - p_1(q_2 p_3 - p_3\hat{x}_3 - \hat{x}_1\hat{x}_2 - p_3 k_2 + k_3\hat{x}_1)$ and $|J_{11_{(3)}}| = (p_1 + k_1)(q_3 + \hat{x}_1^2) - p_1(p_2 q_3 - \hat{p}_3\hat{x}_3 - \hat{x}_1\hat{x}_2 - q_3 k_2 + k_3\hat{x}_1)$.

It is easily derived that our choice of $\mathbf{K} = [30 \quad 30 \quad 0]^T$ in previous example also guarantees $|J_{11_{(i)}}|_{i=1,2,3} > 0$ (Note: it simply demands that k_1 and k_2 are large but k_3 is small. Hence, it is letting that $k_3 = 0$).

Based on (19)–(21), which reflect the convergence rates of the error dynamics under the situations with one single unknown parameter, Γ_1 and Γ_3 are found to be of high order and dependent on their corresponding states (i.e. $\Gamma_i = g(\hat{x}_i)$ where $i = 1, 3$) while Γ_2 in (20) is of the first order and dependent only on $\hat{x}_1$. In order to make all the Γ_i having similar dynamics and convergence rate, it is to set $\mu_1 = \mu_3 = 1$ and $\mu_2 = |\hat{x}_2|$.

The final adaptive observer is hence summarized as follows:

$$\begin{cases} \dot{\hat{x}}_1 = q_1(\hat{x}_2 - \hat{x}_1) + k_1 e_1 \\ \dot{\hat{x}}_2 = q_2\hat{x}_1 - \hat{x}_2 - \hat{x}_1\hat{x}_3 + k_2 e_1 \\ \dot{\hat{x}}_3 = \hat{x}_1\hat{x}_2 - q_3\hat{x}_3 + k_3 e_1 \\ \dot{q}_1 = \delta_1 \mathrm{sgn}(\hat{x}_2 - \hat{x}_1)e_1 \\ \dot{q}_2 = \delta_2|\hat{x}_2|\mathrm{sgn}(\hat{x}_1)e_1 \\ \dot{q}_3 = \delta_3 \mathrm{sgn}(\hat{x}_1\hat{x}_3)e_1 \end{cases} \tag{22}$$

Simulation Results

Figure 1 illustrates the effectiveness of the proposed scheme. In the simulation, it is letting that $\delta_1 = \delta_3 = 10$, $\delta_2 = 2$, $\mathbf{K} = [30 \quad 30 \quad 0]^T$ and the initial conditions are $\mathbf{x}^{(0)} = [1 \quad 1 \quad 1]^T$, $\hat{\mathbf{x}}^{(0)} = [4 \quad 5 \quad 6]^T$, $\mathbf{q}^{(0)} = [13 \quad 25 \quad 2.4]^T$. It can be observed that synchronization is achieved within a short transient time of about 40s and all of the unknown parameters are accurately estimated. At $t = 100s$, one obtains $q_1 = 10 - 2.0262 \times 10^{-6}$, $q_2 = 28 + 5.5468 \times 10^{-6}$, $q_3 = 2.667 - 3.2366 \times 10^{-7}$, which are very close to the corresponding true values, i.e. $p_1 = 10$, $p_2 = 28$, $p_3 = 8/3$.

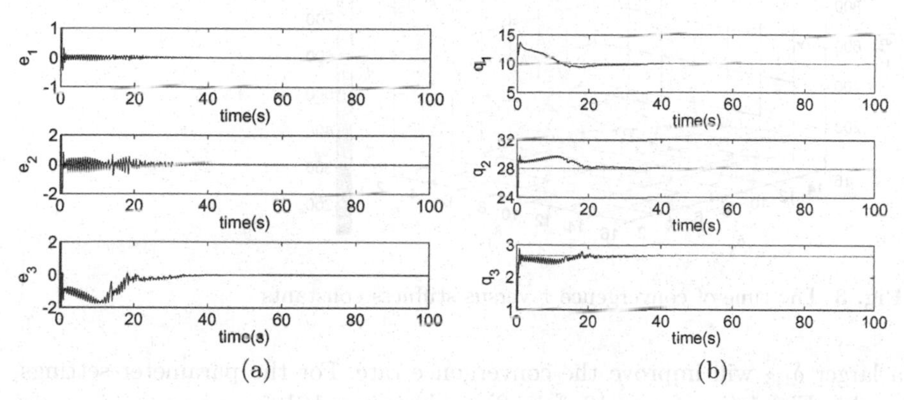

Fig. 1 (a) e_i versus *time* (b) q_i versus *time*, with initial conditions: $\mathbf{x}^{(0)} = [1 \quad 1 \quad 1]^T$, $\hat{\mathbf{x}}^{(0)} = [4 \quad 5 \quad 6]^T$, $\mathbf{q}^{(0)} = [13 \quad 25 \quad 2.4]^T$

To further investigate the effectiveness of the proposed design under a large mismatch in the initial values, another set of initial conditions: $\mathbf{x}^{(0)} = [1 \quad 2 \quad 3]^T$, $\hat{\mathbf{x}}^{(0)} = [-4 \quad -2 \quad -1]^T$ and $\mathbf{q}^{(0)} = [20 \quad 50 \quad 1]^T$ are used. As depicted in Fig. 2, adaptive synchronization is clearly achieved and the parameter estimation errors are just around 10^{-4} after 100s.

To study the effect of the constant δ_i, Fig. 3 depicts the time of convergence, τ[2], with $\delta_{1,3} \in [2, 16]$, $\delta_2 = 2$, $\mathbf{K} = [30 \quad 30 \quad 0]^T$. It is noticed that

[2] τ is defined as the time required for all the estimated parameters converging to their true values within a specified error, such as 10^{-7} in our simulation.

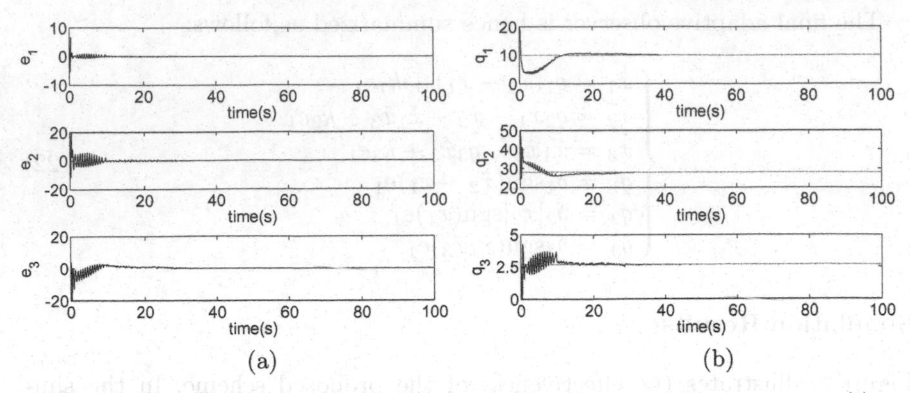

Fig. 2 (a) e_i versus *time* (b) q_i versus *time*, with initial conditions: $\mathbf{x}^{(0)} = [1 \quad 2 \quad 3]^T$, $\hat{\mathbf{x}}^{(0)} = [-4 \quad -2 \quad -1]^T$, $\mathbf{q}^{(0)} = [20 \quad 50 \quad 1]^T$

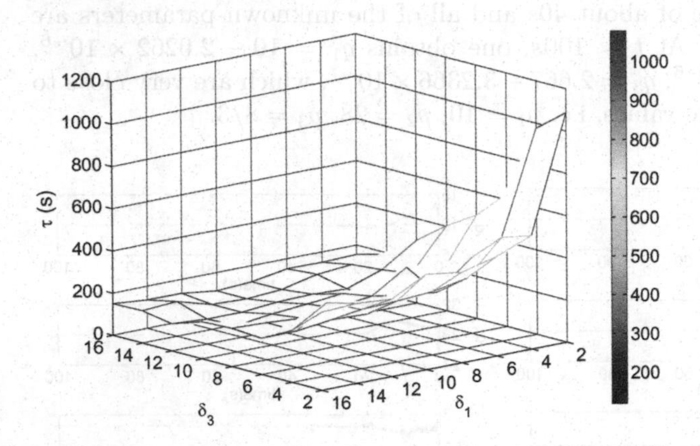

Fig. 3 The time of convergence τ versus stiffness constants

a larger $\delta_{1,3}$ will improve the convergence rate. For the parameter settings used in Fig. 1 (i.e. $\delta_{1,3} = 10$, $\delta_2 = 2$) we have $\tau = 161.6$s.

Design Verification and Stability Analysis

The design isverified by the use of local Lyapunov function [25] and the conditional Lyapunov exponents [1, 2].

Local Lyapunov Function: It is stated in [25] that systems are to be synchronized only if there exists a local Lyapunov function changed smoothly when the estimator, **q**, deviates slightly from its true value, **p**. In practice, the function can be specified by a time average over a period of time T such as,

$$U_i(\tau_0, T) = \frac{1}{T} \int_{\tau_0 - T}^{\tau_0} \dot{q}_i \, d\tau \tag{23}$$

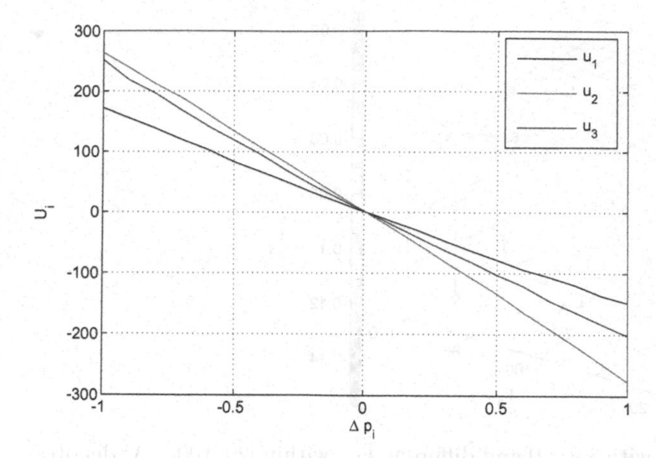

Fig. 4 $U_i, i = 1, 2, 3$ as a function of the relative deviation of the parameter error for q_i

where $q_i \in [p_i - \Delta p_i, p_i + \Delta p_i]$, assuming that the other unknown parameters are set as their nominal values.

Figure 4 depicts the time average of $U_i(\tau_0, T)$ as defined in (23) for each state variable with $T = 500$ (denoted as U_i), where the parameters are the same as those given in Fig. 1. It can be observed that U_i varies smoothly when the parameter deviation Δp_i is small. As illustrated by the slope of the functions, the time average of all parametric updating functions converge to zero in a similar manner under each individual dimension, which is closely agreed with our design concept.

Conditional Lyapunov Exponent: Another common-used approach in justifying the effectiveness of the design is the condition Lyapunov exponent (CLE). It is firstly proposed by Pecora and Carroll [1, 2], stating that synchronization can only be achieved if all the CLE's are negative.

However, it is recently reported that the negativity of the largest CLE is neither a sufficient nor a necessary condition for chaos synchronization due to some unstable invariant sets in the stable synchronization manifold [35, 36]. In the vicinity of zero largest CLE, it may lead to an on-off intermittent desynchronization, vulnerable to perturbation caused by the noise or a slight mismatch in parameter.

Yet, if the largest CLE is small enough and insensitive to the initial parameter differences, there is no on-off intermittent synchronization, and identical synchronization can still be achievable. Therefore, it is still considered to be adequate especially when the studies are carried out with larger initial differences between a given system and the observer [11, 12].

Obviously, the CLE's of our design depend on the choices of the feedback gains $\mathbf{K}$ and the stiffness constants δ. In practice, it is more preferable to

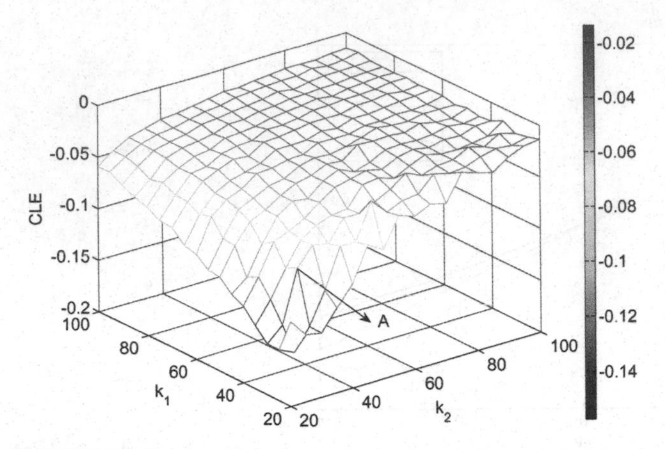

Fig. 5 The largest CLE with $k_3 = 0$ and different $k_{1,2}$ within $[20, 100]$. 'A' denotes the case of $\mathbf{K} = [30 \quad 30 \quad 0]^T$ and the corresponding largest CLE is -0.0815

choose those $\mathbf{K}$ and δ such that the largest CLE associated with the variation equations is more negative.

Since $k_3 = 0$ is recommended in our design, only the effects of gains $k_{1,2}$ on CLE's are analyzed. The result is depicted in Fig. 5, where the stiffness constants are fixed as $\delta = [10 \quad 2 \quad 10]^T$. It is clearly shown that the largest CLE's are negative for all the cases, provided that the control gains are assigned within the regime stabilizing the linear part of the system. Therefore, adaptive synchronization is expected to be achieved.

Remark 1. It is noted that for some smaller feedback gains $\mathbf{K}$, which closes to the boundary of stabilizing the linear part of the system, a negative largest CLE is still obtained. Although it may still achieve adaptive synchronization of the targeted system, however, as discussed in the followings, small stiffness constants should be used. Otherwise, the error dynamics may be diverging.

Figure 6 shows the dependence of the largest CLE with the stiffness constants δ, where the feedback gain is fixed as $\mathbf{K} = [30 \quad 30 \quad 0]^T$. As shown in Fig. 6, negative largest CLE is found within a large domain of δ, such as $0 < \delta_{1,3} \leq 16$ and $0 < \delta_2 \leq 4$.

It is also noticed that the largest CLE is more likely dependent on δ_3 than δ_1. For the cases $\delta_2 = 1, 2, 3$, an increase of δ_3 causes the decrease of the largest CLE. In addition, as depicted in Fig. 6, the largest CLE's is also very sensitive to the value of δ_2. Generally speaking, when δ_2 increases, the largest CLE will also decrease accordingly.

The dependence of CLE on stiffness constant δ_2 and feedback gains k_1 and k_2 are hence further studied as shown in Figs. 7 (a) and (b), respectively. From the simulation results, it is observed that δ_2 should not be too large, especially when $\mathbf{K}$ is small. For example, it is noticed that when $k_1 = 20$, $k_2 = 30$ and $\delta_2 = 5$, the largest CLE may become positive, and hence

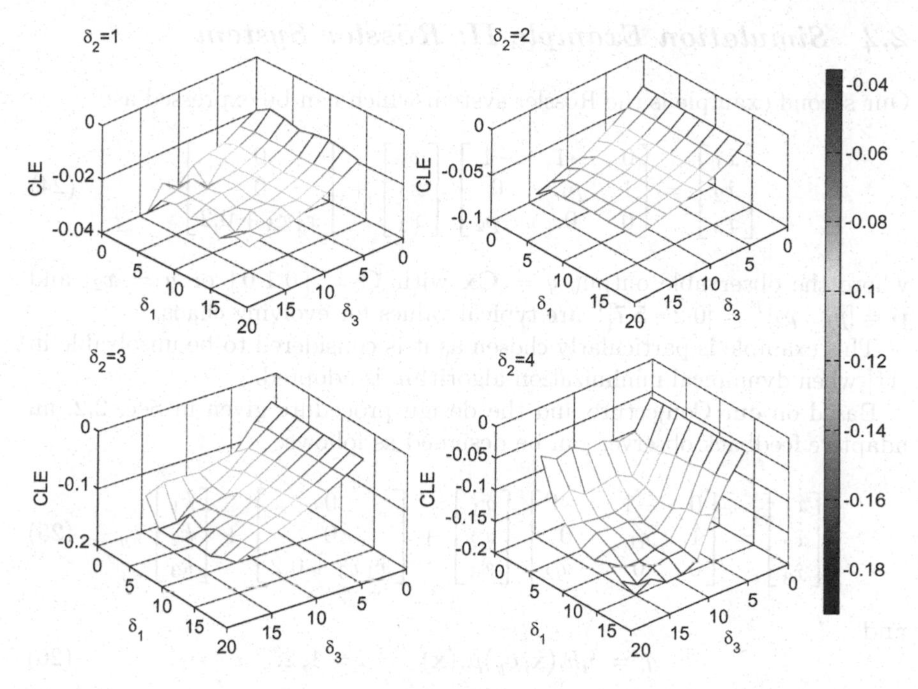

Fig. 6 The largest CLE with different δ. From upper-left to lower-right, $\delta_{1,3} \in [2, 16]$ and $\delta_2 = 1.0,\ 2.0,\ 3.0,\ 4.0$, with $\mathbf{K} = [30 \quad 30 \quad 0]^T$

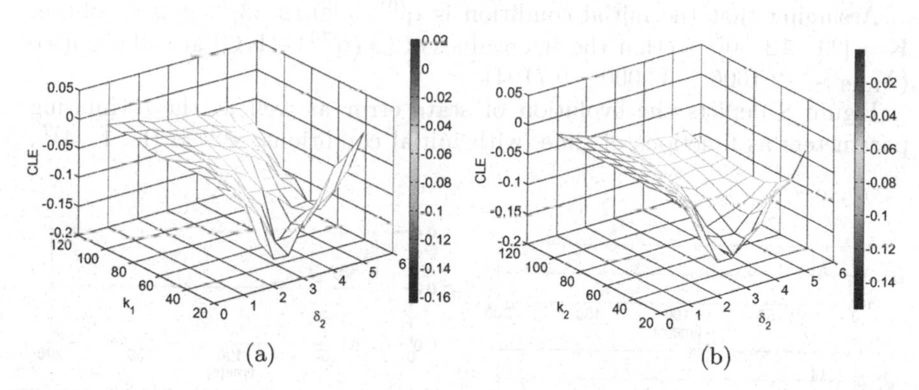

Fig. 7 (a) The dependence of largest CLE on feedback gain k_1 and stiffness constants δ_2 with $k_2 = 30, k_3 = 0,\ \delta_{1,3} = 10,\ k_1 \in [20 \quad 110],\ \delta_2 \in [0.5 \quad 5]$; (b) the dependence of largest CLE on feedback gain k_2 and stiffness constants δ_2 with $k_1 = 30, k_3 = 0,\ \delta_{1,3} = 10,\ k_2 \in [20 \quad 110],\ \delta_2 \in [0.5 \quad 5]$

adaptive synchronization cannot be achieved. Generally speaking, the larger the feedback gains, the larger the upper bound of δ_2.

2.4 Simulation Example II: Rössler System

Our second example is the Rössler system which can be expressed as:

$$\begin{bmatrix} \dot{x}_1 \\ \dot{x}_2 \\ \dot{x}_3 \end{bmatrix} = \begin{bmatrix} 0 & -1 & -1 \\ 1 & p_1 & 0 \\ 0 & 0 & -p_2 \end{bmatrix} \begin{bmatrix} x_1 \\ x_2 \\ x_3 \end{bmatrix} + \begin{bmatrix} 0 \\ 0 \\ x_1 x_3 + 0.2 \end{bmatrix} \tag{24}$$

where the observable output $y = \mathbf{C}\mathbf{x}$ with $\mathbf{C} = \begin{bmatrix} 0\,1\,0 \end{bmatrix}$ or $y = x_2$; and $\mathbf{p} = [p_1 \quad p_2]^T = [0.2 \quad 5.7]^T$ are typical values for evolving chaos.

This example is particularly chosen as it is considered to be unsolvable in [11] when dynamical minimization algorithm is adopted.

Based on our Conjecture and the design procedure given in Sec. 2.2, an adaptive feedback observer can be designed as follows:

$$\begin{bmatrix} \dot{\hat{x}}_1 \\ \dot{\hat{x}}_2 \\ \dot{\hat{x}}_3 \end{bmatrix} = \begin{bmatrix} 0 & -1 & -1 \\ 1 & q_1 & 0 \\ 0 & 0 & -q_2 \end{bmatrix} \begin{bmatrix} \hat{x}_1 \\ \hat{x}_2 \\ \hat{x}_3 \end{bmatrix} + \begin{bmatrix} 0 \\ 0 \\ \hat{x}_1 \hat{x}_3 + 0.2 \end{bmatrix} + \begin{bmatrix} k_1 \\ k_2 \\ k_3 \end{bmatrix} e_y \tag{25}$$

and

$$\dot{q}_i = \delta_i h_i(\hat{\mathbf{x}}, e_y) \mu_i(\hat{\mathbf{x}}) \qquad i = 1, 2; \tag{26}$$

where $e_y = x_2 - \hat{x}_2 \equiv e_2$, $h_1 = \mathrm{sgn}(\hat{x}_2) e_2$, $h_2 = \mathrm{sgn}(\hat{x}_3) e_2$, $\mu_1 = \frac{\mathrm{sgn}(\hat{x}_3)}{1+|\hat{x}_3|}$ and $\mu_2 = 1$.

Assuming that the initial condition is $\mathbf{q}^{(0)} = [0.12 \quad 3]^T$, it is to obtain $\mathbf{K} = [20 \quad 12 \quad 0]^T$ so that the eigenvalues of $(\mathbf{A}(\mathbf{q}^{(0)}) - \mathbf{KC})$ are all negative ($\lambda_{1,2,3} = -2.1606, -3.0000, -9.7194$).

Figure 8 depicts the evolution of state error as well as the estimating parameters as functions of time, with initial conditions: $\mathbf{x}^{(0)} = [1 \quad 1 \quad 1]^T$,

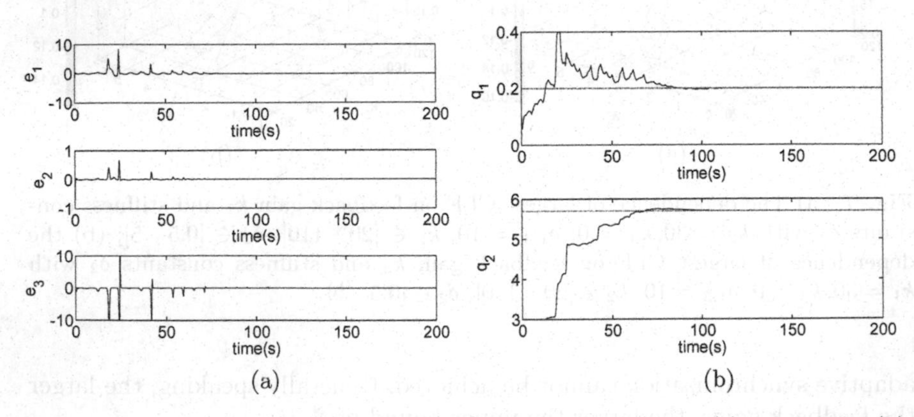

Fig. 8 Convergence of state error e_i and system parameter q_i: (a) e_i vs *time*, (b) q_i vs *time*

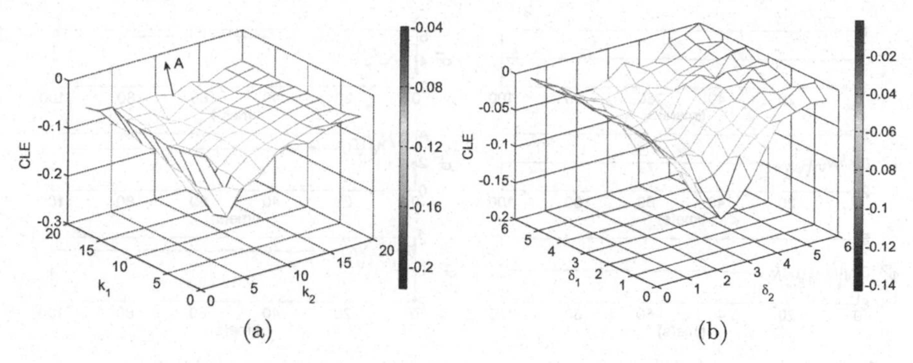

Fig. 9 The largest CLE of Rössler system under different circumstances: (a) The largest CLE when $k_{1,2} \leq 20$, $k_3 = 0$ with stiffness $\delta = [2 \quad 2]^T$. The marked value of is $A = -0.09364$ obtained at $\mathbf{K} = [20 \quad 12 \quad 0]^T$. (b) The largest CLE when $\delta_{1,2} \in [0.5 \quad 6]$ with feedback gain $\mathbf{K} = [20 \quad 12 \quad 0]^T$.

$\hat{\mathbf{x}}^{(0)} = [3 \quad 2 \quad 2]^T$, $\mathbf{K} = [20 \quad 12 \quad 0]^T$, $\delta = [2 \quad 2]^T$. It is clearly shown that both the state and parameter estimation errors converge to zero after the initial transient time of about 100s.

To verify the stability of algorithm, the largest CLE versus the feedback $\mathbf{K}$ and the stiffness gain δ are obtained and shown in Fig. 9 (a) and (b), respectively. It is noticed that a large synchronization regime is available. Both Figs. 9 (a) and (b) have similar profiles, showing that the CLE's of the Rössler system depend more on k_2 and δ_2, instead of k_1 and δ_1. By increasing the values of k_2 or δ_2 from zero, the largest CLE decreases quickly initially, but gradually increases when k_2 or δ_2 are sufficiently large. On the contrary, the largest CLE are less sensitive to the values of k_1 and/or δ_1.

Similar to the Lorenz system, the regime of δ increases with an increase of $\mathbf{K}$ in general. However, in the chosen ranges for both $\mathbf{K}$ and δ, all the nontrivial CLE's are negative. It means that synchronization together with parameter estimation can be achieved by some suitable stiffness constants, as presented in Fig 8.

2.5 Simulation Example III: Genesio System

The last example is the Genesio chaotic system which is expressed as follows:

$$\begin{bmatrix} \dot{x}_1 \\ \dot{x}_2 \\ \dot{x}_3 \end{bmatrix} = \begin{bmatrix} 0 & 1 & 0 \\ 0 & 0 & 1 \\ -p_1 & -p_2 & -p_3 \end{bmatrix} \begin{bmatrix} x_1 \\ x_2 \\ x_3 \end{bmatrix} + \begin{bmatrix} 0 \\ 0 \\ x_1^2 \end{bmatrix} \tag{27}$$

where p_1, p_2, p_3 are the positive real constants satisfying $p_2 p_3 < p_1$.

In our simulation, it is letting that $p_1 = 6$, $p_2 = 2.92$, $p_3 = 1.2$ and the observable output is $y = \mathbf{C}\mathbf{x}$ with $\mathbf{C} = \begin{bmatrix} 1 \, 0 \, 0 \end{bmatrix}$ or simply $y = x_1$.

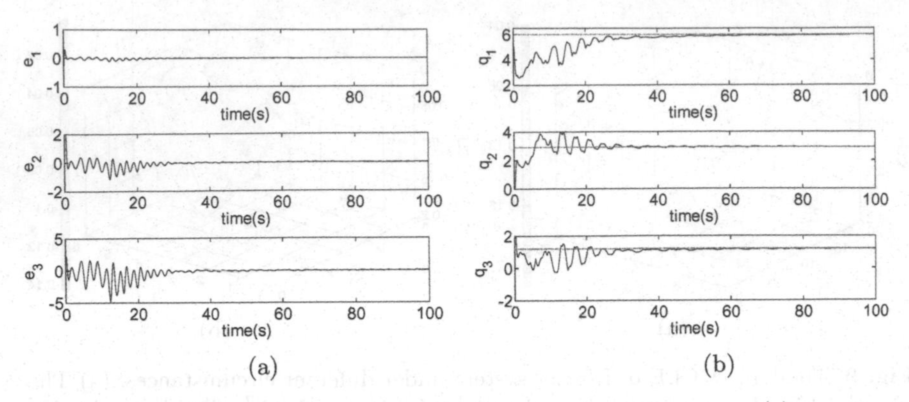

Fig. 10 Convergence of state errors e_i and parameter estimators q_i:(a)e_i vs *time*, (b) q_i vs *time*

Similarly, an adaptive observer system can be designed as follows:

$$\begin{bmatrix} \dot{\hat{x}}_1 \\ \dot{\hat{x}}_2 \\ \dot{\hat{x}}_3 \end{bmatrix} = \begin{bmatrix} 0 & 1 & 0 \\ 0 & 0 & 1 \\ -q_1 & -q_2 & -q_3 \end{bmatrix} \begin{bmatrix} \hat{x}_1 \\ \hat{x}_2 \\ \hat{x}_3 \end{bmatrix} + \begin{bmatrix} 0 \\ 0 \\ \hat{x}_1^2 \end{bmatrix} + \begin{bmatrix} k_1 \\ k_2 \\ k_3 \end{bmatrix} e_y \tag{28}$$

$$\dot{q}_i = \delta_i \text{sgn}(\hat{x}_i) e_1 \qquad \text{for } i = 1, 2, 3 \tag{29}$$

Assuming that $\mathbf{x}^{(0)} = [3 \quad -4 \quad 2]^T$, $\hat{\mathbf{x}}^{(0)} = [4 \quad -6 \quad 3]^T$ and $\mathbf{q}^{(0)} = [4 \quad 0.7 \quad 0.9]^T$, the feedback gains $\mathbf{K} = [10 \quad 40 \quad 40]^T$ are used (the corresponding eigenvalues of $(\mathbf{A}(\mathbf{q}^{(0)}) - \mathbf{KC})$ are $-3.5175 \pm 3.1839i$ and -3.8650). As shown in Fig. 10, both the system states and system parameters will converge to their corresponding values.

2.6 Robustness Analysis

The robustness of an adaptive observer is important, in particular for the practical situation where noise is inevitably existed. Here, the Lorenz system given in (12) is considered as a testbed and two kind of noises are added to the output state. One is a weakly additive uniformly distributed noise with amplitude in a range of $[-0.5 \quad 0.5]$ while another one is a Gaussian noise with zero mean and standard variance 0.5.

Based on the adaptive observer designed in (22), with $\delta = [10 \quad 2 \quad 10]^T$ and $\mathbf{K} = [30 \quad 30 \quad 0]^T$, it is found that generalized synchronization is still possible even if noise is existed. Figures 11 and 12 depict the results for the two different kinds of noises, respectively. It is found that the observer's system states will converge to their true values with some small errors while estimated parameters oscillate around their corresponding true values. It is worth to point out that these errors can be further reduced by taking the

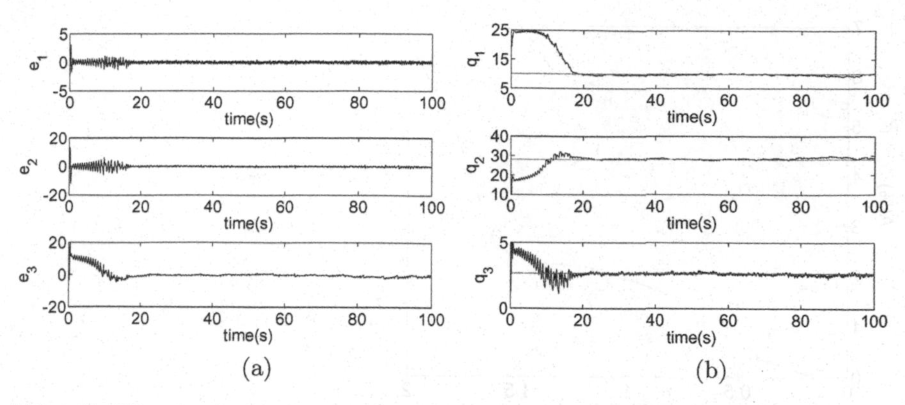

Fig. 11 The simulation result when the system is embedded into uniformly distributed noise; (a) e_i versus *time* (b) q_i versus *time*, with initial conditions: $\mathbf{x}^{(0)} = [1 \quad 1 \quad 1]^T$, $\hat{\mathbf{x}}^{(0)} = [4 \quad 5 \quad 6]^T$, $\mathbf{q}^{(0)} = [20 \quad 10 \quad 2]^T$

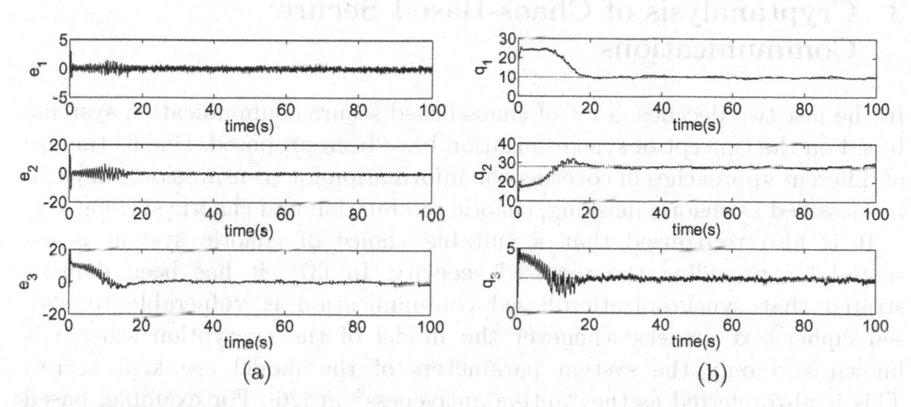

Fig. 12 The simulation result when the system is embedded into random Gaussian noise (a) e_i versus *time* (b) q_i versus *time*, with initial conditions: $\mathbf{x}^{(0)} = [1 \quad 2 \quad 3]^T$, $\hat{\mathbf{x}}^{(0)} = [4 \quad 5 \quad 6]^T$, $\mathbf{q}^{(0)} = [20 \quad 10 \quad 2]^T$

average over the time evolution of the unknown parameters in the asymptotic limit.

Defining the average parameter estimation error as $A = \sqrt{\sum_{i=1}^{3}(p_i - q_i)^2/3}$, Fig. 13 studies the average estimation errors for different noise levels, assuming that a uniformly-distributed noise with strength $[-w, w]$ is added. The error A is found to be increased with noise. A relative error about 7% is noticed for $w = 2$, which may still be good enough for approximating the system parameters in certain circumstances.

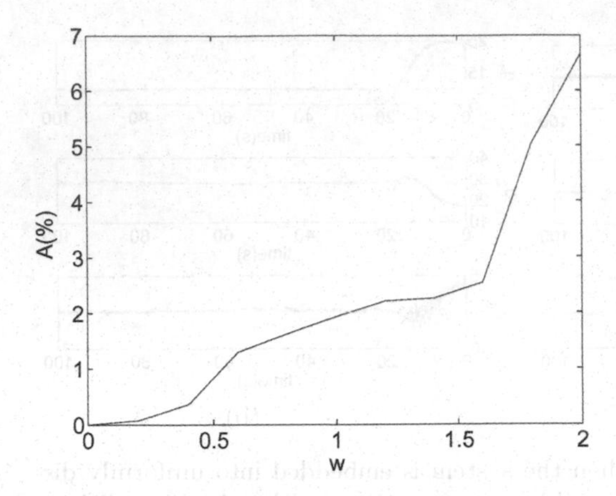

Fig. 13 The plot of parameter errors as a function of strength of noise, w

3 Cryptanalysis of Chaos-Based Secure Communications

In the last two decades, a lot of chaos-based secure communication systems based on the concept of synchronization have been proposed. Due to the use of different approaches in covering the information for transmission, they can be classified as chaotic masking, chaotic modulation and chaotic shifting key.

It is also recognized that a suitable choice of chaotic system is essential for providing the required security. In [37], it has been demonstrated that synchronization-based communication is vulnerable to chosen cipher-text attacks whenever the model of the encryption scheme is known and only the system parameters of the model are kept secret. This is also referred as the "anti-adaptiveness" in [26]. For example, based on the works done in [26], Chua's Circuit and Lur'e systems are neither

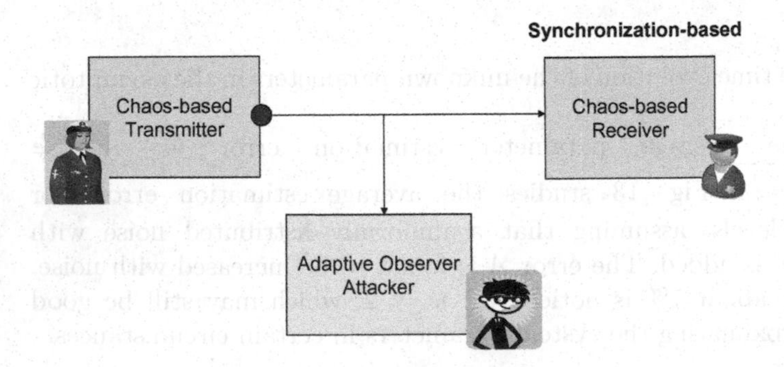

Fig. 14 Attack on a secure communication system using adaptive observer

anti-adaptive nor anti-robust secure, and hence they are not suitable for the use in secure communications. On the other hand, Lorenz system or its unified model are generally considered to be highly anti-adaptive, and many Lorenz-based or Unified-model-based cryptosystems have been proposed.

In this section, the adaptive observer presented in the previous section is used as an attack for various chaos-synchronization based cryptosystems. As shown in Fig. 14, assuming that the transmitted signal has been eavesdropped in an open channel, an adaptive observer can be used by an attacker for revealing the user-specified parameters and the covered messages. This provides as a fundamental cryptanalysis of a particular scheme, justifying whether it is sufficiently anti-adaptive secure.

3.1 Cryptanalysis of Chaotic Masking Scheme

The first scheme to be analyzed is based on chaotic masking. Consider a communication system consisting of a transmitter and a receiver, information signal, $s(t)$, is to be embedded into the output of the transmitter system (also known as the masking carrier), $x_i(t)$, and the resultant signal $y(t)$ is transmitted. Due to the randomness of chaos, the original information is hence masked. At the receiver end, the carrier is obtained via synchronization. It is then subtracted from the received signal $y(t)$ so that $s(t)$ is uncovered [3, 18, 22, 38]. Both theoretical analysis and numerical simulation have shown that synchronization between transmitter and receiver is obtainable if the message is weak enough, typically -20dB to -30dB of the masking carrier $x_i(t)$ [38].

Cryptanalysis of Lorenz-Based Chaotic Masking Scheme

In this section, the security of the Lorenz-based chaotic masking systems proposed in [20] and [39] are investigated.

System I: The transmitter of the cryptosystem given in [20] is formulated as

$$\begin{cases} \dot{\mathbf{x}} = \begin{bmatrix} -p_1 & p_1 & 0 \\ p_2 & -1 & 0 \\ 0 & 0 & -p_3 \end{bmatrix} \mathbf{x} + \begin{bmatrix} 0 \\ -yx_3 \\ yx_2 \end{bmatrix} + \mathbf{K}s \\ y = \mathbf{C}\mathbf{x} + s \end{cases} \tag{30}$$

where $\mathbf{C} = [1 \quad 0 \quad 0]$ and y is the output signal obtained by masking the plaintext signal s with the variable x_1. Similarly to [20], the gain matrix is set as $\mathbf{K} = [30 \quad 28 \quad 0]^T$.

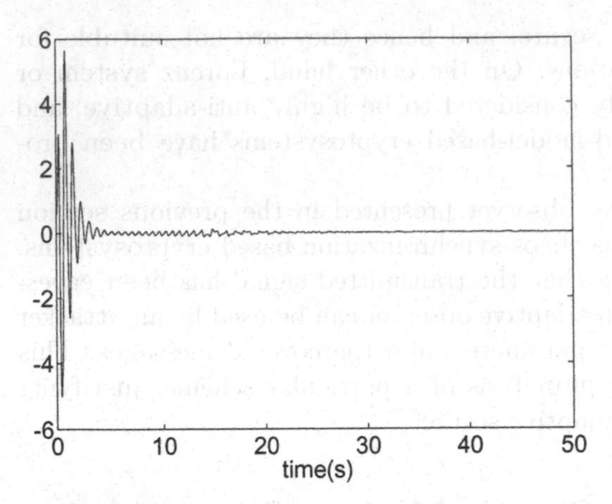

Fig. 15 The signal estimation error obtained by adaptive observer for Case 1 with System I

An adaptive observer can then be constructed as follows:

$$\begin{cases} \dot{\hat{\mathbf{x}}} = \begin{bmatrix} -q_1 & q_1 & 0 \\ q_2 & -1 & 0 \\ 0 & 0 & -q_3 \end{bmatrix} \hat{\mathbf{x}} + \begin{bmatrix} 0 \\ -y\hat{x}_3 \\ y\hat{x}_2 \end{bmatrix} + \mathbf{K}s_o \\ \hat{y} = \mathbf{C}\hat{\mathbf{x}} \end{cases} \tag{31}$$

where $s_o = e_y = y - \hat{y}$ and the unknown parameters are adaptively updated as in (22) with $\delta_1 = \delta_3 = 1.2$, $\delta_2 = 0.35$. Similarly to [40], in the following sections, two cases are considered.

Case 1: In the first case, a low frequency information signal, $s(t) = A\sin(2\pi ft)$, with $A = 0.05$ and $f = 3$Hz, is transmitted. Using the adaptive observer (31), Fig. 15 depicts the estimation errors between the recovered signal and the actual signal. It is noticed that the errors remain as a very small value after a transient time of about 25s.

This secure communication system has also been attacked by a high-pass filter as suggested in [42]. For comparison, both the estimation results using the proposed adaptive observer and a high-pass filter with cut-off frequency of 1 Hz are presented in Fig. 16. It is noticed from Fig. 16 (b) that the absolute estimation error with adaptive observer is about 5×10^{-3}, corresponding to about 10% of the transmitted signal. Although it is slightly worse than that of high-pass filter, the basic form is still closely followed. Moreover, there is no need to do any calibration of phase as required for high-pass filter.

As shown in Fig. 17, the unknown parameters $\mathbf{p}$ in (30) can also be coarsely estimated by the adaptive observer. It is derived that $\bar{\mathbf{q}} = [9.9758 \quad 28.0272 \quad 2.6628]^T$ by taking an average of $\mathbf{q}$ from 25s to 50s. The obtained values are very close to the actual ones which are $\mathbf{p} = [10 \quad 28 \quad 8/3]^T$, and hence the entire system structure is revealed. It should

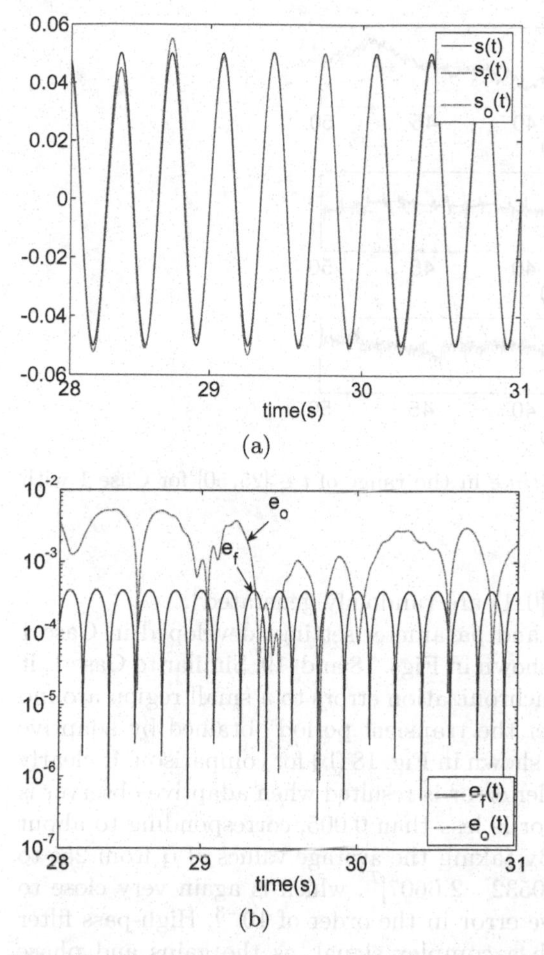

Fig. 16 (a) The original signal ($s(t)$), recovered signal obtained by high-pass filter ($s_f(t)$) and recovered signal obtained by adaptive observer ($s_o(t)$); (b) the estimation errors due to adaptive observer (e_o) and the high-pass filter (e_f) with $t \in [28, 31]$ for Case 1 with System I

be noticed that these major parameters are usually used as the user keys in a chaos-based cryptosystem which are assumed to be kep secret.

Case 2: The message is now assumed to be more complex and a multitonic signal consisting of frequencies from 4 Hz–13 Hz is used:

$$s(t) = \sum_{i=0}^{9} a_i \sin(2\pi(4+i)t + \theta_i) \tag{32}$$

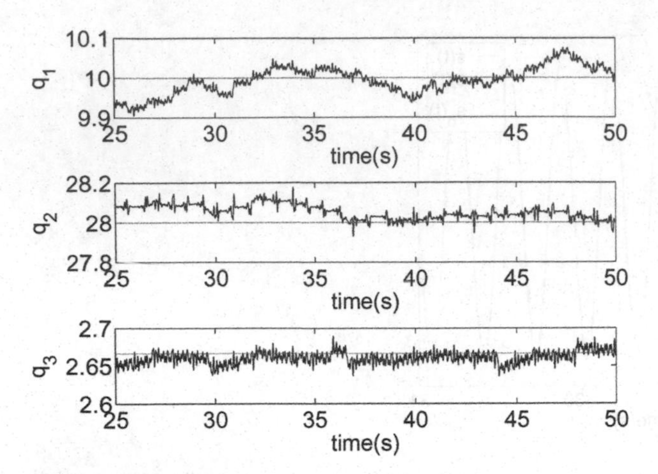

Fig. 17 The estimator **q** against *time* in the range of $t \in [25, 50]$ for Case 1 with System I

where $a_i \in [-0.2, 0.2]$ and $\theta_i \in [0, 1)$ are randomly generated[3].

The same adaptive observer and parameter settings developed in Case 1 are applied, and the results are shown in Figs. 18 and 19. Similar to Case 1, it takes about 25s to drive the synchronization errors to a small region around zero. The estimation errors after the transient period obtained by adaptive observer and high-pass filter are shown in Fig. 18 (b) for comparison. It clearly demonstrates that a much smaller error is resulted when adaptive observer is used. The signal estimation error is less than 0.005, corresponding to about 5% of the transmitted signal. By taking the average values of **q** from 25s to 50s, one gets $\bar{\mathbf{q}} = [9.9939 \quad 28.0532 \quad 2.6607]^T$, which is again very close to the actual values with a relative error in the order of 10^{-3}. High-pass filter becomes inappropriate for such a complex signal, as the gains and phase shifts are in fact frequency dependent.

System II: Now, the system proposed in [39] is considered, for which information is nonlinearly coupled into the Lorenz system. The overall dynamical model can be formulated as:

$$\begin{cases} \dot{\mathbf{x}} = \begin{bmatrix} -p_1 & p_1 & 0 \\ p_2 & -1 & 0 \\ 0 & 0 & -p_3 \end{bmatrix} \mathbf{x} + \begin{bmatrix} -ys \\ -yx_3 - ys \\ yx_2 + ys \end{bmatrix} + \mathbf{K}s \\ y = \mathbf{C}\mathbf{x} + s \end{cases} \tag{33}$$

where the same parameters for System I are now used.

[3] The randomly generated parameters are: $\{a_i\}$ = {-0.0826, 0.0955, 0.0994, 0.1156, 0.1474, -0.0762, 0.0863, -0.1114, -0.1973,-0.1559}; and $\{\theta_i\}$= {0.7850, 0.0420, 0.4878, 0.3726, 0.7462, 0.4500, 0.7034, 0.8726, 0.0268, 0.4083}

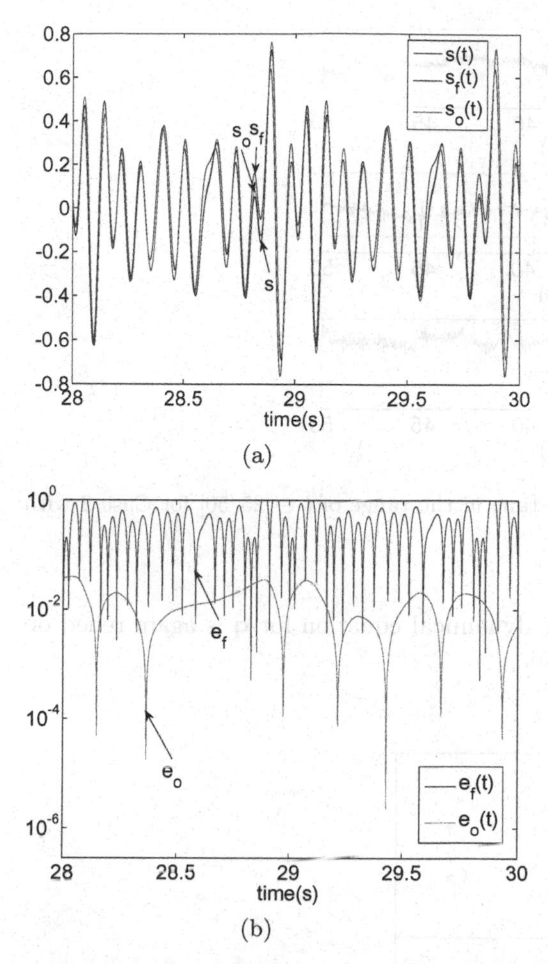

Fig. 18 (a) The original signal ($s(t)$), recovered signal obtained by high-pass filter ($s_f(t)$) and recovered signal obtained by adaptive observer ($s_o(t)$); (b) the estimation errors due to adaptive observer (e_o) and the high-pass filter (e_f) with $t \in [28, 30]$ for Case 2 with System I

Similarly to System I, the high-pass filter attack works fine to recover monotonic information signal. Hence, only the results for multitonic signal are presented.

The adaptive observer is then designed as follows:

$$\begin{cases} \dot{\hat{\mathbf{x}}} = \begin{bmatrix} -q_1 & q_1 & 0 \\ q_2 & -1 & 0 \\ 0 & 0 & -q_3 \end{bmatrix} \hat{\mathbf{x}} + \begin{bmatrix} -ys_o \\ -y\hat{x}_3 - ys_o \\ y\hat{x}_2 + ys_o \end{bmatrix} + \mathbf{K}s_o \\ \hat{y} = \mathbf{C}\hat{\mathbf{x}} \end{cases} \tag{34}$$

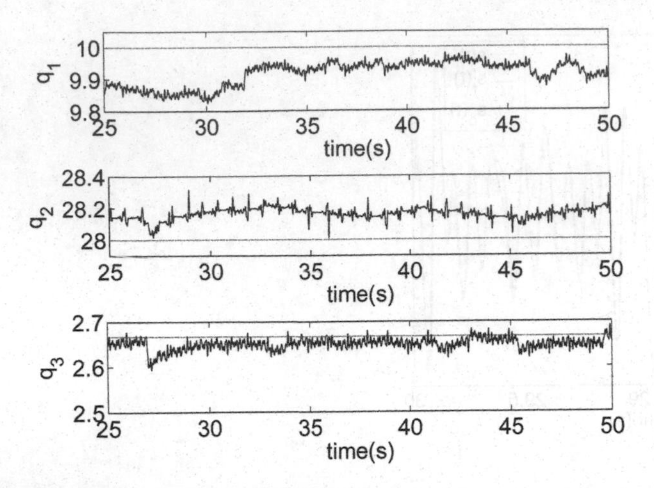

Fig. 19 The estimator $\mathbf{q}$ against time in the range of $t \in [25, 50]$ for Case 2 with System I

where $s_o = e_y = y - \hat{y}$ and the dynamical equation for $\mathbf{q}$ is again relied on (22) with $\delta_1 = \delta_3 = 1.6$, $\delta_2 = 0.6$.

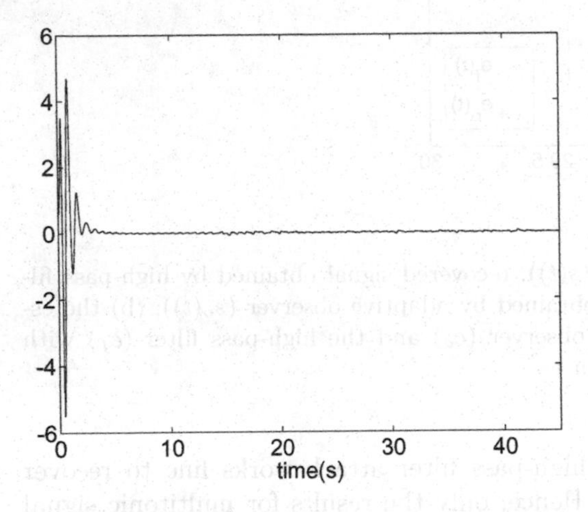

Fig. 20 The signal estimation error obtained by adaptive observer for multitonic signal with System II

As shown in Fig. 20, the estimation error takes about 20s to settle. The corresponding estimation errors obtained by adaptive observer and high-pass filter are compared in Fig. 21. It is noticed that high-pass filter fails to provide

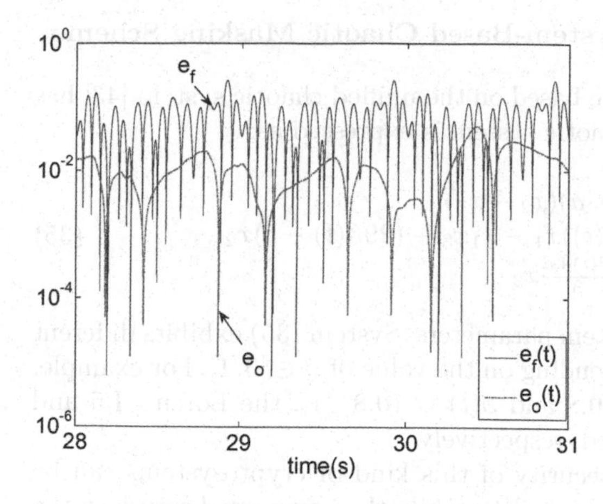

Fig. 21 The estimation errors due to adaptive observer (e_o) and the high-pass filter (e_f) with $t \in [28, 31]$ for multitonic signal with System II

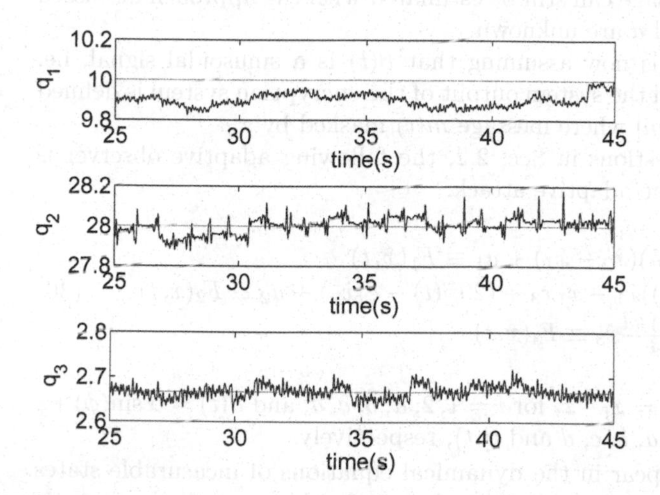

Fig. 22 The estimator $\mathbf{q}$ against time in the range of $t \in [25, 45]$ for multitonic signal with System II

an accurate reconstruction of $s(t)$ while the signal obtained by adaptive observer closely follows $s(t)$. It is also interested to notice that the performance for both Cases I and II are similar when adaptive observer is used.

By taking the average of $\mathbf{q}$ from 25s to 45s in Fig. 22, the estimated system parameter is computed as $\bar{\mathbf{q}} = [9.9015 \quad 28.0046 \quad 2.6669]^T$. The relative error is in the order of 10^{-4}, similarly to the linear case (i.e. System I).

Cryptanalysis of Unified System-Based Chaotic Masking Scheme

In [5], a chaotic masking scheme based on the unified chaotic system [43] has been suggested. The unified chaotic system is expressed as:

$$\begin{cases} \dot{x}_1 = (25\beta(t) + a)(x_2 - x_1) \\ \dot{x}_2 = (b - 35\beta(t))x_1 - x_1x_3 + (29\beta(t) - c)x_2 \\ \dot{x}_3 = x_1x_2 - \frac{\beta(t)+d}{3}x_3 \end{cases} \tag{35}$$

where a, b, c and d are some system parameters. System (35) exhibits different kinds of chaotic attractors, depending on the value of $\beta \in [0,1]$. For example, when $\beta(t) = [0 \quad 0.8)$, $\beta(t) = 0.8$ and $\beta(t) = (0.8 \quad 1]$, the Lorenz, Lü and Chen attractors can be observed, respectively.

It is stated in [5] that the security of this kind of cryptosystems can be granted by applying a time varying $\beta(t)$, while the constant d serves as the key for the authorized receiver. Other constants a, b and c is simply retrieved at the receiver end based on adaptive synchronization. However, as shown in the following, the message can still be estimated with the approach discussed in Sec. 2 even $\beta(t)$ and d are unknown.

Referring to [5], it is now assuming that $\beta(t)$ is a sinusoidal signal, i.e. $\beta(t) = \alpha\sin(\phi)+\varepsilon$, and the system output of the encryption system is defined as $\mathbf{s} = [x_1 \quad x_2 \quad x_3 + m]$ where message $m(t)$ masked by x_3.

Following the suggestions in Sec. 2.2, the following adaptive observer is obtained to perform the adaptive attack:

$$\begin{cases} \dot{\hat{x}}_1 = (25\hat{\beta}(t) + \hat{a})(\hat{x}_2 - \hat{x}_1) + u_1 \equiv F_1(\hat{x}, t) \\ \dot{\hat{x}}_2 = (\hat{b} - 35\hat{\beta}(t))\hat{x}_1 - \hat{x}_1\hat{x}_3 + (29\hat{\beta}(t) - \hat{c})\hat{x}_2) + u_2 \equiv F_2(\hat{x}, t) \\ \dot{\hat{x}}_3 = \hat{x}_1\hat{x}_2 - \frac{\hat{\beta}(t)+\hat{d}}{3}\hat{x}_3 \equiv F_3(\hat{x}, t) \end{cases} \tag{36}$$

where $u_i = k_ie_i$ and $e_i = x_i - \hat{x}_i$ for $i = 1, 2$; $\hat{a}$, $\hat{b}$, $\hat{c}$, $\hat{d}$, and $\hat{\beta}(t) = \hat{\alpha}\sin(\hat{\phi})+\hat{\varepsilon}$ are the estimators for a, b, c, d and $\beta(t)$, respectively.

Since $\hat{a}$, $\hat{b}$ and $\hat{c}$ appear in the dynamical equations of measurable states x_1 or x_2 (i.e. s_1 or s_2), their updating rules can be designed as:

$$\begin{cases} \dot{\hat{a}} = \delta_1(\hat{x}_2 - \hat{x}_1)e_1 \\ \dot{\hat{b}} = \delta_2\hat{x}_1e_2 \\ \dot{\hat{c}} = \delta_3\hat{x}_2e_2 \end{cases} \tag{37}$$

On the other hand, parameters $\hat{\alpha}$, $\hat{\varphi}$ and $\hat{\varepsilon}$ appear in all the three state equations, but only the state information of x_1 and x_2 are now available. By considering both the dynamics of F_1 and F_2, one can obtain:

$$\begin{cases} \dot{\hat{\alpha}} \propto \sum_{k=1}^{2} \frac{\partial F_k(\hat{\mathbf{X}},t)}{\partial \hat{\alpha}} e_k \\ \dot{\hat{\phi}} \propto \sum_{k=1}^{2} \frac{\partial F_k(\hat{\mathbf{X}},t)}{\partial \hat{\phi}} e_k \\ \dot{\hat{\varepsilon}} \propto \sum_{k=1}^{2} \frac{\partial F_k(\hat{\mathbf{X}},t)}{\partial \hat{\varepsilon}} e_k \end{cases} \tag{38}$$

For the parameter $\hat{d}$, which only occurs in the third state equation (state equation of $\hat{x}_3$), while the corresponding state x_3 is not available, a further dependence should be derived and expressed as:

$$\dot{\hat{d}} \propto \frac{\partial F_2(\hat{\mathbf{X}},t)}{\partial \hat{x}_3} \frac{\partial F_3(\hat{\mathbf{X}},t)}{\partial \hat{d}} e_2 \tag{39}$$

Combining (36)–(39), the adaptive observer is finalized as follows:

$$\begin{cases} \dot{\hat{x}}_1 = (25\hat{\beta}(t) + \hat{a})(\hat{x}_2 - \hat{x}_1) + k_1 e_1 \\ \dot{\hat{x}}_2 = (\hat{b} - 35\hat{\beta}(t))\hat{x}_1 - \hat{x}_1\hat{x}_3 + (29\hat{\beta}(t) - \hat{c})\hat{x}_2 + k_2 e_2 \\ \dot{\hat{x}}_3 = \hat{x}_1\hat{x}_2 - \frac{\hat{\beta}(t)+\hat{d}}{3}\hat{x}_3 \\ \dot{\hat{a}} \; = \delta_1(\hat{x}_2 - \hat{x}_1)e_1 \\ \dot{\hat{b}} \; = \delta_2\hat{x}_1 e_2 \\ \dot{\hat{c}} \; = \delta_3\hat{x}_2 e_2 \\ \dot{\hat{d}} \; = \delta_4 \frac{\hat{x}_1\hat{x}_3}{\sigma+|\hat{x}_1|} e_2 \\ \dot{\hat{\alpha}} \; = \delta_5 \sin(\hat{\phi})[25(\hat{x}_2 - \hat{x}_1)e_1 - (35\hat{x}_1 - 29\hat{x}_2)e_2] \\ \dot{\hat{\phi}} \; = \delta_6 \cos(\hat{\phi})[25(\hat{x}_2 - \hat{x}_1)e_1 - (35\hat{x}_1 - 29\hat{x}_2)e_2] \\ \dot{\hat{\varepsilon}} \; = \delta_7[25(\hat{x}_2 - \hat{x}_1)e_1 - (35\hat{x}_1 - 29\hat{x}_2)e_2] \end{cases} \tag{40}$$

where $\delta_i, i = 1, \cdots, 7$ are positive constants.

It should be emphasized that an auxiliary function $\mu = \frac{1}{\sigma+|\hat{x}_1|}$ has been included in the dynamical equation of $\hat{d}$, so that similar convergence rates are obtained for all the error dynamics. Detailed calculation can be referred to [44].

Due to the chaotic properties of system (35), the condition of *persistent excitation*, which is commonly required for system identification, is in general satisfied. However, as reflected in Eqn. (35), $\beta(t)$ is linearly dependent on parameters a, b, c and d and hence direct recovery of $\beta(t)$ and a, b, c and d is impossible [32]. Instead, by defining the parameter vector $\mathbf{q} = [25\hat{\beta} + \hat{a} \quad \hat{b} - 35\hat{\beta} \quad 29\hat{\beta} - \hat{c} \quad \hat{\beta} + \hat{d}]^T$, it is possible to obtain a vector $\mathbf{q}$ being very close to its true value $\mathbf{p} = [25\beta + a \quad b - 35\beta \quad 29\beta - c \quad \beta + d]^T$.

When synchronization is established, i.e. the systems states $\hat{x}_i$ converge to x_i, and the parameter vector $\mathbf{q}$ converges to $\mathbf{p}$ as $t \to \infty$, the information signal can be derived based on the following relationship

$$\hat{m}(t) = s_3(t) - \hat{x}_3(t) \tag{41}$$

where $s_3(t) = m(t) + x_3(t)$ is currently measurable.

Now assuming that $a = 9.5$, $b = 27$, $c = 0.9$ and $d = 8$ and $\beta(t) = \frac{1+\sin(7t)}{2}$, Fig. 23 depicts the simulation results with feedback gains $k_1 = k_2 = 120$,

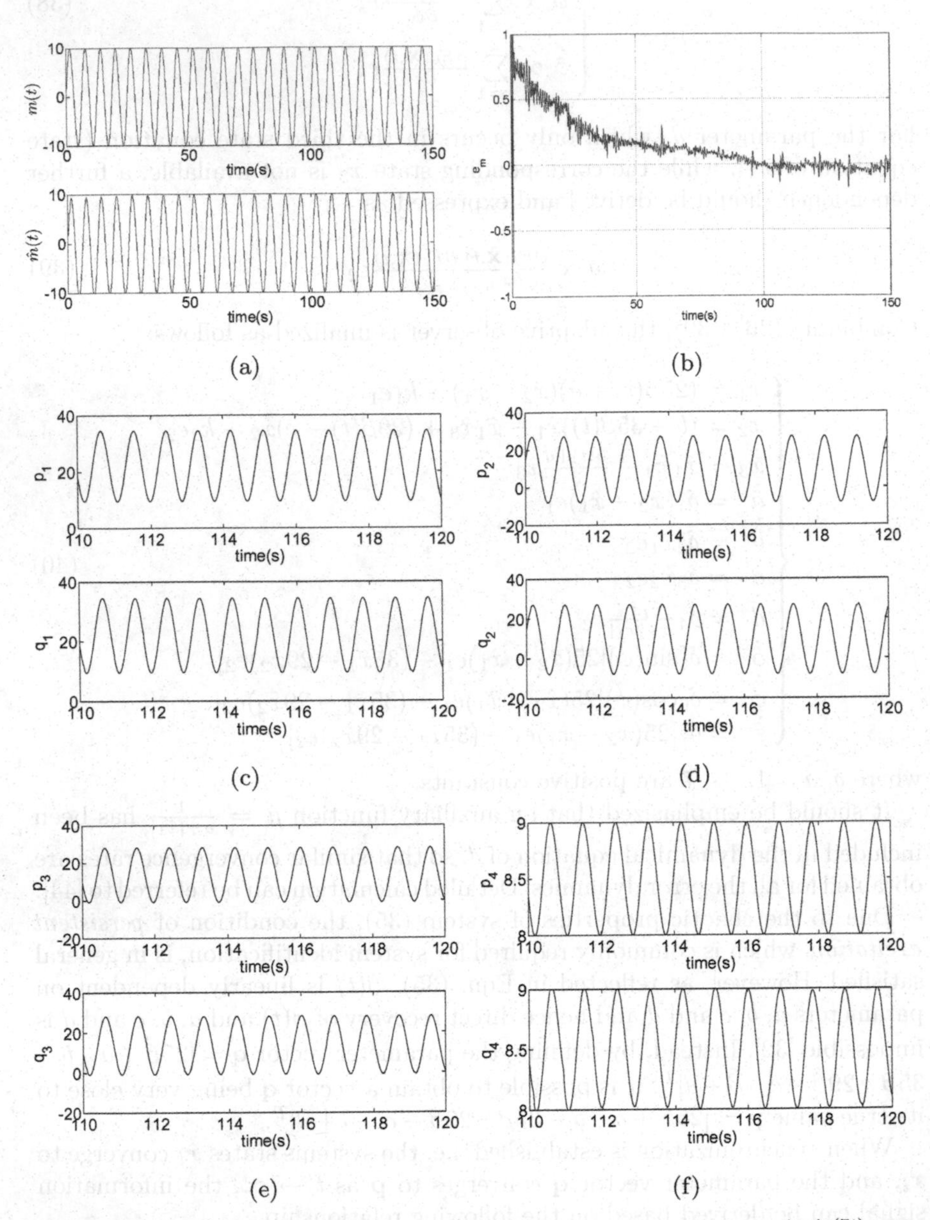

Fig. 23 Simulation results for chaotic masking system with $\beta(t) = \frac{1+\sin(7t)}{2}$ and $m(t) = 10\sin(t)$: (a) The information signal $m(t)$ and recovered signal $\hat{m}(t)$; (b) signal error $e_m = m(t) - \hat{m}(t)$; (c) p_1 and its estimator q_1; (d) p_2 and its estimator q_2;(e) p_3 and its estimator q_3; (f) p_4 and its estimator q_4

$\delta_i = 50$ for $i = 1, \cdots, 7$ and $\sigma = 5$. The initial conditions are arbitrarily chosen as: $\hat{a}^{(0)} = 12$, $\hat{b}^{(0)} = 20$, $\hat{c}^{(0)} = 2$, $\hat{d}^{(0)} = 6$, $\hat{\alpha}^{(0)} = 0.01$, $\hat{\phi}^{(0)} = 1$ and $\hat{\varepsilon}^{(0)} = 0$.

From Fig. 23 (a), it is found that the plaintext $m(t)$ and its recovered signal $\hat{m}(t)$ are very similar. As given in Fig. 23(b), it takes a few seconds for $\hat{m}(t)$ to become closely matched with $m(t)$. The final encrypted message $m(t)$ can be estimated within about 2% relative errors after 100s. As a reference, the elements of $\mathbf{q}$ are also depicted in Figs. 23(c)-(f), respectively, which well represent the true ones. Therefore, the entire structure of the whole cryptosystem can be revealed. It is also worth to point out that, similar performance is obtained for different information signals, since the dynamics of the observer is largely controlled by the error signals, e_1 and e_2, but is irrelevant to the information signal.

Note that the magnitude of the information signal used in the previous example is relatively large. In practical case, it is about -20dB to -30dB of the chaotic carrier [38]. Otherwise, the plaintext presence in the ciphertext may be easily revealed by a simple filter design [45, 42] or even from our naked eyes.

Therefore, another simulation with a weak information signal together with a more complex secret key $\beta(t)$ are carried out. Taking $m(t) = 2\sin(t)$ and $\beta(t) = \frac{3+\sin(t)+\sin(4t)+\sin(7t)}{6}$, the simulation results are presented in Fig. 24. To obtain a faster convergence rate, a larger gain with $\delta_7 = 500$ has been adopted. It is noticed that a similar performance is obtained as compared with our previous simulation. As a conclusive remark, the security of these cryptosystems are found to be questionable.

3.2 Cryptanalysis of Chaotic Modulation Scheme

The second scheme to be investigated is the chaotic modulation scheme. Its design idea is to inject information into a chaotic system or modulate it by means of an invertible transformation so that spread spectrum transmission can be achieved. The transmitter in chaotic modulation is switched among different trajectories of the same chaotic attractor, while at the receiver end, the information is revealed by an inverses process.

In this section, similar to Sec. 3.1, a number of cryptosystems based on Lorenz system [22, 41] and Unified system [5] are studied.

Cryptanalysis of Lorenz-Based Chaotic Modulation Scheme

In the first example, we will focus on those systems proposed in [22, 41], where Lorenz attractor is used in the chaos-based cryptosystem.

System I: The communication scheme proposed in [22] can be expressed as:

$$M : \begin{cases} \dot{x_1} = p_1(x_2 - x_1) \\ \dot{x_2} = p_2 x_1 - x_2 - x_1 x_3 \\ \dot{x_3} = x_1 x_2 - \tilde{p}_3 x_3 \end{cases} \tag{42}$$

where $\tilde{p}_3 = p_3 + m(t) \times \Delta$, $m(t) = \pm 1$ and Δ is a constant.

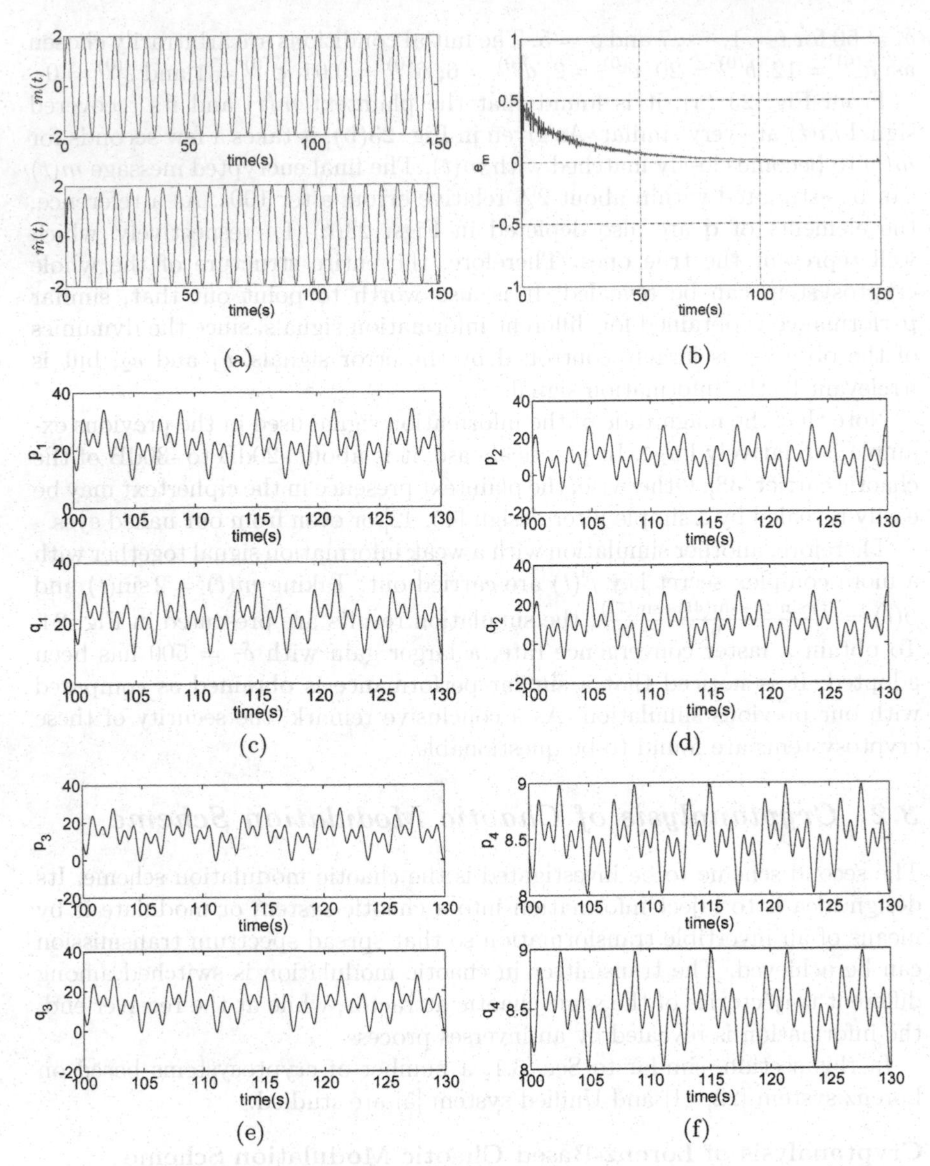

Fig. 24 Simulation results for chaotic masking system with $\beta(t) = \frac{3+\sin(t)+\sin(4t)+\sin(7t)}{6}$ and $m(t) = 2\sin(t)$: (a) The information signal $m(t)$ and recovered signal $\hat{m}(t)$; (b) signal error $e_m = m(t) - \hat{m}(t)$; (c) p_1 and its estimator q_1; (d) p_2 and its estimator q_2;(e) p_3 and its estimator q_3; (f) p_4 and its estimator q_4

The parameters p_i, $i = 1, 2, 3$ are considered to be the secret keys, and p_3 is modulated by the message $m(t)$ while the state x_1 is transmitted. In our simulation, the same parameters set in [22] and [25], i.e. $p_1 = 16$, $p_2 = 45.6$, $p_3 = 4.2$ and $\Delta = 0.2$, are used.

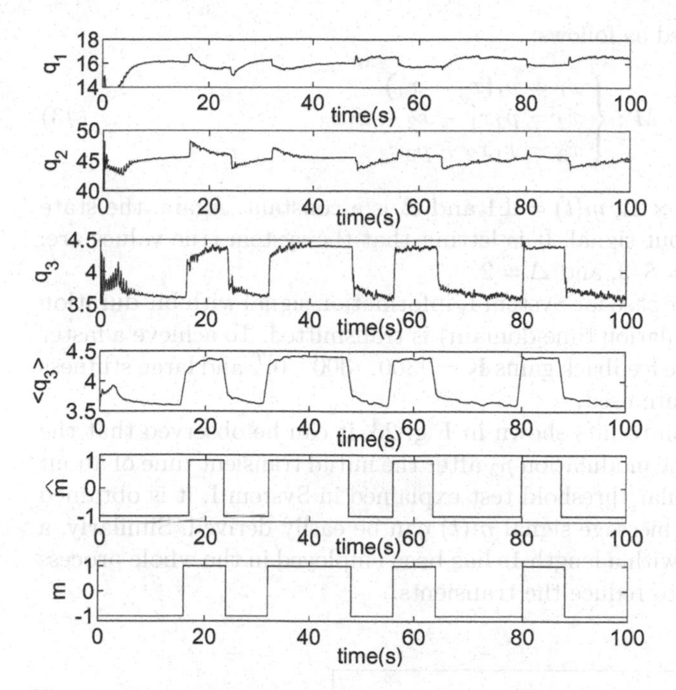

Fig. 25 Recovered parameters against time. From the top to the bottom, q_1,q_2,q_3, $< q_3 >$ obtained by using a moving average filter, recovered message $\hat{m}(t)$ and the plaintext $m(t)$

The same adaptive observer as designed in (22) [46] can be used as an attack of (42), to recover the system parameters and the information signals. To get a fast convergence, larger stiffness constants $\delta_{1,3} = 24$, $\delta_2 = 6$ and feedback gains $\mathbf{K} = [100 \quad 100 \quad 0]^T$ are used.

Figure 25 shows the results based on the bit duration time of 8s (in the simulation time domain, τ). This bit duration time is equivalent to 3.2ms in real system, as a time-scale transformation $t = \tau/T_0$, where $T_0 = 2505$ is required as explained in [22]. Comparing with the attack suggested in [25], where a wrong estimation of $\tilde{p}_3$ has been obtained, the value of q_3 now closely follows the modulated value $\tilde{p}_3$. By taking the average of the maximum and minimum values of q_3 after the initial transient time (in our case, it is taken as 20s), a coarse estimate of $p_3 \approx 4.2$ can be obtained. This is then employed as the threshold to recover the message $m(t)$. Note that the value of the threshold is not exclusive and strict. Generally, it can be chosen as the mean of the maximum and the minimum of the modulated parameters (after the initial transient time). To further improve the performance, a moving average filter with a length of 1s is used.

System II: In our second example, the system described in [41] is studied. The message is used to modulate the parameter p_2 in the Lorenz system, which can be described as follows:

$$M : \begin{cases} \dot{x_1} = p_1(x_2 - x_1) \\ \dot{x_2} = \tilde{p}_2 x_1 - x_2 - x_1 x_3 \\ \dot{x_3} = x_1 x_2 - p_3 x_3 \end{cases} \quad (43)$$

where $\tilde{p}_2 = p_2 + m(t) \times \Delta$, $m(t) = \pm 1$ and Δ is a constant. Again, the state x_1 is used as the output signal. It is letting that the system true values are: $p_1 = 10$, $p_2 = 30$, $p_3 = 8/3$, and $\Delta = 2$.

Similar to the above case for System I, information signal with bit duration time of 8s (in the simulation time domain) is transmitted. To achieve a faster convergence rate, large feedback gains $\mathbf{K} = [300 \quad 300 \quad 0]^T$ and large stiffness constants $\delta_{1,2,3} = 24$ are used.

From the simulation results shown in Fig. 26, it can be observed that the estimator q_2 follows the modulation $\tilde{p}_2$ after the initial transient time of about 20s. By using the similar threshold test explained in System I, it is obtained that $p_2 = 30$ and the message signal $m(t)$ can be easily derived. Similarly, a moving average filter with a length 1s has been employed in the whole process for message recovery to reduce the transients.

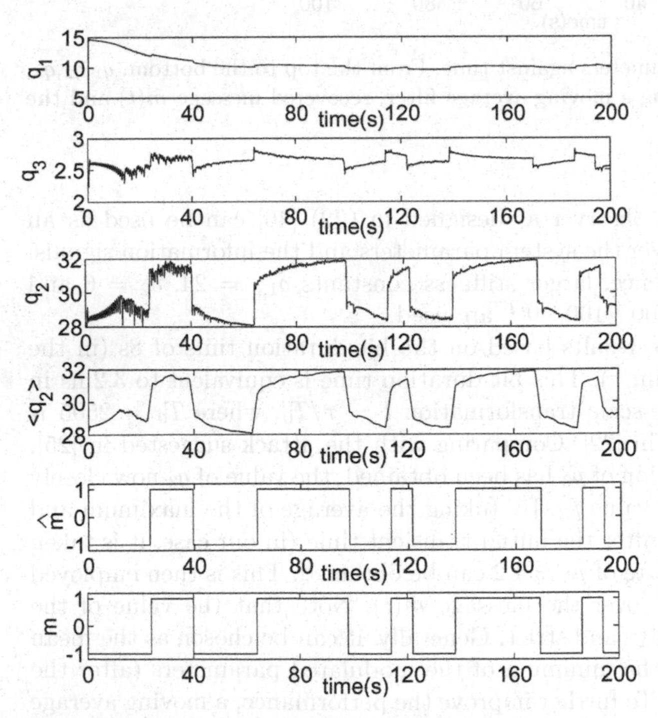

Fig. 26 Recovered parameters against time. From the top to the bottom, q_1,q_3,q_2, $< q_2 >$ obtained by using a moving average filter, recovered message $\hat{m}(t)$ and the plaintext $m(t)$

Cryptanalysis of Unified System-Based Chaotic Modulation Scheme

In [5], a chaotic modulation scheme has also been proposed with the use of a unified system. The transmitter of the cryptosystem is formulated as below:

$$\begin{cases} \dot{x_1} = (25\beta(t) + a)(x_2 - x_1) \\ \dot{x_2} = (b - 35\beta(t))x_1 - x_1x_3 + (29\beta(t) - c)x_2 \\ \dot{x_3} = x_1x_2 - \frac{\beta(t)+d}{3}x_3 + m(t) \end{cases} \tag{44}$$

where $m(t)$ is the plaintext and $\mathbf{s} = [x_1 \quad x_2 \quad x_3 + m]^T$ are the transmitted signals. Similar to the case expressed in (35), a, b, c, d are system parameters and $\beta(t)$ is the secret key. All of them are assumed to be kept secret for the attacker.

An adaptive observer expressed as in (45) is suggested to perform the attack:

$$\begin{cases} \dot{\hat{x}}_1 = (25\hat{\beta}(t) + \hat{a})(\hat{x}_2 - \hat{x}_1) + k_1e_1 \\ \dot{\hat{x}}_2 = (\hat{b} - 35\hat{\beta}(t))\hat{x}_1 - \hat{x}_1\hat{x}_3 + (29\hat{\beta}(t) - \hat{c})\hat{x}_2 + k_2e_2 \\ \dot{\hat{x}}_3 = \hat{x}_1\hat{x}_2 - \frac{\hat{\beta}(t)+\hat{d}}{3}\hat{x}_3 + k_3e_3 \\ \dot{\hat{a}} = \delta_1(\hat{x}_2 - \hat{x}_1)e_1 \\ \dot{\hat{b}} = \delta_2\hat{x}_1e_2 \\ \dot{\hat{c}} = \delta_3\hat{x}_2e_2 \\ \dot{\hat{d}} = \delta_4\frac{\hat{x}_1\hat{x}_3}{\sigma+|\hat{x}_1|}e_2 \\ \dot{\hat{\alpha}} = \delta_5\sin(\hat{\phi})[25(\hat{x}_2 - \hat{x}_1)e_1 - (35\hat{x}_1 - 29\hat{x}_2)e_2] \\ \dot{\hat{\phi}} = \delta_6\cos(\hat{\phi})[25(\hat{x}_2 - \hat{x}_1)e_1 - (35\hat{x}_1 - 29\hat{x}_2)e_2] \\ \dot{\hat{\varepsilon}} = \delta_7[25(\hat{x}_2 - \hat{x}_1)e_1 - (35\hat{x}_1 - 29\hat{x}_2)e_2] \end{cases} \tag{45}$$

Note that $e_i = x_i - \hat{x}_i$ for $i = 1, 2$ and $e_3 = s_3 - \hat{x}_3 = x_3 + m - \hat{x}_3$. Once synchronization is achieved, the plaintext can then be recovered by subtracting $\hat{x}_3$ from the transmitted signal s_3, based on the same calculation given in (41).

In our simulation, a more challenging case as comparing with those used in [5, 44] is considered. It is letting that a weak information signal $m(t) = 2\cos(t)$ is to be transmitted, and the secret key is a multitonic signal expressed as $\beta(t) = \frac{3+\sin(t)+\sin(4t)+\sin(7t)}{6}$. Taking $a = 10$, $b = 28$, $c = 1$ and $d = 8$, Fig. 27 presents the simulation results with $k_1 = k_2 = 120$ and $k_3 = 1$. It clearly shows that the sending information can be retrieved within 1% estimation error after the initial transient time.

Figure 28 depicts another case where information is a frequency rich chaotic signal, that obtained by normalizing a state of a Lorenz system. From the numerical simulation, it is clearly demonstrated that the proposed adaptive observer can also tackle with this case while the secret system parameters are also obtained simultaneously.

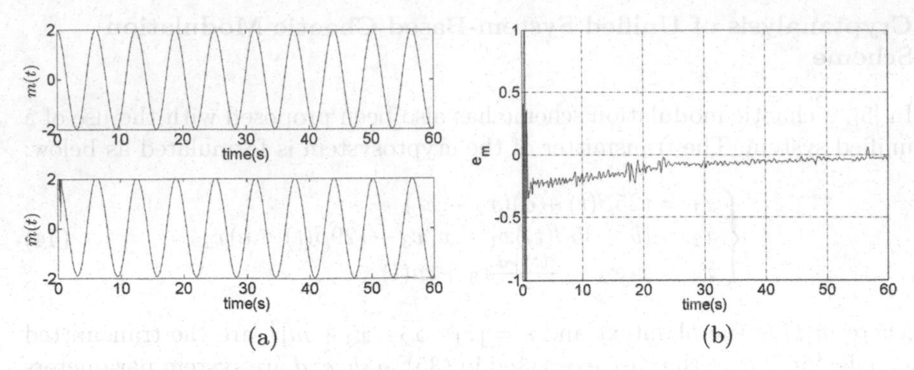

Fig. 27 Simulation results for chaotic modulation system with $\beta(t) = \frac{3+\sin(t)+\sin(4t)+\sin(7t)}{6}$ and $m(t) = 2\cos(t)$: (a) the information signal $m(t)$ and recovered signal $\hat{m}(t)$; (b) signal error $e_m = m(t) - \hat{m}(t)$

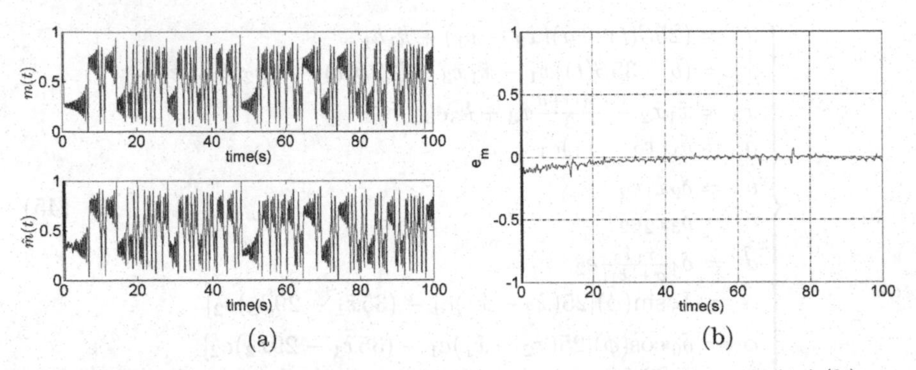

Fig. 28 Simulation results for chaotic modulation system with $\beta(t) = \frac{1+\sin(3t)}{2}$ and $m(t)$ is a chaotic signal obtained from the first state of another Lorenz system:(a)the information signal $m(t)$ and recovered signal $\hat{m}(t)$; (b) signal error $e_m = m(t) - \hat{m}(t)$

3.3 Cryptanalysis of Chaotic Shift Keying Scheme

The last scheme to be testified is the chaotic shift keying scheme. It can be considered as a special case of chaotic modulation, where binary signal is used to switch the parameter(s) of a system between two statistically similar chaotic attractors.

The design given in [5] as shown below is investigated:

$$\begin{cases} \dot{x}_1 = (25\beta(t) + a)(x_2 - x_1) \\ \dot{x}_2 = (b - 35\beta(t))x_1 - x_1x_3 + (29\beta(t) - m(t))x_2 \\ \dot{x}_3 = x_1x_2 - \frac{\beta(t)+d}{3}x_3 \end{cases} \tag{46}$$

where the system output is defined as $\mathbf{s} = [x_1 \quad x_2 \quad x_3]^T$. As compared with the unified system given in (35), it can be noticed that the parameter c in (35) is now modulated by a binary signal $m(t)$.

To perform an effective attack and follow the design procedures given in Sec. 2.2, an adaptive observer is formed as follows:

$$\begin{cases} \dot{\hat{x}}_1 = (25\hat{\beta}(t) + \hat{a})(\hat{x}_2 - \hat{x}_1) + k_1 e_1 \\ \dot{\hat{x}}_2 = (\hat{b} - 35\hat{\beta}(t))\hat{x}_1 - \hat{x}_1\hat{x}_3 + (29\hat{\beta}(t) - \hat{c})\hat{x}_2 + k_2 e_2 \\ \dot{\hat{x}}_3 = \hat{x}_1\hat{x}_2 - \frac{\hat{\beta}(t)+\hat{d}}{3}\hat{x}_3 + k_3 e_3 \\ \dot{\hat{a}} \ = \delta_1(\hat{x}_2 - \hat{x}_1)e_1 \\ \dot{\hat{b}} \ = \delta_2\hat{x}_1 e_2 \\ \dot{\hat{c}} \ = \delta_3\hat{x}_2 e_2 \\ \dot{\hat{d}} \ = \delta_4\hat{x}_3 e_3 \\ \dot{\hat{\alpha}} \ = \delta_5 \sin(\hat{\phi})[25(\hat{x}_2 - \hat{x}_1)e_1 - (35\hat{x}_1 - 29\hat{x}_2)e_2 - \frac{1}{3}\hat{x}_3 e_3] \\ \dot{\hat{\phi}} = \delta_6 \cos(\hat{\phi})[25(\hat{x}_2 - \hat{x}_1)e_1 - (35\hat{x}_1 - 29\hat{x}_2)e_2 - \frac{1}{3}\hat{x}_3 e_3] \\ \dot{\hat{\varepsilon}} \ = \delta_7[25(\hat{x}_2 - \hat{x}_1)e_1 - (35\hat{x}_1 - 29\hat{x}_2)e_2 - \frac{1}{3}\hat{x}_3 e_3] \end{cases} \tag{47}$$

The same parametric settings as that employed in [5] and [44] are used, except that the user-secret key is defined as $\beta(t) = \frac{1+\sin(7t)}{2}$ and $\delta_i = 150$ for $i = 1, \cdots, 7$.

Obviously, if $m(t)$ is a constant, generalized synchronization can be established. Even if it is time varying but sufficiently slow, the synchronization errors can still converge to a small value and coarse estimation of the linear-independent system parameters is possible, as shown in Fig. 29. It is also noticed that the parameter $\hat{c}(t)$ closely follows the modulation information $m(t)$, but with a shift of level. Therefore, $m(t)$ can be obtained by detecting the switching of the corresponding estimator $\hat{c}(t)$, as demonstrated in

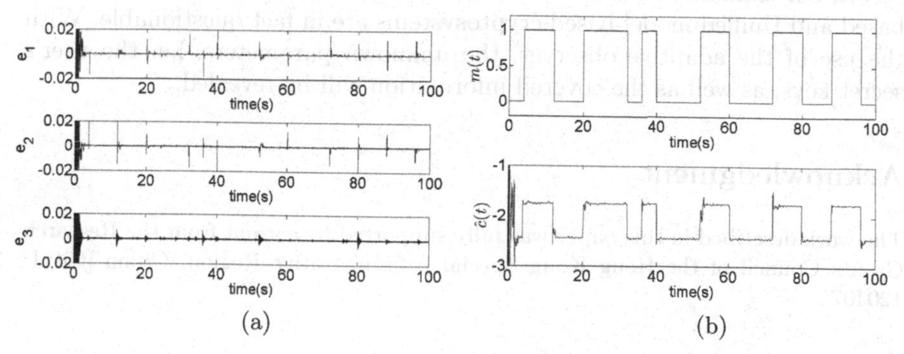

Fig. 29 Simulation results for chaotic shift key scheme with secret key $\beta(t) = \frac{1+\sin(7t)}{2}$, $\delta_i = 150$ for $i = 1, \cdots, 7$: (a) State errors e_i; (b) information signal $m(t)$ and parameter estimator $\hat{c}(t)$

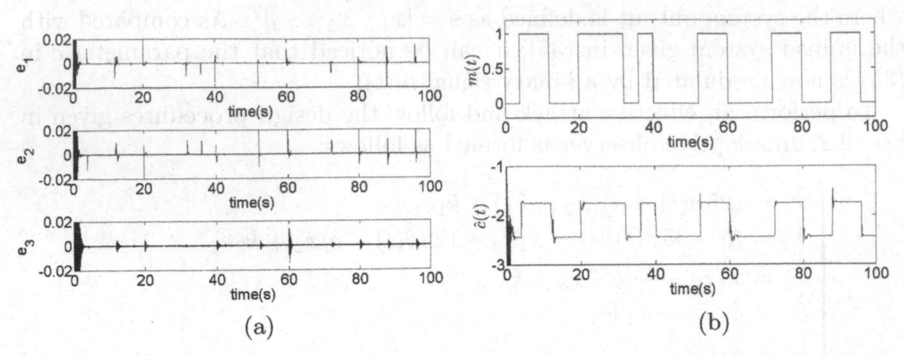

Fig. 30 Simulation results for chaotic shift key scheme with secret key $\beta(t) = \frac{3+\sin(t)+\sin(4t)+\sin(7t)}{6}$, $\delta_i = 150$ for $i = 1, \cdots, 7$: (a) State errors e_i; (b) information signal $m(t)$ and parameter estimator $\hat{c}(t)$

our simulation. Similar results are obtained as shown in Fig. 30 with a more complex multitonic secret key $\beta(t) = \frac{3+\sin(t)+\sin(4t)+\sin(7t)}{6}$.

4 Conclusion

In this chapter, a new design of adaptive observer has been proposed for adaptive synchronization of a class of chaotic sytems with multiple unknown parameters. Its effectiveness is well demonstrated by tackling several difficult synchronization problems found in chaotic systems, such as Lorenz system, Rössler system and Genesio System.

The new adaptive observer is then used as an attack for some chaos-based cryptosystems. By assuming the Kerckhoff's principle, i.e. the basic structure of a cryptosystem is known but some user-specified parameters remain in secret, our simulation results have shown that the security of many Lorenz-based and Unified model-based cryptosystems are in fact questionable. With the use of the adaptive observer, the unknown parameters, i.e. the user's secret keys, as well as the covered information will be revealed.

Acknowledgment

The work described in this paper was fully supported by a grant from the Research Grants Council of the Hong Kong Special Administrative Region, China [CityU 120407].

References

1. Pecora, L.M., Carroll, T.L.: Phys. Rev. Lett. 64, 821 (1990)
2. Carroll, T.L., Pecora, L.M.: IEEE Trans. Circuits Syst. I 38, 453 (1991)

3. Kocarev, L., Halle, K.S., Eckert, K., et al.: Int. J. Bifurcation Chaos Appl. Sci. Eng. 2, 709 (1992)
4. Wu, X.: Chaos 16, 043118 (2006)
5. Yu, W., Cao, J., Wong, K.W., Lü, J.: Chaos 17, 033114 (2007)
6. Besancon, G.: Systems Control Letters 41, 271 (2000)
7. Huang, D.: Phys. Rev. E 73, 066204 (2006)
8. Huang, D., Xing, G., Wheeler, D.: Chaos 17, 023118 (2007)
9. Parlitz, U.: Phys. Rev. Lett. 76, 1232 (1996)
10. Chen, S., Hu, J., Wang, C., et al.: Phys. Lett. A 321, 50 (2004)
11. Maybhate, A., Amritkar, R.E.: Phys. Rev. E 59, 284 (1999)
12. Maybhate, A., Amritkar, R.E.: Phys. Rev. E 61, 6461 (2000)
13. Lü, J., Cao, J.: Chaos 15, 043901 (2005)
14. Koshkouei, A.J., Zinober, A.S.I.: Int. J. Control 77, 118 (2004)
15. Freitas, S.U., Macau, E.E.N., Grebogi, C.: Phys. Rev. E 71, 047203 (2005)
16. Chen, M., Kurths, J.: Physical Review E. 76, 027203 (2007)
17. Isidori, A.: Nonlinear Control Systems. Springer, New York (1995)
18. Itoh, M., Wu, C.W., Chuo, L.O.: Int. J. Bifurcation Chaos Appl. Sci. Eng. 7, 275 (1997)
19. Cuomo, K.M., Oppenheim, A.V., Strogatz, S.H.: IEEE Trans. Circuits Syst. II: Analog Digital Signal Process 40, 626 (1993)
20. Liao, T., Huang, N.: IEEE Trans. Circuits Syst. I 46, 1144 (1999)
21. Liao, T., Tsai, S.: Chaos, Solitons Fractals 11, 1387 (2000)
22. Cuomo, K.M., Oppenheim, A.V.: Phys. Rev. Lett. 71, 65 (1993)
23. Feki, M.: Chaos, Solitons and Fractals 18, 141 (2003)
24. Zhu, F.: Chaos, Solitons & Fractals (in Press); Corrected Proof, Available online (2008)
25. Zhou, C.S., Lai, C.H.: Phys. Rev. E 59, 6629 (1999)
26. vCelikovský, S., Chen, G.R.: IEEE Trans. on Automatic Control 50, 76 (2005)
27. Besancon, G., León-Morales, J., Huerta-Guevara, O.: Int. J. Control 79, 581 (2006)
28. Zhang, Q., Xu, A.: Proc. American Control Conf. 1551 (2006)
29. Liu, Y., Tang, W.K.S., Kocarev, L.: Int. J. Bifurc. Chaos 18, 2415 (2008)
30. John, J.K., Amritkar, R.E.: Int. J. Bifurcation Chaos Appl. Sci. Eng. 4, 1987 (1994)
31. John, J.K., Amritkar, R.E.: Phys. Rev. E 49, 4843 (1994)
32. Yu, W., Chen, G., Cao, J., Lü, J., et al.: Phys. Rev. E 75, 067201 (2007)
33. Hoppensteadt, F.C.: Trans. Amer. Math. Soc., 123, 521 (1966)
34. Hoppensteadt, F.C.: Comm. Pure Appl. Math. 24, 807 (1971)
35. Zhou, C.S., Lai, C.H.: Physica D 135, 1 (2000)
36. Gauthier, D.J., Bienfang, J.C.: Phys. Rev. Lett. 77, 1751 (1996)
37. Hu, G.J., Feng, Z.J., Meng, R.L.: IEEE Trans. Circuits Syst. I 50, 275 (2003)
38. Yang, T.: International Journal of Computational Cognition 2, 81 (2004)
39. Boutayeb, M., Darouach, M., Rafaralahy, H.: IEEE Trans. Circuits Syst. I 49, 345 (2002)
40. Liu, Y., Tang, W.K.S.: IEEE Trans. on Circuits and Systems II 55(11) (2008)
41. Song, Y., Yu, X.: Proc. IEEE Conf. Decision and Control 42 (2000)
42. Alvarez, G., Montoya, F., Romera, M., Pastor, G.: IEEE Trans. on Circuits and Systems II 51(10) (2004)

43. Lü, J., Chen, G., Cheng, D., vCelikovský, S.: Int. J.Bifurcation Chaos Appl. Sci. Eng. 12 (2002)
44. Liu, Y., Mao, Y., Tang, W.K.S.: To be scheduled on Int. J. Bifurcation Chaos Appl. Sci. Eng. 19(7) (2009)
45. Yang, T., Yang, L., Yang, C.: Phys. Rev. A 247, 105 (1998)
46. Liu, Y., Tang, W.K.S.: The First International Workshop on Nonlinear Dynamics and Synchronization 197 (2008)

Trust-Based Collaborative Control for Teams in Communication Networks

P. Ballal, F.L. Lewis, and G.R. Hudas

1 Introduction

Trust establishment in distributed communication networks such as mobile ad hoc networks (MANETs), sensor networks and ubiquitous computing systems is considered to be more difficult than in traditional hierarchical structures such as the Internet and Wireless LANs centered on base-stations and access nodes. Teams in disaster response and elsewhere may be heterogeneous networks consisting of interacting humans, ground sensors, and unmanned airborne or ground vehicles (UAV, UGV). Developed military team scenarios include the War-fighter Information Network-Tactical (WIN-T Figure 1), DARPA Agile Information Control Environment (AICE), C4ISR Architectures for the War-fighter (CAW), Joint Force Air Component Commander (JFACC) Project, etc. Such scenarios should provide intelligent shared services of sensors and mobile nodes to augment the capabilities of the remote-site mission commander and on-site personnel in terms of extended sensing ranges, sensing of modalities such as IR and ultrasound not normally open to humans, and cooperative control of UAV/UGV to extend the range of human team members. Also Automated decision assistance (e.g., handheld PDAs) should be provided to onsite personnel based on algorithms that only depend on local information from nearest neighbor sensor nodes or humans, yet yield network-wide guaranteed performance.

There is a need to provide means for inter-acting teams to develop trust through the extensive use of simulation, scenario-driven games, experiments, and training exercises that challenge leaders and reduce the need to learn "on the job" in actual operations (U.S. Army Training and Doctrine Command (TRADOC) Pamphlet 525-66). As the team members may often be geographically distributed there will be a heightened need for shared conceptualization of teamwork built on trust.

P. Ballal and F.L. Lewis
Fellow, IEEE,
Automation & Robotics Research Institute, University of Texas at Arlington,
7300 Jack Newell Blvd. S., Fort Worth, TX 76118-7115, USA

G. R. Hudas
US Army RDECOM-TARDEC AMSRD-TAR-R
Bldg 200C, Mail Stop 263, 6501 E. 11 Mile Rd., Warren, MI, 48397-5000, USA

K. Kyamakya (Eds.): Recent Adv. in Nonlinear Dynamics and Synchr., SCI 254, pp. 347 – 363.
springerlink.com

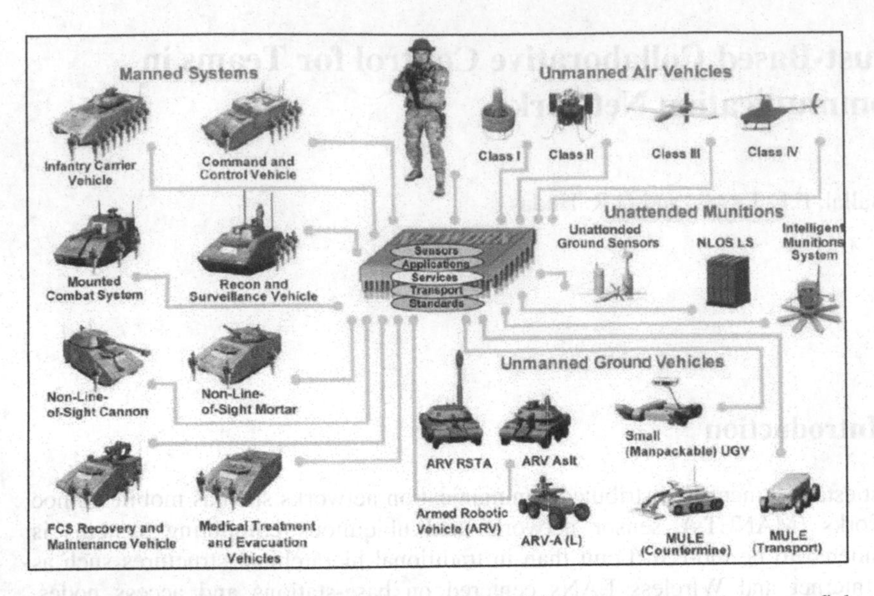

Fig. 1 WIN-T network has ground sensors, UGV, UAV cooperating with the war-fighter

Given the presence of malicious or enemy components and the possibility of node compromise, a *trust consensus* must be reached by the team that determines which nodes to trust, which to disregard, and which to avoid. Trust algorithms for unmanned nodes must be autonomous computationally efficient numerical schemes. However, existing schemes for control of dynamical systems on communications graphs (in the style of work by (Beard and Stepanyan, 2003; Fax and Murray, 2004; Jadbabaie et al., 2003; Lee and Spong, 2007; Olfati-Saber and Murray, 2004; Ren and Beard, 2005; Ren et al., 2005; Saligrama et al., 2006.)) do not take into account trust propagation and maintenance (such as work by (Jiang and Baras, 2006; Theodorakopoulos and Baras, 2006)). Yet it is a fact that biological groups such as flocks, swarms, herds (Reynolds, 1987), do have built-in trust mechanisms to identify team members, team leaders, and enemies to be treated as obstacles or avoided. Cooperative mission planning should involve decisions made in the context of the trust opinions of all nodes, and be based on performance criteria set by human personnel nodes or team leaders. These performance criteria may change with time depending on varying mission objectives in the field.

Recently, many researchers have worked on problems that are essentially different forms of agreement problems with differences in the types of agent dynamics, properties of graphs and the names of the tasks of interest. In (Fax and Murray, 2004), graph Laplacians were used for the task of formation stabilization for groups of agents with linear dynamics. In (Jadbabaie et al., 2003) directed graphs were be used to represent the information exchange between the agents. In (Beard and Stepanyan, 2003), a linear update scheme was introduced for directed graphs. In (Chopra and Spong, 2006) a Lyapunov-based approach was used to consider stability of consensus synchronization for balanced and weakly connected networks. The work by (Olfati-Saber and Murray, 2004) solved the average consensus problem with directed graphs which required the graph to be strongly connected and balanced. In (Ren and Beard, 2005), it was shown that under certain assumptions consensus can be reached asymptotically under dynamically changing interaction topologies if the union of the collection of interaction graphs across some time intervals has a spanning tree frequently enough. The spanning tree requirement is a milder condition than connectedness and is therefore suitable for practical applications. They also allowed the link weighting factors to be time-varying which provides additional flexibility.

In contrast to the aforementioned protocols, this chapter presents a bilinear protocol for trust consensus in directed graphs where the link weights are state-dependent and time-varying. In this chapter, we develop a framework for trust propagation and maintenance in team networks of nodes that yields global consensus of trust under rich enough communication structure graphs. Most of the work in literature considers the graph Laplacian to be static or have time-varying weights which are due to unreliable transmission or limited communication or sensing range. In this chapter we consider the case where the graph Laplacian is a time-varying function of the trusts based on the graph connectivity. This makes the trust consensus protocol bilinear.

There has been a tremendous amount of interest in flocking and swarming that has primarily originated from the pioneering work of Reynolds, 1987. The trust consensus protocols developed in this chapter is incorporated into cooperative control laws that depend on local information from neighboring nodes, yet yield team-wide desired emergent behavior such as flocking and formations.

2 Trust Propagation in Graphs

2.1 Trust Graphs

Motivated by (Jiang and Baras, 2006; Theodorakopoulos and Baras, 2006), consider a network of N agents or nodes $V=\{v_1,\ldots,v_N\}$ who are to engage in cooperative trust evaluation. Define a trust graph $G_T = (V, E)$, where edge $(v_i, v_j)\in E$ if node v_j obtains a direct trust evaluation about node v_i. Note this is backwards from (Jiang and Baras, 2006; Theodorakopoulos and Baras, 2006). Define the direct trust neighborhood of node v_i as $N_i = \{v_j : (v_j, v_i) \in E\}$, i.e. the set of nodes with edges incoming to v_i. The graph is directed since if node j can obtain a direct evaluation of trust about node i, the reverse may not be true. Given the trust graph, define the graph adjacency matrix $A = [a_{ij}]$ where $a_{ij} = 1$ if e_{ji} is an edge, and $a_{ij} = 0$ otherwise. A is a constant matrix defined by the direct trust relations between nodes that determines the communication connectivity and structure in a team. In fact, adjacency matrix A captures the information flow in the trust graph. If there is a directed path, e.g. a sequence of nodes $v_0, v_1, \cdots, v_r$ such that $(v_i, v_{i+1}) \in E, i \in \{0,1,\cdots,r-1\}$, then, node v_r should be able to form an indirect trust opinion about node v_0 based on the opinions of the agents along the path. Likewise, if two paths converge at an agent v_r, each of which contains agent v_0, then v_r has a basis to form a more confident opinion about the trustworthiness of agent v_0 than if there were only a single path.

2.2 Trust Consensus Protocols

We encode the trust opinions an agent i has about other agents in the network as a trust vector $\xi_i \in R^N$ associated with each *node*, with elements indexed by all the nodes about which node i has an opinion. That is $\xi_i = [\xi_{ii1}\ \xi_{ii2} \ldots]^T$ where ξ_{ij} is the trust node i has for node j. Throughout this chapter the trust values ξ_{ij} are assumed to be in [0, 1], with 1 denoting maximum trust, and 0 denoting no opinion.

Consider the following trust protocol in continuous-time.

$$\dot{\xi}_i = u_i \tag{1}$$

$$u_i = \sum_{j \in N_i} w_{ij}(\xi_j - \xi_i) \tag{2}$$

In (Ren and Beard, 2005), w_{ij} was taken as $a_{ij}\sigma_{ij}$ where σ_{ij} is a time-varying weighting factor chosen from any finite set. In (Jiang and Baras, 2006), w_{ij} was taken as $a_{ij}c_{ij}$, where c_{ij} is the confidence node i has in its trust opinion of node j. Hence each node has an associated [ξ,c], i.e. trust and confidence each of which have two operations ($\oplus, \otimes$) which form a semi-group (Theodorakopoulos and Baras, 2006). In (Jiang and Baras, 2006), the weights c_{ij} were kept constant throughout.

In this chapter, we propose the following local voting continuous-time trust protocol,

$$u_i = \sum_{j \in N_i} a_{ij}\xi_{ij}(\xi_j - \xi_i) \tag{3}$$

This protocol is *bilinear* in the trust values. Note that this defines a graph topology that stays constant, yet the edge weights are equal to ξ_{ij}, the trust that node i has for its neighbor node j. The weighted adjacency matrix is defined by $W = [w_{ij}] = [a_{ij}\xi_{ij}]$. This defines a time-varying graph $G(t)$ which has a constant topology given by the adjacency matrix A, yet whose edge weights vary as node i changes its trust opinion about its neighbor nodes, i.e. this is a weighted version of the trust graph defined by the adjacency matrix A.

If ξ_i's are scalars, (3) can be rewritten as,

$$\begin{aligned} u_i &= \sum_{j\in N_i} a_{ij}\xi_{ij}(\xi_j - \xi_i) \\ &= \sum_{j\in N_i} a_{ij}\xi_{ij}\xi_j - \sum_{j\in N_i} a_{ij}\xi_{ij}\xi_i \\ &= -(D(t)-W(t))\xi_i \end{aligned}$$

$$\dot{\xi}_i = -(D(t)-W(t))\xi_i = -L(t)\xi_i \qquad (4)$$

Here, $D(t)$ is the time-varying in-degree matrix defined as $D(t)= diag\{n_i\}$ where $n_i = \sum_{j\in N_i} a_{ij}\xi_{ij}$, and $W(t)$ is a time-varying weighted adjacency matrix. These matrices are functions of node trusts ξ. $L(t)$ is a time-varying Laplacian matrix defined as $D(t)$-$W(t)$ which is also a function of the node trusts. Note that the node trust vectors $\xi_i(t)$ have nonzero entries ξ_{ij} corresponding to the weights of incoming edges e_{ji}, which have $a_{ij} = 1$, but there may also be nonzero entries $\xi_{ij}(t)$ that do not correspond to edges in the graph. Thus, though a node i forms a trust opinion about more and more nodes as trust propagates through the graph, its direct trust neighbors (the graph edges coming into node i) never change, and are defined by the adjacency matrix A.

Since $\xi_i \in R^N$, we must use Kronecker product (Godsil and Royle, 2001) to write,

$$\dot{\xi} = -(L(t)\otimes I_N)\xi \qquad (5)$$

where I_N is an identity matrix of $N\times N$. Here, $\xi = [\xi_1^T \cdots \xi_N^T]^T \in R^{N^2}$ is the overall network trust vector.

The Laplacian $L(t)$ corresponds to a time-varying graph $G(t)$. The initial Laplacian $L(0)$ corresponds to the initial graph $G(0)$. Note that the row sum of $L(t)$ is zero for $\forall t$. Hence, $L(t)$ has a zero eigenvalue corresponding to the right eigenvector of **1**, where **1** is a column vector with all entries equal to one.

Consider the following nonlinear local voting discrete-time trust consensus protocol based on the Vicsek model (Vicsek et al., 1995),

$$\xi_i(k+1) = \xi_i(k) + \frac{1}{n_i+1}\sum_{j\in N_i} a_{ij}\xi_{ij}(\xi_j - \xi_i) \qquad (6)$$

which can be rewritten in the scalar case as,

$$\begin{aligned} \xi_i(k+1) &= (I-(I+D(k))^{-1}L(k))\xi_i(k) \\ \xi_i(k+1) &= F(k)\xi_i(k) \qquad (7) \end{aligned}$$

where

$F(k) = I - (I + D(k))^{-1} L(k) = (I + D(k))^{-1}(I + W(k))$ Since $\xi_i \in R^N$, we must use Kronecker product to write,

$$\xi(k+1) = (F(k) \otimes I_N)\xi(k) \tag{8}$$

Here, $\xi = [\xi_1^T \cdots \xi_N^T]^T \in R^{N^2}$.

Note that $F(k)$ is a time-varying stochastic matrix that depends on the trust values ξ_{ij}. The matrix $F(k)$ corresponds to a time-varying graph $G(k)$ with Laplacian $L(k)$. $F(0)$ corresponds to the initial graph $G(0)$ with initial Laplacian $L(0)$. For each k, $F(k)$ has a eigenvalue of one corresponding to the right eigenvector of **1**, where **1** is a column vector with all entries equal to one. Even if $F(k)$, $F(k\text{-}1)$, $F(k\text{-}2)$,…, $F(0)$ are time-varying, the graph topology remains the same, only the weights in F change, which we prove in Section 3.

3 Convergence of Trust

We say that a protocol achieves (asymptotic) consensus if for every i, j one has $\xi_i(t) \to \xi_j(t) \to \xi_*$ in continuous-time, $\xi_i(k) \to \xi_j(k) \to \xi_*$ in discrete-time, where ξ_* is called the consensus trust vector value. If this occurs, then in the limit one has $\xi_{ip} = \xi_{jp}$ for all i, j so that all nodes arrive at the same trust value for each other at node p. To prove the trust consensus, we need to have the following assumption.

Assumption 1: In the trust graph G_T, $\xi_{ij}(0) > 0$ if $a_{ij} = 1$.

The main result of this chapter is that the bilinear trust protocol (5) for continuous-time and (8) for discrete-time achieve asymptotic consensus for a trust graph G_T if and only if the initial graph $G(0)$ has a spanning tree. Under Assumption 1, this is equivalent to the trust graph G_T containing a spanning tree. We are of course inspired by (Ren and Beard, 2005), which covers the case of linear integrator dynamics.

Two nonnegative matrices are said to be of the same type if their zero elements are in the same locations (Ren and Beard, 2005). We will use the notation $P \sim Q$ to denote that P and Q are of the same type. Two graphs on the same nodes are of the same type if their edge sets are the same.

3.1 Consensus of the Discrete-Time Protocol

In this section, we prove that the trust protocol in (8) achieves asymptotic consensus if and only if the initial graph $G(0)$ has a spanning tree. Assumption 1 means that G_T and $G(0)$ are of the same type, i.e. $G_T \sim G(0)$. For each $F(k)$

associate a set of graphs $\{G(k)\}$. Now, F is a time-varying function of the trusts with the initial trust vectors for each node $\xi_i(0)$ in [0, 1]. Consider the local voting discrete-time trust consensus scheme based on the Vicsek model given in (8). Let $F(0)$ represent the initial directed graph $G(0)$. If $\xi_{ij}(0)$ is an edge in $G(0)$ then $\xi_{ij}(k)$ is an edge for all $G(k)$, for $k \geqslant 0$. This is formalized in the next result.

Lemma 1: Consider a network with initial graph $G(0)$ running the discrete-time consensus scheme in (8) with initial condition $\xi(0)$. Let $\xi_{ij}(k) > 0$ for some time instant $k \geqslant 0$. Then $\xi_{ij}(k+1) > 0$. As a result, $G(k)$ for $k \geqslant 0$ are all of the same type.

Proof: From (8), each updated node trust is a weighted average of its neighboring trust values such that the weights are nonnegative and less than 1, because the row sum of $F(k)$ and $F(k) \otimes I_N$ is 1, i.e. they are stochastic (Wolfowitz, 1963). Protocol in (8) can be rewritten for each state as,

$$\xi_{ij}(k+1) = \sum_l f_{il}(k)\xi_{lj}(k)$$

$$= f_{ii}(k)\xi_{ij}(k) + \sum_{l \neq i} f_{il}(k)\xi_{lj}(k)$$

where $f_{ij}(k)$ is the $(i,j)^{th}$ element of $F(k)$. Then by definition of $F(k)$, we know that, $0 \leqslant f_{ij}(k) < 1$, for $i \neq j$ and $0 < f_{ii}(k) \leqslant 1$. Also, $f_{ii} = \dfrac{1}{1+n_i} > 0$. Hence, if $\xi_{ij}(k) > 0$, the first term is always positive. The second term is a weighted average which once again is always nonnegative for non-zero initial trusts. Therefore, for $k \geqslant 0$, if $\xi_{ij}(k) > 0$, $\xi_{ij}(k+1) > 0$.

Thus, if $\xi_{ij}(0) > 0$ is an edge weight for $G(0)$, then $\xi_{ij}(k) > 0, \forall k > 0$ is an edge weight for $G(k)$. Therefore, $G(k)$, $\forall k \geqslant 0$ are all of the same type. ■

Theorem 1: Let $\xi_{ij}(0) > 0$ if $a_{ij} = 1$. Then the discrete-time trust protocol in (8) achieves a trust consensus for $\xi_{ij}(k)$ if and only if the trust graph G_T has a spanning tree.

Proof: Now $G(0)$ has a spanning tree if and only if $G(k)$, $\forall k > 0$, has a spanning tree by Lemma 1. Under Assumption 1, this is equivalent to the trust graph G_T containing a spanning tree. This is a necessary and sufficient condition for the union of graphs over any finite time interval to have a joint spanning tree. Therefore, Theorem 3.8 in (Ren and Beard, 2005) proves the result. ■

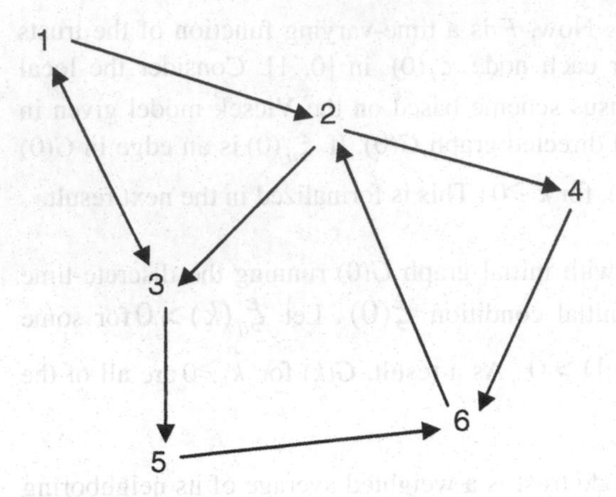

Fig. 2 A Six Node Directed Graph

Example 1: Consider a six node network as shown in Figure 2. Let the initial trust vectors $\xi_i(0) \in R^6$ have elements selected randomly in [0, 1].

Figure 3 shows convergence of trust in a six node network with 6 states using the discrete-time protocol given by (8). The nodes reach a consensus trust value since the graph is strongly connected. ■

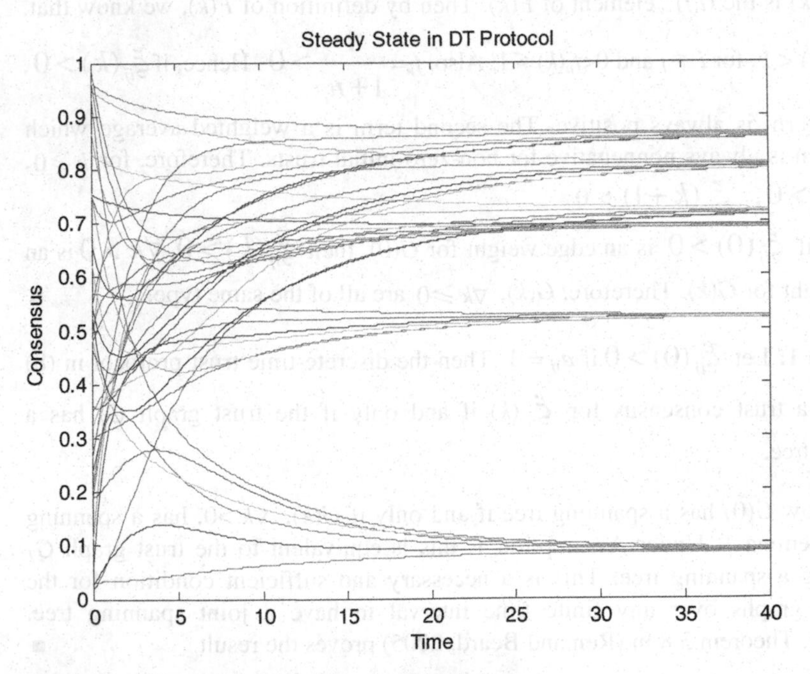

Fig. 3 Trust Consensus in the Discrete-Time Scheme

3.2 *Consensus of the Continuous-Time Trust Protocol*

In this section, we prove that the bilinear trust protocol in (5) achieves asymptotic consensus for a trust graph G_T if and only if the initial graph $G(0)$ has a spanning tree. Assumption 1 means that G_T and $G(0)$ are of the same type, i.e. $G_T \sim G(0)$. For each $L(t)$ associate graph $G(t)$. For the continuous-time scheme, one has $L(t)=[l_{ij}(t)]$, the diagonal elements of $L(t)$ are positive, the off-diagonal elements are negative and $\sum_j l_{ij} = 0$. Let $\phi(t,t_0)$ be the corresponding transition matrix of $L(t)$ defined as $\phi(t,t_0) = I + \int_{t_0}^{t} L(\sigma 1)d\sigma 1 + \int_{t_0}^{\sigma 1} L(\sigma 2)d\sigma 2 d\sigma 1 + \ldots$

From (Ren and Beard, 2005), we know that the transition matrix $\phi_L(t,t_0)$ of $L(t)$ is a nonnegative stochastic matrix with positive diagonal elements. Also, the corresponding transition matrix of $L(t) \otimes I_N$ is $\phi_L(t,t_0) \otimes I_N$ which is once again a nonnegative stochastic matrix with positive diagonal entries. Along the same lines as in Lemma 1, we can prove the following Lemma.

Lemma 2: Consider a network with initial graph $G(0)$ running the continuous-time protocol (5) with initial node trust vectors $\xi_i(0)$. Let $\xi_{ij}(0) > 0$. Then for $\forall t > 0$, $\xi_{ij}(t) > 0$. As a result, $G(t)$ for $t \geqslant 0$ are all of the same type.

Proof: Solution of (5) can be written as $\xi(t) = (\phi_L(t,0) \otimes I_N)\xi(0)$. This can be rewritten for each state as,

$$\xi_{ij}(t) = \phi_{Lii}(t,0)\xi_{ij}(0) + \sum_{l \neq i} \phi_{Lil}(t,0)\xi_{lj}(0) \tag{9}$$

Here, the diagonal elements of $\phi_L(t,0) \otimes I_N$ are always positive and therefore the first term in the RHS of Equation (9) will always be positive for $\xi_{ij}(0) > 0$. The second term in the RHS of Equation (9) is always nonnegative since $\phi_L(t,0) \otimes I_N$ is a nonnegative stochastic matrix with positive diagonal entries (Wolfowitz, 1963). Thus, if $\xi_{ij}(0) > 0$ is an edge weight for $G(0)$, then $\xi_{ij}(t) > 0, \forall t > 0$ is an edge weight for $G(t)$. Therefore, $G(t)$, $\forall t \geqslant 0$ are all of the same type. ■

Theorem 2: Let $\xi_{ij}(0) > 0$ if $a_{ij} = 1$. Then the continuous-time trust consensus protocol in (5) achieves trust consensus for $\xi_{ij}(t)$ if and only if the trust graph G_T has a spanning tree.

Proof: Now $G(0)$ has a spanning tree if and only if $G(t)$, $\forall t$ has a spanning tree by Lemma 2. Under Assumption 1, this is equivalent to the trust graph G_T containing a spanning tree. Also $\varphi_L(t,0)$ is a continuous function of $L(t)$ for the interval [0, t]. This is a necessary and sufficient condition for the union of graphs over any finite time interval to have a joint spanning tree. Therefore, the result (Theorem 3.2) in (Ren et al., 2005) proves the result. ■

Example 2: Consider the same six node network as shown in Figure 2. Let the initial $\xi(0) \in R^6$ be the same as in Example 1. Figure 4 shows convergence of trust in a six node network with six states using the continuous-time protocol given by (5). ■

3.3 Relation of the Continuous and Discrete-Time Protocols

It can be observed that the discrete-time and the continuous-time schemes in Examples 1 and 2 respectively give different consensus values for the same graph and the same initial conditions. Let us examine the relation between the continuous and discrete-time protocols more closely.

The Laplacian L in the continuous-time protocol is related to the stochastic matrix F in the discrete-time protocol at each time instance. As shown in Figures 3 and 4, the trust consensus using (8) and (5) do not converge to the same consensus for the same graph and initial trust values. This is because the graph represented by $F(k)$ is not the same as the graph represented by $L(t)$. In fact,

$$F = I-(I+D)^{-1}L \tag{10}$$

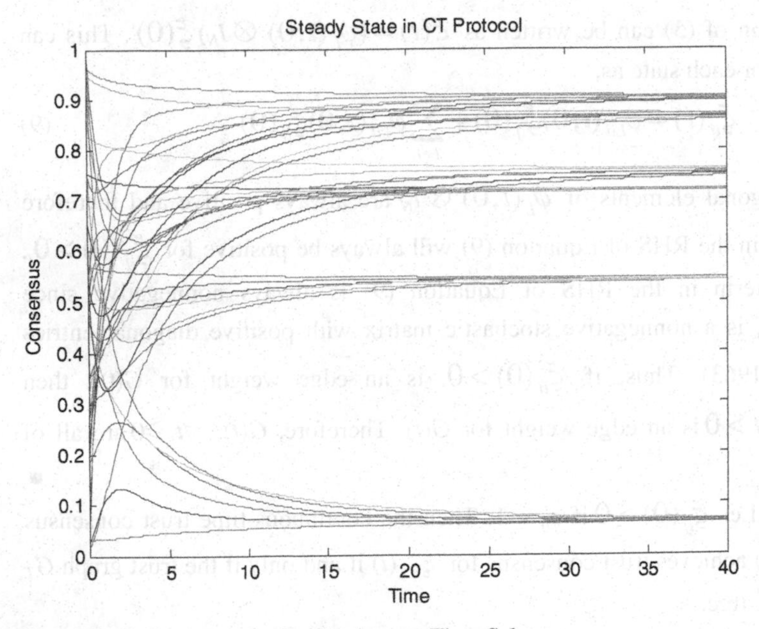

Fig. 4 Trust Consensus in the Continuous-Time Scheme

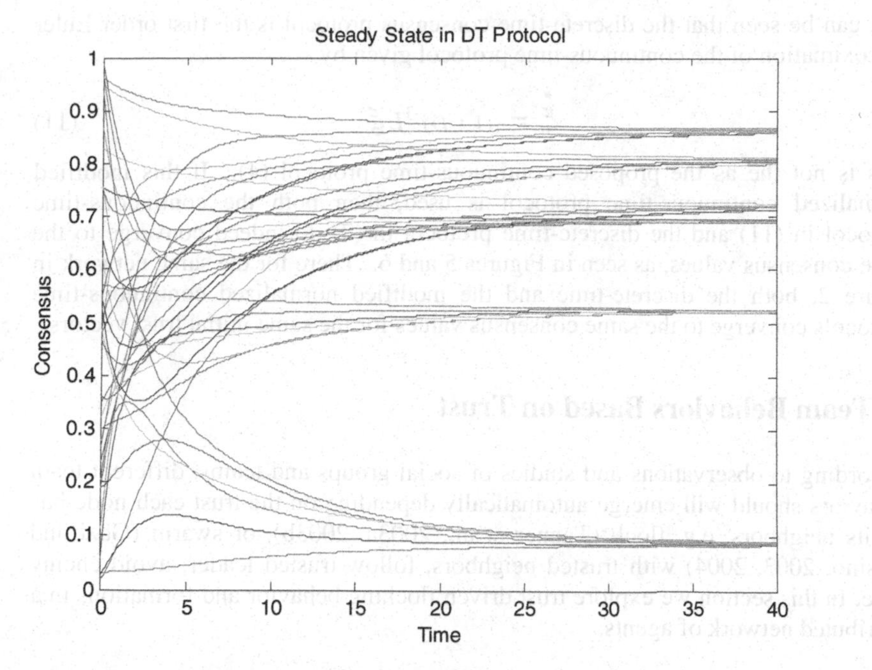

Fig. 5 Trust Consensus in the Discrete-time Scheme

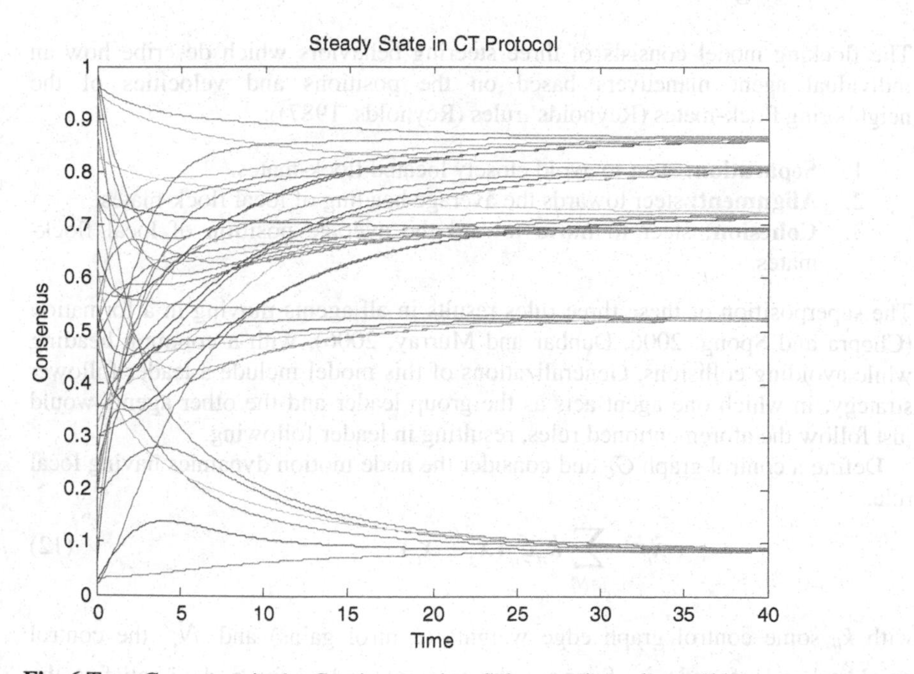

Fig. 6 Trust Consensus in the Continuous-time Scheme using scheme (11)

It can be seen that the discrete-time consensus protocol is the first order Euler approximation of the continuous-time protocol given by

$$\dot{\xi}_i = -(I+D)^{-1} L \xi_i \tag{11}$$

This is not the as the proposed continuous-time protocol (4). If this modified normalized continuous-time protocol is used, then both the continuous-time protocol in (11) and the discrete-time protocol in (7) do indeed converge to the same consensus values, as seen in Figures 5 and 6. There for the same network in Figure 2, both the discrete-time and the modified normalized continuous-time protocols converge to the same consensus values for the same initial trust vectors.

4 Team Behaviors Based on Trust

According to observations and studies of social groups and teams, different team behaviors should will emerge automatically depending on the trust each node has for its neighbors, e.g. flock (Tanner et al., 2003a, 2003b), or swarm (Gazi and Passino, 2003, 2004) with trusted neighbors, follow trusted leader, avoid enemy node. In this section we explore trust driven flocking behavior and formations in a distributed network of agents.

4.1 Flocking

The flocking model consists of three steering behaviors which describe how an individual agent maneuvers based on the positions and velocities of the neighboring flock-mates (Reynolds' rules (Reynolds, 1987)):

1. **Separation:** steer to avoid closely located flock-mates.
2. **Alignment:** steer towards the average heading of local flock-mates.
3. **Cohesion:** steer to move toward the average position of local flock-mates.

The superposition of these three rules results in all agents moving in a formation (Chopra and Spong, 2006, Dunbar and Murray, 2006), with a common heading while avoiding collisions. Generalizations of this model include a leader follower strategy, in which one agent acts as the group leader and the other agents would just follow the aforementioned rules, resulting in leader following.

Define a control graph G_C and consider the node motion dynamics having local rule,

$$\dot{x}_i = \sum_{j \in N_i^c} k_{ij} \xi_{ij} (x_j - x_i) \tag{12}$$

with k_{ij} some control graph edge weights (control gains) and N_i^c the control neighborhood of node i. Suppose the trust of node i for node j satisfies the bilinear trust local voting dynamics,

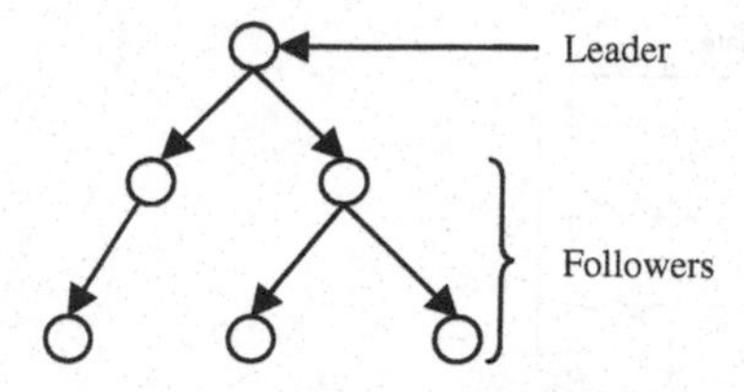

Fig. 7 Tree network with one leader and five followers

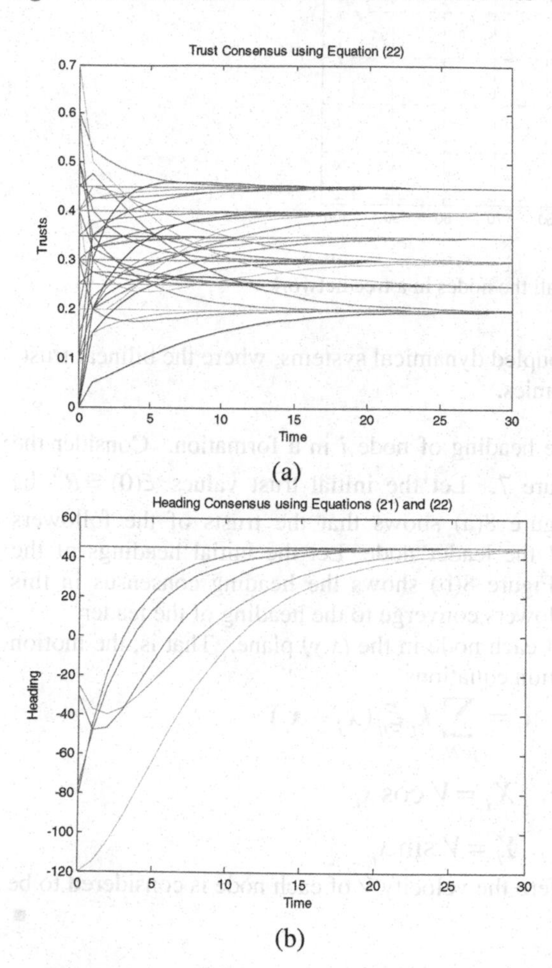

Fig. 8 (a)Convergence of trusts of all the nodes, (b) Convergence of headings of all the nodes in a tree network

$$\dot{\xi}_i = \sum_{j \in N_i^t} a_{ij} \xi_{ij} (\xi_j - \xi_i) \tag{13}$$

with N_i^t the trust neighborhood of node i. The structure of the trust graph G_T is defined by the adjacency matrix $A=[a_{ij}]$.

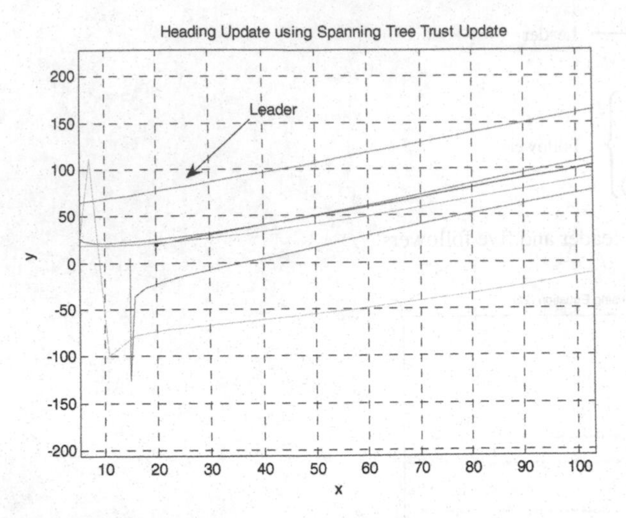

Fig. 9 Convergence of headings of all the nodes in a tree network

Note that (12) and (13) are coupled dynamical systems, where the bilinear trust dynamics drive the motion dynamics.

Example 3: Let x_i represent the heading of node i in a formation. Consider the formation graph shown in Figure 7. Let the initial trust values $\xi(0) \in R^6$ be selected randomly in [0,1]. Figure 8(a) shows that the trusts of the followers converge to the initial trusts of the leader node. Let the initial headings of the nodes be randomly selected. Figure 8(b) shows the heading consensus in this network; the headings of the followers converge to the heading of the leader.

Figure 9 shows the motion of each node in the *(x,y)* plane. That is, the motion of node i is governed by the motion equations

$$\dot{x}_i = \sum_{j \in N_i^c} k_{ij} \xi_{ij} (x_j - x_i)$$

$$\dot{X}_i = V \cos x_i$$

$$\dot{Y}_i = V \sin x_i$$

with x_i the heading of node i. Here the velocity V of each node is considered to be the same. ■

4.2 An Easy Way to Stabilize Formations

A formation refers to a set of spatially distributed vehicles whose dynamic states are coupled through a common control law and which should maintain desired positions relative to each other. This section shows an easy way to assign and maintain formations in a desired configuration. It is not based on potential fields, which can become cumbersome for assignment of precise positions in a formation.

Potential fields also do not allow easy changes in the formation structure. Here, as the desired configuration changes, the formation can quickly be moved into the new desired structure.

Let the communication graph structure be a tree (e.g. as in Figure 7), which is taken as the structure of the trust graph and the control graph. Let the state x_i of node i in (12) be defined as

$$x_i = \begin{bmatrix} p_{i1} \\ p_{i2} \\ \vdots \\ p_{iN} \end{bmatrix}$$

with $p_{ij} = \begin{bmatrix} X_{ij} & Y_{ij} \end{bmatrix}^T$ the estimate (opinion) of node i of the 2-D (x,y) position of node j. (3-D positions can also be accommodated easily.) Let the leader's (e.g. the root node of the tree) initial estimate be

$$x_1(0) = \begin{bmatrix} p_1^d \\ p_2^d \\ \vdots \\ p_N^d \end{bmatrix}$$

the desired positions of each node, which are selected according to the desired formation structure. Let each other node take as its initial motion vector $x_i(0)$ all entries of zero except for entry $p_{ii}(0)$ which is taken as its own initial position.

Now run the coupled trust consensus protocol (13) and motion consensus protocol (12). Since the graph is a tree, all nodes will converge to the motion values of the leader, which means that all nodes will move into the positions assigned by the leader in $x_1(0)$, moving to their desired formation positions.

If the formation changes, the leader only needs to select another value of $x_1(0)$, corresponding to the desired assigned node locations in the new formation. The nodes will then move into their new positions.

Example 4: For the same tree network in Example 3, we want the desired positions of the nodes in the hexagonal formation structure. Let the initial state of the leader $x_l(0)$ contain the desired formation positions of all the other nodes in the network in 2D, i.e. their desired (x,y) positions. If we run the coupled node dynamics and bilinear trust update in (12) and (13), then all nodes converge to the initial state of the leader, i.e. to their desired formation positions, as shown in Figure 10.

If the desired relative positions of all or some of the nodes change, then the leader simply resets $x_{ss} = x_l(0)$, and all nodes will automatically converge to the new consensus trust and positions, as specified by the leader in its initial state vector. ■

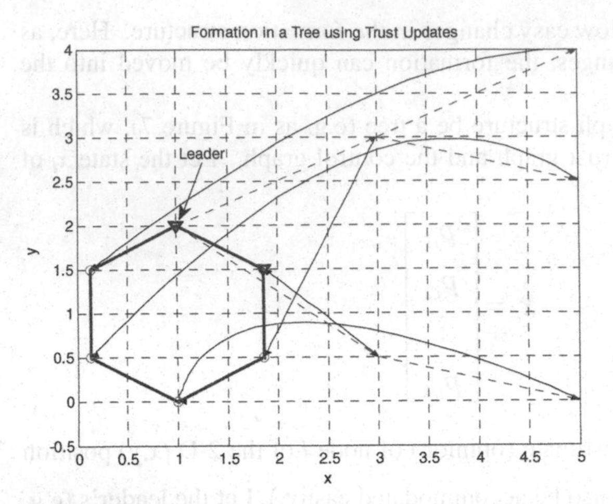

Fig. 10 Convergence of positions of all the nodes in a tree network to a hexagon formation

5 Conclusions

This chapter considered the problem of trust establishment and consensus in a distributed network. Directed graphs were used to represent the information exchange between the nodes. We presented a continuous-time and a discrete-time bilinear trust update scheme for trust consensus. We showed now to take these trust update schemes into account in relation to team motion behaviors such as flocking and formations.

Acknowledgements

This work was supported by ARO grant ARO W91NF-05-1-0314 and the Army National Automotive Center, and NSF grant ECCS-0801330.

References

Beard, R., Stepanyan, V.: Synchronization of information in distributed multiple vehicle coordinated control. In: Proc. IEEE Conf. Decision and Control, pp. 2029–2034 (2003)

Chopra, N., Spong, M.W.: Passivity-based control of multi-agent systems. In: Kawamura, S., Svinin, M. (eds.) Advances in Robot Control: From Everyday Physics to Human-Like Movements, pp. 107–134. Springer, Berlin (2006)

Dunbar, W.B., Murray, R.M.: Distributed receding horizon control for multi-vehicle formation stabilization. Automatica 42, 549–558 (2006)

Fax, J.A., Murray, R.M.: Information flow and cooperative control of vehicle formations. IEEE Trans. Automatic Control 49(9), 1465–1476 (2004)

Gazi, V., Passino, K.M.: Stability analysis of swarms. IEEE Trans. Automatic Control 48(4), 692–697 (2003)

Gazi, V., Passino, K.M.: A class of attractions/repulsion functions for stable swarm aggregations. Int. J. Control 77(18), 1567–1579 (2004)
Godsil, C., Royle, G.: Algebraic Graph Theory, vol. (207). Springer, New York (2001)
Jadbabaie, A., Lin, J., Morse, A.S.: Coordination of groups of mobile autonomous agents using nearest neighbor rules. IEEE Trans. Automatic Control 48(6), 988–1001 (2003)
Jiang, T., Baras, J.S.: Trust evaluation in anarchy: a case study on autonomous networks. In: Proc. Infocom., Barcelona (2006)
Lee, D., Spong, M.W.: Stable flocking of multiple inertial agents on balanced graphs. IEEE Trans. Automatic Control 52(8), 1469–1475 (2007)
Olfati-Saber, R., Murray, R.M.: Consensus problems in networks of agents with switching topology and time-delays. IEEE Trans. Automatic Control 49(9), 1520–1533 (2004)
Ren, W., Beard, R.W.: Consensus seeking in multiagent systems under dynamically changing interaction topologies. IEEE Trans. Automatic Control 50(5), 655–661 (2005)
Ren, W., Beard, R.W., Kingston, D.: Multi-agent Kalman Consensus with Relative Uncertainty. In: Proceedings of ACC (2005)
Reynolds, C.: Flocks, herds and schools: a distributed behavioral model. Computer Graphics 21(4), 25–34 (1987)
Saligrama, V., Alanyali, M., Savas, O.: Distributed detection in sensor networks with packet losses and finite capacity links. IEEE Trans. Signal Proc. 54(11), 4118–4132 (2006)
Tanner, H., Jadbabaie, A., Pappas, G.: Stable flocking of mobile agents, Part i: Fixed Topology. In: Proc. IEEE Conf. Decision and Control, Maui, HI, pp. 2010–2015 (2003a)
Tanner, H., Jadbabaie, A., Pappas, G.: Stable flocking of mobile agents, Part ii: Dynamic Topology. In: Proc. IEEE Conf. Decision and Control, Maui, HI, pp. 2016–2021 (2003b)
Theodorakopoulos, G., Baras, J.S.: On trust models and trust evaluation metrics for ad hoc networks. IEEE J. Selected Areas in Communications 24(2), 318–328 (2006)
US Army Training and Doctrine Command (TRADOC)., March 2008: Pamphlet 525-66, Future Operational Capabilities, Fort Monroe, VA (2008)
Vicsek, T., Czirok, A., Jacob, E., Cohen, I., Schochet, O.: Novel type of phase transitions in a system of self—driven particles. Phys. Rev. Lett. 75, 1226–1229 (1995)
Wolfowitz, J.: Products of indecomposable, aperiodic, stochastic matrices. Proc. Amer. Math. Soc. 15, 733–736 (1963)

Part V
Further Application Areas – Pattern Recognition and Communications Engineering

Coupled Oscillator systems for Microwave Applications: Optimized Design Based on the Study and Control of the Multiple Coexisting Solutions in Systems with Symmetry

Ana Collado and Apostolos Georgiadis

Abstract. Coupled oscillator arrays have attracted a lot of attention due to their intrinsic synchronization properties that make them suitable for a wide range of microwave and radiofrequency applications. Characteristic applications of such arrays are phase shifters and high frequency generation. Their design however is complicated by their nonlinear nature, which, combined with the symmetry properties of these architectures leads to a rich dynamic behavior consisting of multiple operating modes and complicated stability considerations. This work focuses on linear coupled oscillator arrays for beam steering applications, and N-push oscillator topologies for high frequency generation. Both architectures are examined from the point of view of identifying multiple coexisting solutions and their stability. Practical examples are provided.

1 Linear Arrays of Coupled Oscillators

The use of coupled oscillator arrays in communication systems have been widely reported in the literature [1, 4, 7, 13, 14, 15, 16, 19, 27, 29, 32, 33, 40, 46, 30, 23]. One of the most common applications of coupled oscillator chains is feeding antenna arrays and allowing for beam steering or pattern nulling. The use of coupled oscillator arrays allows reducing the size and the cost of the system by avoiding the use of phase shifters and the associated feeding networks.

Ana Collado and Apostolos Georgiadis
Centre Tecnologic de Telecomunicacions de Catalunya, Av. Canal Olimpic, 08860 Castelldefels, Barcelona, Spain
e-mail: acollado@cttc.es, ageorgiadis@cttc.es

K. Kyamakya (Eds.): Recent Adv. in Nonlinear Dynamics and Synchr., SCI 254, pp. 367–398.
springerlink.com

In order to steer a beam in a desired direction it is necessary to feed the antenna elements with a constant phase shift distribution. Due to the intrinsic properties of coupled oscillators, it is possible to synthesize constant phase shift distributions among the array elements by simply detuning the free-running frequencies of the edge elements in the array by the same amount and in opposite directions [46]. This detuning is done by means of a control parameter typically the voltage of a varactor diode.

However the design of coupled oscillator systems presents difficulties associated to the complex nonlinear dynamics present in these circuits. These systems due to their symmetric structure can present multiple coexisting solutions. The desired solution for their use in beam steering applications is the one that establishes a constant phase shift distribution between the oscillator elements in the array. When designing the coupled oscillator system for this type of scanning applications the possible appearance of other undesired solutions, where the phase shift distribution is not constant has to be taken into account.

1.1 Coupled Oscillator System Modelling

Several works have already studied the use of coupled oscillator systems for communications applications [1, 12, 19, 15, 27]. As stated before the design of coupled oscillator systems requires the accurate determination of the multiple coexisting solutions in the system in order to ensure the circuit is working in the desired operation mode.

The calculation of these multiple coexisting solutions can be computationally expensive when performed with a commercial full harmonic balance simulation. Several approaches has been presented in the literature to model oscillators and coupled oscillator systems [44, 28, 11, 15, 8].These approaches allow designing and analyzing coupled oscillator systems in an efficient manner. In the following the formulation presented in [15, 8] is introduced. This formulation will be used in the next sections to obtain the multiple coexisting solutions in linear arrays of coupled oscillator elements and also to check their stability. Coupled oscillator systems formed by a large number of elements require long simulation times in conventional harmonic balance simulators. The definition of a simplified model for this type of systems reduces considerably the simulation time.

The proposed formulation only requires a model for the individual oscillator that form the array. The model of the single oscillator together with the mathematical description of the coupling network allows modeling accurately the complete array of coupled oscillators.

The single oscillator model is obtained using a perturbation analysis of the admittance function of the free-running oscillator. Under the assumption of weak coupling conditions in the system, the admittance is approximated by a first order Taylor series expansion about the free-running solution. The derivatives involved in the single oscillator modelling are obtained numerically at the node where the oscillator is connected to the coupling network [15].

When the individual oscillators are introduced in the array environment, their oscillation amplitude and frequency change due to the influence of the rest of the oscillators of the chain through the coupling network.

The linearization used in the proposed model for the single oscillator is only valid when the oscillators are weakly coupled. If the coupling is weak the effect of introducing the individual oscillators in the system is small and the system can be accurately modelled using a first or second order model of the single oscillator elements.

The proposed formulation uses as inputs the derivatives of the admittance of the free-running oscillator and the coupling network parameters. With these input parameters the model can accurately predict the dynamics of coupled oscillator arrays.

1.1.1 Steady State Solution

In order to model the coupled oscillator system, each of the oscillator elements is described by an amplitude and a phase at their output node. The individual oscillator elements are coupled to the array environment at the same node used to calculate their admittances in the free-running state (Fig. 1). When introduced into the array all the oscillator elements will synchronize to a common oscillation frequency. As stated before, under weak coupling conditions, the amplitude and frequency of the single oscillator elements when introduced into the system, stays close to the free-running values. This implies that the model of the individual oscillator admittance can be approximated by a Taylor series.

As stated before, in order to steer the radiated beam in a desired direction, it is necessary to synthesize a constant phase shift distribution among the oscillator elements that form the array [46, 1, 15]. The value of this constant phase shift determines the direction at which the beam is pointing. The different constant phase shift distributions are obtained by detuning the free-running frequencies of the edge oscillator elements by the same amount and in opposite directions. Depending on the amount of detuning of the edge oscillators different constant phase shift distributions will be achieved. The detuning of the free-running frequencies is done using two control parameters η_1 and η_N that can be the voltages of two varactor diodes V_{T1}, V_{TN}.

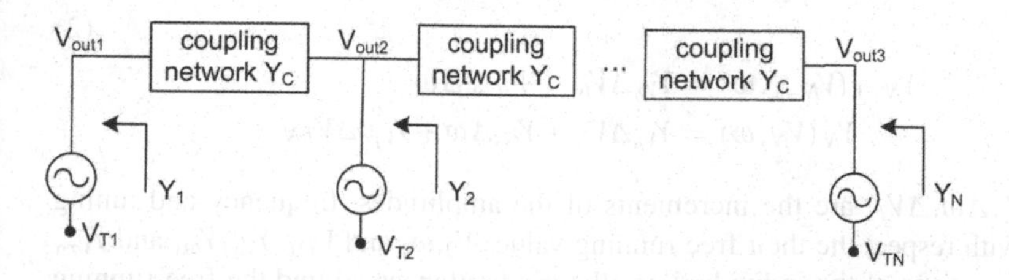

Fig. 1 N coupled oscillator array model. Coupling network defined by its admittance matrix Y_c. The oscillator elements are described by their admittance functions Y_k.

Although the desired solution for using coupled oscillator systems in beam steering applications is the one with a constant phase shift distribution, the proposed model is formulated in a general manner to accommodate arbitrary phase distributions. The phase of the first oscillator is fixed to $\phi_1 = 0$ as a reference. The rest of the oscillator phases take the form ϕ_k with $k = 2, \ldots, N$.

One of the key parameters in the proposed coupled oscillator model is the description of the coupling network. In the proposed formulation the coupling network is described by its admittance matrix Y_c [34]. Global coupling is assumed among all the oscillator elements that form the coupled oscillator array even though the main coupling will be due to the nearest neighbour elements.

The amplitudes and phases of each oscillator are related to the rest of the oscillators in the system by means of the coupling matrix Y_c. The coupled oscillator system model can be written as follows:

$$\begin{aligned}
Y_1 V_1 + \sum_{i=1}^{N} Y_{c1i} V_i e^{j(\phi_i - \phi_1)} &= 0 \\
Y_2 V_2 + \sum_{i=1}^{N} Y_{c2i} V_i e^{j(\phi_i - \phi_2)} &= 0 \\
&\vdots \\
Y_{N-1} V_{N-1} + \sum_{i=1}^{N} Y_{c(N-1)i} V_i e^{j(\phi_i - \phi_{N-1})} &= 0 \\
Y_N V_N + \sum_{i=1}^{N} Y_{cNi} V_i e^{j(\phi_i - \phi_N)} &= 0
\end{aligned} \tag{1}$$

where Y_k kith $k = 1, 2, \ldots, N$ is the admittance function of oscillator k in the connection node to the array and $V_1, \ldots, V_N$ are the amplitudes of the oscillators elements in the steady state. The admittance functions of the oscillators are approximated by a first order Taylor series expansion around the free-running solution:

$$\begin{aligned}
Y_1 (V_1, \omega) &= Y_{V_o} \Delta V_1 + Y_{\omega_o} \Delta \omega + Y_{V_{T0}} \Delta V_{T1} \\
Y_2 (V_2, \omega) &= Y_{V_o} \Delta V_2 + Y_{\omega_o} \Delta \omega \\
&\vdots \\
Y_{N-1} (V_{N-1}, \omega) &= Y_{V_o} \Delta V_{N-1} + Y_{\omega_o} \Delta \omega \\
Y_N (V_N, \omega) &= Y_{V_o} \Delta V_N + Y_{\omega_o} \Delta \omega + Y_{V_{T0}} \Delta V_{TN}
\end{aligned} \tag{2}$$

In (2) $\Delta V_i, \Delta \omega, \Delta V_{Ti}$ are the increments of the amplitudes, frequency and tuning voltages with respect the their free running values V_0, ω_0 and V_{T0}. Y_{V_0}, Y_{ω_0} and $Y_{V_{T0}}$ are the derivatives of the individual oscillator admittances around the free-running solution.

$$Y_{V_0} = Y_{V_0}^R + jY_{V_0}^I = \left.\frac{\partial Y}{\partial V}\right|_{V_0} = Y_V$$

$$Y_{\omega_0} = Y^R_{\omega_0} + jY^I_{\omega_0} = \left.\frac{\partial Y}{\partial \omega}\right|_{\omega_0} = Y_\omega \tag{3}$$

$$Y_{V_{T0}} = Y^R_{V_{T0}} + jY^I_{V_{T0}} = \left.\frac{\partial Y}{\partial V_T}\right|_{V_{T0}} = Y_{V_T}$$

$$\begin{aligned} \Delta V_i &= V_i - V_0 \\ \Delta \omega &= \omega - \omega_0 \\ \Delta V_{Ti} &= V_{Ti} - V_{T0} \end{aligned} \tag{4}$$

Substituting the expressions for the admittance functions (2) in (1) one can write:

$$\begin{aligned} \left[Y_V (V_1 - V_o) + Y_\omega (\omega - \omega_o) + Y_{V_T} (V_{T1} - V_{To})\right] V_1 + \sum_{i=1}^{N} Y_{c1i} V_i e^{j(\phi_i - \phi_1)} &= 0 \\ \left[Y_V (V_2 - V_o) + Y_\omega (\omega - \omega_o)\right] V_2 + \sum_{i=1}^{N} Y_{c2i} V_i e^{j(\phi_i - \phi_2)} &= 0 \\ &\vdots \\ \left[Y_V (V_{N-1} - V_o) + Y_\omega (\omega - \omega_o)\right] V_{N-1} + \sum_{i=1}^{N} Y_{c(N-1)i} V_i e^{j(\phi_i - \phi_{N-1})} &= 0 \\ \left[Y_V (V_N - V_o) + Y_\omega (\omega - \omega_o) + Y_{V_T} (V_{TN} - V_{To})\right] V_N + \sum_{i=1}^{N} Y_{cNi} V_i e^{j(\phi_i - \phi_N)} &= 0 \end{aligned} \tag{5}$$

Expressing (5) in a matrix form, the final system of equations that model the linear array of coupled oscillators can be written as follows:

$$\bar{H}(\bar{V}, \bar{\phi}, \omega) = \begin{pmatrix} Y_V & 0 & \cdots & 0 & Y_\omega V_1 e^{j\phi_1} \\ 0 & Y_V & & \vdots & Y_\omega V_2 e^{j\phi_2} \\ \vdots & & \ddots & 0 & \vdots \\ 0 & \cdots & 0 & Y_V & Y_\omega V_N e^{j\phi_N} \end{pmatrix} \begin{pmatrix} V_1 e^{j\phi_1} (V_1 - V_1^0) \\ \vdots \\ V_N e^{j\phi_N} (V_N - V_N^0) \\ \omega - \omega_0 \end{pmatrix} +$$

$$+ \begin{pmatrix} Y_{V_T} & 0 \\ 0 & 0 \\ \vdots & \vdots \\ 0 & Y_{V_T} \end{pmatrix} \begin{pmatrix} V_1 e^{j\phi_1} \Delta V_{T1} \\ V_N e^{j\phi_N} \Delta V_{TN} \end{pmatrix} +$$

$$+ \begin{pmatrix} Y_{C11}(\omega) & \cdots & Y_{C1N}(\omega) \\ \vdots & & \vdots \\ Y_{CN1}(\omega) & \cdots & Y_{CNN}(\omega) \end{pmatrix} \begin{pmatrix} V_1 e^{j\phi_1} \\ V_2 e^{j\phi_2} \\ \vdots \\ V_N e^{j\phi_N} \end{pmatrix} = \bar{0} \tag{6}$$

When solving the system for a constant phase shift distribution the phase differences in the exponential terms of (5) are substituted for the desired constant phase shift value $\Delta\phi$.

$$\phi_2 - \phi_1 = \phi_3 - \phi_2 = \ldots = \phi_N - \phi_{N-1} = \Delta\phi \tag{7}$$

As stated before, due to their autonomous nature of coupled oscillator systems, one may set one of the phases to zero. Here $\phi_1 = 0$ is taken as a reference phase.

For each of the desired constant phase shift values $\Delta\phi$ the system (6) is solved in terms of V_i, ω, V_{T1} and V_{TN}. This formulation allows designing and analyzing coupled oscillator systems in an efficient manner. The optimum coupling network for each specific coupled oscillator system can also be determined from the proposed formulation.

The presented formulation is only valid for weak coupling conditions when the variations in the amplitude and the frequency of the oscillators when introduced into the array are small in comparison to the free-running state.

In the free-running state the admittance function fulfils $Y_k = 0$. As the N oscillator elements that form the array are typically identical, the derivatives involved in the admittance functions are also equal in all the N oscillators. These derivatives are calculated around the free running solution of the oscillators. They are obtained in a harmonic balance commercial simulator using finite differences. $\left.\frac{\partial Y}{\partial V}\right|_0$ is obtained introducing an increment in V , while keeping the frequency and the tuning parameter in the free running values $\omega = \omega_0$ and $V_T = V_{T0}$. The derivative $\left.\frac{\partial Y}{\partial \omega}\right|_0$ is obtained introducing an increments in ω and keeping the amplitude and the tuning parameter to the free-running values $V = V_0$ and $V_T = V_{T0}$. The derivatives versus the tuning parameter $\left.\frac{\partial Y}{\partial V_T}\right|_0$ are calculated maintaining the amplitude and frequency to the free-running values $V = V_0$ and $\omega = \omega_0$ and introducing an increment in V_T .

The precision of the coupled oscillator model can be increased introducing higher order derivatives in the Taylor series expansion when calculating the admittance functions.

Additionally, in order to consider the frequency dependence of the coupling network admittance matrix Y_c, a Taylor expansion of Y_c around the free-running frequency ω_0 can also be used.

1.1.2 Stability Analysis

In order to study the stability of the steady state solution obtained with the proposed formulation, a small perturbation is introduced around the steady state solution. This perturbation is introduced in all the circuit variables and is time varying.

$$\begin{aligned} V_i^p &= V_i + \Delta V_i(t) \\ \phi_i^p &= \phi_i + \Delta\phi_i(t) \\ \omega_i^p &= \omega_i + \Delta\omega_i(t) = \omega_i + \left[\Delta\dot{\phi}_i - j\frac{\Delta\dot{V}_i}{V_i}\right] \end{aligned} \tag{8}$$

The well known Kurokawa [24, 25, 26] substitution is used to calculate the frequency perturbation. Substituting (8) in (6) the systems of equations necessary to obtain the stability of the coupled oscillator system steady state solutions can be written as follows:

$$\begin{aligned}
&H_{\omega}^{1}\left[\Delta\dot{\phi}_1 - j\frac{\Delta\dot{V}_1}{V_1}\right] + H_{V_1}^{1}\Delta V_1 + \cdots + H_{V_N}^{1}\Delta V_N + H_{\phi_1}^{1}\Delta\phi_1 + \cdots + H_{\phi_N}^{1}\Delta\phi_N = 0\\
&H_{\omega}^{2}\left[\Delta\dot{\phi}_2 - j\frac{\Delta\dot{V}_2}{V_2}\right] + H_{V_1}^{2}\Delta V_1 + \cdots + H_{V_N}^{2}\Delta V_N + H_{\phi_1}^{2}\Delta\phi_1 + \cdots + H_{\phi_N}^{2}\Delta\phi_N = 0\\
&\vdots\\
&H_{\omega}^{N}\left[\Delta\dot{\phi}_N - j\frac{\Delta\dot{V}_N}{V_N}\right] + H_{V_1}^{N}\Delta V_1 + \cdots + H_{V_N}^{N}\Delta V_N + H_{\phi_1}^{N}\Delta\phi_1 + \cdots + H_{\phi_N}^{N}\Delta\phi_N = 0
\end{aligned} \tag{9}$$

where H_{ω}, H_{Vi} and H_{ϕ_1} are the derivatives of the functions H_i in (6). Equation (9) can be rewritten in a compact form as:

$$\begin{aligned}
&H_{\omega}^{1}\left[\Delta\dot{\phi}_1 - j\frac{\Delta\dot{V}_1}{V_1}\right] + \sum_{i=1}^{N}\left[H_{V_i}^{1}\Delta V_i + H_{\phi_i}^{1}\Delta\phi_i\right] = 0\\
&H_{\omega}^{2}\left[\Delta\dot{\phi}_2 - j\frac{\Delta\dot{V}_2}{V_2}\right] + \sum_{i=1}^{N}\left[H_{V_i}^{2}\Delta V_i + H_{\phi_i}^{2}\Delta\phi_i\right] = 0\\
&\vdots\\
&H_{\omega}^{N}\left[\Delta\dot{\phi}_N - j\frac{\Delta\dot{V}_N}{V_N}\right] + \sum_{i=1}^{N}\left[H_{V_i}^{N}\Delta V_i + H_{\phi_i}^{N}\Delta\phi_i\right] = 0
\end{aligned} \tag{10}$$

It should be noted that the proposed formulation allows one to analyze the stability of any coupled oscillator systems independently of the individual oscillator topology.

The stability of the steady state solution is determined form the eigenvalues of (10). The coupled oscillator system is an autonomous circuit, which implies the resulting perturbed system is singular and has an eigenvalue equal to zero. In order for the system to be stable the rest of the eigenvalues must be located on the left hand side of the complex plane, which means all of them must have a negative real part [9, 21, 41, 3, 20, 43].

In the case that a real eigenvalue or a pair of complex conjugate eigenvalues crosses the imaginary axis to the right hand side of the complex plane, the system loses its stability and the analyzed steady state solution is unstable.

In order to use coupled oscillator systems for beam scanning in antenna arrays the solution that establishes a constant phase shift distribution among the elements of the array is desired. To check the stability of this solution (10) is rewritten using (7).

In a weakly coupled linear array of oscillators, a theoretical maximum stable constant phase shift range of [−90°,90°] can be obtained [1, 46] when the coupling phase among elements is a multiple of the wavelength of the system. In the next section these results are confirmed by applying the introduced formulation in a practical design of a three coupled oscillator linear array.

1.2 Practical Example: Three Coupled Oscillator Linear Array

A three element (N=3) coupled oscillator array has been designed using the formulation introduced in the previous sections. Each of the individual oscillator elements is based on a NE3210s01 NEC HJ-FET transistor and has an oscillating frequency around 6 GHz. The tuning of the oscillator frequency is done using a varactor diode MACOM MA46H070. The circuit is implemented in a Cuclad 217 substrate (Fig. 2).

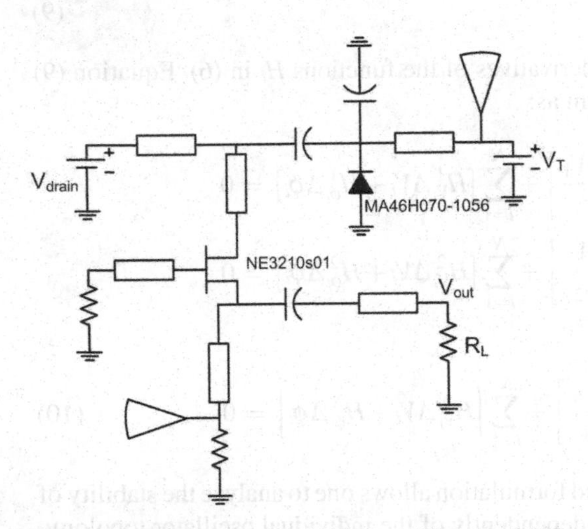

Fig. 2 Schematic of the individual oscillator at 6 GHz

The coupling network used to connect the three oscillator elements is a transmission line with characteristic impedance Z_o (Fig. 3) loaded with two resistors R [46, 15]. The admittance matrix of the coupling network is given by:

$$Y_{ij} = \begin{cases} n_i \dfrac{e^{j\psi} + j\frac{R-Z_0}{Z_0}\sin(\psi)}{2Re^{j\psi} j\frac{(R-Z_0)^2}{Z_0}\sin(\psi)} & , i = j \\ \dfrac{-1}{2Re^{j\psi} j\frac{(R-Z_0)^2}{Z_0}\sin(\psi)} & , |i-j| = 1 \\ 0 & , \text{other} \end{cases} \qquad (11)$$

where $n_i = 1$ for $i = 1$ and N and $n_i = 2$ for the rest of values of i.

The main goal of this example is to design an optimized system of coupled oscillators for beam steering applications. The proposed formulation is used to find

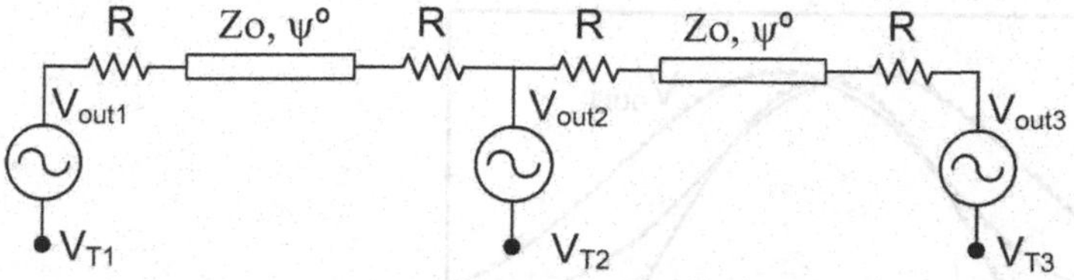

Fig. 3 Three coupled oscillator array. Coupling network formed by a loaded transmission line.

the values of the pairs of tuning voltages (V_{T1}, V_{T3}) needed in order to synthesize the different constant phase shift distributions $\Delta\phi$ necessary to scan the beam in a desired direction. In the next section the multiple coexisting solutions in the designed three elements coupled oscillator system are also obtained and their stability is studied.

As a first step in the design, the formulation is used to determine the optimum coupling network parameters (Z_o, ψ, R) in order to obtain the maximum stable constant phase shift range possible. Weak coupling conditions are ensured by choosing the appropriate value of the coupling resistance R. The stability of the constant phase shift solution for different coupling line lengths ψ has been obtained solving equations (6) and (10) using the conditions in (7). The resolution of (6) is made using a Newton Raphson algoritm. The results are represented in Fig. 4 showing that in

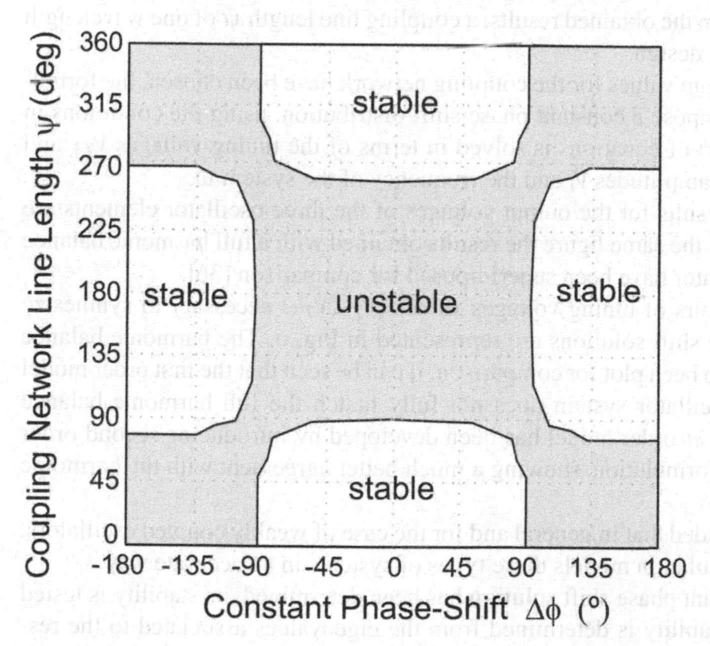

Fig. 4 Stability analysis of the three coupled oscillator array versus the coupling line length ψ

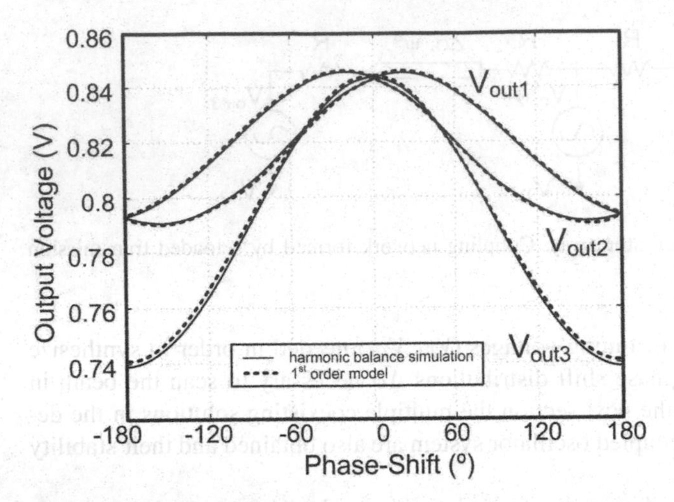

Fig. 5 Output voltage variation of the three oscillator elements versus the imposed constant phase shift

order to obtain the maximum stable constant phase shift range around the in-phase value, the coupling line length has to be a multiple of the wavelength of the system. It is also possible to obtain a maximum stable constant phase shift range around the out of phase value choosing a coupling line length ψ multiple of half the wavelength of the system. From the obtained results, a coupling line length ψ of one wavelength was chosen for the design.

Once the optimum values for the coupling network have been chosen, the formulation is used to impose a constant phase shift distribution, using the conditions in (7). The system (6) of equations is solved in terms of the tuning voltages V_{T1} and V_{T3}, the oscillator amplitudes V_i and the frequency of the system ω.

The obtained results for the output voltages of the three oscillator elements are shown in Fig. 5. In the same figure the results obtained with a full harmonic balance commercial simulator have been superimposed for comparison [36].

The different pairs of tuning voltages values (V_{T1}, V_{T3}) necessary to synthesize the constant phase shift solutions are represented in Fig. 6. The harmonic balance simulation has also been plot for comparison. It can be seen that the first order model of the coupled oscillator system does not fully match the full harmonic balance simulation. A higher order model has been developed by introducing second order derivatives in the formulation, showing a much better agreement with the harmonic balance results.

It can be concluded that in general and for the case of weakly coupled oscillators, the proposed formulation models these types of systems in an accurate way.

Once the constant phase shift solution has been determined, its stability is tested using (10). The stability is determined from the eigenvalues associated to the resolution of the system (10). In Fig. 7 the evolution of the real part of the dominant eigenvalue is plotted versus the positive range of constant phase shift values. It can

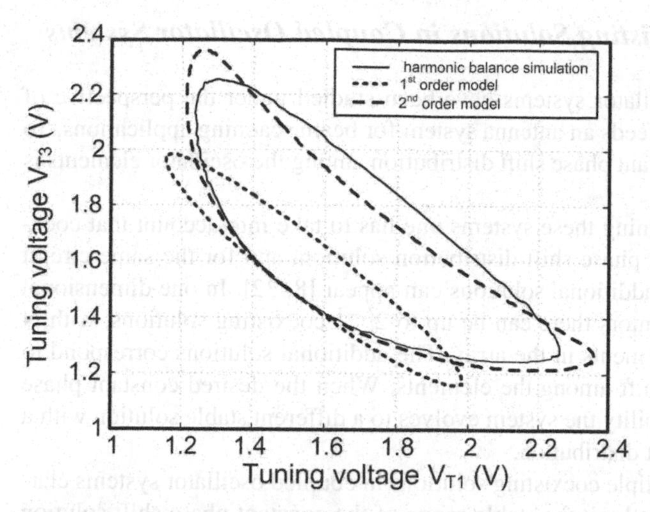

Fig. 6 Pairs of tuning voltages necessary to synthesize the constant phase shift solutions from 0° to 360°

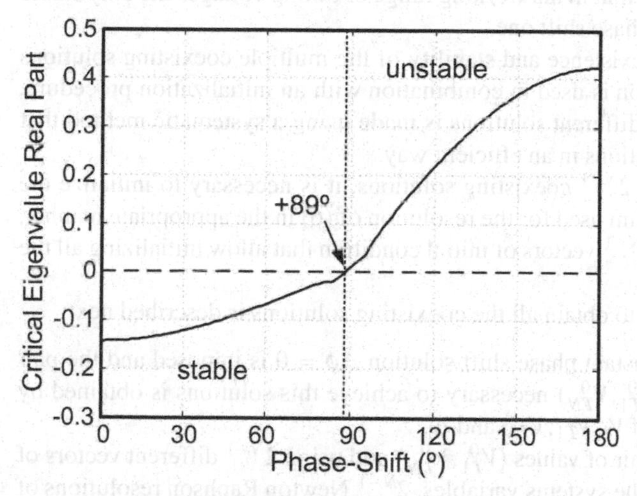

Fig. 7 Evolution of the real part of the critical eigenvalue that determines the stability properties of the constant phase shift solution in the 3x1 coupled oscillator array

be observed that the studied solution is stable in the range (0°, +89°). Due to the symmetry of the system the stability of the negative constant phase shift values will be (−89°,0°). The total stable range is (−89°,+89°) which is approximately the expected theoretical range of (−90°,+90°) [46]. From values of constant phase shifts higher than +89° and lower than −89° the constant phase shift solution becomes unstable and the system evolves to another of the coexisting solutions. A detailed description of the phenomena that lead to the loss of stability is given in the next section.

1.3 Multiple Coexisting Solutions in Coupled Oscillator Systems

Until now coupled oscillator systems have been studied under the perspective of designing an array that feeds an antenna system for beam scanning applications. To achieve this goal a constant phase shift distribution among the oscillator elements is synthesized.

However when designing these systems one has to take into account that coexisting with the constant phase shift distribution solution, and for the same circuit parameters (V_{T1}, V_{TN}), additional solutions can appear [8, 32]. In one dimensional arrays of coupled oscillators there can be up to 2^{N-1} coexisting solutions, with N being the number of elements in the array. The additional solutions correspond to a non constant phase shift among the elements. When the desired constant phase shift solution losses stability the system evolves to a different stable solution with a non-constant phase shift distribution.

The study of the multiple coexisting solutions in coupled oscillator systems clarifies the phenomena that limit the stable range of the constant phase shift solution and how the existence of additional solutions affects the performance of the circuit. The final aim is to ensure that in the working range of tuning voltages the only stable solution is the constant phase shift one.

In order to study the existence and stability of the multiple coexisting solutions the introduced formulation is used in combination with an initialization procedure. The initialization of the different solutions is made using a systematic method that allows obtaining the solutions in an efficient way.

In order to obtain the 2^{N-1} coexisting solutions, it is necessary to initialize the Newton Raphson algorithm used for the resolution of (6) in the appropriate manner. This implies providing 2^{N-1} vectors of initial condition that allow initializing all the coexisting solutions [8].

The process followed to obtain all the coexisting solutions is described next:

1. In a first step, the constant phase shift solution $\Delta\phi = 0$ is imposed and the pair of tuning voltages (V_{T1}^0, V_{TN}^0) necessary to achieve this solutions is obtained by solving (6) in terms of V_i, V_{T1}, V_{TN} and ω.
2. Fixing the obtained pair of values (V_{T1}^0, V_{TN}^0) and using 2^{N-1} different vectors of initial conditions for the systems variables, 2^{N-1} Newton Raphson resolutions of the system (6) in terms of V_i, ϕ_i and ω are performed
3. Once the 2^{N-1} solutions are obtained, and increment in the constant phase shift is introduced leading to $\Delta\phi_1$ and the new pair of values (V_{T1}^1, V_{TN}^1) necessary to achieve it are calculated solving (6) in terms of V_i, V_{T1}, V_{TN} and ω.
4. The additional 2^{N-1} coexisting solutions are obtained solving (6) for the fixed values (V_{T1}^1, V_{TN}^1) and initializing the Newton Raphson algorithm with the results of the variables obtained in step 2.
5. This process is repeated for all the values of constant phase shift using an iterative process.

A key issue for solving (6) for the different coexisting solutions, is how to obtain the 2^{N-1} vectors of initial conditions needed in step 2. For the obtained pair of tuning voltages (V_{T1}^0, V_{TN}^0), the values of the phases of the oscillators ϕ_i in the vector of initial conditions, present a phase relationship given by the 2^{N-1} binary combinations of 0 and π radians. Due to the symmetry conditions in the system the values of the oscillator phases for the 2^{N-1} coexisting solution is close to the values indicated in (12).

$$
\begin{array}{lll}
k = 1, & \bar{\phi}^1 & = [\phi_1^1 \ \phi_2^1 \ldots \ \phi_N^1] = [0,\ldots,0,0] \\
k = 2, & \bar{\phi}^2 & = [\phi_1^2 \ \phi_2^2 \ldots \ \phi_N^2] = [0,\ldots,0,\pi] \\
\vdots & & \\
k = 2^{N-1}, & \bar{\phi}^{2^{N-1}} & = [\phi_1^{2^{N-1}} \ \phi_2^{2^{N-1}} \ \ldots \ \phi_N^{2^{N-1}}] = [\pi,\ldots,\pi,\pi]
\end{array}
\qquad (12)
$$

where k indicate the number of the solution.

The stability analysis of each of the multiple coexisting solutions is performed using the stability equations (10). In the next section the multiple coexisting solutions are obtained and their stability analyzed for a three element coupled oscillator array.

1.4 Practical Example: Multiple Coexisting Solutions in a Three Coupled Oscillator Linear Array

A similar circuit to the one in Fig. 2 also based on the NE3210s01 transistor has been used to illustrate the coexistence of solutions in coupled oscillator systems. In a three oscillator array (N=3) up to $2^{N-1} = 2^{3-1} = 2^2 = 4$ multiple coexisting solutions can appear.

The multiple coexisting solutions have been obtained using the formulation in (6) in combination with the initialization procedure. In Fig. 8a the amplitude of one of the oscillator of the chain for the different coexisting solutions has been plot. The synchronization frequency of the system for the different coexisting solutions has been represented in Fig. 8b. Solution 1 in Fig. 8 corresponds to the constant phase shift solution.

The plots in Fig. 8 are traced versus the constant phase shift distribution $\Delta\phi$. Each constant phase shift value corresponds to a pair of tuning voltages (V_{T1}, V_{T3}). This means coupled oscillator systems used for beam steering applications are two parameter systems. Ideally Fig. 8 should be represented versus the pairs of tuning voltages (V_{T1}, V_{T3}), but in order not to loss perspective with a three dimensional plot, the results have been represented versus the constant phase shift corresponding to each pair (V_{T1}, V_{T3}).

In order to get a broader understanding of the dynamics involved in coupled oscillator arrays, the system has been reduced to a one parameter system and the evolution of the multiple coexisting solutions has been studied versus a single tuning parameter. The value of V_{T1} has been fixed to different values and for each of these values the evolution of the solutions versus V_{T3} has been obtained. When performing this type of analysis on the three coupled oscillator array, for each fixed value of V_{T1}, one obtains two closed curves $C1$ and $C2$ versus V_{T3}.

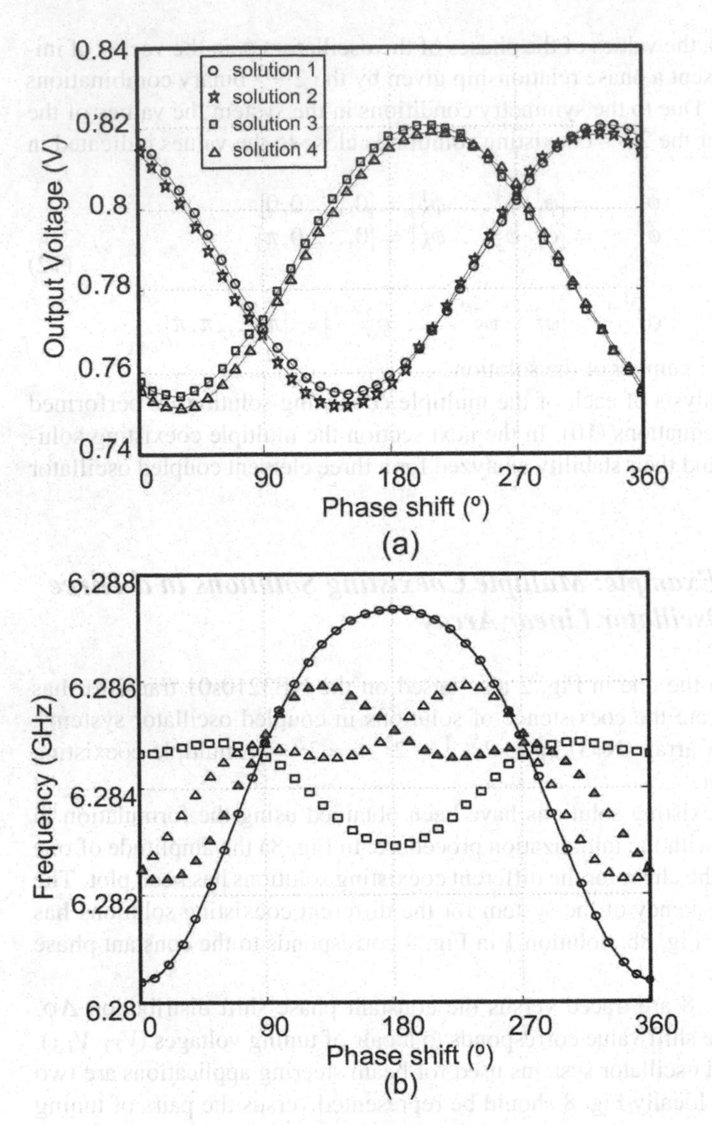

Fig. 8 Four coexisting solutions in the 3x1 coupled oscillator array. Solution 1 correspond to the constant phase shift solution. (a) Output voltage of one of the oscillator elements for the four coexisting solutions versus the constant phase shift. (b) Frequency of the four coexisting modes versus the constant phase shift.

In Fig. 9a the amplitude of the central oscillator of the array has been represented versus V_{T3} for two different values of V_{T1}. The curves $C1$ and $C2$ in Fig. 9a are obtained for a value $V_{T1} = 0.4V$. In these curves there is only one constant phase shift point. This point can be found for the pair of values $(V_{T1} = 0.4V, V_{T3} = 0.6V)$ and correspond to a constant phase shift of 50°. In Fig. 9a it can be seen that, for

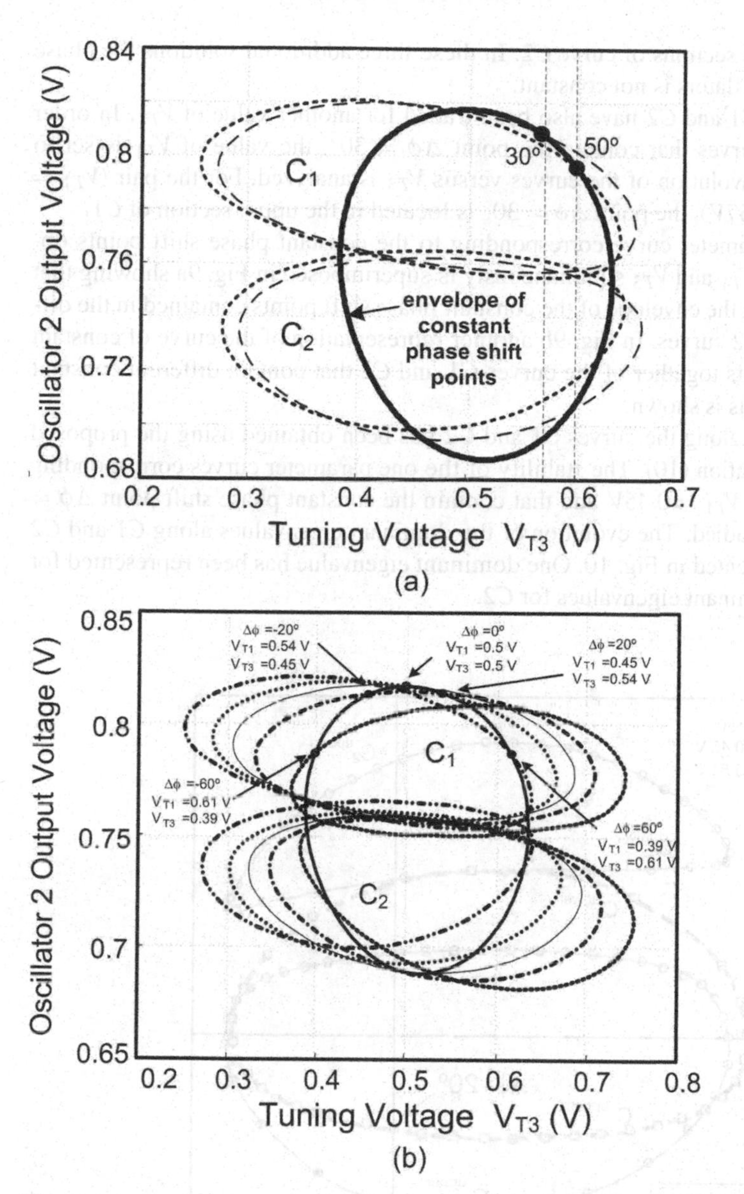

Fig. 9 One parameter and two parameter curves for the 3x1 coupled oscillator array. (a) One parameter curves corresponding to $\Delta\phi = 30^\circ$ and $\Delta\phi = 50^\circ$. (b) Set of one parameter curves both in the positive and negative ranges of constant phase shift.

this case, this point is located in the upper section of curve $C1$. Coexisting with this constant phase shift solution there are three additional ones, as expected. One of them is located in the lower section of curve $C1$ and the other two are located in the upper and lower sections of curve $C2$. In these three additional solutions the phase shift among oscillators is not constant.

The curves $C1$ and $C2$ have also been traced for another value of V_{T1}. In order to obtain the curves that contain the point $\Delta\phi = 30°$, the value of V_{T1} is set to $0.43V$ and the evolution of the curves versus V_{T3} is analyzed. For the pair ($V_{T1} = 0.43V, V_{T3} = 0.57V$), the point $\Delta\phi = 30°$ is located in the upper section of $C1$.

The two parameter curve corresponding to the constant phase shift points obtained varying V_{T1} and V_{T3} simultaneously is superimposed in Fig. 9a showing that this curve forms the envelope of the constant phase shift points contained in the different $C1$ and $C2$ curves. In Fig. 9b another representation of the curve of constant phase shift points together of the curves $C1$ and $C2$ that contain different constant phase shift points is shown.

The stability along the curves $C1$ and $C2$ has been obtained using the proposed stability formulation (10). The stability of the one parameter curves corresponding to a fixed value $V_{T1} = 0.45V$ and that contain the constant phase shift point $\Delta\phi = 20°$ has been studied. The evolution of the dominant eigenvalues along $C1$ and $C2$ has been represented in Fig. 10. One dominant eigenvalue has been represented for $C1$ and two dominant eigenvalues for $C2$.

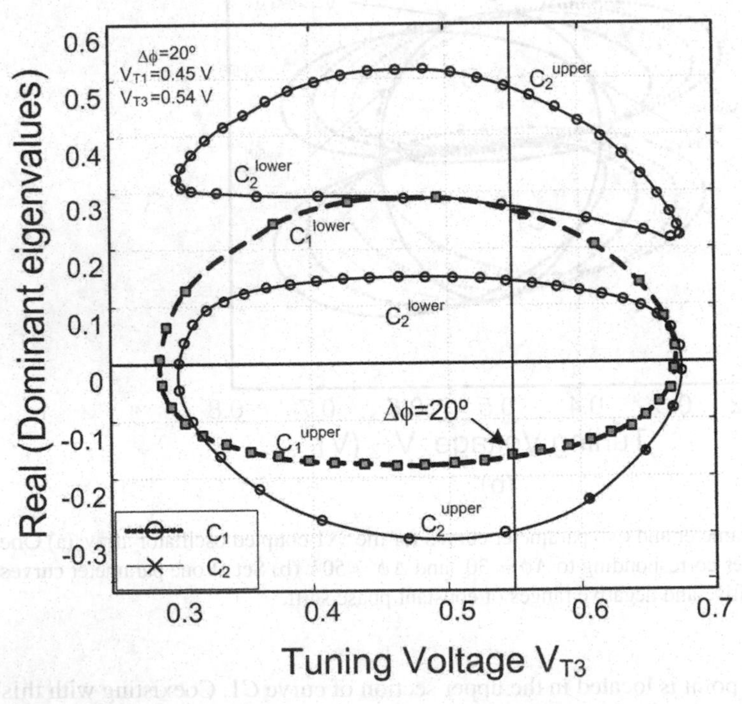

Fig. 10 Evolution of the real part of the dominant eigenvalues corresponding to the constant phase shift solution $\Delta\phi = 20°$. Stability along the one parameter curves.

Observing the obtained results in Fig. 10, it can be seen that for curve $C2$ there is always one eigenvalue with positive real part so it can be concluded that the curve $C2$ is unstable both in its upper and lower sections. In the case of curve $C1$, the dominant eigenvalue has positive real part in the lower section of the curve. This means the lower section of $C1$ is unstable. However, the upper section is stable as the real part of the dominant eigenvalue is negative.

The limit of stability between the upper and lower sections is marked by the infinite slope points (turning points) of the curves in Fig. 9 [41, 18, 43]. These turning points correspond to local/global bifurcations of the system. For values of V_{T3} beyond these turning points the system of coupled oscillators looses synchrony.

When studying the stability of the system versus one parameter the appearance of a turning point means a change of the stability properties of the systems under study. In the present case this parameter corresponds to V_{T3}. However, in a two parameter system a turning point in the solution curves does not imply a change of the stability. This is the case of the envelope curve that only contains the constant phase shift points. In order to obtain this curve both V_{T1} and V_{T3} have been modified. This mean this solution curve is a two parameter curve and its turning points do not imply a change in its stability.

The constant phase shift solution curve (envelope curve) corresponds to a family of points each of which belongs to a different curve $C1$ or $C2$. This means each of the points correspond to a different value of V_{T1}. Observing the evolution of $C1$ in Fig. 9, one can see the constant phase shift points are moving along the different $C1$ curves until the constant phase shift point coincides with the turning point of the curve $C1$. After reaching the turning point, the next point of constant phase shift will fall in the lower section of curve $C2$.

As seen in Fig. 10 the complete curve $C2$ is unstable. This means that the points of constant phase shift that fall in $C2$ will be unstable. At the pair of tuning voltages (V_{T1}, V_{T3}) at which the turning point appears, the constant phase shift solution becomes unstable. This point indicates the maximum achievable stable constant phase shift solution. From that point on, even though the constant phase shift solutions may mathematically exist they will be unstable. The system will evolve to one of the other coexisting solutions with non constant phase shift distributions that now will fall in the upper section of $C1$ and will be the stable solution of the system.

In Fig. 11 it can be seen that the point where the envelope curve losses its stability does not correspond with a turning point. This agrees with the argument presented before as this curve is a two parameter curve. However if looking at the same point in the one parameter curves $C1$ and $C2$, the loss of stability does correspond with a turning point.

In Fig. 11 the envelope curve is represented with a continuous line. In the same figure the curve $C1$ for a value $V_{T1} = 0.38V$ is also represented. Curve $C1$ contains the constant phase shift point $\Delta\phi = 90°$, that is the maximum achievable stable value expected from the theory. Also in Fig. 11 the curve $C2$ corresponding to $V_{T1} = 0.39V$ is represented. This $C2$ curve contains the constant phase shift point $\Delta\phi = 100°$. It can be observed that the point $\Delta\phi = 90°$ corresponds with the right edge turning point of curve $C1$. In this point $\Delta\phi = 90°$ the constant envelope curve also becomes

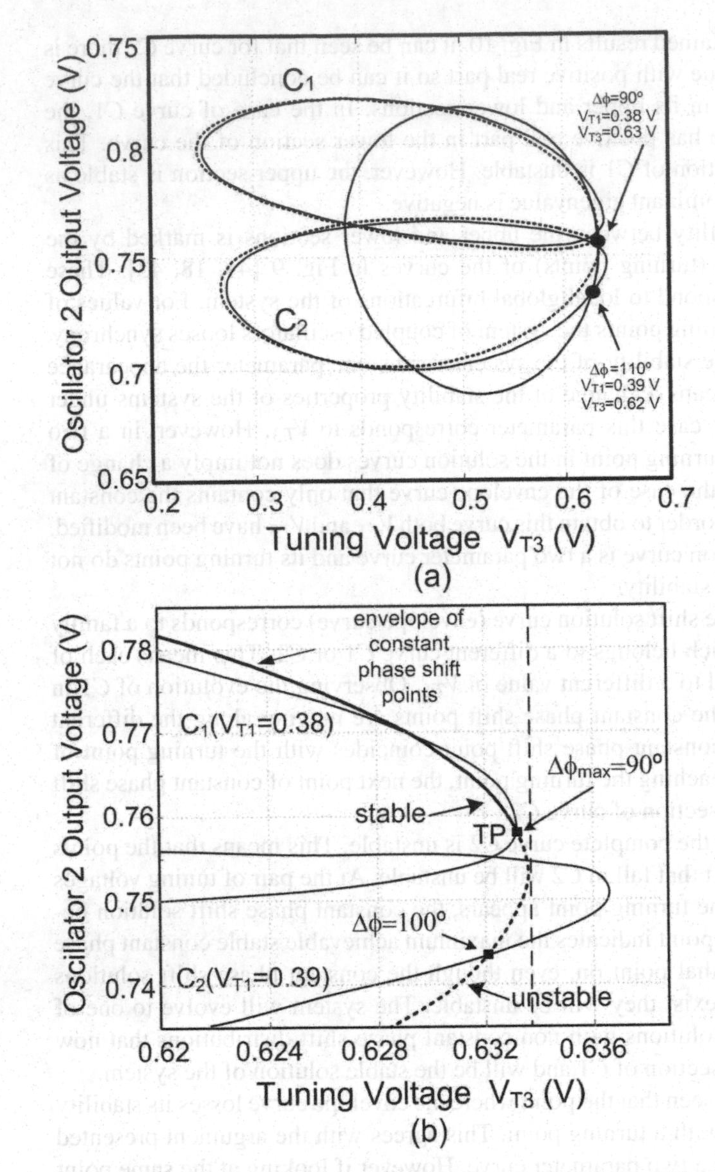

Fig. 11 One parameter and two parameter curves for $\Delta\phi = 90°$ and for $\Delta\phi = 100°$. The stability is lost at $\Delta\phi = 90°$, coinciding with the turning point in the one parameter curves.(a) Complete view (b) Detailed view.

unstable, however it does not present a turning point. This is due to the fact that the constant envelope curve is a two parameter curve.

The points of constant phase shift beyond 90°, as it is the case of 100°, are located in the lower section of $C2$ ($V_{T1} = 0.39V$)(see Fig. 11). As curve $C2$ has been found to be unstable in all its sections, these points are unstable. For the pair of tuning voltages (V_{T1}, V_{T3}) that lead to the mathematical constant phase shift solution of 100°, the systems will evolve to the stable coexisting solution that fall in the upper section of $C1(V_{T1} = 0.39V)$. This solution is a non constant phase shift solution.

1.5 Conclusion

In this section about "Linear coupled oscillator arrays" a formulation for modelling coupled oscillator systems has been introduced. This formulation allows designing these systems in an efficient manner reducing the simulation time compared to full harmonic balance simulations.

The use of coupled oscillator systems for beam steering applications require a constant phase shift distribution among the oscillator elements that form the array. However coexisting with the constant phase shift there are additional solutions. Using the proposed model the multiple coexisting solutions present in coupled oscillator systems have been obtained and their stability studied. From the results obtained it has been concluded that only one of the coexisting solutions is stable at a time under weak coupling conditions. When the coupling line length is a multiple of the wavwlength of the system, the constant phase shift solution is stable in the range [−90° ,+90°] centered around 0° . Outside this range the constant phase shift solution losses stability and the system evolves to stable nonconstant phase shift solutions.

2 N–Push Oscillator Topologies

N-push oscillator topologies have found their main application in high frequency generation. In these topologies N oscillator elements are coupled together by means of a coupling structure forming a ring [37, 38, 6, 5, 42, 45]. The outputs of the oscillating elements are combined at a certain point to maximize the output power at the Nth harmonic. In order to achieve high frequency generation, at the combined output the Nth harmonics of the oscillator elements have to add up while the rest of the harmonics cancel. In order to combine the Nth harmonic components, the phase distribution along the elements of the chain has to be $\frac{2\pi}{N}$, forming a travelling wave mode.

However, due to their symmetry conditions, these N-push topologies can present multiple coexisting solutions for the same values of the circuit parameters [17]. Moreover these coexisting solutions can even be simultaneously stable [10, 22, 35, 2]. This fact can become a problem when there is no control on the appearance of the different solutions or on how to avoid the undesired modes. The determination and control of the multiple coexisting solutions is a key issue when designing these systems. In the following design tips to obtain and control the multiple modes in these systems are introduced.

2.1 Multiple Steady-State Coexisting Solutions

Rings of coupled oscillators are highly symmetric structures. In [6, 17, 10, 2] it has been shown that this symmetry conditions give rise to the appearance of multiple coexisting solutions, some of which could even be simultaneously stable.

The number of solutions (modes) that appear in these N-push oscillators depends directly on the number of oscillators that form the ring. As N increases the number of coexisting modes also increases. The modes in N-push oscillator topologies can be defined by the relationships established among the amplitudes and phases of the oscillator elements that form the chain. The different coexisting modes in triple-push and quadruple-push oscillators are described in the following. A diagram representation of these two structures is shown in Fig. 12.

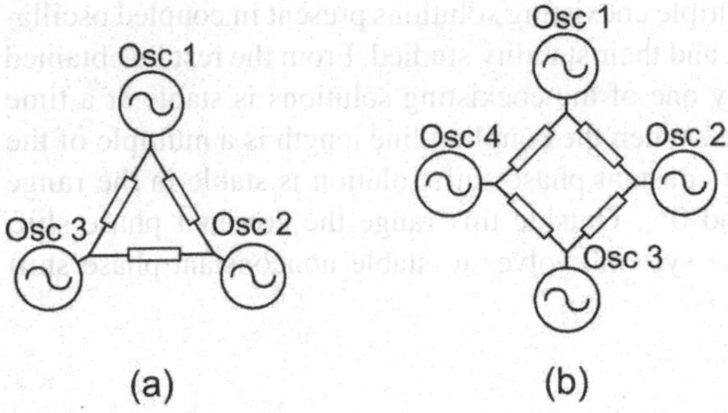

Osc 1
Osc 3
Osc 2
(a)
Osc 1
Osc 4
Osc 2
Osc 3
(b)

Fig. 12 Schematic representation of triple-push and quadruple-push structures. (a) Triple-push schematic. (b) Quadruple-push schematic.

2.1.1 Modes in Triple-Push Oscillators

A triple push oscillator is formed by N=3 oscillator elements. The goal of a triple push oscillator is to combine the 3rd harmonic components of the oscillator elements at the output and cancel the rest of the harmonics. In order to achieve this goal the oscillator elements has to present a phase shit of $\frac{2\pi}{3}$ among them. However in the case of the triple-push oscillators up to four coexisting modes can appear in the system, so three additional undesired modes may coexist with the desired travelling wave mode [17].

A detailed description of the modes is shown in Table 1. These modes are defined by the relationships among the phases and amplitudes of the three oscillators [17] in Fig. 12a and are listed in the following:

1. Mode M1 (in-phase mode). The three oscillator elements in Fig. 12a have the same phase and the same amplitudes. In this mode the 1st harmonic component of the oscillators add up in the combined node.
2. Mode M2 (travelling wave mode). The three oscillator elements in Fig. 12a present a phase-shift of $\frac{2\pi}{3}$ among them. In this mode the three oscillators have

Table 1 Coexisting modes in a triple-push oscillator

Mode	Name	A_1	A_2	A_3	ϕ_1	ϕ_1	ϕ_1
Mode 1	M1	A	A	A	ϕ	ϕ	ϕ
Mode 2	M2	A	A	A	ϕ	$\phi+120$	$\phi+240$
Mode 3	M3	A	A	B	ϕ	$\phi+180$	$\theta+\omega_0 t$
Mode 4	M4	A	A	B	ϕ	ϕ	$\phi+180$

[a] Multiple coexisting modes in a tripe-push oscillator. Relationships between the amplitudes and phases in the different oscillator elements relatives to $\omega_0 t$.

the same amplitude. This mode is the desired one in order to combine the 3rd harmonic components at the output of the circuit.

3. Mode M3 (double frequency mode). Oscillators 1 and 2 in Fig. 12a have the same amplitude and present a phase shift of π radians. Oscillator 3 has an oscillation frequency double to the frequency of oscillators 1 and 2.
4. Mode M4. Oscillators 1 and 2 are in phase and have the same amplitude while oscillator 3 is out of phase and has different amplitude.

2.1.2 Modes in Quadruple-Push Oscillators

A quadruple push oscillator is formed by N=4 oscillator elements. In the case of the quadruple push oscillator the oscillator elements will have a phase relationship of $\frac{2\pi}{4}$ in order to combine the 4th harmonic components of the oscillator elements at the output and cancel the rest of the harmonics. The number of possible synchronized modes in a quadruple-push oscillator is five. The phase and amplitude relationships among the oscillator elements can be seen in Table 2 [17]. A brief explanation of the modes in Table 2 follows:

1. Mode M1 (in-phase mode). The four oscillator elements in Fig. 12b have the same phase and the same amplitudes. In this mode the 1st harmonic component of the oscillators add up in the combined node.
2. Mode M2 (travelling wave mode). The four oscillator elements in Fig. 12b have a phase-shift of $\frac{2\pi}{4}$ among them. The amplitudes are the same at each of the oscillator elements. This mode is the desired one in order to combine the 4th harmonic components at the output of the circuit.
3. Mode M3 (double frequency mode). Oscillators 1 and 3 have the same amplitude and present a phase shift of π radians. Oscillators 2 and 4 are oscillating at a frequency double the one of oscillators 1 and 3.
4. Mode M4. Oscillators 1 and 2 are in phase and have the same amplitude. Oscillators 3 and 4 have the same amplitudes and are in phase. Oscillators 1 and 2 are out of phase with oscillators 3 and 4.
5. Mode M5. Oscillators 1 and 3 are in phase and have the same amplitude. Oscillators 2 and 4 are in phase and have the same amplitude. Oscillators 1 and 3 are out of phase with oscillators 2 and 4.

Table 2 Coexisting modes in a quadruple-push oscillator

Mode	Name	A_1	A_2	A_3	A_4	ϕ_1	ϕ_1	ϕ_1	ϕ_4
Mode 1	M1	A	A	A	A	ϕ	ϕ	ϕ	ϕ
Mode 2	M2	A	A	A	A	ϕ	$\phi+90$	$\phi+180$	$\phi+270$
Mode 3	M3	A	A	B	C	ϕ	$\phi+180$	$\theta+\omega_0 t$	$\psi+\omega_0 t$
Mode 4	M4	A	A	B	B	ϕ	ϕ	$\phi+180$	$\phi+180$
Mode 5	M5	A	B	A	B	ϕ	$\phi+180$	ϕ	$\phi+180$

[a] Multiple coexisting modes in a quadruple-push oscillator. Relationships between the amplitudes and phases in the different oscillator elements relatives to $\omega_0 t$.

2.2 *Stability of the Multiple Coexisting Modes*

As presented in the previous section, the symmetry properties of N-push oscillator circuits lead to the coexistence of several oscillation modes for the same circuit parameters. In general, considering coupling among the oscillators without delay, the multiple solutions, although mathematically coexisting, would not be stable at the same time [17]. This means although the circuit resolution will show several modes, in the final implementation only one of the modes will be stable and consequently observable. The case of quadruple push oscillators is an exception, and it is possible to find multistable modes even when the oscillator elements are coupled without delay [17].

However when introducing delay in N-push oscillator circuits, the complexity of the involved dynamics in these systems increase considerably. The presence of delays gives rise to the multistability phenomenon [10, 2, 22, 35, 6]. For the same circuit parameters not only multiple modes coexist but also can be simultaneously stable. This fact will be shown in the practical example in the next section.

2.3 *Practical Example: Triple-Push Oscillator at 13.8 GHz*

In order to illustrate the introduced concepts, in this section the design of a triple push oscillator circuit will be presented. The circuit is analyzed in order to obtain the multiple coexisting modes. The modes are obtained using harmonic balance based techniques.

Once the modes have been obtained, a stability analysis is performed to determine which of the modes are stable and to study the possible multistability. The stability analysis is performed using a properly initialized envelope transient simulation [31, 39].

Finally a technique to avoid the undesired modes in triple push oscillator circuits is presented. The technique is based on the control of the coupling phase and coupling strength among the oscillator elements that form the system.

2.3.1 Triple-Push Oscillator Design

A triple push oscillator circuit with an output frequency of 13.8GHz has been designed. The designed triple push oscillator is formed by the combination of three single oscillators operating at a frequency of 4.6 GHz. The single oscillators at 4.6GHz are based on the NEC HJ-FET NE3210s01 transistor and are designed to present a high third harmonic. The circuit was fabricated in Cuclad 217 substrate with a thickness of 0.78mm.

The three oscillator elements are connected together to form the triple push oscillator using a star type coupling network. The star network is formed by three 50 Ohm transmission lines. The combined output of the circuit is taken in the middle of the coupling structure.

In Fig. 13 the schematic view of the designed triple push oscillator can be seen. The coupling line length marked as L_c in Fig. 13 controls the phase of the coupling network while the coupling strength for this specific circuit is controlled by introducing attenuation networks at the output of each oscillator element indicated as α in Fig. 13. The combined output is marked with the letter N in Fig. 13.

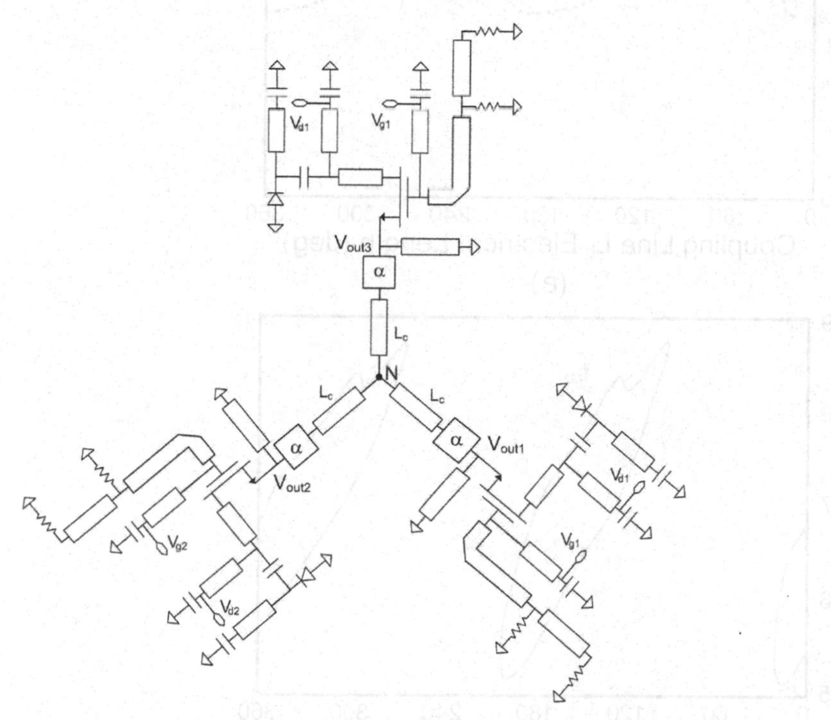

Fig. 13 Schematic of the triple-push oscillator circuit at 13.8 GHz

2.3.2 Multiple Coexisting Modes in the 13.8GHz Triple-Push Oscillator

Due to symmetry conditions triple push oscillator circuits can have up to four coexisting modes for the same values of circuit parameters. This modes were described in detail in Table 1. In this section the different coexisting modes in the designed triple push oscillator are detected following the criteria in Table 1. Once the modes are recognized their behavior versus the coupling strength and phase among the individual oscillator elements is analyzed.

In order to detect the different modes in the triple push oscillator, it is necessary to impose the amplitude and phase relationships that appear in Table 1. The technique used here is based on the combined use of harmonic balance simulation and auxiliary generators [41]. In order to avoid the convergence of the harmonic balance to the DC solution of the circuit, an auxiliary generator is introduced into

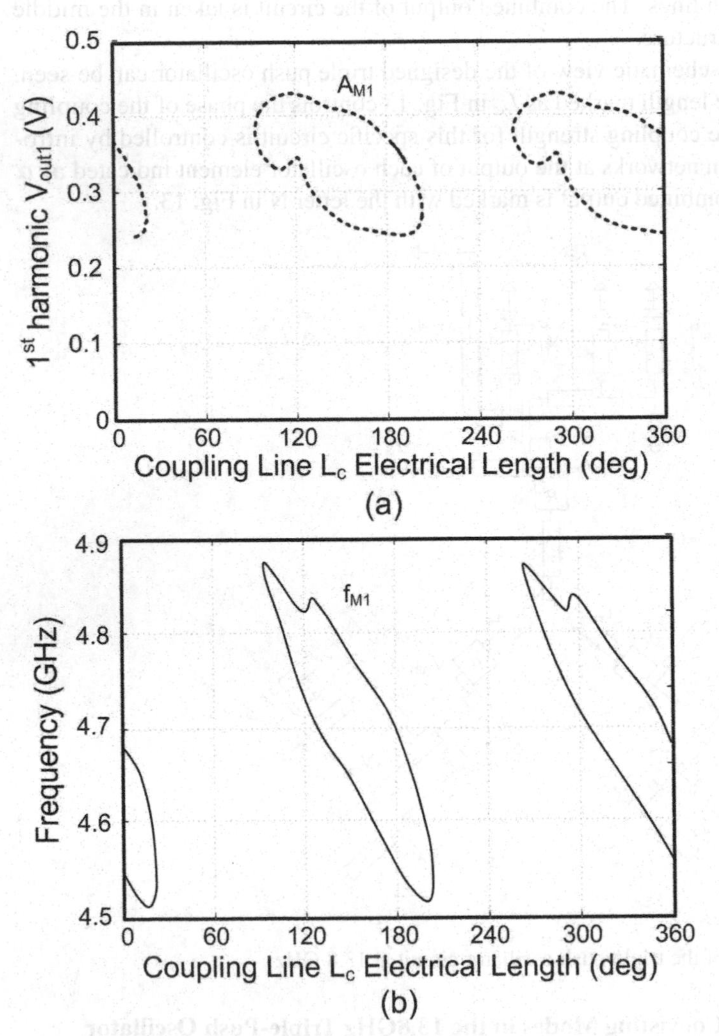

Fig. 14 Evolution of mode M1 versus the coupling line length L_c. (a) Output amplitude A_{M1} of mode M1 (b) Oscillation frequency of mode M1.

the circuit. This auxiliary generator ensures the convergence of the harmonic balance analysis to the oscillating solutions. The auxiliary generator is defined by its amplitude, phase and frequency.

In order to detect the modes it is necessary to use one auxiliary generator at each oscillator element. Imposing the amplitudes and phase relationships from Table 1 using the auxiliary generator it is possible to detect the different coexisting modes. The synchronization of the oscillator elements is ensured by using a common frequency in the three auxiliary generators.

In a first approach an analysis on how the different coexisting modes evolve versus the coupling phase is performed. The coupling phase among the oscillators is modified by changing the coupling line length L_c. In Fig. 14a and Fig. 14b the evolution of the amplitude and the frequency of mode M1 (in-phase mode) versus the coupling line length L_c are shown.

In Fig. 15a and Fig. 15b the evolution of the two different amplitudes in mode M4 versus the coupling line length L_c can be seen. Fig. 15c shows the evolution of the synchronization frequency of mode M4. Finally in Fig. 16 the evolution of the desired mode M2 is shown. The evolution of the synchronization frequency for this mode can be seen in Fig. 16b. Mode M3 was not detected during the analysis of the modes. This mode is more difficult to obtain and is not always observed.

It can be observed that the solution curves of the modes form islands versus the coupling line length L_c. Also, the appearance of thc islands is almost periodic with L_c. The solution curves repeat approximately every half wavelength.

The synchronization frequency of the system for each of the modes is also different. Modes M2 and M4 have similar synchronization frequency while mode M1 appear for lower values of synchronization frequency.

2.3.3 Mode Stability

In order to check the stability of the multiple coexisting modes, an initialized envelope transient analysis was used [31, 39]. In the same way as when detecting the different coexisting modes, an auxiliary generator is used in combination with an envelope transient simulation in order to ensure convergence to the oscillating solution.

The stability analysis was performed along the modes solution curves of Fig. 14-Fig. 16. The results showed that the upper sections of the solutions curves of M1, M2 and M4 were stable. This result illustrates the previously mentioned multistability of solutions in N-push oscillator circuits with time delay. The stability analysis of the rest of the sections in the solution curves showed some of the sections are stable and some unstable. The change in the stability properties was usually delimited by the turning points of the solution curves in Fig. 14- Fig. 16.

2.3.4 Elimination of the Undesired Modes M1 and M4

As the desired mode in order to combine the output power at the 3rd harmonic is the mode M2, it is important to design the coupling structure in such a way that mode M2 exist and at the same time the rest of the coexisting modes does not appear. For

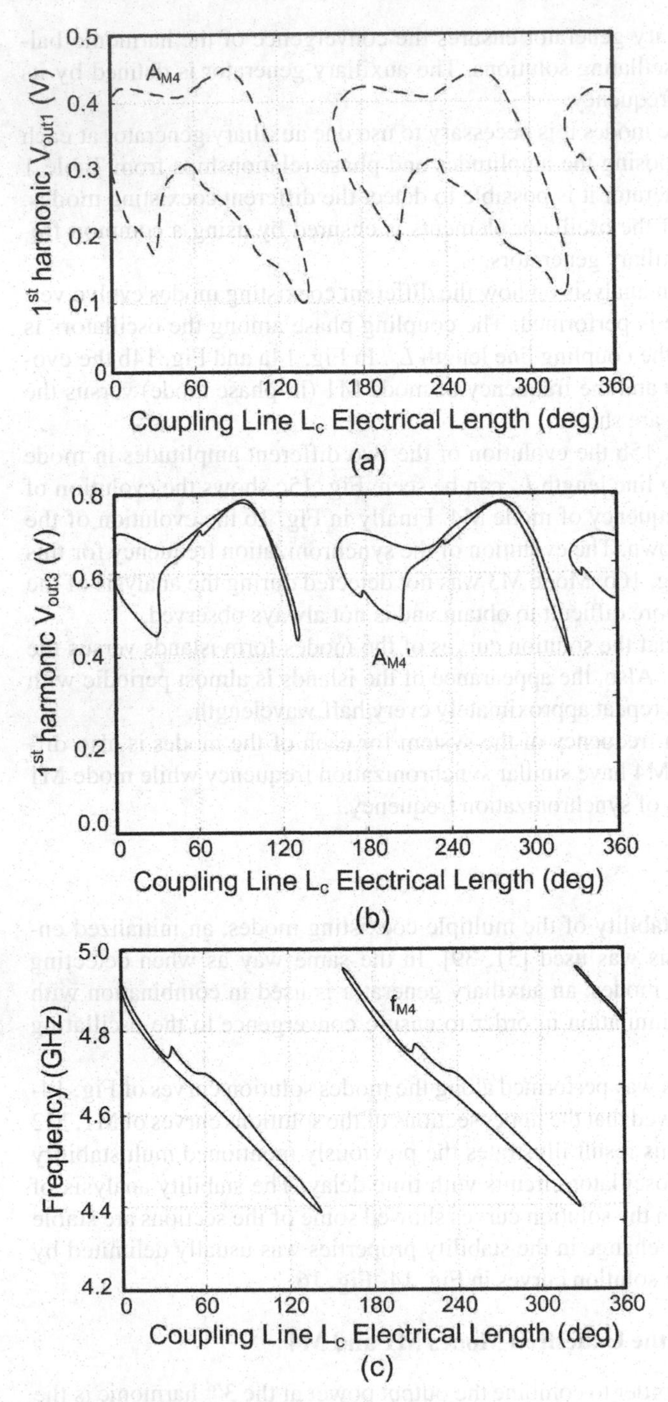

Fig. 15 Evolution of mode M4 versus the coupling line length L_c. (a) Output amplitude A_{M4} of mode M4 (b) Output amplitude A'_{M4} of mode M4 (c) Oscillation frequency of mode M4

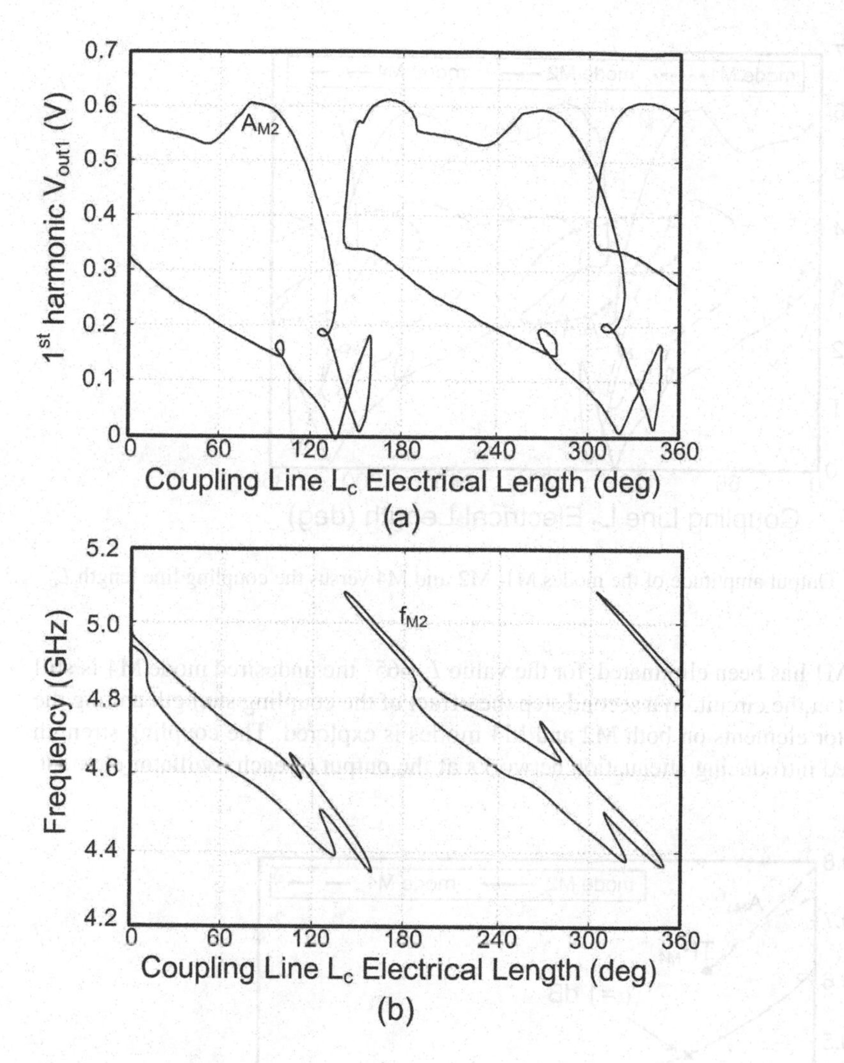

Fig. 16 Evolution of mode M2 versus the coupling line length L_c. (a) Output amplitude A_{M2} of mode M2 (b) Oscillation frequency of mode M2.

the optimized design of the triple push oscillator the objective is to find the optimum design of coupling network that will maintain mode M2 and will eliminate modes M1 and M4.

The solution curves for M1, M2 and M4 have been represented in a single plot (Fig. 17). It can be seen that there are values of the coupling line length L_c for which mode M1 does not exist and however both modes M2 and M4 appear. This means by choosing one of those values of L_c mode M1 will have been eliminated. As a first step in the design of the optimum coupling network a line length L_c=65° was chosen. For this value of L_c mode M1 does not appear. However even though

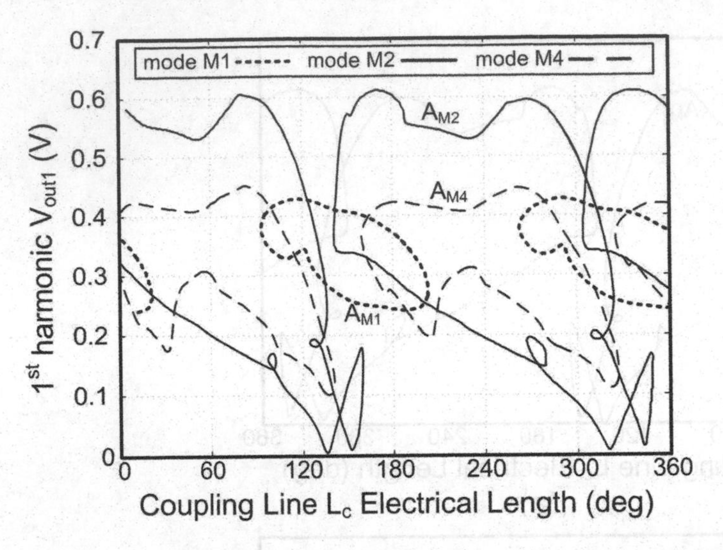

Fig. 17 Output amplitude of the modes M1, M2 and M4 versus the coupling line length L_C

mode M1 has been eliminated, for the value L_C=65° the undesired mode M4 is still present in the circuit. In a second step the effect of the coupling strength among the oscillator elements on both M2 and M4 modes is explored. The coupling strength is varied introducing attenuation networks at the output of each oscillator element.

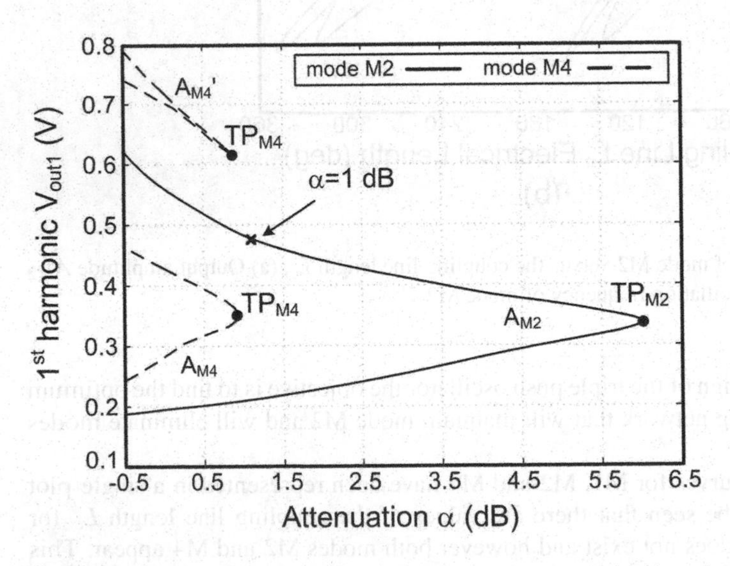

Fig. 18 Evolution of modes M2 and M4 versus the attenuation introduced at the output of each of the oscillator elements of the triple-push oscillator.

The line length is fixed to the value of 65° and the attenuation α at the output of the oscillators is varied.

In Fig. 18 the evolution of modes M2 and M4 versus the attenuation α is shown. From the presented results it can be seen that as the attenuation increases both modes M2 and M4 reach an infinite slope point or turning point (TP_{M2} and TP_{M4} respectively). For values of attenuation beyond these turning points the corresponding mode does not exist. From Fig. 18 it can be concluded that there are values of attenuation for which mode M4 does no exist and however mode M2 is still present in the circuit. These values of attenuation fall in the range [0.9 dB, 6 dB]. In order to minimize the effect of the attenuation on the combined output power, the minimum necessary attenuation that eliminates M4 is chosen. In this case the chosen attenuation was 1dB.

Three versions of the circuit were implemented in order to verify the results obtained in terms of the multistability of the different coexisting modes.

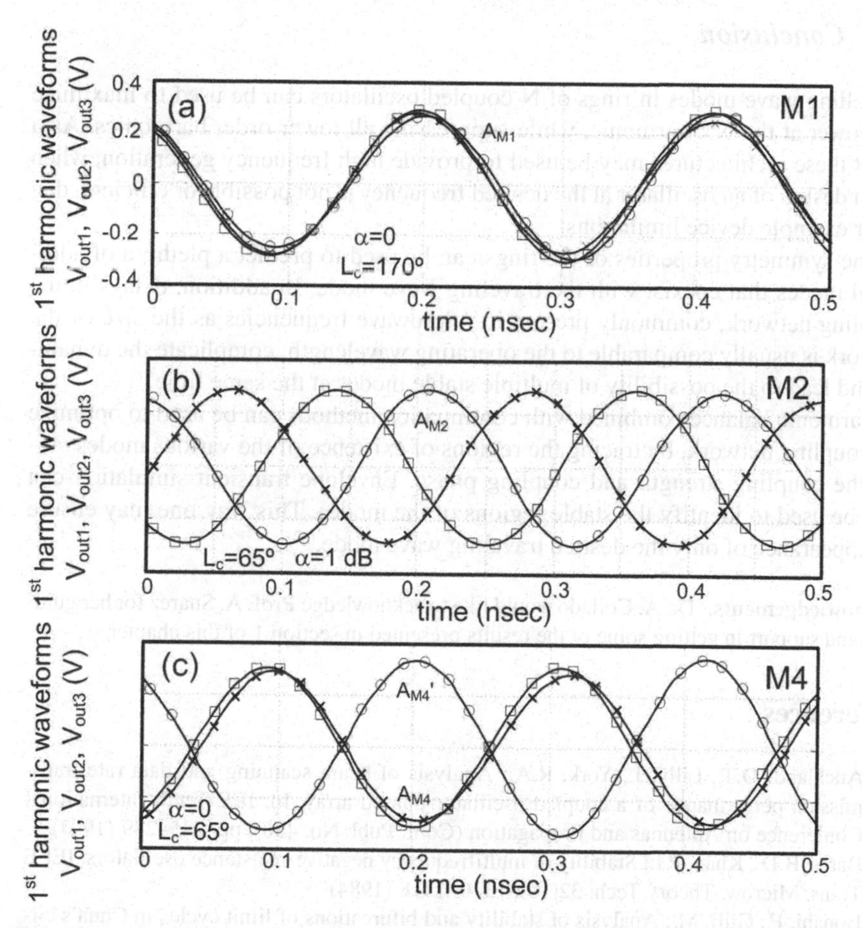

Fig. 19 Measured waveforms of the three observed modes in the 13.8 GHz triple-push oscillator.(a) Mode M2. (b) Mode M4. (c) Mode M1.

A first version of the triple push circuit was implemented using a coupling network with an L_c=65° and an attenuation $\alpha = 1$dB. In this implementation only mode M2 is expected to appear in the circuit. In Fig. 19a the measured waveforms of mode M2 are shown. Small variations in the bias conditions of the circuit were introduced in order to try to excite the other modes. However, only mode M2 was observed.

A second version of the triple push oscillator was implemented using a coupling network with L_c=65° and with no attenuation. Both M2 and M4 were observed in the measurements by slightly changing the initial bias conditions. In Fig. 19b the waveforms of the measured mode M4 can be seen.

Finally a last version of the triple push oscillator with L_c=170° and no attenuation was implemented. For this implementation the three possible modes M1, M2 and M4 were observed, although the system tends to evolve easily to mode M1. In Fig. 19c the waveforms of the observed mode M1 are represented.

2.4 Conclusion

Travelling wave modes in rings of N coupled oscillators can be used to maximize the power at the Nth harmonic, while suppressing all lower order harmonics. As a result these architectures may be used to provide high frequency generation, when direct design of an oscillator at the desired frequency is not possible or efficient, due to for example device limitations.

The symmetry properties of the rings can be used to predict a plethora of additional modes that coexist with the traveling wave mode. In addition, delays in the coupling network, commonly present in microwave frequencies as the size of the network is usually comparable to the operating wavelength, complicate the dynamics and lead to the possibility of multiple stable modes at the same time.

Harmonic balance combined with continuation methods can be used to optimize the coupling network, by tracing the regions of existence of the various modes versus the coupling strength and coupling phase. Envelope transient simulation can then be used to identify the stable regions of the modes. This way one may ensure the appearance of only the desired traveling wave mode.

Acknowledgements. Dr. A. Collado would like to acknowledge Prof. A. Suarez for her guidance and support in getting some of the results presented in section 1 of this chapter.

References

1. Auckland, D.T., Lilly, J., York, R.A.: Analysis of beam scanning and data rate transmission performance of a coupled oscillator phased array. In: IEE Tenth International Conference on Antennas and Propagation (Conf. Publ. No. 436), pp. 245–249 (1997)
2. Bates, B.D., Khan, P.J.: Stability of multifrequency negative-resistance oscillators. IEEE Trans. Microw. Theory Tech. 32(10), 1310–1318 (1984)
3. Bonani, F., Gilli, M.: Analysis of stability and bifurcations of limit cycles in Chua's circuit through a harmonic balance approach. IEEE Trans. Circuits Syst. I, Fundam. Theory Appl. 46(8), 881–890 (1999)

4. Chang, H.C., Cao, X., Mishra, U.K., York, R.A.: Phase noise in coupled oscillators: Theory and experiment. IEEE Trans. Microw. Theory Tech. 45(5), 604–615 (1997)
5. Choi, J., Mortazawi, A.: Design of push-push and triple-push oscillators for reducing 1/f noise upconversion. IEEE Trans. Microw. Theory Tech. 53(11), 3407–3414 (2005)
6. Collado, A., Georgiadis, A.: Nonlinear mode analysis and optimization of a triple-push oscillator. IEEE Microw. Wireless Compon. Lett. 18(8), 545–547 (2008)
7. Collado, A., Ramirez, F., Suarez, A., Pascual, J.: Harmonic-balance analysis and synthesis of coupled-oscillator arrays. IEEE Microw. Wireless Compon. Lett. 14(5), 192–194 (2004)
8. Collado, A., Sancho, S., Suarez, A.: Semi-analytical formulation for the stability analysis of coexisting solutions in coupled-oscillator systems. In: IEEE MTT-S Int. Microw. Symp. Dig., pp. 973–976 (2007)
9. Collantes, J.M., Lizarraga, I., Anakabe, A., Jugo, J.: Stability verification of microwave circuits through floquet multiplier analysis. In: Proc. IEEE Asia-Pacific Circuits Syst. Conf., pp. 997–1000 (2004)
10. Dodla, R., Sen, A., Johnston, G.: Phase-locked patterns and amplitude death in a ring of delay-coupled limit cycle oscillators. Physical Review E 69 (2004)
11. Farzaneh, F., Mehrshahi, E.: A novel noise analysis method in microwave oscillators based on harmonic balance. In: Proc. Eur. Microw. Conf., pp. 255–260 (1998)
12. Fusco, V.F., Drew, S.: Active antenna phase modulator performance. In: Proc. Eur. Microw. Conf., pp. 248–251 (1993)
13. Georgiadis, A.: Design of coupled oscillator arrays for second harmonic radiation. In: IEEE MTT-S Int. Microw. Symp. Dig., pp. 1727–1730 (2007)
14. Georgiadis, A., Collado, A.: Injection locked coupled oscillator arrays. IEEE Microw. Wireless Compon. Lett. 17(12), 900–902 (2007)
15. Georgiadis, A., Collado, A., Suarez, A.: New techniques for the analysis and design of coupled-oscillator systems. IEEE Trans. Microw. Theory Tech. 11(11), 3864–3877 (2006)
16. Georgiadis, A., Collado, A., Suarez, A.: Pattern nulling in coupled oscillator antenna arrays. IEEE Trans. Antennas Propag. 55(5), 1267–1274 (2007)
17. Golubitsky, M., Stewart, I.: Hopf bifurcation with dihedral group symmetry: Coupled nonlinear oscillators. Contemporary Mathematics Multiparameter Bifurcation Theory 56, 131–173 (1986)
18. Guckenheimer, J., Holmes, P.: Nonlinear Oscillations, Dynamic Systems, and Bifurcations of Vector Fields. Springer, New York (1983)
19. Heath, T.: Beam steering of nonlinear oscillator arrays through manipulation of coupling phases. IEEE Trans. Antennas Propag. 52(7), 1833–1842 (2004)
20. Jordan, D.W., Smith, P.: Nonlinear Ordinary Differential Equations: An Introduction to Dynamical Systems, 3rd edn. Oxford Univ. Press, Oxford (1999)
21. Jugo, J., Portilla, J., Anakabe, A., Suarez, A., Collantes, J.M.: Closed-loop stability analysis of microwave amplifiers. Electron. Lett. 37, 226–228 (2001)
22. Kim, S., Park, S.H., Ryu, C.: Multistability in coupled oscillator systems with time delay. Physical Review Letters 79(15) (1997)
23. Kuhn, M.R., Biebl, E.M.: Power combining by means of harmonic injection locking. In: IEEE MTT-S Int. Microw. Symp. Dig., pp. 91–94 (2004)
24. Kurokawa, K.: Noise in synchronized oscillators. IEEE Trans. Microw. Theory Tech. 16(4), 234–240 (1968)
25. Kurokawa, K.: Some basic characteristics of broadband resistance oscillator circuits. Bell Syst. Tech. J., 1937–1955 (1969)

26. Kurokawa, K.: Injection locking of microwave solid-state oscillators. Proc. IEEE 61(10), 1386–1410 (1973)
27. Kykkotis, C., Hall, P.S., Ghafouri-Shiraz, H., Wake, D.: Modulation effects in active integrated locked antenna oscillator arrays. In: IEE Tenth International Conference on Antennas and Propagation (Conf. Publ. No. 436), pp. 510–513 (1997)
28. Lai, X., Roychowdhury, J.: Capturing oscillator injection locking via nonlinear phase-domain macromodels. IEEE Trans. Microw. Theory Tech. 52(9), 2251–2261 (2004)
29. Lynch, J.J., York, R.A.: Synchronization of oscillators coupled through narrowband networks. IEEE Trans. Microw. Theory Tech. 49(2), 237–249 (2001)
30. Meadows, B.K., Heath, T.H., Neff, J.D., Brown, E.A., Fogliatti, D.W., Gabbay, M., In, V., Hasler, P., Deweerth, S.P., Ditto, W.L.: Nonlinear antenna technology. Proc. IEEE 90(5), 882–897 (2002)
31. Ngoya, E., Larcheveque, R.: Envelope transient analysis: A new method for the transient and steady state analysis of microwave communication circuits and systems. In: IEEE MTT-S Int. Microw. Symp. Dig., pp. 1365–1368 (1995)
32. Nogi, S., Lin, J., Itoh, T.: Mode analysis and stabilization of a spatial power combining array with strongly coupled oscillators. IEEE Trans. Microw. Theory Tech. 41(10), 1827–1837 (1993)
33. Pogorzelski, R.J.: On the design of coupling networks for coupled oscillator arrays. IEEE Trans. Antennas Propag. 51(4), 794–801 (2003)
34. Pozar, D.: Microwave Engineering, 3rd edn. Wiley, New York (2004)
35. Redy, D.R., Sen, A., Johnston, G.: Experimental evidence of time-delay-induced death in coupled limit-cycle oscillators. Physical Review Letters 85(16) (2000)
36. Rizzoli, V., Neri, A.: State of the art and present trends in nonlinear microwave CAD techniques. IEEE Trans. Microw. Theory Tech. 36(2), 343–365 (1988)
37. Rohde, U.L., Poddar, A.K., Böck, G.: The Design of Modern Microwave Oscillators for Wireless Applications: Theory and Optimization. Wiley, New York (2005)
38. Rohde, U.L., Poddar, A.K., Schoepf, J., Rebel, R., Patel, P.: Low noise low cost ultra wideband N-push VCO. In: IEEE MTT-S Int. Microw. Symp. Dig., pp. 1171–1174 (2005)
39. Sharrit, D.: Method for simulating a circuit, U.S. Patent 5588142 (1996)
40. Suarez, A., Collado, A., Ramirez, F.: Harmonic-balance techniques for the design of coupled-oscillator systems in both unforced and injection-locked operation. In: IEEE MTT-S Int. Microw. Symp. Dig., pp. 8887–8904 (2005)
41. Suarez, A., Quere, R.: Stabiliry Analysis of Nonlinear Microwave Circuits. Artech House, Norwood (2003)
42. Tang, Y., Wang, H.: Triple-push oscillator approach: Theory and experiments. IEEE J. Solid-State Circuits 36(10), 1472–1479 (2001)
43. Thompson, J.M.T., Stewart, H.B.: Nonlinear Dynamics and Chaos, 2nd edn. Wiley, New York (2002)
44. Vanassche, P., Gielen, G.G.E., Sansen, W.: Behavioral modeling of (coupled) harmonic oscillators. IEEE Trans. Comput.-Aided Design Integr. Circuits Syst. 22(8), 1017–1026 (2003)
45. Yen, S.C., Chu, T.H.: An Nth-harmonic oscillator using an N-push coupled oscillator array with voltage-clamping circuits. In: IEEE MTT-S Int. Microw. Symp. Dig., pp. 2169–2172 (2003)
46. York, R.A., Liao, P., Lynch, J.J.: Oscillator array dynamics with broadband N-port coupling networks. IEEE Trans. Microw. Theory Tech. 42(11), 2040–2042 (1994)

Author Index